中国科协学科发展研究系列报告

中国科学技术协会／主编

2016—2017
化学学科发展报告

中国化学会｜编著

REPORT ON ADVANCE IN CHEMISTRY

中国科学技术出版社
·北京·

图书在版编目（CIP）数据

2016—2017 化学学科发展报告 / 中国科学技术协会主编；中国化学会编著 . —北京：中国科学技术出版社，2018.3

（中国科协学科发展研究系列报告）

ISBN 978-7-5046-7929-1

Ⅰ. ① 2… Ⅱ. ①中… ②中… Ⅲ. ①化学－学科发展－研究报告－中国－ 2016-2017 Ⅳ. ① O6-12

中国版本图书馆 CIP 数据核字（2018）第 055390 号

策划编辑	吕建华　许　慧
责任编辑	韩　颖　冯建刚
装帧设计	中文天地
责任校对	杨京华
责任印制	马宇晨

出　　版	中国科学技术出版社
发　　行	中国科学技术出版社发行部
地　　址	北京市海淀区中关村南大街16号
邮　　编	100081
发行电话	010-62173865
传　　真	010-62179148
网　　址	http://www.cspbooks.com.cn

开　　本	787mm × 1092mm　1/16
字　　数	425千字
印　　张	20
版　　次	2018年3月第1版
印　　次	2018年3月第1次印刷
印　　刷	北京盛通印刷股份有限公司
书　　号	ISBN 978-7-5046-7929-1 / O · 196
定　　价	98.00元

2016—2017

化学
学科发展报告

首席科学家 姚建年

专 家 组（按姓氏笔画排序）

丁元华 丁有钱 卜显和 马 丁 王 冬
王 训 王亚韡 王哲明 王健吉 王强斌
王 磊 毛江高 方维海 石先哲 龙腊生
帅志刚 叶 宁 申有青 田伟生 史 勇
邢 巍 朱广山 朱为宏 朱玉军 朱亚先
朱伟民 朱维良 向 宇 邬 慧 庄 林
刘文剑 刘正平 刘伟生 刘 倩 刘海超
刘琛阳 刘 强 闫文付 关 伟 江桂斌
池振国 安立佳 许 柠 孙为银 孙世刚
阴永光 严以京 严兆华 李 川 李有勇
李攻科 李浩然 李海芳 李景虹 杨开广
杨文胜 杨 帆 杨秀荣 杨国昱 杨国强

序
FOREWORD

党的十八大以来，以习近平同志为核心的党中央把科技创新摆在国家发展全局的核心位置，高度重视科技事业发展，我国科技事业取得举世瞩目的成就，科技创新水平加速迈向国际第一方阵。我国科技创新正在由跟跑为主转向更多领域并跑、领跑，成为全球瞩目的创新创业热土，新时代新征程对科技创新的战略需求前所未有。掌握学科发展态势和规律，明确学科发展的重点领域和方向，进一步优化科技资源分配，培育具有竞争新优势的战略支点和突破口，筹划学科布局，对我国创新体系建设具有重要意义。

2016 年，中国科协组织了化学、昆虫学、心理学等 30 个全国学会，分别就其学科或领域的发展现状、国内外发展趋势、最新动态等进行了系统梳理，编写了 30 卷《学科发展报告（2016—2017）》，以及 1 卷《学科发展报告综合卷（2016—2017）》。从本次出版的学科发展报告可以看出，近两年来我国学科发展取得了长足的进步：我国在量子通信、天文学、超级计算机等领域处于并跑甚至领跑态势，生命科学、脑科学、物理学、数学、先进核能等诸多学科领域研究取得了丰硕成果，面向深海、深地、深空、深蓝领域的重大研究以“顶天立地”之态服务国家重大需求，医学、农业、计算机、电子信息、材料等诸多学科领域也取得长足的进步。

在这些喜人成绩的背后，仍然存在一些制约科技发展的问题，如学科发展前瞻性不强，学科在区域、机构、学科之间发展不平衡，学科平台建设重复、缺少统筹规划与监管，科技创新仍然面临体制机制障碍，学术和人才评价体系不够完善等。因此，迫切需要破除体制机制障碍、突出重大需求和问题导向、完善学科发展布局、加强人才队伍建设，以推动学科持续良性发展。

近年来，中国科协组织所属全国学会发挥各自优势，聚集全国高质量学术资源和优秀人才队伍，持续开展学科发展研究。从 2006 年开始，通过每两年对不同的学科（领域）分批次地开展学科发展研究，形成了具有重要学术价值和持久学术影响力的《中国科协学科发展研究系列报告》。截至 2015 年，中国科协已经先后组织 110 个全国学会，开展了 220 次学科发展研究，编辑出版系列学科发展报告 220 卷，有 600 余位中国科学院和中国工程院院士、约 2 万位专家学者参与学科发展研讨，8000 余位专家执笔撰写学科发展报告，通过对学科整体发展态势、学术影响、国际合作、人才队伍建设、成果与动态等方面最新进展的梳理和分析，以及子学科领域国内外研究进展、子学科发展趋势与展望等的综述，提出了学科发展趋势和发展策略。因涉及学科众多、内容丰富、信息权威，不仅吸引了国内外科学界的广泛关注，更得到了国家有关决策部门的高度重视，为国家规划科技创新战略布局、制定学科发展路线图提供了重要参考。

十余年来，中国科协学科发展研究及发布已形成规模和特色，逐步形成了稳定的研究、编撰和服务管理团队。2016—2017 学科发展报告凝聚了 2000 位专家的潜心研究成果。在此我衷心感谢各相关学会的大力支持！衷心感谢各学科专家的积极参与！衷心感谢编写组、出版社、秘书处等全体人员的努力与付出！同时希望中国科协及其所属全国学会进一步加强学科发展研究，建立我国学科发展研究支撑体系，为我国科技创新提供有效的决策依据与智力支持！

当今全球科技环境正处于发展、变革和调整的关键时期，科学技术事业从来没有像今天这样肩负着如此重大的社会使命，科学家也从来没有像今天这样肩负着如此重大的社会责任。我们要准确把握世界科技发展新趋势，树立创新自信，把握世界新一轮科技革命和产业变革大势，深入实施创新驱动发展战略，不断增强经济创新力和竞争力，加快建设创新型国家，为实现中华民族伟大复兴的中国梦提供强有力的科技支撑，为建成全面小康社会和创新型国家做出更大的贡献，交出一份无愧于新时代新使命、无愧于党和广大科技工作者的合格答卷！

2018 年 3 月

前言
PREFACE

在中国科协学会学术部的指导下，继 2007 年、2009 年、2011 年、2013 年、2015 年先后 5 次出版化学学科发展报告之后，中国化学会再次组织所属各学科委员会和专业委员会对化学学科近 3 年来取得的进展进行调研，撰写完成了《2016—2017 化学学科发展报告》（以下简称“本报告”）。

本报告由综合报告和专题报告两部分组成。在综合报告中，汇集了 24 个化学研究分支的最新重要进展和发展趋势，同时也继续对我国高等化学教育和基础化学教育的发展概况进行整理，由执笔人根据各学科和专业委员会、有关专家以及编写组人员提供的部分资料编撰而成，文中共涉及国内科学家近年发表的论文 800 余篇。本报告的 8 篇专题报告是编写组通过调研有选择地组织的，共引用一千余篇参考文献内容涉及与经济发展和民生相关的近年来化学热门前沿领域。专题报告的内容也尽量避免在综合报告中重复赘述。

本报告的编写得到化学界的多位院士、专家的大力支持和积极响应，他们亲自参与了调研、编写和审稿工作。在此对他们所付出的辛勤劳动表示衷心的感谢。虽经多位专家审阅修改，但本报告一定还会存在材料取舍和编排不当、疏漏等不少的缺陷和瑕疵，也恐难全面反映我国化学研究发展全貌。对于报告的不足，欢迎同仁们批评指正。

中国化学会

2017 年 12 月

目录
CONTENTS

综合报告

专题报告

ABSTRACTS

Comprehensive Report

Reports on Special Topics

综合报告

化学学科研究进展与趋势展望

一、引言

近两年来，中国化学在教育和科研方面均取得显著进展。在各级政府和社会的各种经费的支持下，化学研究的队伍越来越壮大，化学学术论文发表的数量和质量持续处在国际领先地位，中国化学正进入一个由“跟跑”向“领跑”转变的关键时期。

本次化学学科进展报告总结了2015—2017上半年我国化学工作者在科研和教育方面的进展。化学科研涵盖了有机化学、无机化学、物理化学、分析化学、高分子化学、核化学与放射化学等六个主要分支学科，以及计算化学、流变学、环境化学、绿色化学、理论化学、晶体化学、多孔材料、化学生物学等八个交叉学科或研究领域。化学教育涵盖基础化学教育和高等化学教育。本报告共引用参考文献两千余篇，能够比较准确地反映了我国化学学科进展情况。

二、我国化学学科最新重要进展

（一）有机化学

近两年来，我国有机化学研究领域继续保持良好发展势头，在学术论文发表方面成绩突出。中国化学家在《美国化学会志》《德国应用化学》《化学科学》3个杂志上发表论文的数量仅次于美国，居第二位；在《化学通讯》发表论文篇数占总数39.7%，位居第一。

有机化学专业杂志上的统计结果可以部分展现出我国有机化学研究领域的发展状况。在《高等合成与催化》《有机快讯》和《有机化学杂志》3个国际主流杂志上，中国有机化学家发表文章数均居首位，分别占论文总数的40%、34%和28%。这从一个侧面反映出我国有机化学研究的整体水平在国际上已经处于较高的位置。2016年，中国科学院上海有机化学研究所刘国生的研究团队，利用金属催化的自由基接力新策略成功地实现了铜催化的苄基上的碳氢键不对称氰化反应，发表在 *Science* 上，是我国大陆有机化学家在该杂志上发表的首篇研究论文[1]。

在本报告收录的有机化学相关 195 篇学术论文中，《科学》《自然·化学》《自然·通讯》《美国化学会志》《德国应用化学》等共计 177 篇，占 90.8%。在这 177 篇论文中，金属有机化学为主的有机反应和合成方法学论文 107 篇，占 60.5%；与氟等元素有关的有机反应与合成方法学论文 23 篇，占 13.0%；天然产物合成化学论文 45 篇，占 25.4%；天然产物化学 2 篇，占 1.1%。可见，金属有机以及含氟等元素的有机反应和合成方法学仍是研究热点。另外，天然产物合成化学也取得了不俗的进展。

1. 有机反应与合成方法学

不对称有机反应和合成方法学研究，C–H 官能团化的研究，自由基反应等有机化学的热点研究领域都有我国有机化学家的身影。

芳烃或烷烃的选择性碳氢键官能化是高效合成药物、农药、精细化学品以及快速开发它们的新产品对有机合成化学家提出的要求，也是有机合成中的挑战性任务。刘国生等提出将反应中的碳自由基中间体转化为金属有机物种来实现选择性控制的策略，借此来解决烷烃的碳氢键直接不对称官能化的问题，成功地实现了铜催化苄基上的碳氢键不对称氰化反应，提供了合成手性腈类化合物的新方法，具有良好的官能团兼容性和优异的化学、区域和立体选择性[1]。刘国生的团队还发展了铜催化的对映选择性烯烃的分子间三氟甲基氰基化反应、铜催化炔烃的三氟甲基叠氮化反应、钯催化苯乙烯分子间氟磺酰化反应、分子内烯烃胺三氟甲氧基化反应、分子间烯烃的分子间氧 / 胺羰基化反应等[2–8]。

去芳构化反应及其在天然产物合成中应用已经得到有机化学家充分重视。游书力团队使用不同有机催化剂和过渡金属络合物对吲哚、吡咯、苯酚以及萘酚等进行不对称去芳构化反应研究，给出了含有螺环或桥环结构单元化合物的合成方法；他们还使用有机小分子和过渡金属络合物催化一些非活化碳氢键来形成碳碳键和碳杂原子键[9–17]。

冯小明和刘小华等继续发展手性氮氧配体与不同金属络合催化的不对称反应，在多种转化上都实现了优异的立体选择性控制，如与锌络合物催化环状烯醇硅醚和炔酮的［2+2］环加成反应、与金 / 镍双金属接力催化螺环缩酮的合成、与镍催化炔丙醇衍生物的动力学拆分和 Claisen 重排合成手性联烯等[18–25]。还使用新型手性双功能胍 – 酰胺配体，在温和条件下，高效高选择性地实现了多个不对称催化反应[26, 27]。

刘心元和谭斌等在有机催化及烯烃三氟甲基化引发的进一步转化的研究上发展了一系列方法学，如：磷酸催化不对称 Passerini 反应、多组分反应不对称构建螺吲哚酮、有机催化萘酚的芳基化反应合成轴手性联芳基二醇、双重活化策略烯烃的直接不对称自由基胺三氟甲基化反应、以及自由基介导的烯烃 1，2– 甲酰基 / 羰基化反应等[28–33]。

张俊良团队以多官能团化合物的化学反应为中心，致力于新的催化剂开发，催化剂新的活性研究，及其在新反应中的应用；发现过渡金属钯和铑的路易斯酸性是反应的关键，成功解决了基于共轭烯炔类化合物为反应原料的多样性合成和选择性问题；发展了以烯基氮杂环丙烷与烯烃、炔烃以及联烯的催化环加成反应为代表的一系列有机反应[34-39]。

王剑波等专注涉及重氮化合物和金属卡宾的方法学研究，如重氮烷烃与端炔偶联合成三取代联烯、卡宾对硅硅键和锡锡键的插入反应、酰基重氮甲烷的碳氢官能团化和串联交叉偶联反应、铑催化重氮酯对硅氧基烯基环丙烷的碳碳键活化反应等[40-44]。

雷爱文等专注于发展新型的包含小分子活化过程的氧化偶联反应方法学和机理研究，如铜催化下氯乙酸酯促进的杂芳基苄位亚甲基的选择性氧化反应、铜催化的 sp3 碳氢 / 磷氢的自由基交叉偶联反应、端炔与非活化烷烃在多金属催化下的自由基氧化交叉偶联反应等[45-48]。游劲松等在碳氢键活化，尤其是氧化偶联方面做出了创新性和系统性研究，利用活化策略构筑具有 π 共轭结构有机功能材料分子，特别是新颖杂芳环共轭骨架[49-52]。

在高活性配体开发及应用研究方面，汤文军等采用设计深手性口袋膦配体的新思路，发展了有显著结构特征的 P- 手性膦配体，在大位阻偶联以及不对称偶联 / 环化 / 氢化中表现出优异的效率[53-57]。周其林等以螺二氢茚为配体骨架设计的手性双噁唑啉和手性氮膦配体在一系列金属催化的反应中表现出极佳的立体化学控制，如铱催化的 2- 吡啶基环亚胺的不对称氢化、镍催化苯乙烯与醛的氢酰化反应、以及苯胺衍生物对 α- 芳基 α- 重氮乙酸酯的不对称芳基化反应等[58-61]。唐勇等以 BOX 型配体在一些不对称环加成或开环反应中取得很好的结果，如给体 - 受体型环丙烷在水进攻下的亲核开环、环状烯胺与不饱和 α- 酮酸酯在铜催化下形成双环 *N*，*O*- 缩醛、以及双烯与给体 - 受体型环丙烷的不对称环化反应等[62-65]。丁奎岭等将所开发的配体应用到氢气、一氧化碳及二氧化碳活化有关的催化反应研究中，发展清洁的有机合成方法，实现了末端联烯高区域选择性和对映选择性的烷氧羰基胺基化反应以及铑催化以氢气和二氧化碳为原料进行胺的甲酰化反应[66，67]。

在惰性化学键，尤其是碳氢键和碳碳键的活化方面，施章杰等发展了银催化的芳基从碳原子向氮原子远程迁移过程，并进行了初步机理研究[68，69]；史炳峰等以二价钯催化实现非活化亚甲基的碳氢键氟化反应，可立体选择性合成 β- 氟代 α- 氨基酸[70，71]；史壮志等分别使用铜和钯催化实现吲哚 C6 和 C7 位的芳基化[72-75]；李兴伟等使用芳香醛的硝酮无痕导向，实现铑催化芳烃的碳氢键活化[76-78]；宋毛平以混合导向基团策略，用钴催化实现非活化芳烃的氧化碳氢 / 碳氢键芳基化[79]。

在光氧化还原催化剂的存在下，越来越多的转化可以在可见光的激发下进行，有绿色环保的特征。陈以昀、肖文精、俞寿云、李桂根等做出了有益的探索，报道了如高价碘介导的脱羧烯化 / 炔化反应、羧酸的脱羧炔化及羰基炔化反应、应用可见光促进酰基肟形成亚胺自由基合成吡啶 / 喹啉 / 菲啶、以及使用取代的 Hantzsch 酯作为烷基化试剂合成多取代酮等过程[80-86]。

另外，黄正等在设计和合成用于有机转化的金属催化剂方面取得进展，开发的铱、铁催化剂可催化串联的脱氢 / 异构 / 氢硅化反应将烷烃直接转化为高附加值的烷基硅烷[87]。麻生明等专注联烯相关方法学研究外，还发展了铁催化室温下空气氧化伯醇为羧酸的方法[88-90]。马大为等设计的草酸二酰胺配体大大降低了一价铜催化芳基卤化物的反应温度，可在较低温度下实现（杂）芳基氯的羟基化和胺基化[91-93]。龚流柱等利用钯络合物与 Brønsted 酸的协同催化实现吡唑酮对末端烯烃的高对映选择性烯丙位烷基化等[94-96]。焦宁等发展了锰催化烯烃的空气氧化羟基叠氮化反应，能快速制备 β－叠氮醇化合物[97-99]。姜雪峰等在均相金催化中巧用配体效应，实现区域发散性的 π 键活化环化[100, 101]。朱晨等将三价锰催化环丁醇自由基碎裂开环反应与一系列自由基捕获过程结合，实现了远程氰化、炔化、以及氟化等转化[102-104]。

纯不对称有机催化研究也有许多不错的成果。陈应春等使用手性胺催化 α－取代的胺叶立德的［4+1］环化反应，对映选择性地构建螺环吲哚酮[105]。赵刚等基于廉价易得的天然手性源设计合成了一系列新型手性有机催化剂，将其作为手性 Brønsted 酸或 Lewis 碱应用于不对称催化 Strecker 等类型的反应，取得优异的产率和对映选择性；这种双试剂手性离子对的催化策略和概念拓展了手性膦催化反应的范围，可能带来新的反应与应用[106, 107]。

2. 有机合成化学

有机合成化学逐渐加强对新的反应特性的理解和运用，以及对化学转化中的选择性的精确控制，寻求通过有机合成化学满足化合物资源的合理利用和人类社会对有机分子的需要。两年来，我国有机化学家发表与天然产物合成相关的高影响因子文章 40 多篇，合成目标多是具有良好生理活性和独特化学结构的天然产物，热点分子主要为生物碱（吲哚生物碱和石松生物碱居多）和萜类（二萜 / 三萜）等。

李昂等采取基于天然产物生物合成假设及特定核心结构的发散性策略，完成 10 余个具有挑战性的吲哚萜类和氨基酸萜类天然产物全合成[108-115]。

俞飚等利用分子内缩醛环化反应和原创的三氟亚胺酯糖苷化和金催化糖苷化反应为核心，完成了杠柳苷 periploside A、皂甙 Linckosides A 和 B、tunicamycins 的合成[116-118]。马大为等以取代的二氢呋喃酮和香茅醛衍生物的羟醛缩合以及二碘化钐介导的分子内酮－烯自由基环化为关键反应，完成具有拒食作用和抗真菌活性的倍半萜 leucosceptroids A/B 的全合成[119]。洪然等以亚硝基－烯环化反应巧妙构筑 2- 氮杂双环［3.2.1］辛烷结构，完成生物碱 hosieine A 的首次全合成[120]。

雷晓光等继续在石松生物碱方面工作，完成 huperzine Q 和 lycopladines B 与 C 的合成；采用序列碳氢官能团化快速建立骨架，并以仿生串联反应完成 incarviatone A 的对映选择性合成；以区域选择性 1，6- 二烯炔环化为基础快速构筑骨架，完成 jungermannenones B 和 C 的合成；并通过合成修正了对耐药性革兰氏阴性菌有抑制活性的天然产物

aspergillomarasmine A 的结构[121-124]。

受生源过程激发灵感，秦勇等设计了四种二萜生物碱统一的合成策略，关键反应包括碳氢氧化、氮杂频哪醇偶联、以及氮杂 Prins 环化等；还完成吲哚生物碱 lundurine A 的全合成并确定其绝对构型[125, 126]。刘波等罕见地采取以 podocarpane 型邻苯二醌为双烯体的［4+2］环加成反应为关键反应的策略，发散式地完成了多个 atisane 型二萜分子的全合成[127]。徐亮等巧设路线，高效率地完成 5 个 atisane 型二萜 /atisine 二萜生物碱的集体全合成[128]。杨劲松等完成海洋糖脂主要组分 vesparioside B 的首次全合成，主要策略是汇聚式的［4+3］偶联[129]。

涂永强等以异色满衍生的烯丙基硅醚发生串联碳氢氧化 / 环合 / 重排过程高效地构筑三环苯并氧［3.2.1］庚烷化合物为关键，完成了天然产物 brussonol 和 przewalskine E 的不对称合成；还以半频哪醇重排的方式实现可控的区域选择性和非立体选择性扩环，完成四环二萜 conidiogenone、onidiogenol 和 conidiogenone B 的全合成[130, 131]。樊春安等利用对位取代环己二烯酮发生分子内氮杂 Michael 加成反应建立螺环的概念，巧妙设计串联反应完成 2 种 Apocynaceae 生物碱的全合成[132]。邱发洋、翟宏斌等利用有机催化的不对称环加成、分子内跨环羟醛缩合、氧化吲哚的加成偶联及分子内亲核取代等 12 步反应完成勾吻素的对映选择性全合成，是迄今最有效的全合成[133]。

杨玉荣等在单萜吲哚生物碱和石松生物碱的全合成研究方面在国际上率先实现了具有促进神经干细胞生长的吲哚生物碱 alstoscholarisine A 和灯台类单萜吲哚生物碱 aspidophylline A 的首次不对称全合成[134, 135]。高拴虎等基于生源假说设计合成路线，完成含卤天然产物 hamigeran 家族 7 个成员的全合成；他们还发展了可灵活构筑吲嗪酮和喹嗪酮的方法，将其用于天然产物及类天然产物的合成，完成 camptothecin 及相关天然产物的合成[136, 137]。丁寒锋等基于串联反应的复杂分子骨架高效构建新方法及新策略设计，完成二萜 steenkrotin A，以及吲哚生物碱 alsmaphorazine D 等多个具有重要生物活性天然产物的全合成[138, 139]。

汤平平等从易得的商品化原料出发，以自由基环化，自由基加成以及后期碳氢键碘代等为关键反应，高度汇聚式地完成了 schilancitrilactones B 和 C 的首次全合成，为类似的萜类天然产物的合成提供了一种新的思路[140]。陈弓等完成了对多重耐药性革兰氏阳性菌有强抗菌活性且结构独特的糖肽化合物 mannopeptimycins α 和 β 的全合成[141]。

杨震与陈家华等专注 Pauson–Khand 反应在骨架构建和全碳手性中心构建中的独特优势，以 20 步完成复杂天然产物（+）–propindilactone G 分子的不对称全合成工作，并据此将其结构进行了修正。杨震和郝小江等还分离得到柠檬苦素天然产物 perforanoid A，并完成其全合成，初步活性研究显示对一些肿瘤细胞系有细胞毒活性[142, 143]。叶涛和许正双等以汇聚式、全立体化学控制的方式完成 callyspongiolide 的全合成，并指定其立体化学，并完成活性靶点为翻译延长因子 1A 的大环脂肽类天然产物 nannocystin A 的全合成[144, 145]。贾彦兴等以分子内 Larock 型反应为核心高效合成 3，4- 并环的二氢苯并呋喃化合物，实现生物碱 galanthamine 和 lycoramine 的全合成，并设计了一系列的高化学选择性和立体选择性的转化，完成二萜 pallavicinin 和 neopallavicinin 的对映选择性全合成，整个过程无保护基操作[146, 147]。罗拓平等以立体选择性 Cope 重排为核心，完成一系列 ampilectane 和 serrulatane 天然产物的全合成[148]。叶新山等从简单的单糖原料出发，采用前期所发展的基于糖基供体“预活化”的一釜寡糖合成策略，以克级规模实现了寡糖链的大量制备，后经［31+31+30］糖基化反应，顺利实现了高度分支化、由 92 个单糖单元所组成的阿拉伯半乳聚糖的首次全合成；是迄今为止人类所合成的最大、最复杂的均一结构的多糖分子，在糖合成领域具有里程碑式的意义。该研究必将为复杂多糖的合成开启新的篇章。[149]

刘刚和唐叶峰等合作完成肌肉松弛素类天然产物 periconiasins A–F 的全合成，路线高效简洁，一次性解决了该家族所有代表性骨架的构建，为系统研究其生物功能奠定了基础[150]。祖连锁等基于氮杂频哪醇重排合成羟基吲哚的断键策略，实现吲哚啉和假吲哚结构的多样性合成，完成生物碱 calophyline A 的全合成[151]。

黄培强等专注生物碱合成研究，近期完成大环生物碱 haliclonin A 的首次全合成[152]粟武等合成了具有高抗增殖活性的环羧酚酸肽 coibamide A 全合成，并对其立体化学进行修正[153]。

3. 元素有机化学（有机氟化学为主）

有机氟化学已经发展成为国际研究热点，往往围绕着氟化 / 含氟试剂展开。包括开发新型含氟 / 氟化试剂以及为已有试剂找到新的用途等方面，相关领域有氟化、氟烷基化（单氟、二氟、三氟、全氟等）以及二 / 三氟甲硫基化等。

胡金波等将 1，2- 二芳基 -3，3- 二氟环丙烯类化合物发展成一类结构新颖的脱氧氟化试剂，命名为 CpFluors，可用于各类醇的高效氟化，是国内研究者自行研制的首个具有全新分子骨架的氟化试剂[154]；他们还发展了二苯基碘鎓离子催化下芳炔的氟代反应，可用于合成邻碘氟代芳烃[155]。刘国生等发展了钯催化苯乙烯的分子间高立体选择性氟 – 砜化反应，可用于合成一系列的 β – 氟代砜[2]。朱晨等将三价锰催化环丁醇 / 环丙醇自由基碎裂开环反应与氟化结合，可合成一系列 β – 和 γ – 氟代酮[104]。汪舰等发展了手性氮杂环卡宾催化下脂肪醛的氧化对映选择性 α – 氟化反应，可制备 α – 氟代羧酸[156]。

张新刚等以工业品氟代溴甲烷为单氟甲基化试剂，发展了镍催化芳基硼酸的偶联反应用于合成单氟甲基苯衍生物，初步机理研究表明反应经过单电子转移过程[157]。王细胜等以氟代溴甲基苯基砜为试剂，在镍催化下与芳基硼酸偶联，产物脱砜后也可合成单氟甲基苯衍生物[158]。

胡金波等在无过渡金属存在下实现 $TMSCF_3$/$TMSCF_2Br$ 与重氮化合物这两种卡宾前体的交叉偶联，可合成偕二氟末端烯烃[159]；通过超配位硅物种［$Me_3Si(CF_2H)_2$］的中间体，实现了对可烯醇化酮的高效二氟甲基化反应[160]；在可见光氧化还原催化下用氟代砜进行异腈的氟烷基化反应[161]。张新刚等发展了镍催化下（杂）芳基硼酸的二氟烷基化，可用 1- 溴 -1，1- 二氟烷基作为试剂[162]；实现镍催化下烯基酰胺的串联二氟烷基化 - 芳基化反应[163]；还在钯催化下实现芳基硼酸的插羰二氟烷基化[164]。卿凤翎等发现溴二氟甲基鏻盐在可见光诱导下可以现场产生二氟甲基自由基，可进行烯烃的氢化二氟甲基化反应[165]。沈其龙等制备了一种亲电的二氟甲基化硫叶立德，可用于各种醇羟基的直接二氟甲基化[166]。王细胜等发展了镍催化的不饱和羧酸的脱羧二氟甲基化反应，表现出较高的立体选择性[167]。

卿凤翎等以三甲基硅基三氟甲烷为试剂，发展了银介导的苯酚的氧化三氟甲基化反应，可直接合成芳基三氟甲基醚[168]。三氟甲基自由基与烯烃反应是三氟甲基化反应的热门研究领域，刘国生、刘心元、以及王洪银等都做出了有趣的结果[5, 33, 169]。

沈其龙等在二氟甲基硫化方面做了较多有益的探索，发展了铜促进下（杂）芳基重氮盐的 Sandermyer 二氟甲硫化反应，反应条件温和，能兼容各种官能团[170]。他们还开发了两种稳定的二氟甲硫基化试剂，分别为自由基型的苯磺酸二氟甲硫酯[171]和亲电型的 N- 二氟甲硫基苯二酰亚胺[172]，能够实现多种底物的二氟甲硫基化。

汤平平[173]和陈庆云[174]等分别独立报道了一价银介导的非活化碳氢键的氧化三氟甲硫基化。肖吉昌等以三苯基二氟亚甲基羧酸盐在单质硫和氟离子的存在下原位产生二氟卡宾，可在无过渡金属存在下实现脂肪亲电体（卤代物）的三氟甲硫基化，由于其中一个氟来自于外部的氟盐，该反应可用于制备有放射活性的氟 18 标记的三氟甲硫基产物[175]。

4. 天然产物化学

两年间，中国学者报道了 6000 多个新天然产物，在国际学术期刊上发表 1600 多篇论文，在高引用率期刊为 *Journal of the Natural Products*（154 篇）、*RSC Advances*（103 篇）、*Organic Letters*（60 篇）、*Marine Drugs*（22 篇）和 *Scientific Reports*（17 篇），分别占各杂志发表天然产物论文总数的 33%、74%、49%、35% 和 77%。

在所报道的新天然产物中，有 349 个新骨架化合物，占总数的 5.82%，有 1800 多个表现出生物活性，其中肿瘤细胞毒性、抗炎和抗病毒活性化合物的数量最多。其中有 20 个高活性的新骨架化合物。

肿瘤相关活性：张勇慧等从中药元宝草获得带有过氧键的混源萜 hyperisampsin J[176]。徐宏喜和许刚等从近无柄金丝桃中分离鉴定了高活性的新骨架萜类化合物 hypersubone B[177]。邱明华等从升麻中分离得到新骨架三萜类化合物 cimyunnin A[178]。林厚文等从海绵 *Dysidea sp.* 中分离得到新骨架萜类化合物 dysiherbol A[179]。尹胜等从麻疯树中分离鉴定了新骨架二萜 jatrocurcadione A[180]。

高抑菌活性：叶文才与江仁望等从红千层中分离得到新骨架萜类化合物 callistrilone A[181]。王长云和朱华结等从海洋真菌 *Pleosporales* sp. 的代谢产物中分离得到新骨架聚酮类化合物

pleosporalone A[182]。

高抗炎活性：姜勇等从九里香中分离得到新骨架吲哚生物碱 exotine A[183]。胡金锋、张海岩以及赵秋等从银杏叶中分离获得新骨架黄酮苷 biginkgoside E[184]。魏孝义等从土壤真菌的代谢产物中分离鉴定了新骨架混源萜 bisacremine G[185]。林厚文和李玉山等从酥脆倔海绵获得了新骨架喹啉生物碱 dysifragilone A[186]。孔令义等从三椏苦叶中分离得到新骨架聚酮类化合物（+）–melicolone B[187]。柴兴云等从矮紫堇中分离得到新骨架生物碱 hendersine A[188]。

糖尿病相关活性：岳建民与李佳等从桉树中分离得到新骨架倍半萜 eucarobustol A[189]。朱伟明等从浒苔放线菌的代谢产物中分离鉴定了新骨架聚酮类化合物 wailupemycin I[190]。马晓驰等从灵芝中分离鉴定了新骨架甾体 ganoderlactone B[191]。李友宾和吴崇明等从葫芦茶分离鉴定了新骨架的苯丙素苷 tadehaginoside D[192]。

其他高活性：郝小江等从牛筋果中分离鉴定了新骨架降萜类化合物 perforalactone B，具有显著的杀蚜虫活性及黑腹果蝇的烟碱乙酰胆碱受体拮抗活性[193]。彭成等从益母草分离获得新骨架二萜 leonuketal，具有较强的血管舒张活性[194]。岳建民等从海南叶下珠中分离获得新骨架三萜 phainanoid F，具有很强的免疫抑制活性，抑制伴刀豆球蛋白 A 诱导的 T 细胞增殖的 IC_{50} 值为 2.04 nM、抑制脂多糖诱导的 B 细胞增殖的 IC_{50} 值为 1.60 nM[195]。

5. 手性化学

手性是自然界的普遍特征，并与生命现象、人类健康和日常生活休戚相关。研究手性化合物的合成、性能及应用的手性科学与技术近几十年来受到广泛的重视，并已取得重大进展[196]。如 2001 年的诺贝尔化学奖授予了在手性催化合成领域取得杰出成就的三位科学家。我国的手性科学与技术研究虽起步晚但发展较快，并已跻身世界先进行列。据统计，我国在不对称合成领域发表的论文数量、SCI 引文数量和高被引论文数量在国际上均名列前茅[197]，最新数据显示，2011—2015 年我国在不对称合成领域发表的 SCI 论文数量、引文数量和高被引论文数量均居世界首位。近年来，手性相关十余个研究获得国家自然科学、国家科技进步奖二等奖，从侧面彰显了近年来我国手性化学研究取得了飞速的发展。

近三年来，我国科学家在手性合成、手性药物和手性材料等方面均取得了系列重要研究成果。在金属催化手性合成研究方面，我国科学家在新型催化剂设计合成和不对称催化新反应方面均取得了重要进展。设计合成了多类具有自主知识产权的优秀手性配体及手性金属催化剂，发展了系列金属催化的不对称新反应。

在有机小分子催化手性合成研究方面，我国科学家在新型催化剂、催化新反应及新策略方面均取得了重要进展。设计合成了多类新型的手性有机小分子催化剂，发展了系列有机催化的不对称新反应。

在手性药物研究方面，我国科学家开展了具有重要药用价值的天然产物分子及其类似物的手性合成研究，取得了系列重要研究进展。

在手性材料研究方面，刘鸣华等在手性纳米超结构的设计、组装以及手性传感、手性分离以及圆偏振光性能等方面进行了深入的探索，揭示了手性纳米结构的组装规律[198]，他们还实现了基于软材料的旋光逻辑门的有效设计[199]。王为[200]和崔勇[201-203]等分别构筑了系列新型手性共价有机框架材料，实现了系列高效多相不对称催化反应。段春迎[204]和崔勇[205]等分别构筑了一系列新型手性金属－有机框架材料，实现了系列高效的多相不对称串联催化反应和连续催化反应。苏成勇[206]和崔勇[207]等分别组装获得了系列新型手性金属－有机笼状和大环化合物，并系统研究了其对映体分离和手性荧光性能，深入揭示了主客体分子间的手性识别机理。吕小兵等采用手性双金属功能催化剂，成功实现了环氧化合物和环酸酐的不对称共聚合，首次获得了有规立构的主链手性聚酯[208]。车顺爱等实现了系列手性金属、金属氧化物以及有机无机杂化膜的构筑，并揭示了其组装机理[209-211]等。

（二）无机化学

2015—2017 年，无机化学在基础科学问题上取得了突出的新发现，同时在新能源、新材料、催化、癌症治疗等领域取得了丰硕的成果。

1. 基础科学问题的新发现

李隽与赵斌首次提出了［Zn8］和［Mn8］等金属团簇“立方芳香性”的新概念，并揭示了其遵循 6n+2 电子计数规则[212]。赵亮和朱军报道了四核金取代的吲哚与八核金取代的苯并二吡咯的合成，将超共轭芳香性概念拓展到过渡金属取代的杂环体系[213]。夏海平制备了五齿碳链螯合过渡金属的络合物，构筑了迄今最大共轭结构的平面型 Möbius 芳香性体系，产物吸收光谱宽、稳定性优异[214]。

李隽发现相对论效应改变了第六族双原子分子化学键的周期律，对于揭示相对论效应对化学键的周期性变化规律的影响具有重要意义[215]；其通过相对论量子化学理论证实了 Pr 处于 +V 价态，首次确认了稀土金属元素在化学条件下的最高氧化态[216]。程津培和吉鹏举通过对低极性的非质子型离子液体中叶立德前体等盐类化合物 pK_a 的精准测定，发现其 pK_a 与抗衡阴离子无关，对长期争论的“盐在离子液体中是否结为离子对”的问题给出了有力的实验澄清，为今后该类离子液体中溶剂化模型的建立提供了坚实的实验依据[217]。

席振峰和张文雄系统总结了金属杂环戊二烯类化合物的各类合成方法及其与不同小分子底物的反应性，展望了金属杂环戊二烯在金属有机杂环化学、配位化学以及高分子化学领域中的发展趋势[218]。苏成勇和潘梅提出了通过预拆分热力学稳定的手性金属配体进行纯手性金属－有机分子笼分步组装的策略[219]。王新平课题组利用大位阻配体和合适的弱配位阴离子成功合成了首例以锑为自旋中心的阳离子自由基，并证实了自由基的自旋中心主要分布在锑原子上[220]。李隽与 Lai-Sheng Wang 发现 $Au_2I_3^-$ 分子具有奇特的“键弯曲异构（bond-bending isomerism）”的新现象，研究表明这种罕见现象是源于化学成键作用和相对论及电子相关效应引起的亲金性作用的相互竞争[221]。

赵亮利用一系列具有相同绝对构型的手性二胺配体合成了具有非平面 S 型或反 S 型排列的六核手性金簇，并解释了 S 型与反 S 型排列的金簇在最大吸收波长处表现出的相反 Cotton 效应的原因[222]。

2. 新能源与电池

金钟和刘杰制备出多功能的 $CF@TiO_2@MoS_2$ 同轴纳米复合纤维，采用该复合纤维电极构筑了纤维状光电转换与能源存储一体化集成器件，其中太阳能电池部分的光电转换效率达到 9.5%，是目前世界上已发表文献中纤维状太阳能电池效率的最高纪录[223]。他们还设计了“海胆”状锂硫正极材料，在单质硫的面积负载率高达 3.2 mg/cm^2 下，仍然具有稳定的循环性能和较高的面积容量[224]；首次提出了制作完全由无机材料构成的钙钛矿太阳能电池，为发展能够在严苛环境下工作的高性能钙钛矿太阳能电池提供了新思路[225]。金盛烨发展了一种“固 – 固”卤素离子交换的方法，合成了含有卤素梯度的 $MAPbBr_xI_{3-x}$ 单晶纳米线，成功实现了在单个纳米线中光生载流子长距离输运过程的定向调控[226]。

杨清正构筑了基于有机纳米晶的光捕获体系，制备简单，光捕获性能优异，为系统研究光捕获体系中的光物理过程提供了物质基础[227]。陈刚提出了一种全新的基于刃型位错修饰晶面的策略，首次合成出了带有敞开的锂离子通道、大量反应活性位点的氧化钴负极材料，使其在循环稳定性和倍率性能上有了新的突破[228]；此外，采用非金属碳掺杂的方式实现了材料电导率的本征改变，使其电化学性能显著提升[229]。

金属所沈阳材料科学国家（联合）实验室先进炭材料研究部储能材料与器件研究组通过氨气高温处理氧化石墨烯的方法，获得了具有优异电化学性能、高含氮量的氮掺杂石墨烯。将碳纳米材料和具有化学锚定多硫化物功能的高导电金属氮化物相结合，采用一步水热法将氮化钒纳米带负载在三维石墨烯基体上，以多硫化锂作为活性物质填充在石墨烯与氮化钒复合材料集流体的三维孔道中，有效地解决了由“穿梭效应”带来的容量衰减及库伦效率低等问题，获得了优异的电化学性能[230]。

卢锡洪等成功研制了一款柔性可快速充放的水系 Ni//Bi 电池，不仅可以提供更高的比表面积和更多的活性位点，而且可以极大地提升高倍率下的氧化还原反应速率[231]。卢锡洪与吴义强合作共同开发并制备了一种高性能的、循环性能优异的柔性准固态 $Zn–MnO_2$ 电池[232]。卢锡洪与杨世和合作研发出了一种可作为柔性锌空电池新型阴极材料的氮掺杂 Co_3O_4 介孔纳米线阵列，组装了一种全固态柔性锌空电池，在不同弯折程度下仍能保持稳定的性能[233]。

3. 无机催化

郑南峰、傅钢、谷林及张鹏合作，采用乙二醇修饰的超薄二氧化钛纳米片作为载体，成功地制备了钯负载量高达 1.5wt% 的单原子分散钯催化剂[234]。

马丁等发展出一种新的铂 – 碳化钼双功能催化剂，在低温下（150~190℃）的产氢效率较传统铂基催化剂提升了近两个数量级[235]。常春然和瞿永泉设计并制备了一种由镍钴磷组成的三元体系催化剂，通过与石墨烯复合，表现出接近贵金属的优异电解水性能[236]。

傅强和包信和团队发现两维材料限域下的分子插层和催化反应增强现象，并提出两维限域催化效应[237]，发现限域状态下金属表面上分子吸附弱化这一普遍规律[238]。常春然等揭示了氧气在纳米金催化剂上以 OOH 形式活化的新机理，打破了氧气在金 – 载体界面处活化的一贯认识，提出氧气借助含氢底物（R–H）实现活化的新机理[239]。

王训等提出了制备非晶态镍钴配合物和 1T 相 MoS_2 复合全解水催化剂的新方法，在碱性体系中实现了 1.44 V 电化学全解水过电位（10 mA/cm^2 处）[240]。

马丁等设计出低温下即具有极高反应活性的 fcc 相 Ru 纳米粒子催化剂，是目前相同温度下报道的活性最高的费托合成催化剂[241]。

汪骋等设计合成了基于金属 Hf 和 C3 对称性配体的 3.6– 连接 kgd 拓扑结构的二维金属 – 有机层状配合物，为非均相催化剂的设计和合成提供了新的思路[242]。林文斌等利用 Zr 基金属有机框架（MOFs）后修饰和 MOFs 限域效应，原位构筑高混合度和高分散度的超小 Cu/ZnO_x 纳米粒子（NPs），在 CO_2 氢化到甲醇的反应中展现了高活性和 100% 的选择性，为优化催化活性和选择性提供了新思路[243]。

黎占亭团队以 Ru^{2+}/ 联 – 二吡啶配合物（$[Ru(bpy)_3]^{2+}$）为节点，首次实现了均相超分子金属有机框架 SMOF 的构筑，可以在水中均相进行，也可以在有机溶剂中以非均相方式实现，后者效率比文献报道的组成比例相同的 POMÌMOF 催化体系提高 5 倍[244]。

王贵昌等采用角分辨分析技术和理论计算相结合的方法研究了甲酸根在 Cu（110）表面的分解动态学行为，对 CO_2 加氢过程具有重要的指导作用[245]。谢兆雄等首次合成出六方晶系的铂镍合金枝状纳米晶，是 HER 催化剂中质量活性最高的[246]。

苏党生等发现通过水热碳化法可以制备出较高比表面的表面含氧基团富集的功能化碳催化剂材料[247]。苏成勇和张利首次发展了基于铱卟啉金属配体的金属 – 有机框架（Ir–PMOF–1（Zr），提出通过特定催化配位空间的设计实现反常化学反应选择性的学术思想[248]。

苏成勇等采用配位超分子自组装策略，结合配位空间的对称性设计和活性中心精准配置思想，在单一分子笼表面构筑出多个相互独立的能量传递和电子转移通道，实现了多通道电子转移和能量传递的单一性、方向性，极大地促进了产氢效率[249]。

4. 新材料：超级电容器、柔性人工机器、自修复材料、铁电材料

超级电容器是重要的新型储能器件，具有功率密度高、循环寿命长和安全可靠等特点。中科院上海硅酸盐研究所、北京大学和美国宾夕法尼亚大学共同研制新型碳材料所制备成水性电解液的超级电器，制备成水性电解液的超级电容器，行驶里程提高至 25~40km，而且制备成本低、充放电快[250]。

牛志强团队设计出一种全固态、一体化和可弯折的超级电容器与光检测器集成器件，为储能器件与其他电子器件的可弯折化和集成化发展起到积极的推进作用[251]。

胡征等得到了一种新型的多孔三维寡层类石墨烯材料，具有微孔－介孔－大孔相互连通的开放性孔结构，比表面积大，导电性高，在水系和离子液体中均有良好浸润性[252]。

卢锡洪和杨世和等开发了一种氮与低价钼共掺杂的 MoO_3 纳米线基纤维状双功能阳极，在作为微生物燃料电池与超级电容器的双功能阳极应用方面表现出优异的性能，有希望应用于自驱动型能源系统的开发中[253]。

刘忠范等成功在铜箔衬底上制备了不同旋转角度的双层石墨烯，并率先研制了旋转双层石墨烯光电探测器件，再将旋转双层石墨烯与等离子激元纳米结构耦合，可使光电流进一步增强 80 倍[254]。他们在玻璃表面成功地实现了石墨烯的直接生长，石墨烯玻璃防雾视窗同时具备了主动和被动防 / 除雾的能力，将光催化剂负载到石墨烯玻璃表面，还可以有效地提升催化效率[255]。

李承辉和鲍哲南利用配位键设计合成了一种高弹性的自修复材料，具有非常好的拉伸性，该材料制备出的人工肌肉器件动作可以通过外部电压来控制，该人工肌肉受损后可自动修复，修复后的器件仍然可承受与修复前同样高的电压，为人工肌肉走向智能化又迈出了重要的一步[256]。

王颖、陈雪波和江华提出柔性链对分子运动的抑制功能可能与其运动的时间尺度有关的新思路，研究了一系列由刚性框架定子和接枝有烷氧基甲基取代基的苯基转子所组成的分子旋转门，从实验和理论计算上详细研究了这些分子旋转门的动力学行为，特别是其运动的动力学特征与转子上的柔性链特征的相关性，首次在人工分子机器体系上观察到柔性直链对整体分子运动的抑制影响和运动形式的调节作用[257]。

金属所沈阳材料科学国家（联合）实验室先进炭材料研究部石墨烯研究组发展了一种以小分子松香作为转移介质的转移方法，实现了大面积石墨烯的洁净、无损转移；在此基础上与马东阁合作，制备出发光均匀、面积达 $56cm^2$ 的 4 英寸石墨烯基柔性 OLED 原型器件，已达到照明和显示的实用要求，并且数次弯折后性能不衰减[258]。

彭海琳等发现一类同时具有超高电子迁移率、合适带隙、环境稳定和可批量制备特点的全新二维半导体（硒氧化铋，Bi_2O_2Se），在场效应晶体管器件和量子输运方面展现出优异性能，有望解决摩尔定律进一步向前发展的瓶颈问题，给微纳电子器件带来新的技术变革[259]。

邓鹤翔团队采用小角 X 射线衍射原位观察孔材料中的气体吸附行为，首次揭示了小分子在限域空间中的有序自发聚集行为[260]；并通过组合包含 5 类金属的不同次级构筑单元和包含 6 类金属或非金属的不同有机配体，合成了一系列同拓扑的 27 种单组份 MOFs，借助光谱学首次破译了 MTV-MOFs 中的两种金属组分分布类型，并且研究了这两种金属组分分布与 MOFs 能带结构和催化效率的关系[261]。汪成课等提出在 MOF 配体侧端引入刺激响应基团，当负载染料后，经过商业化的环糊精水溶液洗涤，即可得到可控 MOF 药物释放体系的新策略[262]。苏成勇创造性地提出“后合成可调隔层装卸法（PVSI）”，不但可以在 MOF 骨架中插入带有不同官能团的第二桥联配体，改变原有框架的拓扑结构，改善框架孔道的尺寸、形状和表面，而且通过可逆插入 / 卸载不同长度的桥联配体，可以精确控制框架的呼吸效应[263]。

马秀良、许宁生和金传洪采用溶碳量适中的金属铂片作为生长基体，发展出一种基于“析出 – 表面吸附生长”原理的 CVD 方法，仅通过改变析出温度便实现了对石墨烯形核密度的控制，制备出晶粒尺寸在 200 纳米到 1 微米范围内均一可调、且晶界完美拼合的高质量单层多晶石墨烯薄膜[264]。杨敏等以电化学阳极氧化法首次合成了一维有序的 MoO_3 纳米管阵列，通过 H_2S 下的热处理构筑了一维有序的 $MoOx/MoS_2$ 核壳结构，具有协同效应，且电催化析氢和离子嵌入性能出色[265]。

焦丽颖等提出了二维 TMDCs 的化学气相输运合成新方法，通过对生长动力学进行调控，实现了多种二维半导体材料，如 MoS_2、$MoSe_2$、ReS_2 等材料的控制合成，构建了具有高迁移率的场效应晶体管、整流二极管等器件[266]。龙腊生和孔祥建等利用反相微乳法，成功地实现了单分散金属团簇分子 $Gd_{52}Ni_{56}@SiO_2$ 合成，从而实现了金属纳米团簇的单分子化[267]。王泉明和林玉妹等发现卤素作为配体对金属纳米团簇的形成起着重要的作用，并且获得了确定结构的含有 110 核的金属纳米团簇 $[Au_{80}Ag_{30}(C\equiv CPh)_{42}C_{l9}]Cl$[268]。

谢兆雄和匡勤等通过引入聚乙烯吡咯烷酮作为表面调节剂以及调控反应的动力学，成功合成了由高表面能 {110} 面裸露的超薄纳米片互相支撑形成的片状内凹立方体结构的铂锡合金纳米晶，在甲醇电氧化中展现出优异的性能[269]。郑南峰和张文卿等报道了全新手性双金属纳米团簇 $[Ag_{28}Cu_{12}(SR)_{24}]^{4-}$ 的合成及其全结构，成功地实现手性团簇的拆分和直接不对称合成[270]。

邓德会等通过二氧化硅纳米小球作为硬模板，采用一步化学合成法得到了均一的介孔泡沫状硫化钼材料，表现出了优异的酸性电解水制氢性能，展示了替代贵金属催化剂的潜力[271]。

5. 生物成像与癌症的治疗

刘又年团队通过超声剥离的方法制备黑磷纳米片，可将阿霉素的化疗、黑磷的光热和光动力活性有效结合起来，实现三种治疗模式的联合，有效消除肿瘤[272]。

中山大学生物无机化学团队与瑞士苏黎世大学 Gilles Gasser 教授合作，合成了系列带有 8 个正电荷的钌（II）多吡啶配合物，具有暗毒性低，光动力治疗系数高的特点，并成功应用于 3D 肿瘤细胞球的双光子光动力治疗[273]。

巢晖和龙建纲合成了一系列新型三联吡啶环金属化铱（III）配合物，得到了具有不同双光子荧光颜色的配合物，光稳定性好，线粒体靶向性强，不依赖于线粒体膜电位，免洗涤分离，适用于长时间的荧光观察，非常适用于实时示踪活细胞中线粒体的动态重构过程[274]。

尹学博发现选择高配位金属钆离子为金属节点和六羧基钌配位化合物为配体，可以在较宽反应温度范围和时间范围，且无需表面活性剂和辅助配体条件下，制备均匀的纳米级金属有机配位聚合物，很好地保留了钆的磁共振（MR）和配体的红色荧光的性质，成功应用于动物体内肿瘤模型的荧光—MR 双模态成像[275]。

杨凌采用计算机辅助筛选及酶催化位点局部改造的策略成功设计研发了首个细胞色素 P450 1A1 亚酶的高特异性荧光探针底物，不仅适用于单酶、细胞及组织等不同生物体系中 CYP1A1 酶活的实时定量检测及 CYP1A1 抑制剂的高通量筛选与评价，还可用于活细胞及活组织中 CYP1A1 的亚细胞定位及生物成像研究[276]。

（三）物理化学

1. 化学动力学

理论计算研究方面，张东辉等人发展了高效的分子结构筛选方法，大大提高了构建反应体系势能面的效率和精度，对燃烧化学中重要的基元反应 $HO+CO \rightarrow H+CO_2$ 势能面的拟合精度达到几个 meV[277]；该团队发展了基本不变量神经网络算法，成为目前小分子体系高精度势能面拟合最为有效的方法[278]。

气相化学动力学方面，田善喜等用自主研制的负离子速度成像装置研究了 CO_2 分子电子贴附解离动力学过程，发现有稳定 O_2 的产生，从而揭示了地球原始大气中氧气起源的新机制[279]。谢代前等基于苯酚分子高精度的非绝热势能面及其耦合和三维精确的 Chebyshev 量子波包方法，研究了振动基态的苯酚分子第一电子激发态的光解动力学，发现绝热模型对描述锥形交叉附近的动力学过程是不正确的[280]。边文生等在对 $C(^1D)+D_2$ 反应的研究工作中却发现弱的范德华力在这个势阱控速反应的入口谷处形成了范德华鞍，这种结构在低碰撞能下表现出与范德华阱完全不同的动力学行为[281]。刘舒和张东辉首次在理论上对 $H+H_2O$ 的初始基态、第一对称和反对称伸缩振动激发态反应进行了全维态－态量子动力学研究，证实了不反应的 OH 键在反应中作为旁观者的局域模效应[282]。他们还对 $H'+CH_4 \rightarrow CH_3H'+H$ 这一最简单的通过瓦尔登翻转机理实现的反应及其同位素类似物进行了精确的量子动力学研究，发现反应的阈值能量远大于势垒高度，并且显示出不同的同位素效应[283]。

复杂体系反应动力学方面，何圣贵等人在团簇与小分子的反应机理研究方面，开展了钒氧团簇阴离子负载金二聚体（$Au_2VO_{3,4}^-$）氧化 CO 的研究，发现在 Au_2 的辅助下，O_2^{2-} 物种可以直接或间接氧化 CO[284]。周鸣飞研究团队通过低温基质隔离红外光谱和串级飞行时间质谱－红外光解离光谱实验获得了 PrO_2^+ 络合物以及 PrO_4 中性分子的红外振动光谱，结合理论计算证实了其中的 Pr 处于 +V 价态，从而确认了镧系金属元素最高氧化态可以达到 +V 价[285]。

2. 电化学

电催化剂方面，郭少军和黄小青等通过设计出 PtPb/Pt 核壳结构的纳米催化剂，利用

Pb 带来的双轴应变效应提升了 Pt 催化剂的活性及稳定性，在 0.9 V（RHE）时活性达到了 4.3A/mg Pt[286]。孙世刚等提升 M–N–C 型催化剂的电池性能至 1 W/cm^2 以上，并认为活性位点很可能是含铁组分并且在高电位条件下为三价 Fe[287]。陈军等揭示了 $CoMn_2O_4$ 尖晶石立方相、氧缺陷、锰混合价态对于增强氧吸附、降低 O_2 活化能垒、诱导电荷转移、提升氧催化活性的促进作用，构建的可充锌 – 空气电池循环 155 周的容量保持率为 92.6%[288]。包信和和田中群等指出新型过渡金属硫化物、g–C_3N_4/ 石墨烯等均有望成为新型替代贵金属的水电解催化剂[289]。

快速充放方面，黄富强和陈一苇等科研人员合成了一种有序介孔少于 5 个原子层碳的新型材料，碳 sp2 杂化程度高达 98%，氮掺杂后，其电化学储电的比容量高达 855Fg^{-1}[290]，张华民等原创性地提出"离子筛分传导"新概念，合成出不含离子交换基团的高离子选择性、高稳定性、低成本非氟离子传导膜，应用于多项示范工程[291]。郭玉国等提出利用三维纳米集流体来引导金属锂在三维电极内部的均匀沉积与溶解的思想，成功实现了金属锂枝晶的控制[292]。杨勇等通过 ^{15}N 标记的高分辨固体核磁共振谱技术并辅助理论模拟结果，提出了具有大 π 共轭三聚的喹喔啉衍生物有机电极材料的多步逐一锂化的电化学反应机制[293]。张校刚等在 MXene 层间进行酚醛树脂和嵌段共聚物的有机 – 有机限域自组装，经过高温碳化首次实现 MXene 和单层有序介孔碳的复合材料；经过氯化处理，得到 MXene 衍生碳 – 有序介孔碳复合材料；单层有序介孔碳不仅可以阻止 MXene 及其衍生碳材料在制备成电极的过程中发生堆叠，其中介孔孔道还可以为电解液在片层间的传输提供定向的通道[294]。任斌等利用针尖增强拉曼光谱（TERS）以 3nm 的空间分辨率对 Pd/Au（111）双金属模型催化剂表面进行拉曼光谱成像，表征出不同表面位点的电子性质与催化性质，表明 TERS 有望发展为原位表征催化剂表面结构及反应过程与机理的新工具[295]。

3. 化学热力学与热分析

（1）离子液体体系化学热力学研究

张锁江团队发展和完善了网上"IPE ionic liquids database"，现有离子液体 4282 种，阳离子 1539 种，阴离子 360 种，总数据约 19 万条，并据此数据库出版了混合离子液体数据手册和离子液体催化专著[296, 297]，开展了离子液体构效关系研究。尽管氢键属于弱相互作用，但其对离子液体性质，如熔点、粘度及蒸发焓等具有重要影响。如咪唑离子液体，随着阴离子氢键增强这些性质呈现下降趋势，与常规溶剂相比，呈现了相反的变化趋势[298]。基于氢键的特殊性，提出离子液体 Z 键模型，通过引入"离子片"概念和介尺度原理对离子液体进行粗粒化划分，开发了新型的离子液体粗粒化模拟方法；针对离子液体气体分离，提出了吸收分离指数（ASI）新概念，建立了考虑多因素的离子液体设计筛选新方法，被 *AIChE J* 期刊评为"Top Tier" contribution[299]。李浩然等分别应用红外和核磁共振两种谱学方法，结合热力学模型，测定了醋酸正丙铵离子液体的离子率为 93%，为质子型离子液体的离子率测定建立了可靠的方法[300]。

张锁江等对咪唑类离子液体水溶液模拟发现，在低浓度下存在球形团簇，中等浓度下存在棒状团簇，高浓度下则出现囊泡，与实验结果吻合良好[301]。另外，部分离子液体水溶液体系随浓度降低其粘度先上升后下降，分子模拟发现这是由于体系中存在的链状团簇影响了粘度[302]。韩布兴等测定了1-乙基-3-甲基咪唑四氟硼酸盐（$EmimBF_4$）/乙醇、1-丁基-3-甲基咪唑六氟磷酸盐（$BmimPF_6$）/乙醇、1-已基-3-甲基咪唑四氟硼酸盐（$HmimBF_4$）/水二元溶液体系的相行为，用小角X射线散射（SAXS）等技术研究了这些二元体系的微观结构，结果表明当溶液的组成趋于分相点时，在溶液中形成微区，并且其尺寸急剧增大，提出了液-液相分离的机理[303]。尉志武等测定了离子液体/二甲基亚砜和离子液体/醇体系中离子液体与共溶剂的氢键作用以及体系的偏摩尔体积和粘度系数等，发现体系中存在一些具有相对稳定结构的物种[304, 305]。关伟等针对文献中有关离子液体二元体系全组成范围粘度中的问题，提出了两个新概念——混合物的相对体和混合物的相对粘度，推导出计算无量纲的混合物粘滞流动超额活化Gibbs自由能计算公式[306, 307]。

（2）离子液体体系中分子间相互作用对化学反应和材料合成的调控研究

韩布兴等采用相行为测定、SAXS等方法研究了离子液体1-辛基-3-甲基咪唑高氯酸盐（$OmimClO_4$）/三乙基铵硝酸盐/硅酸四乙酯/对苯二甲酸/Ni（NO_3）$_2$体系的相行为和微观结构[308]。该团队还研究了离子液体（$BmimPF_6$）/有机溶剂/水三元体系的微观结构[309]；利用质子型离子液体1，8-二氮杂二环十一碳-7-烯2-甲基咪唑盐作为催化剂和反应介质，实现了常压转化CO_2合成恶唑烷酮类化合物[310]。

（3）CO_2/水/离子液体和CO_2/水/乳化剂体系的热力学研究

王键吉等设计合成了一系列由CO_2驱动的疏水-亲水可逆转变的新型离子液体和由不同分子量聚乙二醇功能化的温度响应离子液体[311, 312]。张建玲等通过分子间相互作用研究，采用与水和CO_2同时具有较强相互作用的金属配合物庚二酮酸铁作为乳化剂，在一定条件时可将CO_2/水两相体系转变为热力学稳定的均相体系[313]；还研究了功能材料金属-有机框架对CO_2/水两相体系相行为的影响，发现在适当压力下，金属-有机框架可有效乳化超临界CO_2和水，形成稳定性很高的乳液[314]。

（4）离子液体水溶液传递-反应过程的非平衡热力学研究

目前沼气提纯为生物甲烷的工艺存在能耗高、投资大等问题。陆小华等在氯化胆碱离子液体体系的基础物性测定、过程强化模型以及工艺装备方面取得进展[315–319]，为新型沼气提纯工艺设计开发奠定了基础。

4. 催化

在创制新催化剂与发展催化反应新体系方面的亮点体现在合成气（CO与H_2的混合物）的催化转化制取烃类（费托合成，FTS）及其相关研究中。包信和等将金属氧化物与沸石分子筛复合，创制了双功能复合“氧化物-沸石”催化剂，发现该催化剂能够直接把合成气高选择性（up to 80%）地转化为C2–C4低碳烯烃[320]。孙予罕等以纳米棱柱结构

Co_2C 为催化剂，在较温和条件下实现了由合成气高选择性（>80%）制烯烃，且产物中烯/烷比高达 30 以上、甲烷选择性低[321]；他们随后还通过双功能催化耦合，实现了 CO_2 加氢转化高选择性（80%）制取汽油组分（C_{5+} 烃），其中异构烃与正构烃之比大于 16、甲烷选择性低至 1%[322]。马丁和石川等在 α-MoC 表面制备了原子级分散的 Pt（Pt/α-MoC）催化剂，实现了高效、低温（120~190℃）甲醇-水蒸气重整制取氢气[323]；他们还发现纳米 Au/α-MoC 催化剂对于活化水分子、以及水-气变换反应中拥有优秀的低温反应活性和催化稳定性[324]。马丁等通过对 Fe_5C_2 的电子性质进行调变，在碳化铁催化剂上显著提高了 FTS 产物中烯烃的选择性[325]。于吉红等采用水热法制得了 Silicalite-1 交叉孔道限域的 0.3 ~ 0.6 nm 的 Pd 金属簇，对于甲酸分解制氢反应表现出高活性和稳定性[326]。苏党生等发现核壳复合结构纳米金刚石的缺陷位等与 Pt 颗粒间有着强相互作用，能够在丙烷脱氢反应中促进丙烯脱附，进而提高催化剂的稳定性[327]。赵凤玉等以纳米晶纤维素（NCC）为模板合成了系列多级孔 ZSM-5 沸石和 Ni/NCC-ZSM-5 催化剂，后者在水体系中可高效催化纤维素氢解制六元醇[328]。刘国生等通过发展金属催化/自由基接力的新策略，发明了苄位碳氢键的不对称氰化反应，实现从简单芳烷烃到手性芳基乙腈的高选择性转化[329]。黄正等发展了高效铱基烷烃脱氢催化剂，利用串联催化反应体系首次实现了从烷烃到高附加值直链烷基硅的高效转化[330]；他们还与加州大学尔湾分校管治斌教授合作，通过建构交叉烷烃复分解反应体系，在温和条件下实现了由聚乙烯废塑料制造燃油的催化可控降解过程[331]。张新刚等以二氟卡宾钯为催化剂，实现了大宗廉价氟化工原料 $ClCF_2H$（8~10 元/千克）对（杂）芳基硼化合物的二氟甲基化反应，为氟烷基化化学提供了新策略[332]。

在催化剂构效关系与催化反应机理研究方面，刘中民等的原位 MAS NMR 研究捕捉到了甲醇制烯烃反应（MTO）中由二甲醚活化形成的类亚甲氧基物种，并提出一个协同反应机理阐释了 C1 物种活化形成初始 C-C 键和烯烃的过程[333]。邓风等建立了 ^{13}C-^{27}Al 双共振固体 NMR 新技术，观测到甲醇制烯烃反应中的“烃池”物种可与沸石分子筛催化剂的骨架作用形成超分子活性中心[334]；他们还发展了灵敏度增强固体 ^{67}Zn MAS NMR 技术，在 Zn 改性 ZSM-5 中观测到 Zn^{2+} 物种与 Bønsted 酸中心作用形成协同催化活性位，为理解烷烃 C-H 键在相关催化剂上的活化机制提供了依据[335]。黄伟新等通过研究甲醇在纳米晶 TiO_2 表面热催化和光催化活化与反应，揭示了晶面效应在表面催化反应中的表现形式[336-339]。

在光催化和光电催化分解水研究方面，李灿团队利用半导体光催化剂不同暴露晶面之间的光生电荷分离效应构建高效的光催化体系[340]，在实验上初步确认了强碱条件下半导体与分子催化剂之间可发生单步两电子转移过程[341]；将构筑光电催化体系的“空穴储存层”的概念用于构筑光阳极体系，使光生电荷分离效率接近 100%，光电流逼近热力学理论极限[342]；构建了国际上最早的自然-人工光合成杂化系统，并实现了完全水分解过程[343]；在国际上创造了多个纪录：在基于宽光谱响应半导体材料构建的 Z- 体系，使光催化全解水制氢过程的量子效率达到 6.8% @420nm[344]，太阳能光电催化分解水制氢效率突破 2.5%[345]；发展了太阳能对液流电池体系的直接充放电储能电池，能量转化效率超过 3.2%[346]。

在应用催化研究方面，中科院山西煤化所和中科合成油公司通过催化剂和反应过程创新，开发了国际领先的高温浆态床费托合成油新技术[347]，已在国内建成了 5 个煤（经合成气）制油工业生产厂，合成油能力达到 700 万吨 / 年；与国外现有技术相比，该新技术不仅催化剂活性更高、稳定性更好，而且显著提高了油品的选择性，总体能量效率提高至 42%~47%（国外为 37%~38%）。中科院大连化物所突破了甲醇羰基化反应沸石催化剂活性低、稳定性差的难题，提出以煤基合成气为原料、经二甲醚羰基化制乙醇新策略，开发了具有自主知识产权的全球首套煤基乙醇工业示范装置并一次投产成功，标志着我国将率先拥有设计和建设百万吨级大型煤基乙醇工厂的能力。宗保宁等成功开发了环已酮绿色生产技术，包括国际上首创原子经济性制备环已酮的新反应途径、创制高效催化苯部分加氢负载型 Ru 催化剂和采用催化蒸馏技术实现苯、环已烯和环已烷的高效反应分离，该套技术已完成千吨级工业示范和 20 万吨 / 年装置工程设计，并获得多项专利授权[348]。鲍晓军等提出了烷基转移脱硫新思路，通过发展适于烷基转移反应的新型固体酸载体，研发出了 FHUDS–5 催化剂以及强化烷基转移功能的催化剂级配技术，有效克服了油品加氢精脱硫难以脱除 4，6- 二甲基二苯并噻吩类组分的世界难题；与国外技术相比，氢耗降低约 15%，运转周期延长 30% 以上。

5. 光化学

在合成光化学方面，可见光促进的多类惰性化学键的活化与官能团化，取得许多重要进展。吴骊珠课题组创造性地提出了放氢交叉偶联反应，使可见光驱使的放氢氧化反应在无外加氧化剂的条件下进行，具有更高的环保价值和经济价值，实现了常温常压下光催化苯和氨气一步合成苯胺以及苯和水一步合成苯酚的交叉偶联放氢反应[349]。罗三中等实现了三级胺 α–C–H 键与酮类化合物 α–C–H 键的不对称交叉偶联放氢反应[350]。雷爱文和吴骊珠等实现了无外加氧化剂条件下硫代苯甲酰苯胺类化合物的分子内氧化环化，生成了苯并噻唑和氢气[351]。雷爱文又实现了光催化芳烃与脂肪环胺放氢交叉偶联制备芳胺[352]以及烯烃和水、醇、胺的放氢交叉偶联反应[353,354]。肖文精利用 CdS 半导体光催化剂[355]，吴骊珠等人工模拟光合作用是实现太阳能的高效转换和利用最直接的方式。吴骊珠等在光催化分解水的氧化、质子还原两个半反应以及水的全分解研究中利用量子点光敏剂和分子催化剂取得有重要影响结果[356-361]。李灿团队利用半导体光催化分解水的研究中提出若干新概念，在国际上产生了重要影响[362-365]。赵进才等针对 α–Fe_2O_3 表面进行质子耦合电子转移的机理做了深入的研究，从原理上和实验室实现了对 α–Fe_2O_3 光电水氧化过程的调控[366]。张铁锐团队制备了富含氮缺陷的石墨相氮化碳纳米片，其光催化产氢速率得到大幅提升[367]；通过调控超薄水滑石纳米材料的堆叠厚度以及氧缺陷掺杂活性位点，在非贵金属光催化还原 CO_2 方面展现出优越的性能[368]；利用 LDH 超薄纳米光催化剂以 N_2 和水为原料合成了 NH_3[369]。

在发光材料制备方面，汪鹏飞等以共轭聚合物为前驱物结合高温碳化技术制备了发红光的碳量子点，这种量子点在近红外激光诱导下具有强的光声响应性和高的光热转化效率，可用于对肿瘤的诊疗一体化[370]。唐波等将卟啉类光敏剂以桥联配体的形式嵌入到 MOFs 材料中，可将硫化氢信号分子特异性激活，从而实现活细胞内活性氧的可控释放并达到有效治疗肿瘤目的[371]。田禾和朱为宏等成功实现了对 β－半乳糖苷酶的实时在体、原位检测，并获取了高分辨三维活体成像信号，实现对肿瘤组织的精确定位[372]。彭晓军和樊江莉等设计合成了比例型检测细胞内极性的荧光探针，首次检测到癌细胞线粒体的极性与正常细胞的差别[373]。

激光材料的发展和应用方面，赵永生等将两种染料的有机单晶纳米线进行轴向耦合，构筑复合谐振腔结构，获得了双波长的单模激光，极大的提高了光子学集成器件的集成度和灵活性[374]。

6. 生物物理化学

理论与计算生物物理化学方面，吴云东等将其开发的残基特异性力场 RSFF 应用于环状多肽的模拟，能准确预测其优势构象并揭示了其靶向蛋白－蛋白相互作用的潜力[375]，并与实验合作研究了稳定 α－螺旋多肽的策略[376, 377]；还将新力场用于研究磷酸化诱导的天然无序蛋白的折叠机理[378]。来鲁华等系统研究了炎症相关花生四烯代谢网络的调控方式[379]；发展了基于残基运动相关性的蛋白质别构位点预测方法[380]；建立了从别构位点预测到别构化合物发现的研究策略，在 15– 脂氧合酶中通过理性设计发现了别构激活剂[381]；在癌细胞代谢网络关键靶标中设计得到了别构抑制剂[382]。葛颢等得到了宏观非线性化学反应动力学模型的非平衡热力学新理论，包括宏观自由能函数的非平衡态推广及其均衡方程，这些亚宏观随机理论在宏观尺度上展示出新性质[383]。葛颢和谢晓亮利用粗粒化的随机模型研究了 DNA 聚合酶的工作机理，阐述了酶分子结合底物之后构象的非平衡态弛豫过程是如何和与催化的 DNA 链延长反应紧密结合在一起的，以及这样的结合又是如何同时显著地降低被催化反应的势垒高度[384]的。

实验生物物理化学方面，徐平勇等发展了活细胞超高分辨成像新方法 SIMBA，具有 50nm 和 1s 的时空分辨率，可直接用于全内反射荧光显微镜、PALM/STORM 等仪器，为活细胞内亚结构的超高分辨动态成像提供了工具[385]。赵新生开发了扫描单分子 FCS 技术，实现从微秒到秒过程的同时探测，为将 FCS 用于更多有生物意义的体系奠定了基础，他还推导出结合二阶和三阶 FCS 唯一确定荧光相对亮度的公式，圆满解决了 FCS 技术应用的两个难题[386, 387]。赵新生和高毅勤研究假尿嘧啶生成酶的催化机理，发现 V149 是催化反应的关键残基，其作用为准确地将尿嘧啶放到接近反应过渡态的位置，酶巧妙地利用与水分子的氢键作用，放大底物与产物的差别[388]。

7. 胶体与界面化学

新型有序分子组合体的构建、调控和功能化方面，张希等构筑了一系列功能性超分子聚合物，实现对聚集体形貌、结构和性质的掌控[389]。黄建滨、阎云等通过设计两亲分子，展示了分子自组装中的别构现象，有望在分子的指纹识别方面发挥重要的作用[390]。杜学忠等首次阐述了构筑的柱芳烃超分子囊泡的微观结构，在肿瘤和相关疾病治疗方面具有潜在的应用前景[391]。王毅琳等近年来研究了寡聚表面活性剂在固体表面的组装行为及规律，拓展了寡聚表面活性剂在生物领域的应用[392]。聚表面活性剂混合体系强的协同作用，可高效地诱导聚集体的转变并优化聚集体结构，为寡聚表面活性剂在实际配方中的应用提供理论依据[393]。寡聚表面活性的高效抗菌活性而低细胞毒性的特性又使其有望成为一种新型抗菌剂[394–396]。吴立新等提出利用阴离子客体诱导阳离子短肽形成多价纳米纤维的新思路，开发了基于多金属氧簇与阳离子短肽共组装构筑多价纳米纤维的新方法[397, 398]。郝京诚等确定了可以实现凝胶化的两亲分子聚集体类型并开展了一系列研究[399–402]，通过设计合成一种具有 pH 和光双重调控能力的非手性的偶氮苯衍生物分子，并以该分子作为一种凝胶剂分子，利用其和 CTAB 在酸性条件下可以在超分子层面上构筑手性凝胶材料[403]。李峻柏等发现二苯丙氨酸有机凝胶中的纳米纤维能转变为高度有序的晶体结构，为利用相转变机理，调控单组分的超分子组装提供了新途径[404]；利用分子组装技术，将光酸分子与 ATP 合酶分子马达共组装，在光照下实现了 ATP 的高效合成；这一组装体系有效地模拟了叶绿体中光系统 II 与 ATP 合酶分子构成的光合作用，为有效利用光能提供了新途径[405]。郭荣等首次以一步混合法批量制备了内相液滴由互不相溶的食物油（VO）和硅油（SO）组成的 Janus 乳液，并研究了该新型乳液的乳化条件及形成机理，以此为模板制备了结构由“雪人状”至“哑铃状”连续转变的高分子 Janus 粒子[406]；还实现了通过油相的组成调控 Cerberus 液滴中各微室的结构，为各向异性粒子的制备奠定了基础[407]。郝京诚等设计合成了含有偶氮基团与磁性反离子的新型刺激响应表面活性剂，用于实现 DNA 分子的远程可控捕捉与释放；以 DNA 作为基本构筑基元，通过与金属离子、磁性表面活性剂以及聚合物组装形成各种形貌的基于 DNA 的纳米结构；还将其作为药物运输的载体，实现了靶向药物运输以及可控释放[408–410]。王毅琳等系统地研究了阳离子表面活性剂 OHAB 和阴离子表面活性剂 SDS 在水溶液中的自组装行为及其对玉米蛋白的增容和与磷脂的相互作用，认识到表面活性剂对皮肤的刺激性，建立具有低皮肤刺激性的表面活性剂体系[411, 412]。

微纳米功能材料方面，郭荣等在导电高分子 – 贵金属复合材料的合成策略、机理及其应用等方面取得了一系列有特色的成果[413]。齐利民等实现了 Ag_2S–Ag 异质结构纳米碗阵列的可控制备[414]，结构新颖的异质结构 TiO_2 纳米棒 @ 纳米碗阵列的制备[415]，大面积高质量的有机金属卤化物钙钛矿纳米网及纳米碗阵列的制备[416]，首次实现由方解石单晶构成的微透镜阵列的人工合成[417]。房喻等利用气相荧光探测技术实现了多种常见毒品的原位、高灵敏气相可逆探测。

农药高效利用方面，杜凤沛等发展了植物叶面性质的表征方法，率先提出了基于植物叶面界面性质和固液界面作用进行农药胶体分散体系精细化研发的理念，深入开展了表面活性剂在提高农药利用率和减少农药用量方面的理论及应用研究，在表面活性剂分子结构、电荷性质与固体界面性质及超咪唑性离子液体分子作用力对液滴对靶润湿、沉积行为的作用规律和机制、降低弹跳等界面行为方面取得了显著进展[418]。

超高分辨荧光成像应用方面，李峻柏等在超高分辨图像采集和数据分析方面发展了实时单分子定位的程序包 SNSMIL，该程序包可广泛应用于高背景成像的数据分析；用此方法获得了被细胞内吞后的外源脂质体在细胞骨架上的分布和运动轨迹，为研究脂质体或外源组装体与细胞作用提供了重要的检测手段[419]。

（四）分析化学

1. 近两年我国分析化学发展概况

2015—2016 年我国在国际分析化学领域的核心期刊发表学术论文近千篇，在不少领域取得重要进展。

功能性核酸的研究方面，谭蔚泓等合成了新的 DNA 探针，可以将瞬时的膜上发生的细胞表面分子与胞外环境相互作用过程转化为累积的、可检测的荧光信号，成功地检测到了细胞膜脂类区域的事件和相互作用[420]；发展了一种基于适配体的外泌体表面纳米修饰技术，实现了对靶细胞来源的外泌体的选择性组装，证明了在纳米尺度的细胞器表面实现 DNA 纳米结构的组装的可行性[421]；对 DMFs 进行了一系列的表征，发现与其他 DNA 纳米结构相比，DMFs 拥有更大的 DNA 负载量，在含有核酶和细胞裂解液的体系中，DMFs 表现出良好的稳定性[423]。交叉学科联用新方法方面，汪劲等基于漏斗能量地貌（Funnel-like energy landscape）理论，提出了特异性理论计算方法，该方法能有效地区分特异性复合物结构与非特异性结构[424]。徐国宝等将青蒿素用于鲁米诺化学发光体系并首次应用于血红素和血迹的检测与成像，用智能手机作为检测器，基于鲁米诺 - 青蒿素化学发光方法可以实现稀释 100000 倍的血液的检测，在某些金属离子和生物分子存在下表现出了较高的选择性，并且能够区别鉴定血迹和其他的类似血迹的污渍如咖啡、红糖、红茶污渍[425]。生物活体分析方面，毛兰群和于萍等提出了通过合理调控离子间相互作用，而实现生理活性高选择性分析的新思路[426]；提出并建立了基于微米管整流的活体分析新原理和新方法[427, 428]，该方法有望发展成为胞内单个囊泡的检测方法和新型的活体分析技术；还提出并建立了基于自触发传感的活体分析新原理，通过有机溶剂有效地解决了由于传统酶在电极表面地取向不确定、效率低的问题，不仅在生物电化学的基础研究中具有重要的意义，而且也为进一步构筑基于生物燃料电池原理的自驱动活体分析与传感奠定了基础[429]。

单分子单细胞检测方面，龙亿涛团队构建了独特的 Aerolysin 生物纳米孔道单分子界面，将 DNA 单链过孔速度降低了 3 个数量级，极大地提高了单分子电化学界面分析的精准度，实现了单个碱基差异 DNA 分子的实时分辨[430]；利用电化学刻蚀技术获得 1.5~20nm 的单个可控固体纳米孔道，限域延长单个生物分子亚稳态结构寿命，可检测单个 DNA 分子的动态构象，发展为单分子动力学研究的新手段[431, 432]；设计了尖端小于 100 nm 的超微玻璃毛细管纳米孔道电极，建立了电化学精准控制单细胞成像新技术[433]；实现了单细胞内 H_2S、CO 和 NO 等内源性气体分子的高选择性、快速、原位动态监测，构建了可特异识别野生和变异 p53 蛋白的纳米囊泡，解决了野生和变异 p53 蛋白难以区分的问题，为研究 p53 蛋白相关信号通路提供了新方法[434–435]；设计了“层层解装”新型复合等离子体纳米探针，对不同肿瘤分子标志物进行成像，研究其相互作用关系和细胞信号通路[436]。王宏达等结合单分子力示踪技术和单颗粒模拟研究单个人类肠病毒 71（HEV71）入侵宿主细胞过程中的细胞膜的动态过程[437]，揭示了细胞膜包被病毒的机理，提供了细胞如何操纵病毒以启动病毒内吞进入细胞的新见解。

纳米和成像技术方面，逯乐慧等发现聚多巴胺黑色素具有非常好的生物兼容性，能够有效地消除导致脑损伤的自由基，从而降低缺血性脑中风导致的脑损伤[438]；利用聚多巴胺（PDA）材料设计了对骨微损伤靶向的 SERS 探针，实现了静脉注射条件下活体内骨微损伤的高效识别[439]。姜秀娥等发展了一种对酸性过氧化氢响应的增氧核壳结构 PDT 纳米平台，利用 MnO_2 壳层作为可控的屏障来阻止 PS 从内核中提前释放，并且还能够提高肿瘤组织中氧气的浓度，具有选择性活体核磁成像（MRI）能力[440]，首次将 MnO_2 同时作为 H_2O_2 敏感的屏障、增氧试剂及 MRI 造影剂。

2. 色谱学

2015 至 2017 年 3 月，我国在该领域发表 SCI 论文数目占到总数的近 26.43%，又有了新的提高。

样品前处理方面，张丽华等发展了基于尿素和离子液体溶解的膜蛋白顺序提取法，实现了对 Hela 细胞膜蛋白质组表达谱的深度分析，建立了目前为止关于 Hela 细胞最为全面的蛋白质组和膜蛋白质组表达谱数据集[441]。严秀平等发展了基于共价有机框架（COFs）材料的固相萃取技术，可实现对双酚 A、双酚 AF 的快速富集（5min 内达到吸附饱和）[442]。李攻科等发展了内置磁性纳米离子的共轭微孔聚合物，并与高效液相色谱和荧光检测相结合，实现了吸烟者尿样中微量羟基多环芳烃的鉴定，检测限可低至 0.01 μg/L[443]。邱洪灯等发展了单层多孔石墨烯，成功实现了 K^+，Na^+，Li^+，Ca^{2+} 以及 Fe^{3+} 的选择性富集与分离[444]。

色谱柱方面，梁鑫淼等开发了“硫醇－烯基”点击聚合的硅球表面修饰技术[445]，并商品化，提高了我国在色谱填料研发领域的地位。叶明亮等发展了基于硫巯－烯点击化学

法制备有机 – 无机杂化整体柱[446-448]，使我国在有机 – 无机整体柱研制领域处于国际领先水平。张丽华等发展了周期性有序介孔硅（PMOs）整体柱[449]并用于蛋白质组学研究，为有机 – 无机杂化整体柱的研制提供了新的思路。

生物体中具有重要生物学价值的低丰度组分的分析方面，张祥民等构建了一个在线阵列二维液相色谱分离平台，并将其成功应用于人血浆高丰度蛋白质的去除，有效提高了低丰度蛋白质的检测[450]；发展了一种集成化蛋白质组分析装置，并应用于 100 个细胞的蛋白质组分析，成功鉴定了 813 个蛋白质，检测灵敏度可达 200 zmol[451]。

色谱仪器关键部件研究方面，关亚风等研制的液相色谱荧光检测器[452]采用具有自主知识产权的光电放大器组件，灵敏度与进口荧光检测器相当，极大降低了仪器成本；全自动在线固相萃取装置[453]能与自动进样器和液相色谱在线联用，实现样品的在线富集、纯化、洗脱和高压进样，解决了传统固相萃取与液相色谱无法在线联用的瓶颈，打破了国外公司的技术垄断。

在应用研究方面，张丽华等发现了一种包含代谢标记、蛋白均衡、蛋白分级和膜辅助的样品预处理含血清培养条件下分泌蛋白定性定量分析的新策略[454]，实现了 575 个分泌蛋白的高灵敏度分析。叶明亮等首次将 SH2 超亲体与实验室前期发展的固定金属离子亲和色谱相结合，用于复杂样品中酪氨酸磷酸化肽段的富集，使酪氨酸磷酸化蛋白质组学的研究达到了一个前所未有的深度和广度[455]。许国旺等发展了基于拟靶向分析技术，QC 样品和信号校正的方法来提高大规模代谢组学研究中数据的稳定性[456]，可以显著的提高多批次间数据的重复性和利用率；还构建了一种代谢组和脂质组同时分离的二维液相色谱系统，实现一次进样同时覆盖代谢组和脂质组信息[457]。尹玉新等发展了基于平行多反应检测模式的大规模靶标代谢组学分析方法，相比 MS1 扫描模式具有更高的重现性和定量准确性[458]。再帕尔等发展了高效提取食管组织中内源性代谢物的样品预处理方案，对提取过程进行了系统优化，提高了低丰度代谢物的覆盖度[459]。

中药分析质量控制及制备方面，果德安等针对经典方剂葛根芩连汤的质量控制难题，发展了同时对 50 种活性物质的快速色谱质谱定量分析方法，实现了不同厂家来源样本的有效区分[460]。岳建明等从三尖杉属的海南粗榧中发现了两个新化合物[461]，具有很好的抗癌活性；梁鑫淼等针对天然生物碱制备难题，发展了基于新型分离材料的多维正交高效液相制备方法，从唐古特山莨菪中发现了 3 个新化合物并进行了构效关系研究[462]。

3. 质谱分析方法与仪器研制

根据 Scopus 数据库不完全统计，2015—2017 年，我国质谱领域在《自然》《科学》《美国化学会会志》等国际一流期刊上共发表文章 600 余篇，占总数的 6.91%。中国的文章总数为美国的 1.47 倍，且高水平文章与美国差距不断缩小。

新型离子源的开发和应用方面，王中林等利用摩擦纳米发电机驱动离子源，实现离子源的精确控制，为质谱分析提供了一个全新的可控参数，也是纳米发电机在大型分析仪器首次应用[463]。杭纬等设计了高功率密度激光电离飞行时间质谱仪（fs-LI-O-TOFMS），实现了实时快速地对固体样品的无标样定性和定量分析；利用该仪器实现了对陨石的三维元素质谱成像和对固体金属表面元素分析[464, 465]。段忆翔等开发了微辉光放电等离子体探针 MPP，可被用于生物组织分析、呼吸气直接分析等生物学应用，扩展了辉光放电等离子体离子源的应用范围[466]。

单细胞研究应用方面，黄光明与熊伟等基于自行开发的单细胞电生理与质谱联合检测平台，在单细胞层次上成功完成了对神经元功能、代谢物组成及其代谢通路的研究[467]。林金明等引进了 Inkjet 喷墨技术，发展出基于 Inkjet 技术的单细胞质谱分析平台，实现了高效率的单细胞液滴生成和快速灵敏细胞代谢物的质谱检测，亦可以应用于多个样品的顺序在线快速质谱检测，简化了复杂样品的前处理步骤和时间[468]。张新荣等开发出了能够快速分离待测样品中基质的新型实用电离源，并将之用于单细胞的组学研究和纸基样品分析[469–471]。

质谱用于蛋白质组学方面，王初等开发了一种基于光交联剂和质谱联用的蛋白质检测检测策略（IMAPP），利用 IMAPP 策略显著增强了蛋白质 - 蛋白质相互作用之间信号，并能同时记录蛋白质 - 蛋白质之间结合位点[472]。

质谱成像方面，李智力等开发了一种基质电喷雾扫描涂层技术（EFASS），有效克服成像时基质干扰，提高了质谱成像的灵敏度[473–476]。聂宗秀等通过质谱成像（MSI）来实现对小鼠体内的碳纳米材料（CNMS）的分形分布检测，克服了以往荧光标记成像的短时间猝灭和拉曼光谱法的检测信号低、速度慢的不足[477–480]。钟鸿英等建立了激光诱导隧道电子俘获软电离方法来进行小分子化合物的鉴定和质谱成像，该方法可应用于人的指纹物理形状成像和揭示指纹内源性代谢物信息[481]。林金明等实现了对细胞表面糖类的原位、多组分分析[482]，拓展了质谱标签技术。

在质谱联用技术开发方面，方群等开发并完善了一种基于半封闭液滴系统的液滴 - 电喷雾质谱脱线联用方式，建立了多功能的微流控液滴 - 电喷雾质谱分析系统，可完成液滴的生成、多步样品前处理和多样品的质谱检测等操作[483–485]。林金明等将微流控芯片联合质谱检测技术应用连续液滴在线成分鉴定和定量分析构建起微流控芯片 - 质谱分析平台（Chip-M S）[486, 487]，这是一种新型全自动的细胞代谢研究平台；应用该系统检测了不同细胞的乳酸外排和在缺氧条件下连续监测药物 α - 氰 -4- 羟基肉桂酸二乙胺盐对不同的细胞种类乳酸外排的抑制，考察了药物的效用[488]。

（五）高分子化学

1. 高分子合成化学

可控聚合反应方面，以有机化合物替代辛酸亚锡作为催化剂，进行丙交酯开环聚合，所获得聚乳酸无毒性残留物，更适于用作生物医用高分子[489, 490]。付雪峰等基于有机钴化合物中 Co–C 键在可见光照射下的可逆均裂，发展了新型温和条件下可见光诱导的活性自由基聚合体系[491]；设计合成了具有普适性的光引发调控剂（salen）Co–CO_2Me，在室温可见光照下，实现了包括丙烯酸酯类、丙烯酰胺类及醋酸乙烯酯的活性自由基聚合[492]。

催化聚合方面，崔冬梅等[493]提出了“烯烃单体上的极性基团可以活化聚合反应”的概念，突破了人们在功能化烯烃配位聚合中极性基团毒化聚合反应的传统观念；通过催化剂的设计实现了亲氧性较强的稀土金属对“未保护”的极性烯烃单体与苯乙烯的高活性高间规选择性共聚合。通过改变甲氧基在苯乙烯单体上的取代位置，可以改变极性单体与中心金属的配位模式，进而获得楔形 / 梯度 / 无规序列结构的共聚物[494]。研究不同稀土金属配合物催化苯乙烯间规聚合的行为，发现 Lewis 碱与中心金属配位，阻碍了单体与中心金属配位，致使配合物的催化活性大幅降低[495]；以弱 Lewis 酸或 Lewis 碱催化可控聚合，通过向催化体系中交替加入 Lewis 酸和 Lewis 碱，调控中心金属周围的空间位阻，实现了对配位聚合的开 / 关控制。陈昶乐等[496]提出了“慢链行走”的概念，通过配体调控二亚胺钯催化剂制备的聚烯烃以及乙烯—丙烯酸甲酯共聚物的支化度只有 20/1000C，与文献相比该催化剂的活性要高十多倍，所制备的聚合物、共聚物的分子量也要高十多倍，所制得的聚合物、共聚物是半结晶的固体，熔点接近 100℃，是之前报道的所有二亚胺钯催化剂所无法实现的，极大地拓宽了二亚胺钯催化剂体系的应用前景。通过催化剂设计与调控，实现了高分子量带功能基团聚乙烯的合成[497]，获得了具有线性结构、高熔点和分子量高达百万的共聚物。凌君采用 Lu（OTf）$_3$ 催化 THF 和 CL 共聚合的时候，发现在引发剂两端分别引发了正离子和负离子活性聚合，形成两嵌段聚合物，且在 CL 单体消耗完后阴阳离子两端自动发生缩合，直接形成了多嵌段热塑性弹性体[498]。张兴宏采用有机化合物受阻 Lewis 酸碱对催化环氧丙烷与羰基硫交替共聚获得分子量高达 92.5 kg/mol 的聚单硫代碳酸酯，分子量分布仅为 1.2[499]。

多组分聚合反应方面，将 3 种以上的非传统单体经反应形成聚合物，例如，Pasarini 三组分聚合反应[500]；炔、醛或单质硫和胺进行的聚合反应[501]；用氨基酸通过简单高效的 Ugi 四组分聚合反应合成获得多肽，该聚合物具有良好生物相容性、抗菌性和药物控制释放功能[502]；通过三组分 Biginelli 反应制备链结构连续变化、具有不同物理性能的系列聚合物[503]，并建立了聚合物结构与其 T_g 相关性，利用所得关系式，可预测该类聚合物特定结构所对应的 T_g。

点击聚合方面，秦安军等发展了一种新的氨 - 炔“点击聚合”反应，活化内炔与芳香胺在 CuCl 的催化下，在无溶剂条件下，于 140℃反应 2 小时便可得到分子量高达 13740 的聚 β - 氨基丙烯酸酯，该反应具有优异的区域选择性和立体选择性，可得到只有马氏加成和 Z 式结构高达 94% 的聚合产物[504]。将活化内炔单体换为活化端炔单体，意外发现与脂肪族仲胺在室温下便可自发进行聚合，无需任何催化剂，且几乎定量地得到高分子量（MW 高达 64400）的 PAAs，只有反马加成和 100% E 式结构的产物生成[505]。

手性聚合物方面，通过在磁场下加入与磁场方向平行或反平行的线偏振紫外光辐照聚合，实现了螺旋聚二乙炔组装体的可控合成，为由非手性单体出发制备螺旋共轭聚合物提供了一条新途径[506]。董泽元设计合成了一种新颖螺旋聚合物，该聚合物可转变为双螺旋超分子结构并带来手性变化，有望具有开关效应[507]。吴宗铨等合成了结构简单、制备容易、空气稳定性高的钯配合物，能高效引发异腈活性聚合，得到分子量可控、分子量分布窄的高立构规整度聚苯基异腈及其嵌段共聚物[508]。基于可再生来源化合物弹性体的合成、聚酯的合成[509, 510]也引起广泛关注。

2. 功能高分子

生物医用高分子方面，高卫平以原位聚合方法在干扰素 C 残基接枝聚合聚丙烯酸（齐聚聚乙二醇）酯，能大大提高产率，抗癌效果突出[511, 512]。汤谷平等证明涂有纳米粒子的细菌可以有效地传递口服 DNA 疫苗，刺激人体自身的免疫系统摧毁癌细胞[513]。颜德岳等发展了聚磷脂超支化聚合物[514]，研究了超支化聚合物的组装及其作为纳米载体应用于药物输送[515]。陈学思等利用聚乙二胺和聚 L- 谷氨酸与治疗基因的复合物通过席夫碱交联构建了一种超灵敏的电荷 / 尺寸双响应的酸敏感基因输送系统，同时解决了表面性能调节和基因释放问题，抗肿瘤效果很好[516]。王浩等用 pH 敏感的聚胺基酯键合自噬调节多肽，通过自组装制备了一种能够诱导肿瘤细胞自噬、肿瘤细胞死亡的纳米组装体[517]。申有青等发展了一类在肿瘤细胞内活性氧或酯酶催化下、能够从带正电荷反转为带负电荷的新型电荷反转高分子，以解决阳离子聚合物在细胞内无法有效释放所携带的基因物质的问题，获得了显著提高的基因转染效率和抗肿瘤性能[518]，并由此实现了肿瘤细胞选择性基因释放，提出了一种全新的肿瘤细胞选择性高效基因治疗方法[519]，总结了肿瘤药物输送五步级联的 *CAPIR* 过程，并提出新型高分子载体必须能够同时实现表面、尺寸和稳定性三种纳米特性在 *CAPIR* 过程中的自适应，深化了肿瘤药物输送的理论[520]。

功能高分子应用于生物成像、光热和光动力治疗以及柔性电子生物器件[521]等方面也取得突破。王振新等利用聚吡咯烷酮保护制备了铁离子与没食子酸配位聚合物的纳米点，利用其 pH 依赖的磁共振成像造影和光热治疗性能，不仅显著增加肿瘤成像的灵敏度，而且能通过光热治疗消融肿瘤[522]。严锋等利用聚噻吩制备了可以高特异性检测酶的柔性晶体管[523]。在高分子组织工程材料方面，丁建东等通过在抗蛋白吸附的聚乙二醇水凝胶表面修饰细胞黏附多肽 RGD，研究了纳米图案和微纳米图案对细胞黏附和去分化的影响，发现细胞黏附配体的空间分布对于细胞的去分化有重要作用[524]，基质的硬度和细胞黏附配体的空间安排都对间叶干细胞的分化都有重要影响[525]。

3. 高分子物理

高分子结晶方面，李良彬等发现在拉伸场下，很小的分子链变形即可导致串晶中 shish 部分的形成，这一重要的实验结果无法用串晶生长的“线团 – 伸展转变”机理或“拉伸网络模型”来解释[526]，需要考虑构象有序化与密度的耦合；还研究了在玻璃化转变温度以上 PET 的拉伸诱导结晶，认为双轴拉伸所做的功可以“储存”在具有高活动性的 PET 链中，体系的总结晶度与外场所做的总功密切相关[527]。邓华和傅强等研究了用高速薄壁注射成型方法制备 PP/PE 的交替多层结构，详细考察了流场、温度场、黏度及剪切时间的影响[528]。李忠明等研究了在剪切场和压力场共同作用下 PLLA 的结晶行为，这一双重外场在一定程度上更接近于实际加工条件，发现通过调节剪切速率和压力，可以得到高含量的 β 型 PLLA，这对 PLLA 的增韧有重要意义[529]。

液晶弹性体研究中，杨洪等设计了一种双层结构的聚硅氧烷液晶变形材料[530]，在一个体系中同时实现了弯曲和卷曲两种运动模式。液晶聚合物的组装结构研究中，沈志豪等利用主 - 客体相互作用将刚 - 柔型液晶嵌段共聚物转变成刚 - 刚型，得到特殊的双六方结构[531]。陈尔强等发现了一种具有超晶格的高级有序手性液晶相，其“四链”晶胞是全新的“受挫结构”[532]，该液晶相向高温手性六方相转变时会发生“受挫”螺旋链的手性反转，可制备双向形变材料。

俞燕蕾等[533]设计合成了非交联液晶高分子，该类材料强而韧，加工性能优异，是具有光致形变和自修复等性能的第二代光致形变材料。在聚电解质研究中，刘光明等制备了既有很强保水能力，又有离子迁移通道的两性聚电解质凝胶，并将之用于石墨烯基固态超级电容器，获得了十分优异的性能[534]。张光照等利用聚离子液体凝胶的高渗透压来吸取纯水，进一步利用其温敏性导致的凝胶体积相变，则可将吸入的纯水放出，有望用于咸水淡化[535]。朱雨田等研究了含弱聚电解质 PAA 的两亲性嵌段共聚物 PS-b-PAA 囊泡对电场的响应行为，观察到囊泡会在电场刺激下发生变形、破裂，最后可分裂而成多个小囊泡，可用于囊泡内含物的可控释放[536]。梁德海等用单链 DNA 和聚氨基酸 PLL 进行复合，制备了原始细胞模型体系[537]，将有助于剖析生命起源的可能机理。

在橡胶弹性机理研究中，吴子良、宋义虎等[538]考察了炭黑 / 天然橡胶黏弹体系的佩恩效应。张立群等用纳米力学成像方法研究了异戊二烯橡胶在形变过程中出现的自增强现象，观察到多层次的纳米纤维结构，认为这些纳米纤维与取向无定形系带分子链共同组成的网络结构是应力增强的主要原因[539]。在新型弹性体的开发上，杨曙光等通过调控分子间氢键相互作用，获得了一种 PAA/PEO 纺丝液，可制取断裂应变达 1200% 以上的高弹性纤维[540]。表面褶皱现象研究方面，鲁丛华等在 PDMS 基底上旋涂上一层偶氮高分子，用紫外光照下偶氮苯的构象异构化来实现应力释放，从而达到了擦除外力作用所产生的表面褶皱[541]。王亚培等观察到了在溶胀的 PDMS 表面上沉积金属银后，溶剂挥发会致使 PDMS 膜表面产生均匀的褶皱[542]。

在高分子理论计算和模拟研究中，李卫华[543]从理论上提出了“大分子冶金学”的新概念——嵌段共聚物自组装制备介观晶体结构，并且成功调控了二元晶体结构的“等效键长”和“等效化学价”，预测了十多种二元离子晶体结构；还预测了多种二维晶体结构（柱状结构），可以直接推广到单元、三元、甚至多元晶体结构的制备，从而在大分子尺度实现无机小分子晶体的结构；提出了多臂嵌段共聚物的“组合熵”概念，指出在 AB 嵌段共聚物的二元共混体系中可得到稳定的 Frank–Kasper 相结构[544]；通过调控 AB 类型两组分多臂星型嵌段共聚物的组合构型熵，可调控体系球状或柱状结构的晶格排列，实现多种非典型有序相[545]；利用聚合物表面刷状聚电解质抗衡离子能够可逆交换的性质，研究了聚合物刷表面抗衡离子调控水成冰的异相成核，多种抗衡离子的该效应符合离子的 Hofmeister 效应序列，异相成核温度窗口可达 7.8℃[546]。

崔树勋等利用单分子力谱研究了无扰条件下高分子侧链的长度和形状对单链焓弹性的影响[547]，研究了聚乙二醇（PEG）在有机溶剂中的单链弹性[548]。在可修复聚氨酯研究方面，利用氨基甲酸酯的交换反应实现形状记忆和可修复功能，突破了对聚氨酯体系的传统认识[549]。植入新形状的过程不是“擦除”之前的永久形状，而是将新形状叠加到其中，称之为形状累积效应，可以获得目前加工方法无法实现的复杂形状[550]。

4. 高分子产业化

武德珍团队研制的高强高模聚酰亚胺纤维制造技术已在江苏建成国内外首条 30 吨生产线并实现小批量稳定生产，具有自主知识产权，其中 3 件核心专利获得美国授权。

Lyocell 纤维从原料、生产过程到产品都是绿色环保，除具有天然纤维本身的特性如吸湿性、透气性、可生物降解性等外，还具有合成纤维的高强度的优点，适用于多种染料，上染率高。2016 年底，年产 1.5 万吨产业化生产线工艺路线一次性全线打通，产品性能达到预期指标，是我国生物基纤维领域“绿色制造”工业化的重要突破。

基于青岛科技大学研究成果,6 万吨聚丁烯 –1 工程在山东开工。聚丁烯 –1 被称为“塑料黄金”，其性能在多方面优于聚乙烯和聚丙烯。董侠等基于生物发酵法原料来源的长碳链尼龙，是我国具有自主知识产权的材料品种，中国科学院化学研究所与山东广垠新材料合作，突破了聚合关键技术，建立了合金化、酰胺交换反应控制、纳米复合等改性技术，建成万吨级产能的生产线。

（六）核化学与放射化学

田国新对酰胺荚醚类的萃取机理进行了系统研究[551, 552]，利用小分子的水溶性在水相体系研究了 TMDGA 与 Nd（Ⅲ）、U（Ⅵ）、Am（III）的配位，合成了萃取剂 N，N–二（2– 乙基己基）二甘酰胺酸（DEHDGA）[553]，在高酸下对镧系和锕系元素具有很强的萃取能力；利用单级萃取设备研究了硝酸肼还原反萃 Np 和 Pu 的过程，得出了 Np（VI）和 Pu（IV）反萃动力学方程和表观活化能、硝酸肼还原反萃 Np（VI）和 Pu（IV）的半反应时间[554]。上海应用物理研究所通过减压蒸馏可从 FLiNaK 中有效除去 SrF_2 和 LaF_3[555]。近代物理研究所用四乙基双三嗪吡啶和各种链长的酰胺荚醚作为萃取剂，研究了硝酸体系中对 Am（III）/Eu（III）的协同萃取行为，发现二者具有良好的协萃效应[556]。

四川大学合成了水溶性的酰胺荚醚类配体 TEE-BisDGA，并研究了其萃取 Am（III）、Eu（III）的行为，结果表明，TEE-BisDGA 对 Am（III）/E（III）选择性分离效果明显优于 TEDGA[557]。

新型分离材料与分离技术方面，中国原子能科学研究院将萃取剂 $CyMe_4$-BTBP 负载到大孔硅基复合材料 SiO_2-P 上获得了 $CyMe_4$-BTBP/SiO_2-P，对 Am（III）和 Eu（III）具有良好的分离效果[558]。上海应用物理研究所制备了含偕胺肟基和羧基的 PP（聚丙烯）纤维吸附剂，对海水中铀的吸附容量可达到 0.81 mg · g^{-1}（吸附时间为 68 天）[559]。中国工程物理研究院合成了用于提取铯的钨掺杂离子筛（W-Cs-IS），对模拟高放废液中的 Cs^+ 和 Sr^{2+} 具有较高的选择性，Cs^+ 的提取率为 88.2%[560]。南华大学制备了磁性氧化石墨烯 / β - 环糊精（MGO/CD）复合材料并研究了对 U（VI）的吸附性能及机理，经过 4 次吸附解吸循环后，MGO/CD 的吸附率仍大于 95%[561]。

核药物化学和标记化合物方面，中国工程物理研究院[562]用 ^{131}I 标记了卟啉化合物 TPPOH 和 $TPPNH_2$，并在荷瘤鼠上考察了它们对肿瘤生长的抑制效果，结果在注射 14 天后仍然滞留在肿瘤中并对肿瘤生长有较强的抑制能力。原子高科股份有限公司将 ^{64}Cu 标记的二硫代氨基甲酸二磷酸（DTCBP）与超顺磁氧化铁纳米粒子（IO-Dex）结合，获得了一种可用于淋巴结显像的 PET/MRI 双模态影像探针 ^{64}Cu（DTCBP）-IO-Dex[563]；还合成了具有靶向叶酸受体阳性肿瘤的 PET/MRI 双模态显像探针 ^{64}Cu-DOTA-SPIONs-PEG-FA，在 PET/CT 及 MRI 模式下均能清晰地显示肿瘤[564]。北京大学用 ^{18}F 标记了肿瘤标志物 microRNA，并研究了该标记探针在血清中的稳定性[565]。北京师范大学合成了可能用于阿尔茨海默病 PET 显像药物的配体 TM-1，并测定了它与 α7 烟碱型乙酰胆碱受体的亲和结合常数[566]。

在环境放射化学领域，北京大学利用毛细管内扩散法系统研究了 pH 值和氧气浓度对 $^{79}SeO_3^{2-}$ 在北山花岗岩中的迁移扩散行为[567]；采用通透扩散实验法及自行研制的高温扩散装置研究了不同温度下 SeO_3^{2-} 在北山花岗岩中的扩散和吸附，分别获得了 26~60℃和 70℃下的扩散系数，为（1.57-6.07）$\times 10^{-13} m^2/s$ 和 $54.5 \times {}^{-13} m^2/s$[568]。西北核技术研究所基于对流弥散方程（ADE）和连续时间随机游走（CTRW）理论的模型，定量分析了锶（模拟 ^{90}Sr）的 non-Fickian 弥散迁移规律，较好地描述了锶的 non-Fickian 迁移（$0 < \beta < 2$）和 Fickian 迁移（$\beta > 2$）行为[569]。中国工程物理研究院研究了接触时间、pH 值、离子强度、共存阳离子（Li^+、Na^+、K^+）/ 阴离子（CO_3^{2-}、SO_4^{2-}、PO_4^{3-}）、腐殖酸对 Th（IV）在高岭土上吸附行为的影响[570]。西南科技大学研究了土壤中的矿化菌对锶离子的矿化固定作用，从土壤中分离出了脱氧硫杆菌，发现该菌对 1.0g/L 模拟 Sr^{2+} 污染的去除率可达 80%，矿化产物为硫酸锶[571]。王殳凹在环境放射化学领域取得一系列进展：初步衍生出基于绿色选择性结晶分离理念的稀土分离及镧锕分离，有望改善我国乏燃料后处理关键流程工艺[572]；合成并解析出能够在强酸溶液中稳定存在的膦酸锆基 MOF 的晶体结构，并将其应用于在高酸度条件下选择性吸附铀酰离子，为设计合成高稳定的功能膦酸锆 MOFs 材料用于乏燃料后处理及环境中放射性核素污染防治开辟了新路径[573]；发现一例能够耐受住高剂量电离辐射的三重互锁铀酰金属有机框架材料[574]；发现铀金属有机骨架材料可以作为高灵敏度传感器用于准确探测低剂量电离辐射，成功开辟了贫铀的又一新应用，也推动了辐射探测技术的发展[575]；发展了一大类新型阳离子 MOF 材料用于高锝酸根的快速、高选择性去除以及高效固定，有望解决核废料中锝的泄漏问题[576, 577]；发展了几类新型固体材料用于阻滞核废料中高毒性放射性核素在环境中的迁移以及在天然水体环境中的准确探测[578, 579]。

放射性废物处理与处置方面，中国原子能科学研究院研究了中、低放固体废物超级压缩饼在 $2m^3$ 废物包装箱内的固定配方，能保证最终废物体的整体性和整体强度满足安全运输、储存和处置的要求，并且能够进行工程应用[580]。中国科学院等离子体物理研究所分别开展了氧化石墨烯与金属 Pb（Ⅱ）、Ni（Ⅱ）和 Sr（Ⅱ），以及 Re（Ⅶ）的吸附行为研究，结果表明氧化石墨烯最大化吸附容量 Pb（Ⅱ）> Ni（Ⅱ）> Sr（Ⅱ）[581]。二氧化锆修饰还原氧化石墨烯对于 Re（Ⅶ）的最大吸附量是 43.55mg/g，优于氧化锆 24.85mg/g 的吸附量，远高于其他吸附剂[582]。在放射性废物处置库场址筛选方面，2016 年 6 月底，已经在甘肃北山完成了 32 个钻孔。同时，还在新疆、内蒙古筛选出 6 个预选地段，并开展了初勘。工程屏障方面，建立了膨润土研究大型实验台架，加大了膨润土的工程性能研究。在地下实验室方面，制定了地下实验室选址准则，2016 年筛选出甘肃北山的新场场址——中国首个地下实验室场址。此外，还在甘肃北山建设了“北山坑探设施”，用于地下实验室安全技术研究[583]。

放射化学应用方面，北京大学利用软件 CHEMSPEC 模拟计算了 Np 和 Pu 在北山地下水中的种态分布以及两者在水合氧化铁上的吸附，Np（V）在水合氧化铁上的表面吸附种态是 $\equiv FeOHNpO_2^+$ 与 $\equiv FeOHNpO_3^-$，Pu（IV）在水合氧化铁上的表面吸附种态为 $\equiv FeOHPuO^{2+}$ 与 $\equiv FeOHPuO_2$[584]。中国核动力研究设计院研究了气载氚标准装置以及气载氚监测仪现场校准装置、现场校准方法、影响校准结果主要因素等，解决了国内气载氚监测仪量值溯源的难题[585]。中国原子能科学研究院研发了 ^3H、^{237}Np、^{90}Sr 等核素的自动化快速化学分离装置，能实现自动定量进样、恒温亚沸蒸馏、冷阱低温冷凝、自动平均分样等功能，蒸馏获取一个 2mL 的样品只需 3min，且全过程一键式控制[586]；建立了闪烁玻璃体为主的放射性在线测量系统，该系统对后处理工艺中 α 射线的本征探测效率为 100%，可测量的 ^{237}Np 溶液最小浓度为 1mg/L[587]；建立了以填充特效吸附材料 SuperLig®620 的塑料闪烁体微型柱为核心的放化传感器系统，对 ^{90}Sr 的探测效率为 21%，探测限和最小可探测活度分别为 0.12cps 和 2.56Bq/L，实现了对低放废水流出物中 ^{90}Sr 的连续监测[588]。

（七）交叉学科或研究领域

1. 计算化学

化学计量学方面，邵学广等发展了以吸附材料为基质的漫反射近红外光谱分析方法，制备了具有低背景、高反射性能的银基质材料，并结合化学计量学手段建立了高灵敏的外光谱分析方法，并在医学检验中得到应用[589]；基于近红外光谱对水结构的高灵敏反映，开展了温控近红外光谱新方法研究，利用多级成分同时分析方法研究了葡萄糖水溶液的温控近红外光谱，实现了血清中葡萄糖含量的定量检测[590]。吴海龙等在提出了采用化学计量学辅助 HPLC-DAD 策略用于处理不同色谱柱和样本基质造成的干扰模式变化的问题[591]；以及多维荧光技术结合三维或四维校正的实时分析方法，用于在包含未校正的光谱干扰的人血浆中对 NADH 的降解反应和 FAD 的生成反应进行化学动力学实时定量分析[592]。

理论计算方面，罗成等首次报道了 TET 蛋白对 3 种 DNA 甲基化衍生物不同催化活性的分子机制，揭示了 TET 蛋白底物偏好性机制，为基因组中 5- 羟甲基胞嘧啶稳定存在提供了分子水平的解释[593]。蔡文生等发展了扩展广义自适应偏置力自由能计算方法，提高了高维空间的采样效率[594]；在 NAMD 的 Colvars 模块基础上提出了新的变量用于精确计算蛋白质 - 配体结合自由能，并用于复杂化学过程的研究，在原子水平上揭示了分子机器中复杂的分子运动机理[595]。侯廷军等提出基于势能面计算的自由能分解算法，并成功应用于手性药物分子结合能力差异性研究[596]；发展 CaFE 自由能计算软件包，可以和国际知名分子模拟软件 VMD 无缝结合，方便高效地实现 MM/GB（PB）SA 和 LIE 结合自由能预测[597]。万坚等发展一种能准确计算预测生物大分子 - 配体结构构象的方法 DOX，其对小分子结合构象和结合自由能均具有较高的预测精度[598]。李晓霞等发展了基于 GPU 并行与化学反应分析结合的 ReaxFF MD 新方法，并利用大规模模拟清晰揭示了木质素热解的三阶段特性、各阶段主要产物及其与不同类型连接键反应行为的对应关系[599]。李浩然等对离子取代基修饰的羟基邻苯二甲酰亚胺的催化性能进行了理论计算，发现离子基团能够有效地提高催化剂的反应活性，改变阴离子比阳离子能更明显地调控反应活性[600]。

在药物设计方面，蒋华良等研究了 GCGR 不同结构域对其活化作用的调控机制，不仅有助于新型Ⅱ型糖尿病治疗新药的开发，而且也有助于深入理解 B 型 GPCR 蛋白结构与功能的关系[601]；针对铜离子伴侣蛋白 Atox1 和 CCS 采用基于结构的药物设计方法并结合生物实验获得了在体内有良好抗肿瘤活性的化合物 DC-AC50，它不仅阻断了铜离子在细胞内的转运，而且能抑制了 Cu/Zn 超氧化物歧化酶 SOD1 的活性[602]。张健等发现了首个作用于 APC 口袋的抑制剂，并证实该抑制剂能有效破坏 APC-Asef 蛋白相互作用阻断结肠癌的恶性转移[603]。许永等首次证实了 RORγ 作为肿瘤靶标的可能性，并获得了可口服给药的候选化合物，为前列腺癌临床耐药问题提供了全新的解决方案[604]。王任小等继

续扩展和更新蛋白 - 配体复合物三维结构和亲合性数据库 PDBbind-CN，使其在国际上处于领先地位[605]。徐峻等建立了国际上第一个蛋白质 - 配体共价结合数据库 cBindingDB，填补了基于结构的共价抑制剂设计研究方面的空白[606]。张健等发展了一系列变构药物设计方法和变构综合专家系统 ASD，并将发展的方法用于多个重要靶标的变构抑制剂的设计[607]。唐赟等发展了结构 - 药物 - 靶标网络推理（SDTNBI）方法，可以对已有网络外的老药和新化学实体的潜在靶标进行预测[608]。朱维良等利用数据库统计等方法发现了蛋白质结构数据库中被明显低估的卤键作用，利用包含卤键作用的分子对接新方法发现了含卤老药的新用途[609, 610]。

在材料模拟和设计方面，李有勇等结合了不同尺度的模拟技术对 OPV 的器件效率进行了预测，并对 OPV 中的分子结构、主链结构、侧链结构进行了优化设计，以及通过 STM 模拟和表征确定了有机无机钙钛矿材料的表面结构以及缺陷结构，对提高其稳定性提出了有效的解决措施[611, 612]。胡文兵等采用理论计算方法深入研究了高分子折叠链片晶的发生机制，分子量多分散性的影响及其有限片晶厚度的动力学原因，对高分子晶体生长微观机理的理解上升到一个新的层次[613, 614]。

2. 流变学

在高分子流变学领域，宋义虎和郑强发表综述论文，总结了高分子纳米复合材料非均质结构与分子弛豫特性的国际研究现状，指出类固体流变学研究的局限性[615]，进一步综述了流体动力学补强理论研究现状[616]，证实了“两相”模型预言的物理图像，并提供了粒子相可逆非线性流变的实验证据[617]，引入受限“本体相”动力学参数，修正“两相”流变模型[618-623]，为高分子动力学行为研究提供了重要、可靠的补充性方法。俞炜发展了纳米复合材料线性黏弹性模型[624]，通过应变速率和应力放大因子等特征参数定量表征了纳米粒子的团聚行为[625, 626]，并给出了纳米粒子在聚合物基体中发生相分离的临界分散指数，发展了傅立叶变换方法，提出了表征多分散多嵌段共聚物相分离行为的 vGP3 图方法[627]，利用广义应力分解算法，建立了用大振幅振荡剪切研究复杂流体边界滑移行为的方法[628]，陈全发展了利用离子间静电能界定离聚物和聚电解质的新方法[629, 630]，构建和发展了缔合高分子可逆凝胶和缔合双蠕动理论模型[631, 632]。刘琛阳发展了高性能石墨烯气凝胶制备技术[633]，研究了聚氧化乙烯在不同结构离子液体中的流变行为[634]，建立了纤维素分子量表征新方法[635]，研究了基于酰脲键的硅橡胶自修复材料的流变行为、力学性能及其自修复机理[636]，使用多种流变学方法研究了聚酰亚胺制备过程中的复杂物理化学变化[637]。

电流变和磁流变流体方面，赵晓鹏利用各向异性纳米颗粒，解决纳米电流变体系零场黏度高、剪切不稳等问题，研制了多种基于一维和二维纳米颗粒的纳米电流变新体系[638, 639]，系统诠释了形貌变化对纳米体系电流变行为的影响规律。将电流变技术与超材料相结合，设计了新型流体基超材料，并研究了负折射、全吸收、隐身等行为的调节规律[640]。龚兴龙将自愈合导电剪切增稠胶与凯夫拉纤维复合，研制出一种可穿戴式的电子传感织物[641, 642]，同时，在磁流变弹性体中引入亚麻编制纤维，其拉伸、剪切强度大幅提升，导电性能也显著增强[643]，研制出基于石墨 / 羰基铁粉 / 硅橡胶三元复合磁流变塑性体[644]，力电耦合性能优良，系统研究了在挤压流状态下，剪切增稠液的挤压应力随挤压速率的变化规律[645]。余淼立足于智能材料与结构系统，以工程应用为牵引，从铁磁颗粒改性入手，研制了系列高性能的磁流变材料[646, 647]。

在表面活性剂、原油开采和原油输运领域，方波和卢拥军重点开展了反应体系过程流变学研究，包括纤维素基凝胶体系[648]和表面活性剂可逆光敏体系[649–654]反应过程流变学，建立相关流变反应动力学方程，为获准实施国家科技重大专项课题“储层改造关键流体研究”奠定了良好基础。

3. 环境化学

我国在新型污染物如《斯德哥尔摩公约》增列污染物（短链氯化石蜡、全氟辛烷磺酸、六氯丁二烯等）的分析方法、环境行为及过程、风险评估以及非目标性污染物的筛查等方面不断推进[655–657]，为保障我国环境安全及履约提供了重要支持。我国大气汞形态行业排放清单[658]及基于物质流的汞流向[659]的建立为我国履行新近生效的《关于汞的水俣公约》提供了良好的支撑。在大气环境化学方面，研究发现了北京霾事件中颗粒物浓度的周期性变化规律以及气象条件的关键作用，指出了挥发性有机物（VOC）、NO_x 和 SO_2 等气态污染物在霾形成中的关键作用，提出了北京霾形成的根本原因是二次颗粒物有效的成核和快速与持续的增长过程，建议对 VOC、NO_x、SO_2 等关键致霾因子进行有效管控[660]。实验室模拟与现场霾事件分析显示，在矿物粉尘表面 NO_x 与 SO_2 存在协同作用，共存 NO_x 可显著降低 SO_2 的环境容量，将 SO_2 快速氧化为吸湿性强的 SO_4^{2-}，因此机动车尾气和燃煤烟气的协同作用可加剧细颗粒的生长[661]。同时，来自农业过程等的大气 NH_3 的存在亦可加速 SO_2 氧化与二次气溶胶成核[662]。在 VOC 光化学氧化机制方面，通过实验室模拟与量子化学计算发现，甲苯氧化以生成甲酚为主，而非之前认为的开环产物与有机过氧自由基，提示应重新评估这一大气氧化过程在臭氧和二次有机气溶胶形成中的作用[663]。

在金属污染物的环境过程方面，采用稳定同位素示踪技术研究了新型农药熏蒸剂碘甲烷对无机汞的光化学甲基化机理，提出了碘甲烷对无机汞的光化学甲基化的两步机制[664]，突破了经典汞甲基化机理认知方面的局限，发现了新的汞光化学甲基化途径，表明亟需对碘甲烷新型农药熏蒸剂的使用进行广泛、审慎的安全评估。最近发现植物叶片对零价汞的吸收是大气汞重要的消除途径，而叶片凋落使得所吸收大气汞向地表迁移[665]。汞同位素指纹及叶面积指数等相关性分析则进一步揭示凋落物较湿沉降对森林地表汞输入更为重要，是森林地表汞输入的主要因素[666, 667]。首次发现了纳米银在自然转化过程中的稳定同位素分馏现象，天然有机质与日光照射条件下纳米银的形成与溶解可导致显著的银稳定同位素（^{107}Ag 与 ^{109}Ag）比例变化，其同位素富集因子可达 0.86‰[668]。

4. 绿色化学

氢气的存储、运输和提纯方面，马丁等构建了一种具有优异结构稳定性的层状金团簇 – 碳化钼负载型催化剂，在低温水煤气变换反应中表现出优异的活性和稳定性，从而实现了氢气的低温制备和纯化[669]。

二氧化碳的高效转化与利用方面，孙予罕和钟良枢等通过耦合双功能催化剂，一步法高选择性得到了 C_{5+} 汽油烃类产物[670]，被认为是二氧化碳转化领域的一大突破。韩布兴课题组设计、构建了一系列催化材料 – 绿色介质体系，提出了多种 CO_2 转化的新原理、新方法和新途径，在较温和条件下将 CO_2 高效转化为醇类、羧酸类重要化学品等[671–673]，并在 CO_2 转化过程中相关化学键演变规律和机理方面取得新的认识。何良年等人提出了一种二氧化碳分级可控的还原策略，实现将二氧化碳可控地还原为不同能量级别的还原产物[674]。江焕峰等人以二氧化碳代替光气，经碱 DBU 促进实现了氨基甲酸酯系列衍生物的高效合成，该合成方法反应原料简单易得、环境友好且底物适应性广[675]；还同时实现了二芳基碘鎓三氟甲磺酸盐或芳基硼酸与胺和二氧化碳的反应，一步得到氨基甲酸芳香酯[676]。赵凤玉等人以二氧化碳和甘油为原料，2– 氰基吡啶为脱水剂，在无金属催化剂和其他添加剂的条件下，甘油碳酸酯的收率可达 18.7%[677]；还以二氧化碳为羰基源，通过与胺类分子缩聚制备了反应合成了一系列聚氨酯 – 脲材料[678，679]。

从合成气直接制烯烃方面，孙予罕等人在温和的反应条件下，实现了高选择性合成气直接制备烯烃，总烯烃选择性高达 80%，烯 / 烷比高达 30 以上[680]。二氧化硫的回收利用方面，吴卫泽等人设计合成了一系列功能化的低共熔溶剂，包括甜菜碱 + 乙二醇、左旋肉碱 + 乙二醇、甜菜碱 + 水、左旋肉碱 + 水、咪唑（或咪唑系列化合物）+ 甘油，考察了这些吸收剂对 SO_2 吸收的条件影响，发现该低共熔溶剂能够高效吸收烟气中 SO_2，并能够再生和回收 SO_2[681–683]。

在生物质催化转化领域，王艳芹等人创新性地构建了多功能的 $Pt/NbOPO_4$ 催化剂，使得原生木质生物质可以在不经过化学预处理或分离的情况下直接加氢脱氧，把纤维素、半纤维素和木质素组分分别转化为己烷、戊烷和烷基环己烷等液态烷烃，质量收率达 28.1 wt%，实现了原生木质生物质在单一催化剂上的一步直接催化转化为液态烷烃[684]。为解决生物质转化中传热、传质困难和催化剂难于发挥作用的问题，胡常伟等发展了溶剂热分级转化方法，通过对反应条件、溶剂和共溶剂的主要、催化剂的调控，实现了原生生物质中不同组分的高选择性转化[685–690]。韩布兴课题组采用可从生物质中直接提取或木质素解聚得到的芳香醚为原料，以溴化物改性的 Pd/C 为催化剂，以 H_2O/CH_2Cl_2 为反应介质，成功实现了芳香醚加氢、水解制备酮类物质，目标产物的产率可高达 96%[691]，开辟了由生物质资源制备酮类物质的新途径；还采用天然植酸为配体，合成了多种具有多级孔结构的金属 – 植酸配合物材料，对乙酰乙酸加氢制备 γ – 戊内酯等反应有良好的催化性能[692]。

绿色有机合成与催化领域，顾彦龙等人以水稳定性良好的超交联多孔有机聚合物为载体，制备了负载酸碱双功能和酸/碱单功能组合催化剂，在水中或者有水存在下，实现了一锅酸碱串联催化合成反应[693, 694]。傅尧等人首次在烷基羧酸脱羧偶联反应中首次引入导向的概念，实现了铜催化的烷基羧酸脱羧偶联反应，表现出了金属催化剂在该类反应中的新的性质[695]。唐勇等人发展了基于推拉电子环丙烷的不对称开环/环化反应，实现了首例共轭双烯与环丙烷的不对称［4+3］反应，为合成手性七元环化合物提供了新途径[696]；采用高氯酸铜水合物既为Lewis酸催化剂又为水的存储器，实现了水对环丙烷的高对映选择性开环[697]；通过催化炔基吲哚与环丙烷的不对称［3+3］反应，发展了合成手性四氢咔唑化合物的新方法[698]。江焕峰等人以简单易得的肟酯为原料，以价廉、绿色的氧气为氧化剂，经钯催化肟酯与烯丙醇的脱氢偶联反应，发展了一种一步法区域选择性合成多取代吡啶的新方法[699]。

绿色化学成果向产业化转化方面，宗保宁团队在2017年开发成功浆态床过氧化氢生产技术，该新技术使得过氧化氢生产成本下降20%、污染物排放下降70%。张锁江团队揭示了离子液体中氢键的特殊性及构效关系，获得了离子液体结构对反应/传递性能的影响规律，突破了离子液体规模制备、工艺创新和系统集成的难题，实现了多项反应/分离技术的工业示范和中试；研发了离子液体固载催化CO_2和环氧乙烷联产碳酸二甲酯/乙二醇新技术，产率接近100%，催化剂易分离，工艺节能达30%。基于该技术，2017年江苏奥克化学有限公司建成3.3万吨/年离子液体催化生产碳酸二甲酯/乙二醇工程示范装置，年消耗二氧化碳1.1万吨，装置总产能可提高2.4倍，产品能量单耗降低30%。该团队还开发了一种新型的高电压锂离子电池电解液，电压上限可提高至5.0V，能量密度有望从原有的200Wh/kg左右提高至300Wh/kg，安全性同时得到大幅度提升。该技术已在河南林州科能材料科技有限公司建成规模生产线，产能达5000吨/年，同时产品被比克电池、国轩高科、多氟多等多家大型动力电池企业和车企所应用。

5. 理论化学

在相对论量子化学方面，刘文剑提出了精确二分量相对论（X2C）、准四分量相对论（Q4C）、有效量子电动力学（eQED）、自旋分离X2C(sf-X2C+so-DKHn）等相对论哈密顿，建立起连续、完整的“哈密顿天梯”[700-702]，提供了解决化学与物理中的相对论问题的整套框架。

针对强关联体系，刘文剑提出了“先静态再动态又静态”（SDS）的思想[703]，发展了多态微扰理论（SDS-MS-MRPT2/3）和高效的迭代组态相互作用（iCI）[704]方法。吴玮在价键自洽场（VBSCF）方法基础上，发展了密度泛函价键方法（DFVB）和活性空间价键自洽场方法［VBSCF（CAS）］等[705, 706]。

大分子体系的线性标度计算方面，黎书华等将基于能量的分片方法推广到周期性体系（如分子晶体），实现了晶格能、结构优化和振动光谱的高精度计算[707]。针对大分子体系的激发态计算，黎书华等发展了基于局域激发近似的含时密度泛函方法[708]，刘文剑等发展了用于激发态计算的基于块局域轨道的方法[709, 710]。

密度泛函理论（DFT）方面，徐昕在绝热近似的框架下进行交换和关联双杂化，提出了包括非占据轨道信息的 XYG3 型系列新一代高精度普适泛函（xDHs），并推导出其解析能量梯度，实现了高精度构型的优化[711, 712]。xDHs 在生成热、键焓、能垒、非键相互作用以及过渡态构型的广泛测试中表现优异[713]，是目前国际上最准确和普适的系列泛函之一，并已在多种专业量化软件中实现了程序化。梁万珍等成功实现了解析计算 TDDFT 激发态能量对核坐标或外电场的二阶导数[714, 715]。这些解析方法的实现有效地克服了有限差分求导带来的误差和计算效率低的问题，使 TDDFT 方法能有效地用于计算大分子的激发态振动频率、特征势能面的稳定点，以及研究大分子的光物理过程[716]，相应计算程序已经在国际主流的量子化学软件包 Q-Chem 中实现。

在小分子反应动力学领域，张东辉等发展了四原子态－态量子动力学方法，首次在全维水平计算了四原子反应的态－态微分截面并与实验取得了高度吻合，被誉为“反应动力学研究的一个里程碑”[717]；针对 H_2O 在 Cu（111）表面解离吸附问题，首次实现了三原子分子在刚性金属表面解离吸附的全维量子动力学研究[718]；结合高分辨交叉分子束实验，在 F+HD（v=1）反应中发现了通过 HD 振动激发才能经历的共振态[719]，在 Cl+HD（v=1）反应中发现了振动激发态反应中广泛存的、由过渡态区域化学键软化引起的共振态[720]，极大提高了们对反应共振态的认识。

方维海等发展了两态和三态势能面锥型交叉结构的优化方法和从头算非绝热动力学方法，并研究了一系列重要的光物理及光化学反应的微观机理[721–723]，包括光诱导多肽折叠和解折叠过程、偶氮苯衍生物激发态失活的手性选择性、光诱导环加成反应的对映体选择性和光诱导沃尔夫重排反应机理等；理论计算模拟预示硫代碱基可能的光物理过程、靛蓝分子光稳定性的分子机理和激发态 Au–Au 共价键形成的时间尺度等，都被随后的实验证实。

在开放量子体系动力学理论和应用领域，严以京等率先构建了费米子库的级联运动方程（HEOM）[724]，以及统一描述了声子库、电子库、激子库的耗散子运动方程（DEOM）[725]，并发展了高效计算方法及程序，用于研究非平衡稳态、多维光谱、电荷转移与输运、强关联电子态性质等[726]。史强等在发展级联方程理论和高效计算方法，及其在二维相干光谱、电荷和能量转移动力学等方面完成多项首创性工作[727, 728]。

生物大分子及溶液中化学反应的模拟方面，高毅勤等人基于前期提出温度积分增强抽样方法 ITS/SITS，发展了结合 QM/MM 的 SITS 方法，实现对化学反应自由能面的快速扫描和不需要反应坐标和预知反应机理的情况下反应动力学信息的快速获得[729, 730]；研究了阴阳离子协同性对水动力学的影响，对长期困扰着研究者的 Hofmeister 序列提供了简单、统一、可由实验直接验证的分子水平的理论模型[731]。

生物发光方面，方维海和刘亚军等人从理论的角度提出了萤火虫发光的新机理[732]，得到实验研究和非绝热分子动力学模拟[733]的印证；首次严格区分了化学激发态、生物发光态和生物荧光态及其辐射的不同[734]，被列为该领域重要研究进展之一；还对哈维氏弧菌生物发光的全过程进行了理论化学计算和模拟[735]，首次阐明了黄素单核苷酸在荧光素酶中荧光猝灭的机理，解决了这个多年悬而未决的问题。

低维纳米体系的生长机理方面，杨金龙和李震宇团队通过发展多尺度模拟方法，发现在 Ir 表面石墨烯生长中的非线性动力学行为是由点阵失配导致的生长前沿非均一性决定的，为调控异相外延生长动力学提供了理论依据[736]；发现铜表面石墨烯生长初期碳原子聚合会引起显著的表面弛豫，C–C 二聚体是石墨烯生长的主要供给单元[737]；发现了石墨烯切割的“吃豆人”机理，为可控制备石墨烯纳米条带提供了理论基础[738]。

有机和碳材料光电功能的理论研究方面，帅志刚等发展了分子激发态衰变的速率理论，通过第一性原理计算发光效率，不仅定量解释了分子聚集诱导发光的异常现象，还可预测发光材料的光物理参数[739]；提出了有机分子半导体电荷传输的核隧穿模型，建立了一套自下而上的多尺度的定量预测电荷迁移率的计算方法和程序[740]，率先基于形变势模型和第一性原理计算开展弱耦合区碳基材料迁移率的理论预测，发现了石墨烯纳米带迁移率随宽度变化的尺寸效应以及石墨二炔的高迁移率特性[741, 742]。该课题组所开发的分子材料光电性能计算程序（MOMAP）已经得到了商业化应用。

6. 晶体化学

据不完全统计，2015—2017 年我国化学家在晶体化学领域发表了约 3000 多篇学术论文。在金属有机框架材料（metal organic frameworks，MOFs）前沿领域，我国化学家的研究成果一直保持着领先地位。

卜显和以离子型 MOFs 为研究对象，制备出一例兼具主体框架刚性和热刺激响应的动态 MOFs 材料[743]，可以用于气体的储存与分离，传感、客体的捕获与释放等。张杰制备了 *N*–（3，5–二羧基苯甲基）–1，2–二（4–吡啶基）乙烯的有机分子晶体[744]，该晶体含有 C=C 双键和吡啶嗡 2 种光活性部位，表现出双波长响应性质。臧双全合成了一种手性的层状 Eu（III）配位聚合物[745]，同时具有光控发光、光控二阶非线性光学、光控压电等性质，显示了联吡啶嗡化合物在多功能开关材料方面的研究潜力。郑寿添获得了一种三重对称性 Eu（III）配合物的分子晶体[746]，具有热致变色性质，而且其特征 Eu（III）发光强度随温度升高呈线性降低，是一种可作为荧光温度计的温度传感材料。高恩庆研究了 *N*，*N*’–二（4–羧苯基）–4，4’–联吡啶嗡两性离子型有机晶体（水合物）的变色性质[747]，该化合物非光致变色，但在千兆帕级高压下发生从黄到绿的颜色变化，释放压力放置后自动恢复原色。

吴新涛得到一种杯状金属有机配位物的分子晶体[748]，该晶态仅呈现 X– 射线光致变色性质，变色机理为金属促进的电子转移，为 X– 射线敏感材料的研究提供了新思路。李巧伟利用拓扑诱导原理构筑了由不同几何金属簇次级单元的介孔 MOF[749]。江海龙发展了种晶诱导合成纯相 MOFs 的方法[750]，即在合成反应体系中加入目标 MOFs 的种晶以导向 MOFs 的结晶过程，并将该方法推广应用于多种 Zr–MOFs 和 Cd–MOFs 合成。金国新利用 *N*，*N'* – 二羟基萘二酰亚胺形成的 Cp*Rh 化合物与线性双吡啶类桥联配体构筑了动态结构的有机金属矩形大环化合物，得到了由有机金属矩形分子构成的 Borromean 环互锁结构[751]。

崔勇利用双吡啶功能化的手性 salalen 和 salen 衍生物配体与 Zn（II）组装了 3 个手性金属大环分子 Zn_6（salalen）$_6$、Zn_6（salalen）$_3$（salen）$_3$ 和 Zn_6（salen）$_6$。这些大环的疏水纳米空腔中分别含有六、三和零个手性 NH 中心，其荧光强度与手性客体分子的对映体组成呈线性相关，因此可用于手性传感[752]。孙頔研究了十九核［$Mn^{II}_{15}Mn^{III}_4$］盘状团簇配合物的逐级组装过程，以 1- 羟甲基 -3，5- 二甲基吡唑和叠氮配体，利用"一锅煮"室温超声方法合成了一种高对称性锰氧簇[753]，在 40K 下为亚铁磁体，是高自旋分子呈现磁有序的罕见例子。

崔勇构筑了一种三维手性 Cd–Cr–MOF，作为多相催化剂用于多种不对称催化反应，获得与均相催化剂相当的立体选择性[754]。董育斌利用含 Pd（II）– 氮杂环卡宾配位单元的羧酸配体构筑了一种含四（氮杂环卡宾）合钯和四核铅单元的三维 MOFs，可作为温和条件下苯炔环化三聚反应和苯炔 – 烯丙基溴化物 – 芳硼酸三组分偶联反应的高效多相催化剂[755]。

赵斌合成了一个基于［Cu_{30}］纳米笼的金属有机框架材料[756]，可以在无溶剂条件下催化氮杂环进行二氧化碳环加成反应，首次报道了基于 MOF 材料催化二氧化碳转化为恶唑烷酮类化合物，进一步加深了 MOFs 在催化领域的应用。钱国栋、陈学元和陈邦林合成了具有一维孔道的锌 MOFs 材料 ZJU–68，利用孔道的限域作用原位封装高度取向的有机激光染料分子 4–［p–（二甲胺基）苯乙烯基］–1– 甲基吡啶嗡，首次利用 MOFs 微单晶获得了高偏振度的多光子泵浦激光[757]。邢华斌等[758]利用无机和有机桥联配体构筑了 SIFSIX–2–Cu–i 和 SIFSIX–1–Cu，材料拥有三维网格结构，SiF_6^{2-} 阴离子通过氢键作用可以在低压下专一性地捕获乙炔分子，获得极高的乙炔吸附容量和迄今为止最佳的乙炔乙烯分离选择性。

张健实现碳纳米点与 MOF 材料的复合，获得分散均一的 CDs@MOF 复合薄膜材料[759]，该材料表现出波段可调的光致发光效应和光限幅效应。赵新构筑了迄今为止结构最为复杂的双孔二维 COFs，结合 X 射线粉末衍射、计算模拟和孔分布分析等方法确定了其结构[760]。洪茂椿与袁大强设计合成了一例多孔的、具有金刚石拓扑的五重穿插的氢键有机框架材料（HOF–TCBP），该材料具有较高的比表面积、优良的水、热稳定性以及易再生性，对低碳烃类气体具有良好的吸附分离效果[761]，是 HOF 材料领域非常罕见的。孙忠明合成了一系列全金属团簇并确定了其单晶结构，发现了多种具有金属芳香性、金属反芳香性的新颖配位模型，如首例全金属三明治化合物，首例全金属反芳香性化合物等，丰富了化学键基础理论[762-764]。

7. 多孔材料研究

于吉红等首次发现羟基自由基（·OH）存在于沸石分子筛的水热合成体系之中[765]，通过紫外照射或 Fenton 反应向沸石分子筛水热合成体系额外引入羟基自由基，能够显著加快沸石分子筛的成核，从而加速其晶化过程。这一发现使人们对沸石分子筛的生成机理有了新认识。赵东元开发了一种界面位点导向的定向化、无试剂包裹、无表面活性剂辅助的生长方法来直接制备超小（小于 5 nm）多孔纳米晶[766]。赵东元和邓勇辉以聚环氧乙烷 -block- 聚苯乙烯（PEO-b-PS）两亲性嵌段共聚物为模板剂、酚醛树脂为前驱体，开发了一种全新的用以合成具有高度有序二维六方介观结构的介孔碳材料的胶束融合 - 聚集组装方法[767]。陈小明和张杰鹏提出了“控制柔性客体分子构型可反转吸附选择性”的概念，常温常压下将 C4 碳氢化合物的混合物通过 MAF-23 填充的固定床吸附装置后，丁二烯最先流出而且纯度很容易达到 99.9%，同时避免了常规蒸馏和吸附纯化过程中因加热而产生的丁二烯自聚问题[768]。

洪茂椿利用噻吩二羧酸配体与主族金属钙离子构筑了一例结构独特的 MOF 框架—FJI-H9[769]，该材料对镉离子具有超高的选择性吸附性能，不仅可以重复利用，而且多次使用后的材料可以原位重构生成新鲜的晶体材料，实现了晶体材料的原位重构。同时，利用该材料可以快速识别溶液中浓度低至 10ppm（工业上镉污染的浓度级别）的镉离子。

裘式纶等发展了一种在固体表面生长 MOF 膜，COF 膜或者 COF-MOF 混合膜的方法[770, 771]，具有一定的普适性，通过该方法制备的多孔聚苯胺具有高比表面，优异的电活性及较高的导电性。通过这种方法生长的 COF 膜，以及进一步进行二次生长得到的 COF-MOF 复合膜材料均有优异的气体通量和高选择性，这些数值均超过了 2008 年修改后的氢气 - 二氧化碳吸附罗布森上限。而 COF-MOF 复合膜相对于单独的 COF 以及 MOF 分离膜显示出更加优异的选择性。此外，裘式纶还提出了一种基于双重链接的新策略合成三维 COF[772]，利用该方法获得的三维 COF 显示出高的比表面积和大的气体吸收能力，同时含有酸性和碱性位点，可作为级联反应的优秀双功能催化剂；合成出了多种三维多孔结晶聚酰亚胺共价有机骨架（称为 PI-COF）[773]，在所合成的 PI-COF 中，有两种 PI-COF 显示出高的热稳定性（大于 450℃）和表面积（高达 $2403m^2/g$），有较高的药物担载量和良好的可控释放性。

陈接胜等研究了多级孔 silicalite-1 纳米晶晶内镶嵌 Pd 纳米颗粒的合成及其择形催化性能，他们将 Pd 纳米颗粒成功地镶嵌在多级孔 silicalite-1 分子筛纳米晶中（Pd@mnc-S1）[774]，在加氢还原、催化氧化以及偶联反应中都展现出优异的择形选择性。

卜显和等制备出一例兼具主体框架刚性和热刺激响应性的动态 MOFs 材料[775]，展现出刚性的嵌套型结构特征，所有窗口均为三角形且被作为抗衡离子的硝酸根所封堵，他们认为这种“带塞瓶子”构型的形成得益于模板效应的诱导。左景林等通过引入具有氧化还原活性的四硫富瓦烯配体〔TTF-（py）4〕和自旋交叉的二价铁中心构建了两例三维MOFs，使得该类材料同时具备热力学和光诱导自旋交叉性（SCO）和氧化还原性[776]，通过这种氧化性掺杂使得来自于框架电子态调制的导电性明显加强（高达 3 个数量级）。施剑林等形成一种新颖的基于空心介孔普鲁士蓝的核壳结构纳米粒子（HMPB-Mn）[777]，该纳米粒子不仅具有超高的 pH 响应弛豫率，而且可以实现 MRI 对药物 DOX 释放的实时监控，可以作为一种高效的 T1-MRI 造影剂。

邢华斌等提出了离子杂化多孔材料分离乙炔和乙烯的新方法[778]，不仅为乙烯和乙炔的高效分离与节能降耗提供解决方法，而且也为其他气体的分离提供新的思路。杨维慎和李砚硕制备了单分子层厚度（~1nm）的金属有机框架纳米片，并通过热组装方法得到超薄金属有机框架膜[779]，该纳米片金属有机框架膜的氢气 / 二氧化碳分离系数达到 200 以上，氢气透量达到 2000 GPUs 以上，远高于迄今报道的有机和无机膜的氢气 / 二氧化碳分离性能。

唐智勇等将 Pt 纳米颗粒封装到 MOF 层间或孔道内构建了具有高效选择性催化的反应容器[780]，对催化 α，β - 不饱和醛的加氢反应具有高活性和高选择性。臧双全等用简单桥联配体同十二核的银硫簇组装得到了具有优异稳定性和荧光性质的金属 - 有机框架材料（MOFs）[781]，其稳定性和荧光强度大大提高，并被应用于荧光传感检测 O_2 和挥发性有机物 VOCs，阐明了荧光开 - 关的机理。

朱广山等报道了一类新型的芳香有机多孔材料，通过简单的离子交换方法高效的对 5 种工业用气体（氢气，氧气，氮气，甲烷和二氧化碳）进行逐一筛分纯化，实现多组份气体全分离[782]。苏成勇等提出了“后合成可调隔层装卸法（PVSI）”合成 MOF，可以插入带有不同官能团的第二桥联配体来改变原来框架的拓扑结构，改善框架孔道的尺寸、形状和表面，精确控制框架的呼吸效应[783]。

车顺爱等设计了一系列新型有机两亲性模板分子，在水热条件合成了多种具有有序介孔结构的 MFI 型沸石分子筛[784]，突破在于将常规的表面活性剂疏水链端引入芳香基团，通过分子间的 π - π 堆积增强疏水作用力，来进一步稳定胶束和沸石骨架结构。王建国等采用同晶取代硼掺杂的方法，对 H-MCM-22 分子筛三种孔道结构中的骨架铝和酸分布在原子尺度进行了有效调控[785]。钱国栋合成了具有一维孔道的 MOFs 材料 ZJU-68，并利用孔道的限域作用原位组装且高度取向的有机激光染料分子，使其量子效率从溶液状态的 0.45% 大幅提高到 24.28%，同时利用 MOFs 单晶的天然晶面作为谐振腔的反射镜面，获得了高偏振度（大于 99.9%）的三光子泵浦激光，其品质因子 Q 值高达 1700[786]。

孙剑和葛庆杰等通过设计一种新型 Na-Fe_3O_4/HZSM-5 多功能复合催化剂，实现了 CO_2 直接加氢制取高辛烷值汽油[787]，在接近工业生产的条件下，该催化剂实现了 CH_4 和 CO 的低选择性，烃类产物中汽油馏分烃类的选择性达到 78%，还具有较好的稳定性，可连续稳定运转 1000 小时以上。孙予罕等采用氧化铟 / 分子筛双功能催化剂，实现了 CO_2 加氢一步转化高选择性得到液体燃料[788]，对汽油组分碳氢化合物选择性高达78.6%，而甲烷选择性仅为1%。

肖丰收等合成了沸石分子筛封装的过渡金属氧化锰纳米催化材料，并实现了一步C–H键活化制备腈类化合物的过程[789]，多种表征手段证明了氧化锰颗粒完美地封装于沸石晶体内部，实现了通过沸石孔道控制金属表面催化反应的微环境，在多种烷烃的C–H键直接氧化腈化反应中，该催化材料均给出了优异的产率。

宋卫国等制备了N/P/S共掺杂的中空碳壳非金属催化剂，实现了水体系下芳香烷烃的高效选择性催化氧化[790]，该材料在水体系中显示出了高效的催化芳香烷烃选择氧化性能。

此外，国内同行们还在新型微孔、介孔、大孔、复合多孔材料的合成、制备、表征、功能化以及性能研究，新合成方法的开发等方面取得了重要进展。

8. 化学生物学

在基于天然产物、化学合成及生物合成的小分子探针的开发与应用方面，杨财广与周翔合作，设计发现双功能化学探针，实现对FTO蛋白的"可视化"跟踪[791]。李艳梅和陈永湘通过多功能分子促进细胞内Tau蛋白的多聚泛素化修饰，实现Tau蛋白的降解，为阿尔兹海默症的治疗提供潜在新策略[792]。郭子建发展了基于过氧化氢激活的药物输运体系，实现了光动力学药物在肿瘤细胞的靶向性释放[793]。李笑宇与刘磊合作，发现市场上广泛销售的α1肾上腺素受体阻滞剂特拉唑嗪可以结合新靶点，在中风和败血症治疗中发挥新用途[794]。肖有利化学修饰青蒿素分子，基于活性蛋白谱策略研究青蒿素与靶点蛋白的相互作用，为天然产物作用机制研究提供新思路[795]。胡有洪等设计的小分子探针NP3用于ONOO–的特异灵敏性检测，实现了硝化应激损伤过程的活体实时示踪[796]。周翔发现丙肝病毒和埃博拉病毒均存在G–四链体结构，通过特异性靶向G–四链体结构的小分子探针来稳定结构可以抑制病毒RNA复制，为抗病毒治疗提供新思路[797]，并开发了偶氮苯类小分子探针，它可以通过光来调控蛋白酶的活性[798]。席真和周传政发展了N–取代–5–亚甲基吡咯酮作为一类新的巯基特异性、可控脱除的生物共轭标记分子[799]。

刘文等发现两种小分子硫醇麦角硫因（EGT）和放线硫醇（MSH）在林可链霉菌中相互配合精确有序地介导了林可霉素这一临床上广泛应用的含硫抗生素的生物合成[800]；还报道了螺环乙酰乙酸内酯/内酰胺抗感染抗生素的生物合成中两种不同的酶，可以极大地促进［4+2］D–A环化反应的发生，有助于螺环乙酰乙酸内酯类抗生素的合成[801]。赵宗保通过构建杂合酶扩展脂肪酸合成机器产物谱，利用脂肪酸代谢途径进行生物合成[802]。

化学生物学特色技术和前沿分析方法的发展方面，陈鹏设计了一种特殊的逆电子需求的狄尔斯–阿尔德反应（invDA）用来介导"化学脱笼"反应，在活细胞内及活体动物上实现了激酶的特异激活与靶向研究[803]。陈兴开发了一项基于碳纳米管的蛋白近红外光激活技术，用以在活细胞及小鼠活体中调控转化生长因子–β（TGF–β）信号转导，对活细胞内蛋白质含量的化学控制是调控其功能与活性的另一重要方法[804]；还开发了一系列基于生物正交标记的聚糖标记新方法，应用于小鼠活体中，实现了肿瘤组织及大脑组织中聚糖的靶向标记与成像[805]。董甦伟通过脯氨酸内硫酯中间体的设计实现了多肽在脯氨酸位点的高效连接，有望用于蛋白质的化学合成[806]。刘磊课题组设计并合成了第一例靶向免疫细胞、并可通过双光子激活趋化因子探针 hCCL5**，利用双光子良好的组织穿透性，实现了在活体组织中（小鼠耳朵及淋巴结）对免疫细胞定向运动的控制[807]；发现了第一例针对 B 细胞的光控蛋白抗原 HEL–K96NPE，揭示 B 细胞活化早期信号传导动力学[808]。王江云开发了一种在活细胞内观察 RNA 分子的特异标记技术[809]。席真与易龙合作，发展了多反应 – 多淬灭基探针，实现对肠癌细胞中内源性硫化氢的灵敏检测和肿瘤近红外成像，并揭示双氧水通过 CBS 途径诱导内源硫化氢合成机制[810]。张艳与张辰宇合作，通过光点击化学反应的四氮唑基团对 microRNA 进行化学修饰，实现细胞内 microRNA 的富集和分离[811]。

方晓红等为获得细胞信号转导相关蛋白的高分辨、多参数成像提供了新的研究平台，可以更准确有效地获取和分析细胞上单分子动态变化的信息[812]。毛兰群等针对脑神经化学过程，建立并发展了活体微透析取样和活体原位检测技术，在活体层次研究耳鸣过程中与谷氨酸毒性相关信号转导通路以及水杨酸钠诱导耳鸣过程中抗坏血酸变化的规律，这些活体原位定量分析新技术促进了对实际生物体系的研究[813]。杨朝勇提出了基于气压的免疫检测新原理，实现了痕量生物分子的便携定量检测[814]。黄岩谊发展了微流控微液滴单细胞全基因组扩增技术，实现对单个细胞高覆盖和高均衡的全基因组测序[815]。鞠熀先通过循环杂交酶切手段、单激发 – 双发光共振能量转移[816]等策略发展了一系列细胞表面聚糖糖型的原位检测方法，构建了基于外源酶底物的 FRET 系统和唾液酸酶响应型纳米探针，提出原位检测活细胞内唾液酸转移酶和唾液酸酶活性的新策略[817]。樊春海、师咏勇和晁洁发展了一整套 DNA 折纸结构并作为纳米力学成像探针，实现了原子力显微镜下对基因组 DNA 的直读检测和高分辨单倍型分析[818]。林金明等构建了在线微流控芯片 – 质谱联用装置，用于细胞代谢研究[819]。苏循成通过顺磁核磁共振测定了金黄色葡萄球菌 Sortase A 与底物多肽形成硫酯中间体的三维结构[820]，并与以色列威兹曼研究所合作报道了测定顺磁间高精度距离的信息以检测细胞内蛋白质动态变化方法[821]，与哈佛大学合作通过顺磁核磁共振技术揭示了线粒体 MCU 的离子转运机制[822]。

曲晓刚发现阿尔海默症病变蛋白 Aβ 可抑制端粒酶活性[823]，基于纳米材料开展多功能人工酶的设计和应用研究[824]，并通过稀土上转换材料或光响应来实现细胞行为的精确控制[825]。赵劲通过对金离子解毒蛋白的晶体结构解析和机械强度测定，揭示了金属转运蛋白特异识别的分子机制[826]。季泉江基于 CRISPR/Cas9 系统构建了能够在金黄色葡萄球菌中进行快速高效基因组编辑的 pCasSA 质粒体系[827]。

生物问题导向的多学科交叉研究方面，针对这一胆固醇代谢领域的重要问题，宋保亮等筛选鉴定出300多个胆固醇运输相关的基因，并揭示了一种全新的胆固醇转运途径、发现了过氧化物酶体细胞器的新功能[828]。姚雪彪等揭示了一个调控真核细胞染色体稳定性的CDK1–TIP60–Aurora B信号轴，并详尽地阐明了蛋白质磷酸化与乙酰化修饰动态调控Aurora B激酶活性的新机制[829]。吴乔等以前期发现的核受体Nur77为靶标，探讨其通过NF–kB信号通路抑制炎症的分子机制，利用上游因子和Nur77的关系，建立独特的筛选平台，在自己构建的靶向Nur77的小分子化合物库中寻找和确定能够通过Nur77介导抑制炎症的小分子化合物，并在不同模型的炎症小鼠中进一步证实化合物的生理功能[830]。刘勇揭示了内质网应激感应蛋白IRE1α 通路与机体肥胖及代谢炎症之间的关系，发现IRE1α 缺失能够抵御高脂诱导下肥胖的发生，同时明显改善代谢紊乱症状，特异性抑制巨噬细胞的IPE1通路将在肥胖、胰岛素抵抗和II型糖尿病等代谢疾病的防治方面具有重要的转化潜力[831]。张立新等从真菌中鉴定了单核非血红素铁酶FtmOx1，通过晶体结构分析阐明了FtmOx1催化环内过氧桥键合成的催化机理，推动了青蒿素等复杂天然产物的生物合成[832]，并开展链霉菌生物元件的调控用于激活阿维链霉菌中天然产物的表达[833]。梁宏和杨峰等合作解析了ACF7蛋白结构域，阐述了细胞移动中的重要信号传导过程，并证明ACF7蛋白通过肠上皮细胞紧密连接的动态协调的微管进行调节肠道细胞运动和伤口愈合[834]。元英进首次化学设计合成了五号真核生物染色体，开发了定制酵母环形染色体的方法，为染色体研究提供研究思路和模型，并通过化学设计合成酵母十号染色体，创建一种高效定位生长缺陷靶点的方法，解决了合成型基因组导致细胞失活的难题[835]。

药物发现的早期探索方面，蒋华良与芝加哥大学何川等合作，对铜离子伴侣蛋白的可靶性进行了化学生物学研究，不仅为肿瘤的靶向治疗提供了优秀的先导化合物，同时也为抗肿瘤策略的研究开辟了全新的领域[836]。蒋华良、杨财广和刘江发现了能够破坏SPOP和PTEN相互作用的小分子化合物[837]，证明了破坏SPOP与底物蛋白相互作用有可能成为特异性靶向肾癌治疗的新策略。蓝乐夫和李剑等通过表型筛选老药分子库，发现萘替芬显著地抑制金黄色葡萄球菌金黄色色素的产生，发现并验证了CrtN蛋白是萘替芬的作用靶标，为抗生素替代品如抗致病力药物的研究提供了新的药物作用靶点[838]。杨财广与贾桂芳等从FTO与ALKBH5识别甲基化修饰核酸的机制不同的特点出发设计高通量筛选方案，发现靶向FTO的选择性小分子抑制剂－非甾体抗炎药甲氯芬那酸[839]。徐天乐发现缺血导致的氧气、葡萄糖缺失引起脑组织酸化，在质子作用下，酸敏感离子通道ASIC1a蛋白与RIP1结合，引发后者磷酸化，进而激活神经细胞程序性坏死进程，为基于ASIC1a通道的脑卒中治疗药物研究提供了全新的理论指导[840]。沈旭建立了一套完整高效的分子及细胞水平FXR拮抗剂筛选平台，从商业数据库Enamine中筛选出了50个化合物用于生物活性测试后，得到5个有活性的化合物，表明筛选策略的有效性[841]。张健通过大规模肿瘤基因组在蛋白结构上的变构映射精准识别各类型肿瘤的全新靶标，并利用该方法发现非小细胞肺癌的全新靶标 PDE10A[842]。

周德敏和张礼和以甲型流感病毒为模型，应用基因密码子扩展技术成功创制了复制缺陷型活病毒疫苗[843]，是一项革命性进展。章晓联筛选了能特异性结合卡介苗的单链 DNA 配体，可显著提高卡介苗免疫原性，增强宿主对结核菌感染的免疫力和保护作用[844]。杨光富开发了一种用于药效团连接的碎片虚拟筛选方法的自动化计算程序，为新医药和农药创制提供信息技术支撑[845]。

9. 纳米催化

2016 年我国在纳米催化方面取得了一些具有国际影响力的原创性研究成果。

（1）研制出将二氧化碳高效清洁转化为液体燃料的新型钴基电催化剂

谢毅和孙永福团队制备了四原子厚的钴金属层和钴金属 / 氧化钴杂化层，评估金属和金属氧化物 2 种不同催化位点的作用，发现在低过电位下，相对于块材表面的钴原子，原子级薄层表面的钴原子具有更高的生成甲酸盐的本征活性和选择性，而部分氧化的原子层进一步提高了它们的本征催化活性，在过电位仅为 0.24V 条件下实现了 $10mA/cm^2$ 的电流输出超过 40 小时，且其甲酸盐选择性接近 90%，这超过此前报道的金属或金属氧化物电极在同等条件下得到的结果[846]。

（2）催化功能 MOFs 的精准构筑：选择性加氢反应调控器

α，β－不饱和醇是香料、香精、医药等的原料和中间体，因天然可利用资源有限，其主要由 α，β－不饱和醛选择性加氢制得。然而，热力学上易于 C=C 双键加氢，对应的不饱和醇的选择性较低。唐智勇、李国栋和赵惠军团队提出用 MOFs 作为选择性加氢反应的调控器，设计和构筑了三明治结构 MIL–101@Pt@MIL–101 催化剂，该结构催化剂可以显著提高肉桂醇的选择性（~95.6%），同时可实现肉桂醛的高转化率（~99.8%），并证实了这种催化剂设计理念具有一定的普适性[847]。

（3）开创煤制烯烃新捷径

针对传统烯烃制备副产物多水消耗量大灯缺点，包信和及潘秀莲团队开发出了一种过渡金属氧化物和有序孔道分子筛复合催化剂，成功实现了煤基合成气一步法高效生产烯烃，C_2 到 C_4 低碳烯烃单程选择性突破了费托过程的极限，一跃超过 80%。同时，反应过程完全避免了水分子的参与，在纳米尺度上实现了对分别控制反应活性和产物选择性的两类催化活性中心的有效分离，使在氧化物催化剂表面生成的碳氢中间体在分子筛的纳米孔道中发生受限偶联反应，成功实现了目标产物随分子筛结构的可控调变[848]，被产业界同行誉为“煤转化领域里程碑式的重大突破”。

（4）棱柱形 Co_2C 纳米晶助合成气直接制烯烃

孙予罕和钟良枢团队发现棱柱形状的碳化钴（Co_2C）的纳米粒子能够实现合成气到低碳烯烃的高效直接转化，取得了超过 60% 的低碳烯烃选择性，同时把甲烷的选择性保持在 5% 左右，反应的条件很温和（大气压即可），温度为 250℃，并且可以用来自生物质

的 H_2/CO 比例较低的合成气作为原料生产低碳烯烃[849]。

（5）单原子分散贵金属催化剂的新突破

郑南峰和傅钢团队采用光化学方法，在乙二醇修饰的超薄二氧化钛纳米片上制备了稳定的单原子分散钯－二氧化钛催化剂（Pd_1/TiO_2），其中钯负载量高达 1.5wt.%；Pd_1/TiO_2 在碳碳双键和碳氧双键加氢反应中都表现出了出色的稳定性和极高的催化活性，苯乙烯加氢反应中 Pd_1/TiO_2 的催化活性是商业化 Pd/C 催化剂的 9 倍以上（以单位表面原子计算），而且历经 20 个循环依然保持高水平；该研究工作不仅制备了高负载的单原子分散钯催化剂，为其工业应用打下坚实基础，还能帮助化学家在分子层面理解非均相催化，在均相和非均相催化之间搭起桥梁[850]。

（八）化学教育发展

1. 基础化学教育

高考改革与高中课程改革方面，为配合 2014 年我国全面启动的新一轮考试招生制度改革[851]，围绕中国学生发展核心素养，华东师范大学王祖浩教授、北京师范大学王磊教授、东北师范大学郑长龙教授、陕西师范大学周青教授等研究团队自 2015 年开始，逐步完成教育部组织的 2003 版全国高中化学课程标准修订工作。在课程目标方面，提出化学学科核心素养，为课程内容选取和评价设计提供依据；在课程结构方面，课程标准调整为必修、选修Ⅰ和选修Ⅱ三类课程；在课程实施方面，构建“素养为本”的实施建议，突出化学学科大概念的统领作用；在化学课堂教学开展方面，提供基于主题的教学策略建议和学习情境素材建议，提高对教学实践的具体指导等。

教学改革及其研究方面，主要围绕“核心素养”“化学核心素养”。除理论研究层面提出的 5 个化学学科核心素养要素外，还有围绕对化学学科核心素养的概念分析和层次分析的一系列研究[852-856]。毕华林团队在《美国化学教育》期刊发文，分析、比较了我国化学课程标准与美国新一代课程标准对“化学物质和化学反应”核心概念的呈现方式[857]；在《化学教育研究与实践》刊发利用 Rasch 测评模型研发的中学生电化学概念理解测试工具[858]。郑长龙团队在《化学教育研究与实践》介绍了如何利用 Rasch 测评模型验证中学化学课堂教学质量评估工具[859]。周青团队在《化学教育研究与实践》分享了对中学生在使用流程图学习醋酸相关知识时的核心障碍点和认知结构探查[860]。王磊和朱玉军等在《亚洲科学教育研究与实践》一书中撰写“中国大陆的科学教育研究”一章[861]；参与《东亚科学教育研究与实践：现状与展望》书稿撰写，介绍我国高中科学教育教师的职前培训经验[862]。

学术交流方面，中国化学会化学教育委员会、《化学教育》编辑部、北京师范大学化学教育研究所等单位举办了多个国内会议，如“中国化学会关注中国西部地区中学化学教学发展论坛”、“中国化学会全国中学化学教育高峰论坛”等。并推动国际交流会议在我国境内举办，如第四届东亚科学教育国际会议暨 EASE2015 国际交流会等。

2. 高等化学教育

截止 2016 年底，我国共开办化学类专业 800 个，厦门大学于 2016 年申请创设第一个能源化学专业。三家国家级出版社共出版、再版化学类教材 137 部。2016 年国内期刊共发表大学化学教育论文 574 篇，论文内容主要涉及教学理念、教学方法、教学技术和课程建设等，体现了大学化学教学的改革热点。召开的教师教学交流类会议十余场，为教师们搭建了交流、切磋教学的平台，如“高等学校应用化学专业教学研讨会”、“第六届全国高等学校物理化学（含实验）课程教学研讨会”、“高校化学化工课程教学系列报告会”等。

教学质量国家标准和专业认证标准的制订和完善方面，教指委根据教育部的要求，进一步完善了《化学类专业教学质量国家标准》，受教育部评估中心的委托，起草了《化学类专业（理科）学生学习成果要求》。完成《化学类专业化学理论教学建议内容》《化学类专业化学实验教学建议内容》《高等学校化学类专业物理化学相关教学内容与教学要求建议》《浅议基于学科思维的化学类专业课程体系设计与课程内容编排》等重要成果。这些将对未来一段时间我国化学类专业的教学改革起到重要的指导作用。化学专业认证试点方面，教育部评估中心对武汉大学化学专业开展了认证试点。理科化学专业认证在世界上还是首次，相关标准的建立和认证的开展，为今后我国理科化学类专业认证提供了有益经验。

三、我国化学学科发展趋势和展望

（一）有机化学

我国有机化学在学术论文发表方面取得了显著成绩。但是，还缺乏真正意义上的原创性学术成果，还缺少真正解决工业生产需求的应用成果。

有机反应研究应不能仅仅局限于在热点研究工作跟踪或专注于扩展反应适用范围等补缺式研究，特别需要开辟对一些重要的、深藏闺阁的资源型化合物反应性能研究，以发现原始性反应。有机合成化学的目标分子选择则不应以首次合成为目标，而应该追求有学术或应用价值的目标分子的高效合成。天然产物化学领域，不再局限于新天然有机分子的发现和结构，而是侧重新活性分子的发现和开发，希望不要因为论文导向的考核标准影响天然有机化学的正常发展。手性化学方面，目前的许多不对称催化反应仍然存在催化剂用量大、效率和选择性低等问题，真正能够用于工业生产的不对称催化反应仍然屈指可数。手性合成作为手性化学研究的基础，发展具有原创性、实用性的手性合成方法学的任务仍十分艰巨。因此，设计与合成新型、高效、高选择性的手性配体及催化剂，发展不对称催化合成的新策略与新方法，发展高效的不对称催化新反应，实现手性物质的精准、高效创制将是今后的重要发展方向之一。另一方面发展精准的不对称催化合成反应和方法、以及发展手性片段的精准集成策略，从而实现复杂天然产物分子合成中手性的精准控制是未来发展的挑战性方向之一，同时将推动手性化学与生命科学、环境科学等其他学科的深度交叉与融合。此外，手性物质在材料、信息等领域中的应用我们还知之甚少，但手性液晶材料在显示方面所展现出的特殊性能，足以使我们相信手性物质未来在信息科学、材料科学等领域将大有作为，甚至会带来革命性的变化。因此，及早重视光学活性手性物质在信息科学、材料科学等领域的潜在用途，深入探索宏观尺度的手性物质的可控有序构筑，精准创造出更丰富的手性物质和材料是我们的目标。

（二）物理化学

化学动力学领域的研究者们从理论和实验两方面努力将研究方向扩展到了激发态反应、多势能面的非绝热过程、表界面化学反应、大分子（包括团簇和生物分子）体系动态学等方面，并取得了令人瞩目的研究成果。今后，随着化学动力学的实验技术和理论方法以及计算机能力的进一步提高和扩展应用，特别是国家重大科研仪器设备专项“基于可调极紫外相干光源的综合实验研究装置”项目目前已经顺利出光，将于 2018 年投入科研工作，我国的化学动力学的研究水平将得到极大的提高，使我们有能力面对化学动力学领域来自学科自身发展和国家能源、环境以及国防需求的挑战。

化学电源总的趋势朝着高比能量和高安全性方向发展，一方面为提高电池的能量密度，发展高电压，大容量的正极材料及高容量的负极材料，包括硅碳复合材料，以及以金属锂负极的锂硫，锂空电池等；另一方面发展高安全性的固态和水系的锂离子电池。我国学者在上述领域材料设计、制备方面有一定特色，另外在钠离子、液流电池和水系离子电池等方面处于国际领先水平，但在电池体系和材料的原创性工作、机理研究及其新型现场表征技术创制等方面跟国外有一定的距离，这是我国研究工作的发展方向和重点应该关注的问题。未来固体聚合物电解质膜燃料电池能量转换的核心科学与技术问题仍将集中在开发高性能且稳定的低贵金属 - 非贵金属催化剂上，通过研究反应过程机理与构效关系入手，理性设计新型催化剂是重要发展方向之一；同时，催化反应器现场中存在的实际问题确认与催化剂性能的有效表达是决定催化剂可用性的先决条件，必将获得更多关注。发展现代原位表征方法，深入研究反应过程仍是一个重点研究课题。分子电催化在美国化学会揭晓的目前化学领域十大热门研究课题中位居第三，国内孙立成教授、杜平武教授已经开展了分子电催化的研究，制备出一系列高性能析氢、析氧催化剂，并在光催化系统中实现了使用分子催化剂催化的水分解反应。但是国内分子电催化的研究仍未引起广泛重视，未来还需重点关注难度更大的反应，比如氮气还原和甲烷氧化。

我国催化科学基础与应用开发研究取得了一些国际领先、有重大学术影响的成果，在一些重要研究方向已经或正在实现从跟踪到引领国际潮流的转变。预计我国催化学科的这种发展趋势将会延续下去，在研究工作内容上，基于传统化石资源（包括合成气）的高效选择性转化制取基础化学品的创新工作还会进一步强化，另一方面基于太阳能、可再生碳资源（如生物质）等制取清洁燃料、高附加值化学品和材料的催化研究将被更加重视。在研究方法上将强调在催化反应的原位、实时条件下研究催化剂的构效关系和理解催化过程，实验测试和表征催化剂的工作将会更多地吸纳或参考理论计算化学研究的成果。催化剂加快反应和掌控反应方向（选择性）的属性还会使更多工作在化学其他分支领域的化学家和工程师自觉从事催化剂和催化活性研究。

胶体与界面化学方面，工农业进一步发展中将会更广泛地运用其基本原理和研究方法，特别是石油的开采和炼制，油漆、印染和选矿，新能源、甚至土壤改良和人工降雨等。工农业生产实践中所提出来的问题，又进一步推动着胶体化学理论的发展。胶体体系近年来也经常被称为“软凝聚态物质”或“软物质”，表明它们的结构与动态性质受弱的物理作用所掌控。随着热力学、统计力学和液体理论的不断发展，计算机模拟的进步，均对此领域产生重要的影响，也越来越成为研究的热点。

（三）分析化学

发展实时、原位、在线、活体、高灵敏的分析技术依然是分析化学面临着的挑战。研究者对新型纳米结构的研究发展给予了高度关注，基于纳米孔、等离激元纳米颗粒、荧光纳米探针的分析应用以及纳米传感器等纳米分析的研究依然是目前分析化学的热点俺就领域；成像分析尤其是活体的成像分析是近年来关注度很高的领域，如何在复杂的机体中实现对低浓度物质的灵敏、时空精确的成像是今后的研究重点；活体分析化学正在成为分析化学重要的前沿领域之一。然而活体分析目前在临床上应用的并不多，需要解决复杂的体液成分对传质和信号的影响，更重要的是需要避免对机体尤其是脑和神经系统的干扰；单分子和单细胞分析，迫切需要新的分析材料和高灵敏的检测技术；新的蛋白质组学和多肽组学、代谢组学研究需要高通量的灵敏的分离和检测技术，提取和分析完整、全部的蛋白质或代谢组分是终极的目标，尤其是新的质谱技术和数据处理和分析方法。针对生命体系的分析以及环境分析为主要的研究，始终是分析化学的研究的主要内容，发展新的分析原理、仪器和装置，解决涉及生命、环境和材料科学等有影响的重大科学问题为主要任务。

我国分析化学基础研究已经取得了长足的进步，然而，在发展和建立原创性的分析化学原理和方法方面，还有相当大的提升空间。分析方法的建立，离不开新的分析材料和分析仪器，目前的高精尖的分析仪器依然被国外的公司所垄断，未来必须必须加强分析仪器装置研究的原始创新性工作。

生命科学、材料科学和环境科学的发展要求色谱的功能更为齐全、分析速度更快、分离能力更高，以及自动化程度更高。在色谱填料和色谱柱方面，传统的色谱填料和色谱柱技术仍然是主要分离介质，整体柱、开管毛细管柱将会有新的研究和突破。另外手性固定相、生物亲和色谱材料、纳米多孔材料将有较大发展；在色谱新技术和新方法方面，超高效色谱、快速、高通量分离技术和方法、生物标志物发现和多目标化合物的分离鉴定技术、在线实时监测等技术将有新的技术突破和应用，毛细管电泳与微流控芯片技术也将在生命物质分离分析方面发挥重要作用；在色谱仪器方面，气相色谱、高效液相色谱将保持一个持续增长态势。超高压液相色谱类仪器在技术和应用方面会得到快速发展；气相色谱类仪器在多维色谱、现场检测等方面得到较大发展应用。在应用领域方面，色谱与质谱联用技术在生命科学、环境分析、食品安全、公共安全等方面也会越来越成为主流手段。

国际质谱领域正朝着开发新型实用的敞开式质谱、小型便携的小质谱、高灵敏度高分辨率的高端质谱三大方向发展。在敞开式质谱的领域里，我国的研发进度与国际相差不大，并与国际顶尖团队有广泛合作。国外最先发明了 DART，后相继发明了 ELDL、ASAP EESI、DAPCL、DBDI 等多种敞开式离子源。国内近几年也相继发展出 EESI、DAPCI、LIP、AFAI、PAMLDI、SALDI 等多种实用型敞开式离子源，其种类和离子化效率都在追赶国外先进水平。小型化质谱领域里，我国与国际差距不断缩小，国外已经推出 Torion、Inficon、908device 等不少商业化小型化质谱，并应用于消防和航天等领域。国内的企业也推出 Mars400、mini12 等小型化质谱，并成功应用于疾控中心、防化院等场所里。在高端质谱研制领域，我国有明显的进步，但与国外还有较大的差距。目前市场上国产高端质谱产品屈指可数。国内 2015 年推出了首台国产化的 MADLI-TOF-MS，并通过了专家组鉴定，后续的性能需要用户的进一步体验和确认。按照目前质谱学术领域的潮流，敞开式质谱、质谱成像、质谱的组学应用研究将继续成为未来的研究热点。敞开式质谱研究将重点解决定量问题和小型化问题。质谱成像将会朝着提高成像分辨率和、灵敏度和采集速度上方向努力。质谱的组学应用研究将着重于不同组学间数据库大平台的构建和对接，检测进一步标准化。

（四）交叉学科或研究领域

1. 流变学

中国流变学研究起步较晚，与先进国家相比，尚存在一定的差距：在国际著名流变学刊物上发表的论文数量偏少，由我国学者提出的具有原创性的理论成果尚不多见，没有主办过世界级的国际流变学学术会议。现在国内流变学研究的优势领域在高分子纳米复合材料流变学研究、电流变和磁流变流体研究和采油输油相关流变学研究等方面。国内从事流变学研究的核心课题组不到 50 个，针对流变学作为交叉学科的特点，建议进一步加强工程过程流变学研究，优化流变学在国家自然科学基金委学科目录中的布局。进一步强化流变学应用研究，拓展流变学应用行业领域，为相关行业发展和转型升级提供流变学指导。吸引国内相关学科研究人员进入流变学研究领域，吸引国外的年青学者回国工作，尽快扩大研究队伍该，拓宽研究领域，加强合作，快速突破，从而实现国内流变学研究的后发优势。

2. 环境化学

在环境化学的未来研究中，应继续围绕国家环境保护重大需求，提高我国环境化学领域的创新能力，为解决我国面临的环境问题、履行国际公约、促进可持续发展提供保障。在大气环境化学研究方面，针对我国大气污染的独特性，大气自由基未知来源及其机制、VOC 对臭氧及二次有机气溶胶的影响、大气非均相反应机制等是未来的重要研究内容；继续做好履约相关持久性有机污染物和汞的排放、环境过程与效应、削减技术及管控化学品的替代问题研究，为履约提供有力保障；深入研究微塑料和纳米塑料这一新型污染物的环境多介质定性 / 定量识别技术，评价其自身及所载带污染物的环境与健康风险；进一步利用代谢组学等先进技术，揭示环境污染物的毒性效应及其作用的分子机制。此外，新化学品合成数量与速度的快速增长为面向化学品风险评价的计算（预测）毒理学及高通量毒性测试平台的建立提出了新的需求与研究方向；应结合流行病学手段开展大气污染物与持久性有毒污染物的健康影响研究，尤其强调污染长期暴露（一年或更长时间）的健康影响。

3. 绿色化学

在当前以及今后一段时间里，设计高效价廉的新型催化剂，发展新的合成路线，缩短反应步骤，提高反应原子经济性，仍将是绿色化学的一个重要研究方向。绿色化学的研究特点包括：采用清洁、无污染、可再生的化学原料，代替不可再生的或者有毒的原料；探索新的反应条件，如使用绿色的反应介质（水、超临界流体、离子液体等），甚至使用无溶剂反应体系；设计更安全的、毒性更低或更环保的化学产品等。在绿色能源领域，太阳能、氢能、生物质能等清洁能源的开发难度仍然较大，经济效益偏低，因而，急需研发新的技术方法，实现对这些清洁能源的高效利用。其中解决氢的经济性制取和大规模输尤为重要；人工固碳是减少二氧化碳排放的有效途径，同时可为工业生产提供碳源，减少石油资源的消耗。但目前人工固碳的技术较匮乏、效率较低、经济性较差，是绿色化学研究领域中亟待解决的一个难题。

4. 多孔材料研究

目前，虽然合成的多孔材料种类以及很多，但是实际用于工业应用的却非常有限，仅有十多种类型的分子筛用于石油化工的催化、吸附和分离等领域，金属 – 有机框架材料以及共价 – 框架材料目前还没有进入工业应用阶段。因此，在多孔材料研究领域，应该更加注重以功能为导向的研究，实现多孔材料的多种应用，为能源环境和健康等问题提供解决途径。

四、总结

与美国等西方国家在化学研究领域的经费投入减少和研究队伍缩水不同，我国化学研究队伍不断壮大，有各级政府的研究经费支持，学术论文发表篇数连年保持第一，展现出良好的发展势头。但是，学术论文导向促使我国化学成为学术论文发表大国，并不代表我国化学已经成为学术论文发表强国。我国化学在纯学术研究方面原创性研究成果还很少见，主动解决化学化工产业发展中重大问题的理念也有待建立。中国化学正在处在转型期。

参考文献

［1］Zhang W, Wang F, McCann S D, et al. Enantioselective cyanation of benzylic C–H bonds via copper-catalyzed radical relay［J］. Science, 2016, 353（6303）: 1014–1018.

［2］Yuan Z L, Wang H Y, Mu X, et al. Highly selective Pd-catalyzed intermolecular fluorosulfonylation of styrenes［J］. Journal of the American Chemical Society, 2015, 137（7）: 2468–2471.

［3］Yuan Z L, Cheng R, Chen P H, et al. Efficient pathway for the preparation of aryl（isoquinoline）iodonium（III）salts and synthesis of radiofluorinated isoquinolines［J］. Angewandte Chemie International Edition, 2016, 55（39）: 11882–11886.

［4］Wang F, Zhu N, Chen P H, et al. Copper-catalyzed trifluoromethylazidation of alkynes: Efficient access to CF_3-substituted azirines and aziridines［J］. Angewandte Chemie International Edition, 2015, 54（32）: 9356–9360.

［5］Wang F, Wang D H, Wan X L, et al. Enantioselective copper-catalyzed intermolecular cyanotrifluoromethylation of alkenes via radical process［J］. Journal of the American Chemical Society, 2016, 138（48）: 15547–15550.

［6］Li M, Yu F, Qi X X, et al. A cooperative strategy for the highly selective intermolecular oxycarbonylation reaction of alkenes using a palladium catalyst［J］. Angewandte Chemie International Edition, 2016, 55（44）: 13843–13848.

［7］Cheng J S, Qi X X, Li M, et al. Palladium-catalyzed intermolecular aminocarbonylation of alkenes: Efficient access of β-amino acid derivatives［J］. Journal of the American Chemical Society, 2015, 137（7）: 2480–2483.

［8］Chen C H, Chen P H, Liu G S. Palladium-catalyzed intramolecular aminotrifluoromethoxylation of alkenes［J］. Journal of the American Chemical Society, 2015, 137（50）: 15648–15651.

［9］Zhuo C X, Zhou Y, Cheng Q, et al. Enantioselective construction of spiroindolines with three contiguous stereogenic centers and chiral tryptamine derivatives via reactive spiroindolenine intermediates［J］. Angewandte Chemie International Edition, 2015, 54（47）: 14146–14149.

［10］Zheng J, Wang S-B, Zheng C, et al. Asymmetric dearomatization of naphthols via a Rh-catalyzed C（sp^2）-H functionalization/annulation reaction［J］. Journal of the American Chemical Society, 2015, 137（15）: 4880–4883.

［11］Zheng J, Cui W J, Zheng C, et al. Synthesis and application of chiral Spiro Cp ligands in rhodium-catalyzed asymmetric oxidative coupling of biaryl compounds with alkenes［J］. Journal of the American Chemical Society, 2016, 138（16）: 5242–5245.

［12］Yang Z P, Wu Q F, Shao W, et al. Iridium-catalyzed intramolecular asymmetric allylic dearomatization reaction of pyridines, pyrazines, quinolines, and isoquinolines［J］. Journal of the American Chemical Society, 2015, 137（50）: 15899–15906.

［13］Wang S-G, Yin Q, Zhuo C-X, et al. Asymmetric dearomatization of β-naphthols through an amination reaction catalyzed by a chiral phosphoric acid［J］. Angewandte Chemie International Edition, 2015, 54（2）: 647–650.

［14］Shao W, Li H, Liu C, et al. Copper-catalyzed intermolecular asymmetric propargylic dearomatization of indoles［J］. Angewandte Chemie International Edition, 2015, 54（26）: 7684–7687.

［15］Huang L, Dai L X, You S L. Enantioselective synthesis of indole-annulated medium-sized rings［J］. Journal of the

American Chemical Society, 2016, 138（18）: 5793–5796.

[16] Gao D W, Gu Q, You S L. An enantioselective oxidative C–H/C–H cross–coupling reaction: Highly efficient method to prepare planar chiral ferrocenes [J]. Journal of the American Chemical Society, 2016, 138（8）: 2544–2547.

[17] Cheng Q, Wang Y, You S L. Chemo–, diastereo–, and enantioselective iridium–catalyzed allylic intramolecular dearomatization reaction of naphthol derivatives [J]. Angewandte Chemie International Edition, 2016, 55（10）: 3496–3499.

[18] Kang T F, Ge S L, Lin L L, et al. A Chiral *N, N'*–dioxide–ZnII complex catalyzes the enantioselective [2+2] cycloaddition of alkynones with cyclic enol silyl ethers[J]. Angewandte Chemie International Edition, 2016, 55(18): 5541–5544.

[19] Li J, Lin L L, Hu B W, et al. Bimetallic gold（I）/chiral *N,N'*–dioxide nickel（II）asymmetric relay catalysis: Chemo– and enantioselective synthesis of spiroketals and spiroaminals [J]. Angewandte Chemie International Edition, 2016, 55（20）: 6075–6078.

[20] Li W, Tan F, Hao X Y, et al. Catalytic asymmetric intramolecular homologation of ketones with alpha–diazoesters: Synthesis of cyclic α–aryl/alkyl beta–ketoesters [J]. Angewandte Chemie International Edition, 2015, 54（5）: 1608–1611.

[21] Xia Y, Lin L L, Chang F Z, et al. Asymmetric ring opening/cyclization/retro–mannich reaction of cyclopropyl ketones with aryl 1,2–diamines for the synthesis of benzimidazole derivatives [J]. Angewandte Chemie International Edition, 2016, 55（40）: 12228–12232.

[22] Xia Y, Lin L L, Chang F Z, et al. Asymmetric ring–opening of cyclopropyl ketones with thiol, alcohol, and carboxylic acid nucleophiles catalyzed by a chiral N,N'–dioxide–scandium（III）complex [J]. Angewandte Chemie International Edition, 2015, 54（46）: 13748–13752.

[23] Xia Y, Liu X H, Zheng H F, et al. Asymmetric synthesis of 2,3–dihydropyrroles by ring–opening/cyclization of cyclopropyl ketones using primary amines [J]. Angewandte Chemie International Edition, 2015, 54（1）: 227–230.

[24] Zhao J N, Fang B, Luo W W, et al. Enantioselective construction of vicinal tetrasubstituted stereocenters by the mannich reaction of silyl ketene imines with isatin–derived ketimines [J]. Angewandte Chemie International Edition, 2015, 54（1）: 241–244.

[25] Zhao X H, Liu X H, Mei H J, et al. Asymmetric dearomatization of indoles through a michael/friedel–crafts–type cascade to construct polycyclic spiroindolines [J]. Angewandte Chemie International Edition, 2015, 54（13）: 4032–4035.

[26] Chen Q G, Tang Y, Huang T Y, et al. Copper/guanidine–catalyzed asymmetric alkynylation of isatins [J]. Angewandte Chemie International Edition, 2016, 55（17）: 5286–5289.

[27] Tang Y, Chen Q G, Liu X H, et al. Direct synthesis of chiral allenoates from the asymmetric C–H insertion of α–diazoesters into terminal alkynes [J]. Angewandte Chemie International Edition, 2015, 54（33）: 9512–9516.

[28] Zhang J, Lin S X, Cheng D J, et al. Phosphoric acid–catalyzed asymmetric classic passerini reaction [J]. Journal of the American Chemical Society, 2015, 137（44）: 14039–14042.

[29] Wu M Y, He W W, Liu X Y, et al. Asymmetric construction of spirooxindoles by organocatalytic multicomponent reactions using diazooxindoles [J]. Angewandte Chemie International Edition, 2015, 54（32）: 9409–9413.

[30] Lin J S, Dong X Y, Li T T, et al. A dual–catalytic strategy to direct asymmetric radical aminotrifluoromethylation of alkenes [J]. Journal of the American Chemical Society, 2016, 138（30）: 9357–9360.

[31] Chen Y H, Cheng D J, Zhang J, et al. Atroposelective synthesis of axially chiral biaryldiols via organocatalytic arylation of 2–naphthols [J]. Journal of the American Chemical Society, 2015, 137（48）: 15062–15065.

[32] Yu P, Zheng S–C, Yang N–Y, et al. Phosphine–catalyzed remote β–C–H functionalization of amines triggered by trifluoromethylation of alkenes: One–pot synthesis of bistrifluoromethylated enamides and oxazoles [J]. Angewandte

Chemie International Edition, 2015, 54 (13) : 4041–4045.

[33] Li Z–L, Li X–H, Wang N, et al. Radical–mediated 1,2–formyl/carbonyl functionalization of alkenes and application to the construction of medium–sized rings [J] . Angewandte Chemie International Edition, 2016, 55 (48) : 15100–15104.

[34] Zhou W, Su X, Tao M N, et al. Chiral sulfinamide bisphosphine catalysts: Design, synthesis, and application in highly enantioselective intermolecular cross–rauhut–currier reactions [J] . Angewandte Chemie International Edition, 2015, 54 (49) : 14853–14857.

[35] Yu Z Z, Li Y F, Shi J M, et al. (C_6F_5)$_3$B catalyzed chemoselective and ortho–selective substitution of phenols with alpha–aryl alpha–diazoesters [J] . Angewandte Chemie International Edition, 2016, 55 (47) : 14807–14811.

[36] Su X, Zhou W, Li Y Y, et al. Design, synthesis, and application of a chiral sulfinamide phosphine catalyst for the enantioselective intramolecular rauhut–currier reaction [J] . Angewandte Chemie International Edition, 2015, 54 (23) : 6874–6877.

[37] Feng J J, Lin T Y, Zhu C Z, et al. The divergent synthesis of nitrogen heterocycles by rhodium (I) –catalyzed intermolecular cycloadditions of vinyl aziridines and alkynes [J] . Journal of the American Chemical Society, 2016, 138 (7) : 2178–2181.

[38] Feng J–J, Lin T–Y, Wu H–H, et al. Transfer of chirality in the rhodium–catalyzed intramolecular formal hetero–[5+2] cycloaddition of vinyl aziridines and alkynes: Stereoselective synthesis of fused azepine derivatives [J] . Journal of the American Chemical Society, 2015, 137 (11) : 3787–3790.

[39] Chen P, Yue Z T, Zhang J Y, et al. Phosphine–catalyzed asymmetric umpolung addition of trifluoromethyl ketimines to morita–baylis–hillman carbonates [J] . Angewandte Chemie International Edition, 2016, 55 (42) : 13316–13320.

[40] Ye F, Qu S L, Zhou L, et al. Palladium–catalyzed C–H functionalization of acyldiazomethane and tandem cross–coupling reactions [J] . Journal of the American Chemical Society, 2015, 137 (13) : 4435–4444.

[41] Xia Y, Feng S, Liu Z, et al. Rhodium (I) –catalyzed sequential C (sp) _C (sp^3) and C (sp^3) _C (sp^3) bond formation through migratory carbene insertion [J] . Angewandte Chemie International Edition, 2015, 54 (27) : 7891–7894.

[42] Liu Z X, Tan H C, Fu T R, et al. Pd (0) –catalyzed carbene insertion into Si–Si and Sn–Sn bonds [J] . Journal of the American Chemical Society, 2015, 137 (40) : 12800–12803.

[43] Feng S, Mo F Y, Xia Y, et al. Rhodium (I) –catalyzed C–C bond activation of siloxyvinylcyclopropanes with diazoesters [J] . Angewandte Chemie International Edition, 2016, 55 (49) : 15401–15405.

[44] Chu W–D, Zhang L, Zhang Z K, et al. Enantioselective synthesis of trisubstituted allenes via Cu (I) –catalyzed coupling of diazoalkanes with terminal alkynes [J] . Journal of the American Chemical Society, 2016, 138 (44) : 14558–14561.

[45] Wang H M, Lu Q Q, Qian C H, et al. Solvent–enabled radical selectivities: controlled syntheses of sulfoxides and sulfides [J] . Angewandte Chemie International Edition, 2016, 55 (3) : 1094–1097.

[46] Tang S, Wang P, Li H R, et al. Multimetallic catalysed radical oxidative C (sp^3) –H/C (sp) –H cross–coupling between unactivated alkanes and terminal alkynes [J] . Nature Communications, 2016, 7: 11676.

[47] Liu J M, Zhang X, Yi H, et al. Chloroacetate–promoted selective oxidation of heterobenzylic methylenes under copper catalysis [J] . Angewandte Chemie International Edition, 2015, 54 (4) : 1261–1265.

[48] Ke J, Tang Y L, Yi H, et al. Copper–catalyzed radical/radical Csp_3–H/P–H cross–coupling: α –phosphorylation of aryl ketone O–acetyloximes [J] . Angewandte Chemie International Edition, 2015, 54 (22) : 6604–6607.

[49] Wu Q, Luo Y, Lei A W, et al. Aerobic copper–promoted radical–type cleavage of coordinated cyanide anion: Nitrogen transfer to aldehydes to form nitriles [J] . Journal of the American Chemical Society, 2016, 138 (9) : 2885–2888.

[50] Li K Z, Wu Q, Lan J B, et al. Coordinating activation strategy for C (*sp*^{3}) –H/C (*sp*^{3}) –H cross–coupling to access beta–aromatic alpha–amino acids [J]. Nature Communications, 2015, 6: 8404.

[51] Huang X L, Wang Y, Lan J B, et al. Rhodium (III) –catalyzed activation of C_{sp3}–H bonds and subsequent intermolecular amidation at room temperature [J]. Angewandte Chemie International Edition, 2015, 54 (32) : 9404–9408.

[52] Cheng Y Y, Wu Y M, Tan G Y, et al. Nickel catalysis enables oxidative C (sp^2) –H/C (sp^2) –H cross–coupling reactions between two heteroarenes [J]. Angewandte Chemie International Edition, 2016, 55 (40) : 12275–12279.

[53] Du K, Guo P, Chen Y, et al. Enantioselective palladium–catalyzed dearomative cyclization for the efficient synthesis of terpenes and steroids [J]. Angewandte Chemie International Edition, 2015, 54 (10) : 3033–3037.

[54] Hu N F, Li K, Wang Z, et al. Synthesis of chiral 1,4–benzodioxanes and chromans by enantioselective palladium–catalyzed alkene aryloxyarylation reactions [J]. Angewandte Chemie International Edition, 2016, 55 (16) : 5044–5048.

[55] Hu N F, Zhao G Q, Zhang Y Y, et al. Synthesis of chiral α–amino tertiary boronic esters by enantioselective hydroboration of α–arylenamides [J]. Journal of the American Chemical Society, 2015, 137 (21) : 6746–6749.

[56] Huang L W, Zhu J B, Jiao G J, et al. Highly enantioselective rhodium–catalyzed addition of arylboroxines to simple aryl ketones: Efficient synthesis of escitalopram [J]. Angewandte Chemie International Edition, 2016, 55 (14) : 4527–4531.

[57] Li C X, Zhang Y Y, Sun Q, et al. Transition–metal–free stereospecific cross–coupling with alkenylboronic acids as nucleophiles [J]. Journal of the American Chemical Society, 2016, 138 (34) : 10774–10777.

[58] Guo C, Sun D–W, Yang S, et al. Iridium–catalyzed asymmetric hydrogenation of 2–pyridyl cyclic imines: A highly enantioselective approach to nicotine derivatives [J]. Journal of the American Chemical Society, 2015, 137 (1) : 90–93.

[59] Xiao L J, Fu X N, Zhou M J, et al. Nickel–catalyzed hydroacylation of styrenes with simple aldehydes: Reaction development and mechanistic insights [J]. Journal of the American Chemical Society, 2016, 138 (9) : 2957–2960.

[60] Xu B, Li M L, Zuo X D, et al. Catalytic asymmetric arylation of α–aryl– α–diazoacetates with aniline derivatives [J]. Journal of the American Chemical Society, 2015, 137 (27) : 8700–8703.

[61] Bao D H, Wu H L, Liu C L, et al. Development of chiral spiro P–N–S ligands for iridium–catalyzed asymmetric hydrogenation of β–alkyl– β–ketoesters [J]. Angewandte Chemie International Edition, 2015, 54 (30) : 8791–8794.

[62] Kang Q K, Wang L J, Liu Q J, et al. Asymmetric H_2O–nucleophilic ring opening of D–A cyclopropanes: Catalyst serves as a source of water [J]. Journal of the American Chemical Society, 2015, 137 (46) : 14594–14597.

[63] Liu Q–J, Wang L J, Kang Q–K, et al. Cy–SaBOX/copper (II) –catalyzed highly diastereo and enantioselective synthesis of bicyclic N,O acetals [J]. Angewandte Chemie International Edition, 2016, 55 (32) : 9220–9223.

[64] Wang P, Feng L–W, Wang L, et al. Asymmetric 1,2–perfluoroalkyl migration: Easy access to enantioenriched α–hydroxy– α–perfluoroalkyl esters [J]. Journal of the American Chemical Society, 2015, 137 (14) : 4626–4629.

[65] Xu H, Hu J L, Wang L J, et al. Asymmetric annulation of donor–acceptor cyclopropanes with dienes [J]. Journal of the American Chemical Society, 2015, 137 (25) : 8006–8009.

[66] Liu J W, Han Z B, Wang X M, et al. Highly regio– and enantioselective alkoxycarbonylative amination of terminal allenes catalyzed by a spiroketal–based diphosphine/Pd (II) complex [J]. Journal of the American Chemical Society, 2015, 137 (49) : 15346–15349.

[67] Zhang L, Han Z B, Zhao X Y, et al. Highly efficient ruthenium–catalyzed N–formylation of amines with H_2 and CO_2 [J]. Angewandte Chemie International Edition, 2015, 54 (21) : 6186–6189.

[68] Lei Z-Q, Pan F, Li H, et al. Group exchange between ketones and carboxylic acids through directing group assisted Rh-catalyzed reorganization of carbon skeletons [J]. Journal of the American Chemical Society, 2015, 137 (15): 5012-5020.

[69] Zhou T G, Luo F X, Yang M Y, et al. Silver-catalyzed long-distance aryl migration from carbon center to nitrogen center [J]. Journal of the American Chemical Society, 2015, 137 (46): 14586-14589.

[70] Liao G, Yin X-S, Chen K, et al. Stereoselective alkoxycarbonylation of unactivated C (sp^3) -H bonds with alkyl chloroformates via Pd (II) /Pd (IV) catalysis [J]. Nature Communications, 2016, 7: 12901.

[71] Zhang Q, Yin X S, Chen K, et al. Stereoselective synthesis of chiral β -fluoro α -amino acids via Pd (II) -catalyzed fluorination of unactivated methylene C (sp^3) -H bonds: Scope and mechanistic studies [J]. Journal of the American Chemical Society, 2015, 137 (25): 8219-8226.

[72] Gao P, Guo W, Xue J J, et al. Iridium (III) -catalyzed direct arylation of C-H bonds with diaryliodonium salts [J]. Journal of the American Chemical Society, 2015, 137 (38): 12231-12240.

[73] Yang Y Q, Li R R, Zhao Y, et al. Cu-catalyzed direct C6-arylation of indoles [J]. Journal of the American Chemical Society, 2016, 138 (28): 8734-8737.

[74] Yang Y Q, Qiu X D, Zhao Y, et al. Palladium-catalyzed C-H arylation of indoles at the C7 position [J]. Journal of the American Chemical Society, 2016, 138 (2): 495-498.

[75] Zhu C D, Liang Y, Hong X, et al. Iodoarene-catalyzed stereospecific intramolecular sp^3 C-H amination: Reaction development and mechanistic insights [J]. Journal of the American Chemical Society, 2015, 137 (24): 7564-7567.

[76] Wang H, Tang G D, Li X W. Rhodium (III) -catalyzed amidation of unactivated C (sp^3) -H bonds [J]. Angewandte Chemie International Edition, 2015, 54 (44): 13049-13052.

[77] Xie F, Yu S J, Qi Z S, et al. Nitrone directing groups in rhodium(III)-catalyzed C-H activation of arenes: 1,3-dipoles versus traceless directing groups [J]. Angewandte Chemie International Edition, 2016, 55 (49): 15351-15355.

[78] Yu S J, Liu S, Lan Y, et al. Rhodium-catalyzed C-H activation of phenacyl ammonium salts assisted by an oxidizing C-N bond: A combination of experimental and theoretical studies [J]. Journal of the American Chemical Society, 2015, 137 (4): 1623-1631.

[79] Du C, Li P-X, Zhu X J, et al. Mixed directing-group strategy: Oxidative C-H/C-H bond arylation of unactivated arenes by cobalt catalysis [J]. Angewandte Chemie International Edition, 2016, 55 (43): 13571-13575.

[80] Chen W X, Liu Z, Tian J Q, et al. Building congested ketone: Substituted hantzsch ester and nitrile as alkylation reagents in photoredox catalysis [J]. Journal of the American Chemical Society, 2016, 138 (38): 12312-12315.

[81] Zhou Q Q, Guo W, Ding W, et al. Decarboxylative alkynylation and carbonylative alkynylation of carboxylic acids enabled by visible-light photoredox catalysis [J]. Angewandte Chemie International Edition, 2015, 54 (38): 11196-11199.

[82] Jiang H, An X D, Tong K, et al. Visible-light-promoted iminyl-radical formation from acyl oximes: A unified approach to pyridines, quinolines, and phenanthridines [J]. Angewandte Chemie International Edition, 2015, 54 (13): 4055-4059.

[83] Huang H C, Jia K F, Chen Y Y. Hypervalent iodine reagents enable chemoselective deboronative/decarboxylative alkenylation by photoredox catalysis [J]. Angewandte Chemie International Edition, 2015, 54 (6): 1881-1884.

[84] Huang H C, Zhang G J, Chen Y Y. Dual hypervalent iodine (III) reagents and photoredox catalysis enable decarboxylative ynonylation under mild conditions [J]. Angewandte Chemie International Edition, 2015, 54 (27): 7872-7876.

[85] Jia K F, Zhang F Y, Huang H C, et al. Visible-light-induced alkoxyl radical generation enables selective C (sp^3) -C (sp^3) bond cleavage and functionalizations [J]. Journal of the American Chemical Society, 2016, 138 (5): 1514-

1517.

[86] Zhang J, Li Y, Zhang F Y, et al. Generation of alkoxyl radicals by photoredox catalysis enables selective C（sp^3）–H functionalization under mild reaction conditions [J]. Angewandte Chemie International Edition, 2016, 55（5）: 1872–1875.

[87] Jia X Q, Huang Z. Conversion of alkanes to linear alkylsilanes using an iridium–iron–catalysed tandem dehydrogenation–isomerization–hydrosilylation [J]. Nature Chemistry, 2016, 8（2）: 157–161.

[88] Dai J X, Wang M Y, Chai G B, et al. A practical solution to stereodefined tetrasubstituted olefins [J]. Journal of the American Chemical Society, 2016, 138（8）: 2532–2535.

[89] Huang X, Wu S Z, Wu W T, et al. Palladium–catalysed formation of vicinal all–carbon quaternary centres via propargylation [J]. Nature Communications, 2016, 7: 12382.

[90] Jiang X G, Zhang J S, Ma S M. Iron catalysis for room–temperature aerobic oxidation of alcohols to carboxylic acids [J]. Journal of the American Chemical Society, 2016, 138（27）: 8344–8347.

[91] Fan M Y, Zhou W, Jiang Y W, et al. CuI/oxalamide catalyzed couplings of (hetero) aryl chlorides and phenols for diaryl ether formation [J]. Angewandte Chemie International Edition, 2016, 55（21）: 6211–6215.

[92] Xia S H, Gan L, Wang K L, et al. Copper–catalyzed hydroxylation of (hetero) aryl halides under mild conditions [J]. Journal of the American Chemical Society, 2016, 138（41）: 13493–13496.

[93] Zhou W, Fan M G, Yin J L, et al. CuI/oxalic diamide catalyzed coupling reaction of (hetero) aryl chlorides and amines [J]. Journal of the American Chemical Society, 2015, 137（37）: 11942–11945.

[94] Ling H C, Wang P S, Tao Z L, et al. Highly enantioselective allylic C–H alkylation of terminal olefins with pyrazol-5–ones enabled by cooperative catalysis of palladium complex and bronsted acid [J]. Journal of the American Chemical Society, 2016, 138（43）: 14354–14361.

[95] Tao Z–L, Li X–H, Han Z–Y, et al. Diastereoselective carbonyl allylation with simple olefins enabled by palladium complex–catalyzed C–H oxidative borylation [J]. Journal of the American Chemical Society, 2015, 137（12）: 4054–4057.

[96] Wang P S, Liu P, Zhai Y J, et al. Asymmetric allylic C–H oxidation for the synthesis of chromans [J]. Journal of the American Chemical Society, 2015, 137（40）: 12732–12735.

[97] Li X Y, Li X W, Jiao N. Rh–catalyzed construction of quinolin–2（1*H*）–ones via C–H bond activation of simple anilines with CO and alkynes [J]. Journal of the American Chemical Society, 2015, 137（29）: 9246–9249.

[98] Shen T, Zhang Y Q, Liang Y–F, et al. Direct tryptophols synthesis from 2–vinylanilines and alkynes via C≡C triple bond cleavage and dioxygen activation [J]. Journal of the American Chemical Society, 2016, 138（40）: 13147–13150.

[99] Sun X, Li X Y, Song S, et al. Mn–catalyzed highly efficient aerobic oxidative hydroxyazidation of olefins: A direct approach to β–azido alcohols [J]. Journal of the American Chemical Society, 2015, 137（18）: 6059–6066.

[100] Ding D, Mou T, Feng M H, et al. Utility of ligand effect in homogenous gold catalysis: Enabling regiodivergent π–bond–activated cyclization [J]. Journal of the American Chemical Society, 2016, 138（16）: 5218–5221.

[101] Feng M H, Tang B Q, Wang N Z, et al. Ligand controlled regiodivergent C_1 insertion on arynes for construction of phenanthridinone and acridone alkaloids [J]. Angewandte Chemie International Edition, 2015, 54（49）: 14960–14964.

[102] Ren R G, Wu Z, Xu Y, et al. C–C bond–forming strategy by manganese–catalyzed oxidative ring–opening cyanation and ethynylation of cyclobutanol derivatives [J]. Angewandte Chemie International Edition, 2016, 55（8）: 2866–2869.

[103] Ren R G, Zhao H J, Huan L T, et al. Manganese–catalyzed oxidative azidation of cyclobutanols: Regiospecific synthesis of alkyl azides by C–C bond cleavage [J]. Angewandte Chemie International Edition, 2015, 54（43）:

12692–12696.

[104] Zhao H J, Fan X F, Yu J J, et al. Silver–catalyzed ring–opening strategy for the synthesis of β – and γ –fluorinated ketones [J]. Journal of the American Chemical Society, 2015, 137 (10) : 3490–3493.

[105] Zheng P F, Ouyang Q, Niu S L, et al. Enantioselective [4+1] annulation reactions of α –substituted ammonium ylides to construct spirocyclic oxindoles [J]. Journal of the American Chemical Society, 2015, 137 (29) : 9390–9399.

[106] Wang H–Y, Zhang K, Zheng C–W, et al. Asymmetric dual–reagent catalysis: Mannich–type reactions catalyzed by ion pair [J]. Angewandte Chemie International Edition, 2015, 54 (6) : 1775–1779.

[107] Wang H–Y, Zheng C–W, Chai Z, et al. Asymmetric cyanation of imines via dipeptide–derived organophosphine dual–reagent catalysis [J]. Nature Communications, 2016, 7: 12720.

[108] Li H L, Chen Q F, Lu Z H, et al. Total syntheses of aflavazole and 14–hydroxyaflavinine [J]. Journal of the American Chemical Society, 2016, 138 (48) : 15555–15558.

[109] Li Y, Zhu S G, Li J, et al. Asymmetric total syntheses of aspidodasycarpine, lonicerine, and the proposed structure of lanciferine [J]. Journal of the American Chemical Society, 2016, 138 (12) : 3982–3985.

[110] Lu Z H, Li H L, Bian M, et al. Total synthesis of epoxyeujindole A [J]. Journal of the American Chemical Society, 2015, 137 (43) : 13764–13767.

[111] Meng Z C, Yu H X, Li L, et al. Total synthesis and antiviral activity of indolosesquiterpenoids from the xiamycin and oridamycin families [J]. Nature Communications, 2015, 6: 6096.

[112] Yang M, Li J, Li A. Total synthesis of clostrubin [J]. Nature Communications, 2015, 6: 6445.

[113] Yang M, Yang X W, Sun H B, et al. Total synthesis of ileabethoxazole, pseudopteroxazole, and seco–pseudopteroxazole [J]. Angewandte Chemie International Edition, 2016, 55 (8) : 2851–2855.

[114] Yang P, Yao M, Li J, et al. Total synthesis of rubriflordilactone B [J]. Angewandte Chemie International Edition, 2016, 55 (24) : 6964–6968.

[115] Zhou S P, Chen H, Luo Y J, et al. Asymmetric total synthesis of mycoleptodiscinA [J]. Angewandte Chemie International Edition, 2015, 54 (23) : 6878–6882.

[116] Zhang X H, Zhou Y, Zuo J P, et al. Total synthesis of periploside A, a unique pregnane hexasaccharide with potent immunosuppressive effects [J]. Nature Communications, 2015, 6: 5879.

[117] Zhu D P, Yu B. Total synthesis of linckosides A and B, the representative starfish polyhydroxysteroid glycosides with neuritogenic activities [J]. Journal of the American Chemical Society, 2015, 137 (48) : 15098–15101.

[118] Li J K, Yu B. A modular approach to the total synthesis of tunicamycins [J]. Angewandte Chemie International Edition, 2015, 54 (22) : 6618–6621.

[119] Guo S, Liu J, Ma D W. Total synthesis of leucosceptroids A and B [J]. Angewandte Chemie International Edition, 2015, 54 (4) : 1298–1301.

[120] Ouyang J, Yan R, Mi X W, et al. Enantioselective total synthesis of (–) –hosieineA [J]. Angewandte Chemie International Edition, 2015, 54 (37) : 10940–10943.

[121] Hong B K, Li H H, Wu J B, et al. Total syntheses of (–) –huperzine Q and (+) –lycopladines B and C [J]. Angewandte Chemie International Edition, 2015, 54 (3) : 1011–1015.

[122] Hong B K, Li C, Wang Z, et al. Enantioselective total synthesis of (–) –incarviatone A [J]. Journal of the American Chemical Society, 2015, 137 (37) : 11946–11949.

[123] Liao D H, Yang S Q, Wang J Y, et al. Total synthesis and structural reassignment of aspergillomarasmine A [J]. Angewandte Chemie International Edition, 2016, 55 (13) : 4291–4295.

[124] Liu W L, Li H H, Cai P J, et al. Scalable total synthesis of rac–jungermannenones B and C [J]. Angewandte Chemie International Edition, 2016, 55 (9) : 3112–3116.

[125] Jin S J, Gong J, Qin Y. Total synthesis of (–) –lundurine A and determination of its absolute configuration [J]. Angewandte Chemie International Edition, 2015, 54 (7) : 2228–2231.

[126] Li X–H, Zhu M, Wang Z–X, et al. Synthesis of atisine, ajaconine, denudatine, and hetidine diterpenoid alkaloids by a bioinspired approach [J]. Angewandte Chemie International Edition, 2016, 55 (50) : 15667–15671.

[127] Song L Q, Zhu G L, Liu Y J, et al. Total synthesis of atisane–type diterpenoids: Application of diels–alder cycloadditions of podocarpane–type unmasked ortho–benzoquinones [J]. Journal of the American Chemical Society, 2015, 137 (42) : 13706–13714.

[128] Cheng H, Zeng F H, Yang X, et al. Collective total syntheses of atisane–type diterpenes and atisine–type diterpenoid alkaloids: (±) –spiramilactone B, (±) –spiraminol, (±) –dihydroajaconine, and (±) –spiramines C and D [J]. Angewandte Chemie International Edition, 2016, 55 (1) : 392–396.

[129] Gao P C, Zhu S Y, Cao H, et al. Total synthesis of marine glycosphingolipid vesparioside B [J]. Journal of the American Chemical Society, 2016, 138 (5) : 1684–1688.

[130] Jiao Z W, Tu Y Q, Zhang Q, et al. Tandem C–H oxidation/cyclization/rearrangement and its application to asymmetric syntheses of (–) –brussonol and (–) –przewalskine E [J]. Nature Communications, 2015, 6: 7332.

[131] Hou S H, Tu Y Q, Wang S H, et al. Total syntheses of the tetracyclic cyclopiane diterpenes conidiogenone, conidiogenol, and conidiogenone B [J]. Angewandte Chemie International Edition, 2016, 55 (14) : 4456–4460.

[132] Du J–Y, Zeng C, Han X–J, et al. Asymmetric total synthesis of Apocynaceae hydrocarbazole alkaloids (+)–deethylibophyllidine and (+)–limaspermidine[J]. Journal of the American Chemical Society, 2015, 137(12) : 4267–4273.

[133] Chen X M, Duan S G, Tao C, et al. Total synthesis of (+) –gelsemine via an organocatalytic Diels–Alder approach [J]. Nature Communications, 2015, 6: 7204.

[134] Liang X, Jiang S Z, Wei K, et al. Enantioselective total synthesis of (–) –alstoscholarisine A [J]. Journal of the American Chemical Society, 2016, 138 (8) : 2560–2562.

[135] Jiang S Z, Zeng X Y, Liang X, et al. Iridium–catalyzed enantioselective indole cyclization: Application to the total synthesis and absolute stereochemical assignment of (–) –AspidophyllineA [J]. Angewandte Chemie International Edition, 2016, 55 (12) : 4044–4048.

[136] Li X J, Xue D S, Wang C, et al. Total synthesis of the hamigerans [J]. Angewandte Chemie International Edition, 2016, 55 (34) : 9942–9946.

[137] Li K, Ou J J, Gao S H. Total synthesis of camptothecin and related natural products by a flexible strategy [J]. Angewandte Chemie International Edition, 2016, 55 (47) : 14778–14783.

[138] Zhu C, Liu Z, Chen G, et al. Total synthesis of indole alkaloid alsmaphorazine D [J]. Angewandte Chemie International Edition, 2015, 54 (3) : 879–882.

[139] Pan S Y, Xuan J, Gao B L, et al. Total synthesis of diterpenoid steenkrotin A [J]. Angewandte Chemie International Edition, 2015, 54 (23) : 6905–6908.

[140] Wang L, Wang H T, Li Y H, et al. Total synthesis of schilancitrilactones B and C [J]. Angewandte Chemie International Edition, 2015, 54 (19) : 5732–5735.

[141] Wang B, Liu Y P, Jiao R, et al. Total synthesis of mannopeptimycins α and β [J]. Journal of the American Chemical Society, 2016, 138 (11) : 3926–3932.

[142] Lv C, Yan X H, Tu Q, et al. Isolation and asymmetric total synthesis of perforanoid A [J]. Angewandte Chemie International Edition, 2016, 55 (26) : 7539–7543.

[143] You L, Liang X–T, Xu L–M, et al. Asymmetric total synthesis of propindilactone G [J]. Journal of the American Chemical Society, 2015, 137 (32) : 10120–10123.

[144] Zhou J J, Gao B W, Xu Z S, et al. Total synthesis and stereochemical assignment of callyspongiolide [J]. Journal of

the American Chemical Society, 2016, 138 (22): 6948–6951.

[145] Liao L P, Zhou J J, Xu Z S, et al. Concise total synthesis of nannocystin A [J]. Angewandte Chemie International Edition, 2016, 55 (42): 13263–13266.

[146] Li L, Yang Q, Wang Y, et al. Catalytic asymmetric total synthesis of (–) -galanthamine and (–) -lycoramine [J]. Angewandte Chemie International Edition, 2015, 54 (21): 6255–6259.

[147] Huang B, Guo L, Jia Y X. Protecting-group-free enantioselective synthesis of (–) -pallavicinin and (+) -neopallavicinin [J]. Angewandte Chemie International Edition, 2015, 54 (46): 13599–13603.

[148] Yu X R, Su F, Liu C, et al. Enantioselective total syntheses of various amphilectane and serrulatane diterpenoids via cope rearrangements [J]. Journal of the American Chemical Society, 2016, 138 (19): 6261–6270.

[149] Wu Y, Xiong D-C, Chen S-C, et al. Total synthesis of mycobacterial arabinogalactan containing 92 monosaccharide units [J]. Nature Communications, 2017, 8: 14851.

[150] Tian C, Lei X Q, Wang Y H, et al. Total syntheses of periconiasins A–E [J]. Angewandte Chemie International Edition, 2016, 55 (24): 6992–6996.

[151] Li G, Xie X N, Zu L S. Total synthesis of calophyline A [J]. Angewandte Chemie International Edition, 2016, 55 (35): 10483–10486.

[152] Guo L D, Huang X Z, Luo S P, et al. Organocatalytic, asymmetric total synthesis of (–) -haliclonin A [J]. Angewandte Chemie International Edition, 2016, 55 (12): 4064–4068.

[153] Yao G Y, Pan Z Y, Wu C L, et al. Efficient synthesis and stereochemical revision of coibamide A [J]. Journal of the American Chemical Society, 2015, 137 (42): 13488–13491.

[154] Li L C, Ni C F, Wang F, et al. Deoxyfluorination of alcohols with 3,3-difluoro-1,2-diarylcyclopropenes [J]. Nature Communications, 2016, 7: 13320.

[155] Zeng Y W, Li G Y, Hu J B. Diphenyliodonium-catalyzed fluorination of arynes: Synthesis of *ortho*-fluoroiodoarenes [J]. Angewandte Chemie International Edition, 2015, 54 (37): 10773–10777.

[156] Li F Y, Wu Z J, Wang J. Oxidative enantioselective α-fluorination of aliphatic aldehydes enabled by N-heterocyclic carbene catalysis [J]. Angewandte Chemie International Edition, 2015, 54 (2): 656–659.

[157] An L, Xiao Y L, Min Q Q, et al. Facile access to fluoromethylated arenes by nickel-catalyzed cross-coupling between arylboronic acids and fluoromethyl bromide [J]. Angewandte Chemie International Edition, 2015, 54 (31): 9079–9083.

[158] Su Y-M, Feng G-S, Wang Z-Y, et al. Nickel-catalyzed monofluoromethylation of aryl boronic acids [J]. Angewandte Chemie International Edition, 2015, 54 (20): 6003–6007.

[159] Hu M Y, Ni C F, Li L C, et al. gem-difluoroolefination of diazo compounds with $TMSCF_3$ or $TMSCF_2Br$: Transition-metal-free cross-coupling of two carbene precursors [J]. Journal of the American Chemical Society, 2015, 137 (45): 14496–14501.

[160] Chen D B, Ni C F, Zhao Y C, et al. Bis (difluoromethyl) trimethylsilicate Anion: A key intermediate in nucleophilic difluoromethylation of enolizable ketones with Me_3SiCF_2H [J]. Angewandte Chemie International Edition, 2016, 55 (41): 12632–12636.

[161] Rong J, Deng L, Tan P, et al. Radical fluoroalkylation of isocyanides with fluorinated sulfones by visible-light photoredox catalysis [J]. Angewandte Chemie International Edition, 2016, 55 (8): 2743–2747.

[162] Xiao Y L, Min Q Q, Xu C, et al. Nickel-catalyzed difluoroalkylation of (hetero) arylborons with unactivated 1-bromo-1,1-difluoroalkanes [J]. Angewandte Chemie International Edition, 2016, 55 (19): 5837–5841.

[163] Gu J-W, Min Q-Q, Yu L-C, et al. Tandem difluoroalkylation-arylation of enamides catalyzed by nickel [J]. Angewandte Chemie International Edition, 2016, 55 (40): 12270–12274.

[164] Zhao H-Y, Feng Z, Luo Z J, et al. Carbonylation of difluoroalkyl bromides catalyzed by palladium [J]. Angewandte

Chemie–International Edition, 2016, 55（35）: 10401–10405.

[165] Lin Q Y, Xu X H, Zhang K, et al. Visible–light–induced hydrodifluoromethylation of alkenes with a bromodifluoromethylphosphonium bromide [J]. Angewandte Chemie International Edition, 2016, 55（4）: 1479–1483.

[166] Zhu J S, Liu Y F, Shen Q L. Direct difluoromethylation of alcohols with an electrophilic difluoromethylated sulfonium ylide [J]. Angewandte Chemie International Edition, 2016, 55（31）: 9050–9054.

[167] Li G, Wang T, Fei F, et al. Nickel–catalyzed decarboxylative difluoroalkylation of α, β –unsaturated carboxylic acids [J]. Angewandte Chemie International Edition, 2016, 55（10）: 3491–3495.

[168] Liu J B, Chen C, Chu L L, et al. Silver–mediated oxidative trifluoromethylation of phenols: Direct synthesis of aryl trifluoromethyl ethers [J]. Angewandte Chemie International Edition, 2015, 54（40）: 11839–11842.

[169] Lv W–X, Zeng Y–F, Li Q J, et al. Oxidative difunctionalization of alkenyl MIDA boronates: A versatile platform for halogenated and trifluoromethylated α –boryl ketones [J]. Angewandte Chemie International Edition, 2016, 55（34）: 10069–10073.

[170] Wu J, Gu Y, Leng X B, et al. Copper–promoted sandmeyer difluoromethylthiolation of aryl and heteroaryl diazonium salts [J]. Angewandte Chemie International Edition, 2015, 54（26）: 7648–7652.

[171] Zhu D H, Shao X X, Hong X, et al. $PhSO_2SCF_2H$: A shelf–stable, easily scalable reagent for radical difluoromethylthiolation [J]. Angewandte Chemie International Edition, 2016, 55（51）: 15807–15811.

[172] Zhu D H, Gu Y, Lu L, et al. N–difluoromethylthiophthalimide: A shelf–stable, electrophilic reagent for difluoromethylthiolation [J]. Journal of the American Chemical Society, 2015, 137（33）: 10547–10553.

[173] Guo S, Zhang X F, Tang P P. Silver–mediated oxidative aliphatic C–H trifluoromethylthiolation [J]. Angewandte Chemie International Edition, 2015, 54（13）: 4065–4069.

[174] Wu H, Xiao Z W, Wu J H, et al. Direct trifluoromethylthiolation of unactivated C（sp^3）–H using silver（I）trifluoromethanethiolate and potassium persulfate [J]. Angewandte Chemie International Edition, 2015, 54（13）: 4070–4074.

[175] Zheng J, Wang L, Lin J H, et al. Difluorocarbene–derived trifluoromethylthiolation and [^{18}F] trifluoromethylthiolation of aliphatic electrophiles [J]. Angewandte Chemie International Edition, 2015, 54（45）: 13236–13240.

[176] Zhu H C, Chen C M, Tong Q Y, et al. Hyperisampsins H–M, cytotoxic polycyclic polyprenylated acylphloroglucinols from *Hypericum sampsonii* [J]. Scientific Reports, 2015, 5: 14772.

[177] Liao Y, Liu X, Yang J, et al. Hypersubones A and B, new polycyclic acylphloroglucinols with intriguing adamantane type cores from *Hypericum subsessile* [J]. Organic Letters, 2015, 17（5）: 1172–1175.

[178] Nian Y, Yang J, Liu T–Y, et al. New anti–angiogenic leading structure discovered in the fruit of *Cimicifuga yunnanensis* [J]. Scientific Reports, 2015, 5: 9026.

[179] Jiao W–H, Shi G–H, Xu T–T, et al. Dysiherbols A–C and dysideanone E, cytotoxic and NF– κ B inhibitory tetracyclic meroterpenes from a *Dysidea* sp. marine sponge [J]. Journal of Natural Products, 2016, 79（2）: 406–411.

[180] Bao J–M, Su Z–Y, Lou L–L, et al. Jatrocurcadiones A and B: Two novel diterpenoids with an unusual 10,11–seco–premyrsinane skeleton from Jatropha curcas [J]. RSC Advances, 2015, 5（77）: 62921–62925.

[181] Cao J–Q, Huang X–J, Li Y–T, et al. Callistrilones A and B, triketone–phloroglucinol–monoterpene hybrids with a new skeleton from *Callistemon rigidus* [J]. Organic Letters, 2016, 18（1）: 120–123.

[182] Cao F, Yang J–K, Liu Y–F, et al. Pleosporalone A, the first azaphilone characterized with aromatic A–ring from a marine–derived *Pleosporales* sp. fungus [J]. Natural Product Research, 2016, 30（21）: 2448–2452.

[183] Liu B–Y, Zhang C, Zeng K–W, et al. Exotines A and B, two heterodimers of isopentenyl–substituted indole and coumarin derivatives from *Murraya exotica* [J]. Organic Letters, 2015, 17（17）: 4380–4383.

[184] Ma G-L, Xiong J, Yang G-X, et al. Biginkgosides A-I, unexpected minor dimeric flavonol diglycosidic truxinate and truxillate esters from ginkgo *Biloba leaves* and their antineuroinflammatory and neuroprotective activities [J]. Journal of Natural Products, 2016, 79 (5): 1354-1364.

[185] Wu P, Xue J, Yao L, et al. Bisacremines E-G, three polycyclic dimeric acremines produced by acremonium persicinum SC0105 [J]. Organic Letters, 2015, 17 (19): 4922-4925.

[186] Jiao W-H, Xu T-T, Zhao F, et al. Dysifragilones A-C, unusual sesquiterpene aminoquinones and inhibitors of NO production from the south China sea sponge dysidea fragilis [J]. European Journal of Organic Chemistry, 2015, 2015 (5): 960-966.

[187] Xu J-F, Zhao H-J, Wang X-B, et al. (±)-Melicolones A and B, rearranged prenylated acetophenone stereoisomers with an unusual 9-oxatricyclo [$3.2.1.1^{3,8}$] nonane core from the leaves of melicope ptelefolia [J]. Organic Letters, 2015, 17 (1): 146-149.

[188] Yin X, Bai R F, Guo Q, et al. Hendersine A, a novel isoquinoline alkaloid from *Corydalis hendersonii* [J]. Tetrahedron Letters, 2016, 57 (43): 4858-4862.

[189] Yu Y, Gan L-S, Yang S-P, et al. Eucarobustols A-I, conjugates of sesquiterpenoids and acylphloroglucinols from eucalyptus robusta [J]. Journal of Natural Products, 2016, 79 (5): 1365-1372.

[190] Chen Z B, Hao J J, Wang L P, et al. New α-glucosidase inhibitors from marine algae-derived *Streptomyces* sp. OUCMDZ-3434 [J]. Scientific Reports, 2016, 6: 20004.

[191] Zhao X-R, Huo X-K, Dong P-P, et al. Inhibitory effects of highly oxygenated lanostane derivatives from the fungus *Ganoderma lucidum* on P-glycoprotein and α-glucosidase [J]. Journal of Natural Products, 2015, 78 (8): 1868-1876.

[192] Zhang X P, Chen C Y, Li Y H, et al. Tadehaginosides A-J, phenylpropanoid glucosides from *Tadehagi triquetrum*, enhance glucose uptake via the upregulation of PPARγ and GLUT-4 in C2C12 myotubes [J]. Journal of Natural Products, 2016, 79 (5): 1249-1258.

[193] Fang X, Di Y T, Zhang Y, et al. Unprecedented quassinoids with promising biological activity from *Harrisonia perforata* [J]. Angewandte Chemie International Edition, 2015, 54 (19): 5592-5595.

[194] Xiong L, Zhou Q-M, Zou Y K, et al. Leonuketal, a spiroketal diterpenoid from *Leonurus japonicus* [J]. Organic Letters, 2015, 17 (24): 6238-6241.

[195] Fan Y-Y, Zhang H, Zhou Y, et al. Phainanoids A-F, a new class of potent immunosuppressive triterpenoids with an unprecedented carbon skeleton from *Phyllanthus hainanensis* [J]. Journal of the American Chemical Society, 2015, 137 (1): 138-141.

[196] 谢建华，周其林．手性物质创造的昨天、今天和明天［J］．科学通报，2015, 60（S2）：2679-2696.

[197] 国家自然科学基金委．化学十年：中国与世界［R］．杭州：国家自然科学基金委，2012.

[198] Zhang L, Wang T Y, Shen Z C, et al. Chiral nanoarchitectonics: Towards the design, self-assembly, and function of nanoscale chiral twists and helices [J]. Advanced Materials, 2016, 28 (6): 1044-1059.

[199] Liu C X, Yang D, Jin Q X, et al. A chiroptical logic circuit based on self-assembled soft materials containing amphiphilic Spiropyran [J]. Advanced Materials, 2016, 28 (8): 1644-1649.

[200] Xu H S, Ding S Y, An W K, et al. Constructing crystalline covalent organic frameworks from chiral building blocks [J]. Journal of the American Chemical Society, 2016, 138 (36): 11489-11492.

[201] Wang X R, Han X, Zhang J, et al. Homochiral 2D porous covalent organic frameworks for heterogeneous asymmetric catalysis [J]. Journal of the American Chemical Society, 2016, 138 (38): 12332-12335.

[202] Zhang J, Han X, Wu X W, et al. Multivariate chiral covalent organic frameworks with controlled crystallinity and stability for asymmetric catalysis [J]. Journal of the American Chemical Society, 2017, 139 (24): 8277-8285.

[203] Han X, Xia Q C, Huang J J, et al. Chiral covalent organic frameworks with high chemical stability for heterogeneous

asymmetric catalysis [J]. Journal of the American Chemical Society, 2017, 139 (25): 8693–8697.

[204] Han Q X, Qi B, Ren W M, et al. Polyoxometalate-based homochiral metal-organic frameworks for tandem asymmetric transformation of cyclic carbonates from olefins [J]. Nature Communications, 2015, 6: 10007.

[205] Xia Q C, Li Z J, Tan C X, et al. Multivariate metal-organic frameworks as multifunctional heterogeneous asymmetric catalysts for sequential reactions [J]. Journal of the American Chemical Society, 2017, 139 (24): 8259–8266.

[206] Wu K, Li K, Hou Y, et al. Homochiral D_4-symmetric metal-organic cages from stereogenic Ru (II) metalloligands for effective enantioseparation of atropisomeric molecules [J]. Nature Communications, 2016, 7: 10487.

[207] Dong J Q, Tan C X, Zhang K, et al. Chiral NH-controlled supramolecular metallacycles [J]. Journal of the American Chemical Society, 2017, 139 (4): 1554–1564.

[208] Li J, Liu Y, Ren W, et al. Asymmetric alternating copolymerization of meso-epoxides and cyclic anhydrides: Efficient access to enantiopure polyesters [J]. Journal of the American Chemical Society, 2016, 138 (36): 11493–11496.

[209] Duan Y Y, Han L, Zhang J J, et al. Optically active nanostructured Zno films [J]. Angewandte Chemie International Edition, 2015, 54 (50): 15170–15175.

[210] Cao Y Y, Kao K C, Mou C Y, et al. Oriented chiral DNA-silica film guided by a natural mica substrate [J]. Angewandte Chemie International Edition, 2016, 55 (6): 2037–2041.

[211] Ma L G, Cao Y Y, Duan Y Y, et al. Silver films with hierarchical chirality [J]. Angewandte Chemie International Edition, 2017, 56 (20): 8657–8662.

[212] Hu H C, Hu H S, Zhao B, et al. Metal-organic frameworks (MOFs) of a cubic metal cluster with multicentered Mn^I–Mn^I bonds [J]. Angewandte Chemie International Edition, 2015, 54 (40): 11681–11685.

[213] Yuan J, Sun T T, He X, et al. Synthesis of tetra- and octa-aurated heteroaryl complexes towards probing aromatic indoliums [J]. Nature Communications, 2016, 7: 11489.

[214] Zhu C Q, Yang C X, Wang Y H, et al. CCCCC pentadentate chelates with planar Möbius aromaticity and unique properties [J]. Science Advances, 2016, 2 (8): e1601031.

[215] Wang Y L, Hu H S, Li W L, et al. Relativistic effects break periodicity in group 6 diatomic molecules [J]. Journal of the American Chemical Society, 2016, 138 (4): 1126–1129.

[216] Hu S X, Jian J W, Su J, et al. Pentavalent lanthanide nitride-oxides: NPrO and NPrO− complexes with N ≡ Pr triple bonds [J]. Chemical Science, 2017, 8 (3): 4035–4043; Zhang Q N, Hu S X, Qu H, et al. Pentavalent lanthanide compounds: Formation and characterization of praseodymium (V) oxides [J]. Angewandte Chemie International Edition, 2016, 55 (24): 6896–6900.

[217] Mao C, Wang Z D, Wang Z, et al. Weakly polar aprotic ionic liquids acting as strong dissociating solvent: a typical "ionic liquid effect" revealed by accurate measurement of absolute pK_a of ylide precursor salts [J]. Journal of the American Chemical Society, 2016, 138 (17): 5523–5526.

[218] Ma W Y, Yu C, Chen T Y, et al. Metallacyclopentadienes: Synthesis, structure and reactivity [J]. Chemical Society Reviews, 2017, 46 (4): 1160–1192.

[219] Wu K, Li K, Hou Y J, et al. Homochiral D_4-symmetric metal-organic cages from stereogenic Ru (II) metalloligands for effective enantioseparation of atropisomeric molecules [J]. Nature Communications, 2016, 7: 10487.

[220] Li T, Wei H J, Fang Y, et al. Elusive antimony-centered radical cations: Isolation, characterization, crystal structures, and reactivity studies [J]. Angewandte Chemie International Edition, 2017, 56 (2): 632–636.

[221] Li W l, Liu H T, Jian T, et al. Bond-bending isomerism of $Au_2I_3^-$: Competition between covalent bonding and aurophilicity [J]. Chemical Science, 2016, 7 (1): 475–481.

[222] He X, Wang Y C, Jiang H, et al. Structurally well-defined sigmoidal gold clusters: Probing the correlation between metal atom arrangement and chiroptical response [J]. Journal of the American Chemical Society, 2016, 138 (17):

5634–5643.

[223] Liang J, Zhu G Y, Wang C X, et al. MoS2–Based All–Purpose Fibrous Electrode and Self–Powering Energy Fiber for Efficient Energy Harvesting and Storage [J].Advanced Energy Materials, 2017, 7 (3): 1601208

[224] Chen T, Cheng B R, Zhu G Y, et al. Highly efficient retention of polysulfides in "sea urchin" –like carbon nanotube/ nanopolyhedra superstructures as cathode material for ultralong–life lithium–sulfur batteries [J]. Nano Letters, 2017, 17 (1): 437–444.

[225] Liang J,Wang C, Wang Y, et al. All–Inorganic Perovskite Solar Cells [J]. Journal of American Chemical Society, 2016, 138 (49) , 15829–15832.

[226] Tian W M, Leng J, Zhao C Y, et al. Long–distance charge carrier funneling in perovskite nanowires enabled by built–in halide gradient [J]. Journal of the American Chemical Society, 2017, 139 (2): 579–582.

[227] Chen P Z, Weng Y X, Niu L Y, et al. Light–harvesting systems based on organic nanocrystals to mimic chlorosomes [J]. Angewandte Chemie International Edition, 2016, 55 (8): 2759–2763.

[228] Yan C S, Chen G, Sun J X, et al. Edge dislocation surface modification: A new and efficient strategy for realizing outstanding lithium storage performance [J]. Nano Energy, 2015, 15: 558–566.

[229] Yan C S, Chen G, Zhou X, et al. Template - based engineering of carbon–doped Co_3O_4 hollow nanofibers as anode materials for lithium - ion batteries [J]. Advanced Functional Materials, 2016, 26 (9): 1428–1436.

[230] Yin L–C, Liang J, Zhou G–M, et al. Understanding the interactions between lithium polysulfides and N–doped graphene using density functional theory calculations [J]. Nano Energy, 2016, 25: 203–210;Sun Z H, Zhang J Q, Yin L C, et al. Conductive porous vanadium nitride/graphene composite as chemical anchor of polysulfides for lithium–sulfur batteries [J]. Nature Communications, 2017, 8: 14627.

[231] Zeng Y X, Lin Z Q, Meng Y, et al. Flexible ultrafast aqueous rechargeable Ni//Bi battery based on highly durable single–crystalline bismuth nanostructured anode [J]. Advanced Materials, 2016, 28 (41): 9188–9195.

[232] Zeng Y X, Zhang X Y, Meng Y, et al. Achieving Ultrahigh energy density and long durability in a flexible rechargeable quasi–solid–state Zn–MnO_2 battery [J]. Advanced Materials, 2017, 29 (16): 1700274.

[233] Yu M H, Wang Z K, Hou C, et al. Nitrogen–doped Co_3O_4 mesoporous nanowire arrays as an additive–free air–cathode for flexible solid–state zinc–air batteries [J]. Advanced Materials, 2017, 29 (15): 1602868.

[234] Liu P X, Zhao Y, Qin, R X, et al. Photochemical route for synthesizing atomically dispersed palladium catalysts [J]. Science, 2016, 352 (6287): 797–800; Chen G X, Xu C F, Huang X Q, et al. Interfacial electronic effects control the reaction selectivity of platinum catalysts [J]. Nature Materials, 2016, 15 (5): 564–569.

[235] Lin L L, Zhou W, Gao R, et al. Low–temperature hydrogen production from water and methanol using Pt/ α –MoC catalysts [J]. Nature, 2017, 544 (7648): 80–83.

[236] Liang J, Wang C X, Wang Y R, et al. All–inorganic perovskite solar cells [J]. Journal of the American Chemical Society, 2016, 138 (49): 15829–15832.

[237] Li H B, Xiao J P, Fu Q, et al. Confined catalysis under two–dimensional materials [J]. Proceedings of the National Academy of Sciences of the United States of America, 2017, 114 (23): 5930–5934.

[238] Deng D H, Novoselov K S, Fu Q, et al. Catalysis with two–dimensional materials and their heterostructures [J]. Nature Nanotechnology, 2016, 11 (3): 218–230.

[239] Chang C–R, Huang Z–Q, Li J. Hydrogenation of molecular oxygen to hydroperoxyl: An alternative pathway for O_2 activation on nanogold catalysts [J]. Nano Research, 2015, 8 (11): 3737–3748.

[240] Li H Y, Chen S M, Jia X F, et al. Amorphous nickel–cobalt complexes hybridized with 1T–phase molybdenum disulfide via hydrazine–induced phase transformation for water splitting [J]. Nature Communications, 2017, 8: 15377.

[241] Li W Z, XLiu J X, G J, et al. Chemical Insights into the Design and Development of Face–Centered Cubic

Ruthenium Catalysts for Fischer-Tropsch Synthesis [J]. Journal of American Society, 2017, 139 (6) : 2261-2276.

[242] Cao L Y, Lin Z K, Peng F, et al. Self-supporting metal-organic layers as single-site solid catalysts [J]. Angewandte Chemie International Edition, 2016, 55 (16) : 4962-4966.

[243] An B, Zhang J Z, Cheng K, et al. Confinement of ultrasmall Cu/ZnOx nanoparticles in metal-organic frameworks for selective methanol synthesis from catalytic hydrogenation of CO_2 [J]. Journal of the American Chemical Society, 2017, 139 (10) : 3834-3840.

[244] Tian J, Xu Z Y, Zhang D W, et al. Supramolecular metal-organic frameworks that display high homogeneous and heterogeneous photocatalytic activity for H_2 production [J]. Nature Communications, 2016, 7: 11580.

[245] Quan J M, Kondo T, Wang G C, et al. Energy transfer dynamics of formate decomposition on Cu (110) [J]. Angewandte Chemie, 2017, 129 (13) : 3550-3554.

[246] Cao Z M, Chen Q L, Zhang J W, et al. Platinum-nickel alloy excavated nano-multipods with hexagonal close-packed structure and superior activity towards hydrogen evolution reaction [J]. Nature Communications, 2017, 8: 15131.

[247] Wen G D, Wang B L, Wang C X, et al. Hydrothermal carbon enriched with oxygenated groups from biomass glucose as an efficient carbocatalyst [J]. Angewandte Chemie International Edition, 2017, 56 (2) : 600-604.

[248] Wang Y X, Cui H, Wei Z W, et al. [J]. Chemical Science, 2017, 8 (1) : 775-780.

[249] Chen S, Li K, Zhao F, et al. A metal-organic cage incorporating multiple light harvesting and catalytic centres for photochemical hydrogen production [J]. Nature Communications, 2016, 7: 13169.

[250] Lin T Q, Chen I W, Liu F X, et al. Nitrogen-doped mesoporous carbon of extraordinary capacitance for electrochemical energy storage [J]. Science, 2015, 350 (6267) : 1508-1513.

[251] Chen C, Cao J, Lu Q Q, et al. Foldable all - solid - state supercapacitors integrated with photodetectors [J]. Advanced Functional Materials, 2017, 27 (3) : 1604639.

[252] Zhao J, Jiang Y F, Fan H, et al. Porous 3D few-layer graphene-like carbon for ultrahigh-power supercapacitors with well-defined structure-performance relationship [J]. Advanced Materials, 2017, 29 (11) : 1604569.

[253] Yu M H, Cheng X Y, Zeng Y X, et al. Dual-doped molybdenum trioxide nanowires: A bifunctional anode for fiber-shaped asymmetric supercapacitors and microbial fuel cells [J]. Angewandte Chemie International Edition, 2016, 128 (23) : 6874-6878.

[254] Yin J B, Wang H, Peng H, et al. Selectively enhanced photocurrent generation in twisted bilayer graphene with van Hove singularity [J]. Nature Communications, 2016, 7: 10699.

[255] Chen Y B, Sun J Y, Gao J F, et al. Growing uniform graphene disks and films on molten glass for heating devices and cell culture [J]. Advanced Materials, 2015, 27 (47) : 7839-7846; Hennessy J. Nanomedicine: Polymeric vaccines [J]. Nature Materials, 2015, 14 (12) : 1186.

[256] Li C H, Wang C, Keplinger C et al. A highly stretchable autonomous self-healing elastomer [J]. Nature Chemistry, 2016, 8 (6) : 618-624.

[257] Yu C Y, Ma L S, He J J, et al. Flexible, linear chains act as baffles to inhibit the intramolecular rotation of molecular turnstiles [J]. Journal of the American Chemical Society, 2016, 138 (49) : 15849-15852.

[258] Zhang Z K, Du J H, Zhang D D, et al. Rosin-enabled ultraclean and damage-free transfer of graphene for large-area flexible organic light-emitting diodes [J]. Nature Communications, 2017, 8: 14560.

[259] Wu J X, Yuan H T, Meng M M, et al. High electron mobility and quantum oscillations in non-encapsulated ultrathin semiconducting Bi_2O_2Se [J]. Nature Nanotechnology, 2017, 12 (6) : 530-534.

[260] Cho H S, Deng H X, Miyasaka K, et al. Extra adsorption and adsorbate superlattice formation in metal-organic frameworks [J]. Nature, 2015, 527 (7579) : 503-507.

[261] Liu Q, Cong H J, Deng H X. Deciphering the spatial arrangement of metals and correlation to reactivity in multivariate metal–organic frameworks [J]. Journal of the American Chemical Society, 2016, 138 (42): 13822–13825.

[262] Meng X S, Gui B, Yuan D Q, et al. Mechanized azobenzene–functionalized zirconium metal–organic framework for on–command cargo release [J]. Science Advances, 2016, 2 (8): e1600480.

[263] Chen C X, Wei Z W, Jiang J H, et al. Precise modulation of the breathing behavior and pore surface in Zr–MOFs by reversible post–synthetic variable–spacer installation to fine–tune the expansion magnitude and sorption properties [J]. Angewandte Chemie International Edition, 2016, 55 (34): 9932–9936.

[264] Ma T, Liu Z B, Wen J X, et al. Tailoring the thermal and electrical transport properties of graphene films by grain size engineering [J]. Nature Communications, 2017, 8: 14486.

[265] Jin B W, Zhou X M, Huang L, et al. Aligned MoO_x/MoS_2 core–shell nanotubular structures with a high density of reactive sites based on self - ordered anodic molybdenum oxide nanotubes [J]. Angewandte Chemie International Edition, 2016, 55 (40): 12252–12256.

[266] Hu D K, Xu G C, Xing L, et al. Two–dimensional semiconductors grown by chemical vapor transpor [J]. Angewandte Chemie International Edition, 2017, 56 (13): 3611–3615; Zheng J Y, Yan X X, Lu Z X, et al. High–mobility multilayered MoS_2 flakes with low contact resistance grown by chemical vapor deposition [J]. Advanced Materials, 2017, 29 (13): 1604540.

[267] Liu D P, Lin X P, Zhang H, et al. Magnetic properties of a single–molecule lanthanide–transition–metal compound containing 52 gadolinium and 56 nickel atoms [J]. Angewandte Chemie International Edition, 2016, 55 (14): 4532–4536.

[268] Zeng J L, Guan Z J, Du Y, et al. Chloride–promoted formation of a bimetallic nanocluster $Au_{80}Ag_{30}$ and the total structure determination [J]. Journal of the American Chemical Society, 2016, 138 (25): 7848–7851.

[269] Chen Q L, Yang Y N, Cao Z M, et al. Excavated cubic platinum–tin alloy nanocrystals constructed from ultrathin nanosheets with enhanced electrocatalytic activity [J]. Angewandte Chemie International Edition, 2016, 55 (31): 9021–9025.

[270] Yan J Z, Su H F, Yang H Y, et al. Asymmetric synthesis of chiral bimetallic $[Ag_{28}Cu_{12}(SR)_{24}]^{4-}$ nanoclusters via ion pairing [J]. Journal of the American Chemical Society, 2016, 138 (39): 12751–12754.

[271] Deng J, Li H B, Wang S H, et al. Multiscale structural and electronic control of molybdenum disulfide foam for highly efficient hydrogen production [J]. Nature Communications, 2017, 8: 14430.

[272] Chen W S, Ouyang J, Liu H, et al. Black phosphorus nanosheet–based drug delivery system for synergistic photodynamic/photothermal/chemotherapy of cancer [J]. Advanced Materials, 2017, 29 (5): 1603824.

[273] Huang H Y, Yu B L, Zhang P Y, et al. Highly charged ruthenium (II) polypyridyl complexes as lysosome–localized photosensitizers for two–photon photodynamic therapy [J]. Angewandte Chemie International Edition, 2015, 54 (47): 14049–14052.

[274] Huang H Y, Yang L, Zhang P Y, et al. Real–time tracking mitochondrial dynamic remodeling with two–photon phosphorescent iridium (III) complexes [J]. Biomaterials, 2016, 83: 321–331.

[275] Wang Y M, Liu W, Yin X B. Self - limiting growth nanoscale coordination polymers for fluorescence and magnetic resonance dual - modality imaging [J]. Advanced Functional Materials, 2016, 26 (46): 8463–8470.

[276] Dai Z–R, Feng L, Jin Q L, et al. A practical strategy to design and develop an isoform–specific fluorescent probe for a target enzyme: CYP1A1 as a case study [J]. Chemical Science, 2017, 8 (4): 2795–2803.

[277] Chen J, Xu X, Zhang D H. Communication: An accurate global potential energy surface for the OH + CO → H + CO_2 reaction using neural networks [J]. The Journal of Chemical Physics, 2013, 138 (22): 221104.

[278] Shao K J, Chen J, Zhao Z Q, et al. Communication: Fitting potential energy surfaces with fundamental invariant

neural network [J]. The Journal of Chemical Physics, 2016, 145 (7): 071101.

[279] Wang X–D, Gao X–F, Xuan C–J, et al. Dissociative electron attachment to CO_2 produces molecular oxygen [J]. Nature Chemistry, 2016, 8 (3): 258–263.

[280] Xie C J, Ma J Y, Zhu X L, et al. Nonadiabatic tunneling in photodissociation of phenol [J]. Journal of the American Chemical Society, 2016, 138 (25): 7828–7831.

[281] Shen Z T, Ma H T, Zhang C F, et al. Dynamical importance of van der Waals saddle and excited potential surface in C (^1D) + D_2 complex–forming reaction [J]. Nature Communications, 2017, 8: 14094.

[282] Liu S, Zhang D H. A local mode picture for H atom reaction with vibrationally excited H_2O: A full dimensional state–to–state quantum dynamics investigation [J]. Chemical Science, 2016, 7 (1): 261–265.

[283] Zhao Z Q, Zhang Z J, Liu S, et al. Dynamical barrier and isotope effects in the simplest substitution reaction via Walden inversion mechanism [J]. Nature Communications, 2017, 8: 14506.

[284] Wang Li–N, Li Z–Y, Liu Q–Y, et al. CO oxidation promoted by the gold dimer in $Au_2VO_3^-$ and $Au_2VO_4^-$ Clusters [J]. Angewandte Chemie International Edition, 2015, 54 (40): 11720–11724.

[285] Zhang Q N, Hu S–X, Qu H, et al. Pentavalent lanthanide compounds: Formation and characterization of praseodymium (V) oxides [J]. Angewandte Chemie International Edition, 2016, 55 (24): 6896–6900.

[286] Bu L Z, Zhang N, Guo S J, et al. Biaxially strained PtPb/Pt core/shell nanoplate boosts oxygen reduction catalysis [J]. Science, 2016, 354 (6318): 1410–1414.

[287] Wang Y C, Lai Y J, Song L, et al. S–doping of an Fe/N/C ORR catalyst for polymer electrolyte membrane fuel cells with high power density [J]. Angewandte Chemie International Edition, 2015, 54 (34): 9907–9910; Wang Q, Zhou Z Y, Lai Y J, et al. Phenylenediamine–based FeN_x/C catalyst with high activity for oxygen reduction in acid medium and its active–site probing [J]. Journal of the American Chemical Society, 2014, 136 (31): 10882–10885.

[288] Li C, Han X P, Cheng F Y, et al. Phase and composition controllable synthesis of cobalt manganese spinel nanoparticles towards efficient oxygen electrocatalysis [J]. Nature Communications, 2015, 6: 7345.

[289] Deng D H, Novoselov K S, Fu Q, et al. Catalysis with two–dimensional materials and their heterostructures [J]. Nature Nanotechnology, 2016, 11 (3): 218–230.

[290] Lin T Q, Chen I W, Liu F X, et al. Nitrogen–doped mesoporous carbon of extraordinary capacitance for electrochemical energy storage [J]. Science, 2015, 350 (6267): 1508–1513.

[291] Zhang H Z, Zhang H M, Zhang F X, et al. Advanced charged membranes with highly symmetric spongy structures for vanadium flow battery application [J]. Energy & Environmental Science, 2013, 6 (3): 776–781.

[292] Yang C P, Yin Y X, Zhang S F, et al. Accommodating lithium into 3D current collectors with a submicron skeleton towards long–life lithium metal anodes [J]. Nature Communications, 2015, 6: 8058.

[293] Peng C X, Ning G H, Su J, et al. Reversible multi–electron redox chemistry of π–conjugated N–containing heteroaromatic molecule–based organic cathodes [J]. Nature Energy, 2017, 2: 17074.

[294] Wang J, Tang J, Ding B, et al. Hierarchical porous carbons with layer–by–layer motif architectures from confined soft–template self–assembly in layered materials [J]. Nature Communications, 2017, 8: 15717.

[295] Zhong J H, Jin X, Meng L Y, et al. Probing the electronic and catalytic properties of a bimetallic surface with 3 nm resolution [J]. Nature Nanotechnology, 2017, 12 (2): 132–136.

[296] Zhang S J, Zhou Q, Lu X M, et al. Physicochemical properties of ionic liquid mixtures [M]. Netherlands: Springer, 2016.

[297] Ozokwelu D, Zhang S J, Okafor O C, et al. Novel catalytic and separation processes based on ionic liquids [M]. Amsterdam: Elsevier, 2017.

[298] Dong K, Zhang S J, Wang J J. Understanding the hydrogen bonds in ionic liquids and their roles in properties and

reactions [J]. Chemical Communications, 2016, 52 (41): 6744–6764.

[299] Zhao Y S, Gani R, Afzal M, et al. Ionic liquids for absorption and separation of gases: An extensive database and a systematic screening method [J]. AIChE Journal, 2017, 63 (4): 1353–1367.

[300] Shen M M, Zhang Y Y, Chen K Z, et al. Ionicity of protic ionic liquid: Quantitative measurement by spectroscopic methods [J]. Journal of Physical Chemistry B, 2017, 121 (6): 1372–1376.

[301] Liu X M, Zhou G H, Huo F, et al. Unilamellar vesicle formation and microscopic structure of ionic liquids in aqueous solutions [J]. Journal of Physical Chemistry C, 2016, 120 (1): 659–667.

[302] Zhou J, Liu X M, Zhang S J, et al. Effect of small amount of water on the dynamics properties and microstructures of ionic liquids [J]. AIChE Journal, 2017, 63 (6): 2248–2256.

[303] Kang X C, Ma X X, Zhang J L, et al. Formation of large nanodomains in liquid solutions near the phase boundary [J]. Chemical Communications, 2016, 52 (99): 14286–14289.

[304] Zheng Y Z, Zhou Y, Deng G, et al. Hydrogen–bonding interactions between a nitrile–based functional ionic liquid and DMSO [J]. Journal of Molecular Structure, 2016, 1124: 207–215.

[305] Wang H Y, Zhang S L, Wang J J, et al. Standard partial molar volumes and viscosity *B*-coefficients of ionic liquids [$C_n$mim] Br (n=4,6,8) in alcohols at 298.15 K [J]. Journal of Molecular Liquids, 2015, 209: 563–568.

[306] Xing N N, Zheng L, Pan Y, et al. The thermodynamics of the activation for viscous flow and prediction of dynamic viscositiy of aqueous [C_3py] [DCA] [J]. Colloids and Surfaces A: Physicochemical and Engineering Aspects, 2017, 518: 1–6.

[307] Tong J, Zhang D, Li K, et al. The thermodynamics of the activation for viscous flow of aqueous [C_6mim] [Ala] (1–hexyl–3–methylimidazolium alanine salt) [J]. The Journal of Chemical Thermodynamics, 2016, 101: 356–362.

[308] Kang X C, Liu H Z, Hou M Q, et al. Synthesis of supported ultrafine non–noble subnanometer–scale metal particles derived from metal–organic frameworks as highly efficient heterogeneous catalysts [J]. Angewandte Chemie International Edition, 2016, 55 (3): 1080–1084.

[309] Zhu Q G, Ma J, Kang X C, et al. Efficient reduction of CO_2 into formic acid on a lead or tin electrode using an ionic liquid catholyte mixture [J]. Angewandte Chemie International Edition, 2016, 55 (31): 9012–9016.

[310] Hu J Y, Ma J, Zhu Q G, et al. Transformation of atmospheric CO_2 catalyzed by protic ionic liquids: Efficient synthesis of 2–oxazolidinones [J]. Angewandte Chemie International Edition, 2015, 54 (18): 5399–5403.

[311] Xiong D Z, Cui G K, Wang J J, et al. Reversible hydrophobic–hydrophilic transition of ionic liquids driven by carbon dioxide [J]. Angewandte Chemie International Edition, 2015, 54 (25): 7265–7269.

[312] Yao W H, Wang H Y, Cui G K, et al. Tuning the hydrophilicity and hydrophobicity of the respective cation and anion: Reversible phase transfer of ionic liquids [J]. Angewandte Chemie International Edition, 2016, 55 (28): 7934–7938.

[313] Luo T, Zhang J L, Tan X N, et al. Water–in–supercritical CO_2 microemulsion stabilized by a metal complex [J]. Angewandte Chemie International Edition, 2016, 55 (43): 13533–13537.

[314] Liu C C, Zhang J L, Zheng L R, et al. Metal–organic framework for emulsifying carbon dioxide and water [J]. Angewandte Chemie International Edition, 2016, 55 (38): 11372–11376.

[315] Chen Y F, Zhang Y Y, Yuan S J, et al. Thermodynamic study for gas absorption in choline–2–pyrrolidine–carboxylic acid+polyethylene glycol [J]. Journal of Chemical & Engineering Data, 2016, 61 (10): 3428–3437.

[316] Ma C Y, Guo Y H, Li D X, et al. Molar enthalpy of mixing for choline chloride/urea deep eutectic solvent+water system [J]. Journal of Chemical & Engineering Data, 2016, 61 (12): 4172–4177.

[317] Ma C Y, Guo Y H, Li D X, et al. Molar enthalpy of mixing and refractive indices of choline chloride–based deep eutectic solvents with water [J]. The Journal of Chemical Thermodynamics, 2017, 105: 30–36.

[318] Xie W L, Ji X Y, Feng X, et al. Mass transfer rate enhancement for CO_2 separation by ionic liquids: Effect of film

thickness [J]. Industrial & Engineering Chemistry Research, 2016, 55 (1): 366–372.

[319] Xie W L, Ji X Y, Feng X, et al. Mass–transfer rate enhancement for CO_2 separation by ionic liquids: Theoretical study on the mechanism [J]. AIChE Journal, 2015, 61 (12): 4437–4444.

[320] Jiao F, Li J J, Pan X L, et al. Selective conversion of syngas to light olefins [J]. Science, 2016, 351 (6277): 1065–1068.

[321] Zhong L S, Yu F, An Y L, et al. Cobalt carbide nanoprisms for direct production of lower olefins from syngas [J]. Nature, 2016, 538 (7623): 84–87.

[322] Gao P, Li S G, Bu X N, et al. Direct conversion of CO_2 into liquid fuels with high selectivity over a bifunctional catalyst [J]. Nature Chemistry, 2017, 9 (10): 1019–1024.

[323] Lin L L, Zhou W, Gao R, et al. Low–temperature hydrogen production from water and methanol using Pt/ α –MoC catalysts [J]. Nature, 2017, 544 (7648): 80–83.

[324] Yao S Y, Zhang X, Zhou W, et al. Atomic–layered Au clusters on α –MoC as catalysts for the low–temperature water–gas shift reaction [J]. Science, 2017, 357 (6349): 389–393.

[325] Zhai P, Xu C, Gao R, et al. Highly tunable selectivity for syngas–derived alkenes over zinc and sodium–modulated Fe_5C_2 catalyst [J]. Angewandte Chemie International Edition, 2016, 55 (34): 9902–9907.

[326] Wang N, Sun Q M, Bai R S, et al. In situ confinement of ultrasmall Pd clusters within nanosized silicalite–1 zeolite for highly efficient catalysis of hydrogen generation [J]. Journal of the American Chemical Society, 2016, 138 (24): 7484–7487.

[327] Liu J, Yue Y Y, Liu H Y, et al. Origin of the robust catalytic performance of nanodiamond–graphene–supported Pt nanoparticles used in the propane dehydrogenation reaction [J]. ACS Catalysis, 2017, 7 (5): 3349–3355.

[328] Zhang B, Li X R, Wu Q F, et al. Synthesis of Ni/mesoporous ZSM–5 for direct catalytic conversion of cellulose to hexitols: Modulating the pore structure and acidic sites via a nanocrystalline cellulose template [J]. Green Chemistry, 2016, 18 (11): 3315–3323.

[329] Zhang W, Wang F, McCann S D, et al. Enantioselective cyanation of benzylic C–H bonds via copper–catalyzed radical relay [J]. Science, 2016, 353 (6303): 1014–1018.

[330] Jia X Q, Huang Z. Conversion of alkanes to linear alkylsilanes using an iridium–iron–catalysed tandem dehydrogenation–isomerization–hydrosilylation [J]. Nature Chemistry, 2016, 8 (2): 157–161.

[331] Jia X Q, Qin C, Friedberger T, et al. Efficient and selective degradation of polyethylenes into liquid fuels and waxes under mild conditions [J]. Science Advances, 2016, 2 (6): e1501591.

[332] Zhang F, Min Q Q, Fu X P, et al. Chlorodifluoromethane–triggered formation of difluoromethylated arenes catalysed by palladium [J]. Nature Chemistry, 2017, 9 (9): 918–923.

[333] Wu X Q, Xu S T, Zhang W N, et al. Direct mechanism of the first carbon–carbon bond formation in methanol–to–hydrocarbons process [J]. Angewandte Chemie International Edition, 2017, 56 (31): 9039–9043.

[334] Wang C, Wang Q, Xu J, et al. Direct detection of supramolecular reaction centers in the methanol–to–olefins conversion over zeolite H–ZSM–5 by ^{13}C–^{27}Al solid–state NMR spectroscopy [J]. Angewandte Chemie International Edition, 2016, 55 (7): 2507–2511.

[335] Qi G D, Wang Q, Xu J, et al. Synergic effect of active sites in zinc–modified ZSM–5 zeolites as revealed by high–field solid–state NMR spectroscopy [J]. Angewandte Chemie International Edition, 2016, 55 (51): 15826–15830.

[336] Xiong F, Yu Y Y, Wu Z F, et al. Methanol conversion into dimethyl ether on the anatase TiO_2 (001) surface [J]. Angewandte Chemie International Edition, 2016, 55 (2): 623–628.

[337] Xiong F, Yin L L, Wang Z M, et al. Surface reconstruction–induced site–specific charge separation and photocatalytic reaction on anatase TiO_2 (001) surface [J]. The Journal of Physical Chemistry C, 2017, 121 (18):

9991–9999.

[338] Yuan Q, Wu Z F, Jin Y K, et al. Photocatalytic cross-coupling of methanol and formaldehyde on a rutile TiO_2 (110) surface [J]. Journal of the American Chemical Society, 2013, 135 (13): 5212–5219.

[339] Huang W X. Oxide nanocrystal model catalysts [J]. Accounts of Chemical Research, 2016, 49 (3): 520–527.

[340] (a) Li R G, Zhang F X, Wang D E, et al. Spatial separation of photogenerated electrons and holes among {010} and {110} crystal facets of $BiVO_4$ [J]. Nature Communications, 2013, 4: 1432; (b) Li R G, Han H X, Zhang F X, et al. Highly efficient photocatalysts constructed by rational assembly of dual-cocatalysts separately on different facets of $BiVO_4$ [J]. Energy & Environmental Science, 2014, 7 (4): 1369–1376; (c) Mu L C, Zhao Y, Li A L, et al. Enhancing charge separation on high symmetry $SrTiO_3$ exposed with anisotropic facets for photocatalytic water splitting [J]. Energy & Environmental Science, 2016, 9 (7): 2463–2469.

[341] Xu Y X, Ye Y, Liu T F, et al. Unraveling a single-step simultaneous two-electron transfer process from semiconductor to molecular catalyst in a CoPy/CdS hybrid system for photocatalytic H_2 evolution under strong alkaline conditions [J]. Journal of the American Chemical Society, 2016, 138 (34): 10726–10729.

[342] (a) Liu G J, Shi J Y, Zhang F X, et al. A tantalum nitride photoanode modified with a hole-storage layer for highly stable solar water splitting[J]. Angewandte Chemie International Edition, 2014, 53(28): 7295–7299; (b) Liu G J, Ye S, Yan P L, et al. Enabling an integrated tantalum nitride photoanode to approach the theoretical photocurrent limit for solar water splitting [J]. Energy & Environmental Science, 2016, 9 (4): 1327–1334.

[343] (a) Wang W Y, Chen J, Li C, et al. Achieving solar overall water splitting with hybrid photosystems of photosystem II and artificial photocatalysts [J]. Nature Communications, 2014, 5: 4647; (b) Wang W Y, Wang H, Zhu QJ, et al. Spatially separated photosystem II and a silicon photoelectrochemical cell for overall water splitting: A natural-artificial photosynthetic hybrid [J]. Angewandte Chemie International Edition, 2016, 55 (32): 9229–9233.

[344] Chen S S, Qi Y, Hisatomi T, et al. Efficient visible-light-driven Z-scheme overall water splitting using a $MgTa_2O_{6-}xNy$/TaON heterostructure photocatalyst for H_2 evolution[J]. Angewandte Chemie International Edition, 2015, 54(29): 8498–8501.

[345] Ding C M, Qin W, Wang N, et al. Solar-to-hydrogen efficiency exceeding 2.5% achieved for overall water splitting with all earth-abundant dual-photoelectrode [J]. Physical Chemistry Chemical Physics, 2014, 16 (29): 15608–15614.

[346] Liao S C, Zong X, Seger B, et al. Integrating a dual-silicon photoelectrochemical cell into a redox flow battery for unassisted photocharging [J]. Nature Communications, 2016, 7: 11474.

[347] (a) Li Y W, Hao X, Cao L r, et al. A gas-liquid-solid three-phase suspension bed reactor for Fischer-Tropsch synthesis and the use thereof: Australian Patent, AU2007359708 [P]. 2011-08-04; (b) 温晓东，杨勇，相宏伟，等. 费托合成铁基催化剂的设计基础：从理论走向实践 [J]. 中国科学：化学，2017, 47 (11): 1298–1311.

[348] (a) 宗保宁，马东强，温朗友，等. 一种联产环己醇和乙醇的方法：中国，CN201310001152.9 [P]. 2014-07-09; (b) 宗保宁，谭怀山，马东强，等. 一种联产环己醇和链烷醇的方法：中国，CN201310512160.X [P]. 2015-04-29.

[349] Zheng Y W, Chen B, Ye P, et al. Photocatalytic hydrogen-evolution cross-couplings: Benzene C-H amination and hydroxylation [J]. Journal of the American Chemical Society, 2016, 138 (32): 10080–10083.

[350] Yang Q, Zhang L, Ye C, et al. Visible-light-promoted asymmetric cross-dehydrogenative coupling of tertiary amines to ketones by synergistic multiple catalysis [J]. Angewandte Chemie International Edition, 2017, 129 (13): 3748–3752.

[351] Zhang G T, Liu C, Yi H, et al. External oxidant-free oxidative cross-coupling: A photoredox cobalt-catalyzed aromatic C-H thiolation for constructing C-S bonds[J]. Journal of the American Chemical Society, 2015, 137(29): 9273–9280.

[352] Niu L B, Yi H, Wang S C, et al. Photo-induced oxidant-free oxidative C-H/N-H cross-coupling between arenes and azoles [J]. Nature Communications, 2017, 8: 14226.

[353] Zhang G T, Hu X, Chiang C W, et al. Anti-markovnikov oxidation of β-alkyl styrenes with H_2O as the terminal oxidant [J]. Journal of the American Chemical Society, 2016, 138 (37): 12037-12040.

[354] Yi H, Niu L B, Song C L, et al. Photocatalytic dehydrogenative cross-coupling of alkenes with alcohols or azoles without external oxidant [J]. Angewandte Chemie International Edition, 2017, 56 (4): 1120-1124.

[355] Chai Z G., Zeng T T, Li Q, et al. Efficient visible light-driven splitting of alcohols into hydrogen and corresponding carbonyl compounds over a Ni-modified CdS photocatalyst [J]. Journal of the American Chemical Society, 2016, 138 (32): 10128-10131.

[356] Li X B, Gao Y J, Wang Y, et al. Self-assembled framework enhances electronic communication of ultra-small sized nanoparticles for exceptional solar hydrogen evolution [J]. Journal of the American Chemical Society, 2017, 139 (13): 4789-4796.

[357] Jian, J X, Ye C, Wang X Z, et al. Comparison of H_2 photogeneration by [FeFe] -hydrogenase mimics with CdSe QDs and Ru (bpy) $_3Cl_2$ in aqueous solution [J]. Energy & Environmental Science, 2016, 9 (6): 2083-2089.

[358] Yang B, Jiang X, Guo Q, et al. Self-assembled amphiphilic water oxidation catalysts: Control of O—O bond formation pathways by different aggregation patterns[J]. Angewandte Chemie International Edition, 2016, 55(21): 6229-6234.

[359] Li J X, Ye C, Li X B, et al. A redox shuttle accelerates O_2 evolution of photocatalysts formed in situ under visible light [J]. Advanced Materials, 2017, 29 (17): 1606009.

[360] Liu B, Li X B, Gao Y J, et al. A Solution-processed, mercaptoacetic acid-engineered CdSe quantum dot photocathode for efficient hydrogen production under visible light irradiation [J]. Energy & Environmental Science, 2015, 8 (5): 1443-1449.

[361] Li J, Gao X, Liu B, et al. Graphdiyne: A metal-free material as hole transfer layer to fabricate quantum dot-sensitized photocathodes for hydrogen production [J]. Journal of the American Chemical Society, 2016, 138 (12): 3954-3957.

[362] Xu Y, Ye Y, Liu T, et al. Unraveling a single-step simultaneous two-electron transfer process from semiconductor to molecular catalyst in a CoPy/CdS hybrid system for photocatalytic H_2 evolution under strong alkaline conditions [J]. Journal of the American Chemical Society, 2016, 138 (34): 10726-10729.

[363] Wang S Y, Gao Y Y, Miao S, et al. Positioning the water oxidation reaction sites in plasmonic photocatalysts [J]. Journal of the American Chemical Society, 2017, 139 (34): 11771-11778.

[364] Yao T T, Chen R T, Li J J, et al. Manipulating the interfacial energetics of n-type silicon photoanode for efficient water oxidation [J]. Journal of the American Chemical Society, 2016, 138 (41): 13664-13672.

[365] Mu L C, Zhao Y, Li A L, et al. Enhancing charge separation on high symmetry $SrTiO_3$ exposed with anisotropic facets for photocatalytic water splitting [J]. Energy & Environmental Science, 2016, 9 (7): 2463-2469.

[366] Zhang Y C, Zhang H N, Ji H W, et al. Pivotal role and regulation of proton transfer in water oxidation on hematite photoanodes [J]. Journal of the American Chemical Society, 2016, 138 (8): 2705-2711.

[367] Yu H J, Shi R, Zhao Y X, et al. Alkali-assisted synthesis of nitrogen deficient graphitic carbon nitride with tunable band structures for efficient visible-light-driven hydrogen evolution [J]. Advanced Materials, 2017, 29 (16): 1605148.

[368] Zhao Y F, Chen G B, Bian T, et al. Defect-rich ultrathin ZnAl-layered double hydroxide nanosheets for efficient photoreduction of CO_2 to CO with water [J]. Advanced Materials, 2015, 27 (47): 7824-7831.

[369] Zhao Y F, Zhao Y X, Waterhouse G I N, et al. Layered-double-hydroxide nanosheets as efficient visible-light-driven photocatalysts for dinitrogen fixation [J]. Advanced Materials, 2017, 29 (42): 1703828.

[370] Ge J C, Jia Q Y, Liu W M, et al. Red-emissive carbon dots for fluorescent, photoacoustic, and thermal theranostics in living mice [J]. Advanced Materials, 2015, 27 (28): 4169–4177.

[371] Ma Y, Li X Y, Li A J, et al. H_2S–activable MOF nanoparticle photosensitizer for effective photodynamic therapy against cancer with controllable singlet–oxygen release [J]. Angewandte Chemie International Edition, 2017, 129 (44): 13940–13944.

[372] Gu K Z, Xu Y S, Li H, et al. Real–time tracking and *in vivo* visualization of β –galactosidase activity in colorectal tumor with a ratiometric near–infrared fluorescent probe [J]. Journal of the American Chemical Society, 2016, 138 (16): 5334–5340.

[373] Jiang N, Fan J L, Xu F, et al. Ratiometric fluorescence imaging of cellular polarity: decrease in mitochondrial polarity in cancer cells [J]. Angewandte Chemie International Edition, 2015, 54 (8): 2510–2514.

[374] Zhang C H, Zou C L, Dong H Y, et al. Dual–color single–mode lasing in axially coupled organic nanowire resonators [J]. Science Advances, 2017, 3 (7): e1700225.

[375] Geng H, Jiang F, Wu Y D. Accurate structure prediction and conformational analysis of cyclic peptides with residue–specific force fields [J]. Journal of Physical Chemistry Letters, 2016, 7 (10): 1805–1810.

[376] Hu K, Geng H, Zhang Q Z, et al. An in–tether chiral center modulates the helicity, cell permeability, and target binding affinity of a peptide [J]. Angewandte Chemie International Edition, 2016, 55 (28): 8013–8017.

[377] Zhao H, Liu Q S, Geng H, et al. Crosslinked aspartic acids as helix–nucleating templates [J]. Angewandte Chemie International Edition, 2016, 55 (39): 12088–12093.

[378] Zeng J, Jiang F, Wu Y D. Mechanism of phosphorylation–induced folding of 4E–PB2 revealed by molecular dynamics simulations [J]. Journal of Chemical Theory and Computation, 2017, 13 (1): 320–328.

[379] Meng H, Liu Y, Lai L H. Diverse ways of perturbing the human arachidonic acid metabolic network to control inflammation [J]. Accounts of Chemical Research, 2015, 48 (8): 2242–2250.

[380] Ma X M, Meng H, Lai L H. Motions of allosteric and orthosteric ligand–binding sites in proteins are highly correlated [J]. Journal of Chemical Information and Modeling, 2016, 56 (9): 1725–1733.

[381] Meng H, McClendon C L, Dai Z W, et al. Discovery of novel 15–lipoxygenase activators to shift the human arachidonic acid metabolic network toward inflammation resolution [J]. Journal of Medicinal Chemistry, 2016, 59 (9): 4202–4209.

[382] Wang Q, Liberti M V, Liu P, et al. Rational design of selective allosteric inhibitors of PHGDH and serine synthesis with anti–tumor activity [J]. Cell Chemical Biology, 2017, 24 (1): 55–65.

[383] Ge H, Qian H. Mathematical formalism of nonequilibrium thermodynamics for nonlinear chemical reaction systems with general rate law [J]. Journal of Statistical Physics, 2017, 166 (1): 190–209.

[384] Zhao Z Q, Xie X S, Ge H. Nonequilibrium relaxation of conformational dynamics facilitates catalytic reaction in an elastic network model of T7 DNA polymerase [J]. The Journal of Physical Chemistry B, 2016, 120 (11): 2869–2877.

[385] Xu F, Zhang M S, He W T, et al. Live cell single molecule–guided Bayesian localization super resolution microscopy [J]. Cell Research, 2017, 27 (5): 713–716.

[386] Wu Z Q, Bi H M, Pan S C, et al. Determination of equilibrium constant and relative brightness in fluorescence correlation spectroscopy by considering third–order correlations [J]. Journal of Physical Chemistry B, 2016, 120 (45): 11674–11682.

[387] Bi H M, Yin Y D, Pan B L, et al. Scanning Single–molecule fluorescence correlation spectroscopy enables kinetics study of DNA hairpin folding with a time window from microseconds to seconds [J]. Journal of Physical Chemistry Letters, 2016, 7 (10): 1865–1871.

[388] Wang P, Yang L J, Gao Y Q, et al. Accurate placement of substrate RNA by Gar1 in H/ACA RNA–guided

pseudouridylation [J]. Nucleic Acids Research, 2015, 43 (15): 7207–7216.
[389] Qin B, Zhang S, Song Q, et al. Supramolecular interfacial polymerization: A controllable method of fabricating supramolecular polymeric materials [J]. Angewandte Chemie International Edition, 2017, 56 (26): 7639–7643.
[390] Liu S, Zhao L, Xiao Y L, et al. Allostery in molecular self-assemblies: metal ions triggered self-assembly and emissions of terthiophene [J]. Chemical Communications, 2016, 52 (27): 4876–4879.
[391] Huang X, Wu S S, Ke X K, et al. Phosphonated pillar[5]arene–valved mesoporous silica drug delivery systems[J]. Angewandte Chemie International Edition, 2017, 56 (10): 2655–2659.
[392] Kampf N, Wu C X, Wang Y L, et al. A trimeric surfactant: Surface micelles, hydration–lubrication, and formation of a stable, charged hydrophobic monolayer [J]. Langmuir, 2016, 32 (45): 11754–11762.
[393] Qiao F L, Wang M N, Liu Z, et al. Transitions in the molecular configuration and aggregates for mixtures of a star-shaped hexameric cationic surfactant and a monomeric anionic surfactant [J]. Chemistry – An Asian Journal, 2016, 11 (19): 2763–2772.
[394] Zhou C C, Wang F Y, Chen H, et al. Selective antimicrobial activities and action mechanism of micelles self-assembled by cationic oligomeric surfactants [J]. ACS Applied Materials & Interfaces, 2016, 8 (6): 4242–4249.
[395] Zhou C C, Wang D, Cao M W, et al. Self-aggregation, antibacterial activity, and mildness of cyclodextrin/cationic trimeric surfactant complexes [J]. ACS Applied Materials & Interfaces, 2016, 8 (45): 30811–30823.
[396] Jia Y, Li J B. Molecular assembly of Schiff base interactions: Construction and application [J]. Chemical Reviews, 2015, 115 (3): 1597–1621.
[397] Li B, Li W, Li H L, et al. Ionic complexes of metal oxide clusters for versatile self-assemblies [J]. Accounts of Chemical Research, 2017, 50 (6): 1391–1399.
[398] Li J F, Chen Z J, Zhou M C, et al. Polyoxometalate-driven self-assembly of short peptides into multivalent nanofibers with enhanced antibacterial activity [J]. Angewandte Chemie International Edition, 2016, 55 (7): 2592–2595.
[399] Wang X L, Hao J C. Ionogels of sugar surfactant in ethylammonium nitrate: Phase transition from closely packed bilayers to right-handed twisted ribbons [J]. The Journal of Physical Chemistry B, 2015, 119 (42): 13321–13329.
[400] Sui J F, Wang L H, Zhao W R, et al. Iron-naphthalenedicarboxylic acid gels and their high efficiency in removing arsenic (V) [J]. Chemical Communications, 2016, 52 (43): 6993–6996.
[401] Li G H, Hu Y Y, Sui J F, et al. Hydrogelation and crystallization of sodium deoxycholate controlled by organic acids[J]. Langmuir, 2016, 32 (6): 1502–1509.
[402] Li G H, Wang Y T, Wang L, et al. Hydrogels of superlong helices to synthesize hybrid Ag-helical nanomaterials[J]. Langmuir, 2016, 32 (46): 12100–12109.
[403] Zhao W R, Wang D, Lu H S, et al. Self-assembled switching gels with multiresponsivity and chirality [J]. Langmuir, 2015, 31 (8): 2288–2296.
[404] Liu X C, Fei J B, Wang A H, et al. Transformation of dipeptide-based organogels into chiral crystals by cryogenic treatment [J]. Angewandte Chemie International Edition, 2017, 56 (10): 2660–2663.
[405] Xu Y Q, Fei J B, Li G L, et al. Enhanced photophosphorylation of a chloroplast-entrapping long-lived photoacid [J]. Angewandte Chemie International Edition, 2017, 56 (42): 12903–12907.
[406] Wei D, Ge L L, Lu, S H, et al. Janus particles templated by Janus emulsions and application as a pickering emulsifier [J]. Langmuir, 2017, 33 (23): 5819–5828.
[407] Ge L L, Li J J, Zhong S T, et al. Single, Janus, and Cerberus emulsions from the vibrational emulsification of oils with significant mutual solubility [J]. Soft Matter, 2017, 13 (5): 1012–1019.
[408] Xu L, Wang Y T, Wei G C, et al. Ordered DNA-surfactant hybrid nanospheres triggered by magnetic cationic

surfactants for photon- and magneto-manipulated drug delivery and release [J]. Biomacromolecules, 2015, 16 (12): 4004-4012.

[409] Wang L, Wang Y T, Hao J C, et al. Magnetic fullerene-DNA/hyaluronic acid nanovehicles with magnetism/reduction dual-responsive triggered release [J]. Biomacromolecules, 2017, 18 (3): 1029-1038.

[410] Wang Y T, Yan M M, Xu L, et al. Aptamer-functionalized DNA microgels: A strategy towards selective anticancer therapeutic systems [J]. Journal of Materials Chemistry B, 2016, 4 (32): 5446-5454.

[411] Chen Y, Ji X L, Han Y C, et al. Self-assembly of oleyl bis (2-hydroxyethyl) methyl ammonium bromide with sodium dodecyl sulfate and their interactions with zein [J]. Langmuir, 2016, 32 (32): 8212-8221.

[412] Chen Y, Qiao F L, Fan Y X, et al. Interactions of cationic/anionic mixed surfactant aggregates with phospholipid vesicles and their skin penetration ability [J]. Langmuir, 2017, 33 (11): 2760-2769.

[413] Han J, Wang M G, Hu Y M, et al. Conducting polymer-noble metal nanoparticle hybrids: Synthesis mechanism application [J]. Progress in Polymer Science, 2017, 70: 52-91.

[414] Li Y, Ye X Z, Ma Y R, et al. Interfacial nanosphere lithography toward Ag_2S-Ag heterostructured nanobowl arrays with effective resistance switching and enhanced photoresponses [J]. Small, 2015, 11 (9-10): 1183-1188.

[415] Ye X Z, Zhang F, Ma Y R, et al. Brittlestar-inspired microlens arrays made of calcite single crystals [J]. Small, 2015, 11 (14): 1677-1682.

[416] Wang W H, Dong J Y, Ye X Z, et al. Heterostructured TiO_2 nanorod@nanobowl arrays for efficient photoelectrochemical water splitting [J]. Small, 2016, 12 (11): 1469-1478.

[417] Wang W H, Ma Y R, Qi L M. High-performance photodetectors based on organometal halide perovskite nanonets [J]. Advanced Functional Materials, 2017, 27 (12): 1603653.

[418] Cao C, Zhou Z L, Zhang L, et al. Dilational rheology of different globular protein with imidazolium-based ionic liquid surfactant adsorption layer at the decane/water interface [J]. Journal of Molecular Liquids, 2017, 233: 344-351.

[419] Fu M F, Wang A H, Li J B, et. al. Direct observation of the distribution of gelatin in calcium carbonate crystals by super-resolution fluorescence microscopy [J], Angew. Chem. Int. Ed, 2016, 55: 908-911.

[420] You M X, Lyu Y F, Han D, et al. DNA probes for monitoring dynamic and transient molecular encounters on live cell membranes [J]. Nature Nanotechnology, 2017, 12 (5): 453-459.

[421] Wan S, Zhang L Q, Wang S, et al. Molecular recognition-based DNA nanoassemblies on the surfaces of nanosized exosomes [J]. Journal of the American Chemical Society, 2017, 139 (15): 5289-5292.

[422] Wang Y Y, Wu C C, Chen T, et al. DNA micelle flares: A study of the basic properties that contribute to enhanced stability and binding affinity in complex biological systems [J]. Chemical Science, 2016, 7 (9): 6041-6049.

[423] Yan Z Q, Wang J. SPA-LN: A scoring function of ligand-nucleic acid interactions via optimizing both specificity and affinity [J]. Nucleic Acids Research, 2017, 45 (12): e110.

[424] Gao W Y, Wang C, Muzyka K, et al. Artemisinin-luminol chemiluminescence for forensic bloodstain detection using a smart phone as a detector [J]. Analytical Chemistry, 2017, 89 (11): 6160-6165.

[425] Deng J J, Wang K, Wang M, et al. Mitochondria targeted nanoscale zeolitic imidazole framework-90 for ATP imaging in live cells [J]. Journal of the American Chemical Society, 2017, 139 (16): 5877-5882.

[426] He X L, Zhang K L, Li T, et al. Micrometer-scale ion current rectification at polyelectrolyte brush-modified micropipets [J]. Journal of the American Chemical Society, 2017, 139 (4): 1396-1399.

[427] Li T, He X L, Zhang K L, et al. Observing single nanoparticle events at the orifice of a nanopipet [J]. Chemical Science, 2016, 7 (10): 6365-6368.

[428] Wu F, Su L, Yu P, et al. Role of organic solvents in immobilizing fungus laccase on single-walled carbon nanotubes for improved current response in direct bioelectrocatalysis [J]. Journal of the American Chemical Society, 2017, 139 (4): 1565-1574.

[429] Cao C, Ying Y L, Hu Z L, et al. Discrimination of oligonucleotides of different lengths with a wild–type aerolysin nanopore [J]. Nature Nanotechnology, 2016, 11 (8) : 713–718.

[430] Ying Y L, Long Y T. Single–molecule analysis in an electrochemical confined space [J]. Science China Chemistry, 2017, 60 (9) : 1184–1190.

[431] Lin Y, Shi X, Liu S C, et al. Characterization of the DNA duplex unzipping through a sub–2 nm solid–state nanopore [J]. Chemical Communications, 2017, 53 (25) : 3539–3542.

[432] Lv J, Qian R C, Hu Y X, et al. A precise pointing nanopipette for single–cell imaging via electroosmotic injection [J]. Chemical Communications, 2016, 52 (96) : 13909–13911.

[433] Li D W, Qu L L, Hu K, et al. Monitoring of endogenous hydrogen sulfide in living cells using surface–enhanced Raman scattering [J]. Angewandte Chemie, 2015, 127 (43) : 12949–12952.

[434] Cao Y, Li D W, Zhao L J, et al. Highly selective detection of carbon monoxide in living cells bypalladacycle carbonylation–based SERS nanosensors [J]. Analytical Chemistry, 2015, 87 (19) : 9696–9701.

[435] Qian R C, Cao Y, Long Y T. Dual–targeting nanovesicles for in situ intracellular imaging of and discrimination between wild–type and mutant p53 [J]. Angewandte Chemie International Edition, 2016, 55 (2) : 719–723.

[436] Qian R C, Cao Y, Zhao L J, et al. A two–stage dissociation system for multilayer imaging of cancer biomarker–synergic networks in single cells [J]. Angewandte Chemie International Edition, 2017, 56 (17) : 4802–4805.

[437] Pan Y G, Zhang F X, Zhang L Y, et al. The process of wrapping virus revealed by a force tracing technique and simulations [J]. Advanced Science, 2017, 4 (9) : 1600489.

[438] Liu Y L, Ai K L, Ji X Y, et al. Comprehensive insights into the multi–antioxidative mechanisms of melanin nanoparticles and their application to protect brain from injury in ischemic stroke [J]. Journal of the American Chemical Society, 2017, 139 (2) : 856–862.

[439] Jiang C C, Wang Y, Wang J W, et al. Achieving ultrasensitive *in vivo* detection of bone crack with polydopamine–capsulated surface–enhanced Raman nanoparticle [J]. Biomaterials, 2017, 114: 54–61.

[440] Ma Z J, Jia X D, Bai J, et al. MnO_2 gatekeeper: An intelligent and O_2–evolving shell for preventing premature release of high cargo payload core, overcoming tumor hypoxia, and acidic H_2O_2–sensitive MRI [J]. Advanced Functional Materials, 2017, 27 (4) : 1604258.

[441] Zhao Q, Fang F, Shan Y C, et al. In–depth proteome coverage by improving efficiency for membrane proteome analysis [J]. Analytical Chemistry, 2017, 89 (10) : 5179–5185.

[442] Li Y, Yang C X, Yan X P. Controllable preparation of core–shell magnetic covalent–organic framework nanospheres for efficient adsorption and removal of bisphenols in aqueous solution [J]. Chemical Communications, 2017, 53 (16) : 2511–2514.

[443] Zhou L J, Hu Y L, Li G K. Conjugated microporous polymers with built–in magnetic nanoparticles for excellent enrichment of trace hydroxylated polycyclic aromatic hydrocarbons in human urine [J]. Analytical Chemistry, 2016, 88 (13) : 6930–6938.

[444] Li Z, Liu Y Q, Zhao Y, et al. Selective separation of metal ions via monolayer nanoporous graphene with carboxyl groups [J]. Analytical Chemistry, 2016, 88 (20) : 10002–10010.

[445] Yu D P, Shen A J, Guo Z M, et al. A controlled thiol–initiated surface polymerization strategy for the preparation of hydrophilic polymer stationary phases [J]. Chemical Communications, 2015, 51 (79) : 14778–14780.

[446] Zhang H Y, Ou J J, Liu Z S, et al. Preparation of hybrid monolithic columns via "one–pot" photoinitiated thiol–acrylate polymerization for retention–independent performance in capillary liquid chromatography [J]. Analytical Chemistry, 2015, 87 (17) : 8789–8797.

[447] Liu Z S, Liu J, Liu Z Y, et al. Functionalization of hybrid monolithic columns via thiol–ene click reaction for proteomics analysis [J]. Journal of Chromatography A, 2017, 1498: 29–36.

[448] Bai J Y, Liu Z S, Wang H W, et al. Preparation and characterization of hydrophilic hybrid monoliths via thiol-ene click polymerization and their applications in chromatographic analysis and glycopeptides enrichment [J]. Journal of Chromatography A, 2017, 1498: 37-45.

[449] Wu C, Liang Y, Yang K G, et al. Clickable periodic mesoporous organosilica monolith for highly efficient capillary chromatographic separation [J]. Analytical Chemistry, 2016, 88 (3): 1521-1525.

[450] Huang Z, Yan G Q, Gao M X, et al. Array-based online two dimensional liquid chromatography system applied to effective depletion of high-abundance proteins in human plasma [J]. Analytical Chemistry, 2016, 88 (4): 2440-2445.

[451] Chen Q, Yan G Q, Gao M X, et al. Ultrasensitive proteome profiling for 100 living cells by direct cell injection, online digestion and nano-LC-MS/MS analysis [J]. Analytical Chemistry, 2015, 87 (13): 6674-6680.

[452] 关亚风，耿旭辉，刘洪鹏，等. 一种用于液相色谱荧光检测器的大体积流通池：中国，201310728584.X[P]. 2016-08-24.

[453] 关亚风，沈铮，张健，等. 一种全自动阵列固相萃取装置：中国，201510920927.1 [P]. 2015-12-11.

[454] Weng Y J, Sui Z G, Shan Y C, et al. In-depth proteomic quantification of cell secretome in serum-containing conditioned medium [J]. Analytical Chemistry, 2016, 88 (9): 4971-4978.

[455] Bian Y Y, Li L, Dong M M, et al. Ultra-deep tyrosine phosphoproteomics enabled by a phosphotyrosine superbinder [J]. Nature Chemical Biology, 2016, 12 (11): 959-966.

[456] Zhao Y N, Hao Z Q, Zhao C X, et al. A novel strategy for large-scale metabolomics study by calibrating gross and systematic errors in gas chromatography-mass spectrometry [J]. Analytical Chemistry, 2016, 88 (4): 2234-2242.

[457] Wang S Y, Zhou L N, Wang Z C, et al. Simultaneous metabolomics and lipidomics analysis based on novel heart-cutting two-dimensional liquid chromatography-mass spectrometry [J]. Analytica Chimica Acta, 2017, 966: 34-40.

[458] Zhou J T, Liu H Y, Liu Y, et al. Development and evaluation of a parallel reaction monitoring strategy for large-scale targeted metabolomics quantification [J]. Analytical Chemistry, 2016, 88 (8): 4478-4486.

[459] Wang H Q, Xu J, Chen Y H, et al. Optimization and evaluation strategy of esophageal tissue preparation protocols for metabolomics by LC-MS [J]. Analytical Chemistry, 2016, 88 (7): 3459-3464.

[460] Wang Q, Song W, Qiao X, et al. Simultaneous quantification of 50 bioactive compounds of the traditional Chinese medicine formula Gegen-Qinlian decoction using ultra-high performance liquid chromatography coupled with tandem mass spectrometry [J]. Journal of Chromatography A, 2016, 1454: 15-25.

[461] Ni G, Zhang H, Fan Y Y, et al. Mannolides A-C with an intact diterpenoid skeleton providing insights on the biosynthesis of antitumor *Cephalotaxus* troponoids [J]. Organic Letters, 2016, 18 (8): 1880-1883.

[462] Du N N, Liu Y F, Zhang X L, et al. Discovery of new muscarinic acetylcholine receptor antagonists from *Scopolia tangutica* [J]. Scientific Reports, 2017, 7: 46067.

[463] Li A, Zi Y L, Guo H Y, et al. Triboelectric nanogenerators for sensitive nano-coulomb molecular mass spectrometry [J]. Nature Nanotechnology, 2017, 12 (5): 481-487.

[464] He M H, Meng Y F, Yan S S, et al. Three-dimensional elemental imaging of nantan meteorite via femtosecond laser ionization time-of-flight mass spectrometry [J]. Analytical Chemistry, 2017, 89 (1): 565-570.

[465] Li W F, Yin Z B, Cheng X L, et al. Pulsed microdischarge with inductively coupled plasma mass spectrometry for elemental analysis on solid metal samples [J]. Analytical Chemistry, 2015, 87 (9): 4871-4878.

[466] Zhao Z J, Wang B, Duan Y X. Exploration of microplasma probe desorption/ionization mass spectrometry (MPPDI-MS) for biologically related analysis [J]. Analytical Chemistry, 2016, 88 (3): 1667-1673.

[467] Zhu H, Zou G, Huang G, et al. Single-neuron identification of chemical constituents, physiological changes, and

metabolism using mass spectrometry [J]. Proceedings of the National Academy of Sciences of the United States of America, 2017, 114 (10): 2586–2591.

[468] Chen F M, Lin L Y, Zhang J, et al. Single-cell analysis using drop-on-demand inkjet printing and probe electrospray ionization mass spectrometry [J]. Analytical Chemistry, 2016, 88 (8): 4354–4360.

[469] Zhang X C, Wei Z W, Gong X Y, et al. Integrated droplet-based microextraction with ESI-MS for removal of matrix interference in single-cell analysis [J]. Scientific Reports, 2016, 6: 24730.

[470] Zhao Y Y, Wei Z W, Zhao H S, et al. *In situ* ion-transmission mass spectrometry for paper-based analytical devices [J]. Analytical Chemistry, 2016, 88 (22): 10805–10810.

[471] He Q, Xing Z, Wei C, et al. Rapid screening of copper intermediates in Cu (I) -catalyzed azide-alkyne cycloaddition using a modified ICP-MS/MS platform [J]. Chemical Communications, 2016, 52 (69): 10501–10504.

[472] Yang Y, Song H P, He D, et al. Genetically encoded protein photo crosslinker with a transferable mass spectrometry-identifiable label [J]. Nature Communications, 2016, 7: 12299.

[473] Guo S, Wang Y M, Zhou D, et al. Electric field-assisted matrix coating method enhances the detection of small molecule metabolites for mass spectrometry imaging [J]. Analytical Chemistry, 2015, 87 (12): 5860–5865.

[474] Zhang D, Chen B C, Wang Y M, et al. Disease-specific IgG Fc N-glycosylation as personalized biomarkers to differentiate gastric cancer from benign gastric diseases [J]. Scientific Reports, 2016, 6: 25957.

[475] Liu Y J, Liu Y J, Zhang D, et al. Kapok fiber: A natural biomaterial for highly specific and efficient enrichment of sialoglycopeptides [J]. Analytical Chemistry, 2016, 88 (2): 1067–1072.

[476] He M W, Guo S, Li Z L, et al. *In situ* characterizing membrane lipid phenotype of breast cancer cells using mass spectrometry profiling [J]. Scientific Reports, 2015, 5: 11298.

[477] Chen S M, Xiong C Q, Liu H H, et al. Mass spectrometry imaging reveals the sub-organ distribution of carbon nanomaterials [J]. Nature Nanotechnology, 2015, 10 (2): 176–182.

[478] Zhang N, Hou J, Chen S M, et al. Rapidly probing antibacterial activity of graphene oxide by mass spectrometry-based metabolite fingerprinting [J]. Scientific Reports, 2016, 6: 28045.

[479] Xiong C Q, Zhou X Y, He Q, et al. Development of visible-wavelength MALDI cell mass spectrometry for high-efficiency single-cell analysis [J]. Analytical Chemistry, 2016, 88 (23): 11913–11918.

[480] Wang J N, Qiu S L, Chen S M, et al. MALDI-TOF MS imaging of metabolites with a N- (1-naphthyl) ethylenediamine dihydrochloride matrix and its application to colorectal cancer liver metastasis [J]. Analytical Chemistry, 2015, 87 (1): 420–430.

[481] Tang X M, Huang L L, Zhong H Y, et al. Chemical imaging of latent fingerprints by mass spectrometry based on laser activated electron tunneling [J]. Analytical Chemistry, 2015, 87 (5): 2693–2701.

[482] He Z Y, Chen Q S, Chen F M, et al. DNA-mediated cell surface engineering for multiplexed glycan profiling using MALDI-TOF mass spectrometry [J]. Chemical Science, 2016, 7 (8): 5448–5452.

[483] Wu Z Q, Du W B, Li J Y, et al. Establishment of a finite element model for extracting chemical reaction kinetics in a micro-flow injection system with high throughput sampling [J]. Talanta, 2015, 140: 176–182.

[484] Jin D Q, Zhu Y, Fang Q. Non-tapered PTFE capillary as robust and stable nanoelectrospray emitter for electrospray ionization mass spectrometry [J]. Rapid Communications in Mass Spectrometry, 2016, 30 (S1): 62–67.

[485] Liang Y R, Zhu L N, Gao J, et al. 3D-printed high-density droplet array chip for miniaturized protein crystallization screening under vapor diffusion mode [J]. ACS Applied Materials & Interfaces, 2017, 9 (13): 11837–11845.

[486] Liu W, Chen Q S, Lin X X, et al. Online multi-channel microfluidic chip-mass spectrometry and its application for quantifying noncovalent protein-protein interactions [J]. The Analyst, 2015, 140 (5): 1551–1554.

[487] Wu J, He Z Y, Chen Q S, et al. Biochemical analysis on microfluidic chips [J]. TrAC Trends in Analytical

Chemistry, 2016, 80: 213–231.

[488] Liu W, Lin J M. Online monitoring of lactate efflux by multi–channel microfluidic chip–mass spectrometry for rapid drug evaluation [J]. ACS Sensors, 2016, 1 (4) : 344–347.

[489] 中国发明专利，201510319853.6.

[490] Zhi X, Liu J J, Li Z J, et al. Ionic hydrogen bond donor organocatalyst for fast living ring–opening polymerization [J]. Polymer Chemistry, 2016, 7 (2) : 339–349.

[491] Zhao Y G, Zhang S L, Wu Z Q, et al. Visible–light–induced living radical polymerization (LRP) mediated by (salen) Co (II) /TPO at ambient temperature [J]. Macromolecules, 2015, 48 (15) : 5132–5139.

[492] Zhao Y G, Liu X, Liu Y C, et al. When CMRP met alkyl vinyl ketone: Visible light induced living radical polymerization (LRP) of ethyl vinyl ketone (EVK) [J]. Chemical Communications, 2016, 52 (81) : 12092–12095.

[493] Liu D T, Yao C G, Wang R, et al. Highly isoselective coordination polymerization of ortho–methoxystyrene with β–diketiminato rare–earth–metal precursors [J]. Angewandte Chemie International Edition, 2015, 54 (17) : 5205–5209.

[494] Liu D T, Wang M Y, Wang Z C, et al. Stereoselective copolymerization of unprotected polar and nonpolar styrenes by an yttrium precursor: Control of polar–group distribution and mechanism [J]. Angewandte Chemie International Edition, 2017, 56 (10) : 2714–2719.

[495] Liu B, Cui D M, Tang T. Stereo–and temporally controlled coordination polymerization triggered by alternating addition of a lewis acid and base [J]. Angewandte Chemie International Edition, 2016, 55 (39) : 11975–11978.

[496] Dai S Y, Sui X L, Chen C L. Highly robust Pd (II) α–diimine catalysts for slow–chain–walking polymerization of ethylene and copolymerization with methyl acrylate [J]. Angewandte Chemie International Edition, 2015, 54 (34) : 9948–9953.

[497] Dai S Y, Chen C L. Direct synthesis of functionalized high–molecular–weight polyethylene by copolymerization of ethylene with polar monomers [J]. Angewandte Chemie International Edition, 2016, 55 (42) : 13281–13285.

[498] Tao X F, Deng Y W, Shen Z Q, et al. Controlled polymerization of N–substituted glycine N–thiocarboxyanhydrides initiated by rare earth borohydrides toward hydrophilic and hydrophobic polypeptoids [J]. Macromolecules, 2014, 47 (18) : 6173–6180.

[499] Yang J L, Wu H L, Li Y, et al. Perfectly alternating and regioselective copolymerization of carbonyl sulfide and epoxides by metal–free lewis pairs [J]. Angewandte Chemie International Edition, 2017, 56 (21) : 5774–5779.

[500] Zhang J, Zhang M, Du F S, et al. Synthesis of functional polycaprolactones via passerini multicomponent polymerization of 6–oxohexanoic acid and isocyanides [J]. Macromolecules, 2016, 49 (7) : 2592–2600.

[501] Hu R R, Li W Z, Tang B Z. Recent advances in alkyne–based multicomponent polymerizations [J]. Macromolecule Chemistry and Physics, 2016, 217 (2) : 213–224.

[502] Zhang X J, Wang S X, Liu J, et al. Ugi reaction of natural amino acids: A general route toward facile synthesis of polypeptoids for bioapplications [J]. ACS Macro Letters, 2016, 5 (9) : 1049–1054.

[503] Xue H D, Zhao Y, Wu H B, et al. Multicomponent combinatorial polymerization via the biginelli reaction [J]. Journal of the American Chemical Socity, 2016, 138 (28) : 8690–8693.

[504] He B Z, Zhen S J, Wu Y W, et al. Cu (I) –catalyzed amino–yne click polymerization [J]. Polymer Chemistry, 2016, 7 (48) : 7375–7382.

[505] He B Z, Su H F, Bai T W, et al. Spontaneous amino–yne click polymerization: A powerful tool toward regio– and stereospecific poly (β–aminoacrylate) s [J]. Journal of the American Chemical Socity, 2017, 139 (15) : 5437–5443.

[506] Xu Y Y, Yang G, Xia H Y, et al. Enantioselective synthesis of helical polydiacetylene by application of linearly

polarized light and magnetic field [J]. Nature Communications, 2014, 5: 5050.

[507] Zhu J Y, Dong Z Y, Lei S B, et al. Design of aromatic helical polymers for STM visualization: Imaging of single and double helices with a pattern of $\pi-\pi$ stacking [J]. Angewandte Chemie International Edition, 2015, 54 (10): 3097-3101.

[508] Jiang Z Q, Xue Y X, Chen J L, et al. One-pot synthesis of brush copolymers bearing stereoregular helical polyisocyanides as side chains through tandem catalysis [J]. Macromolecules, 2015, 48 (1): 81–89.

[509] Zhou X X, Wang R G, Lei W W, et al. Design and synthesis by redox polymerization of a bio-based carboxylic elastomer for green tire [J]. Science China-Chemistry, 2015, 58 (10): 1561–1569.

[510] Gao S H, Wang R G, Fang B W, et al. Preparation and properties of a novel bio-based and non-crystalline engineering elastomer with high low-temperature and oil resistance [J]. Journal of Appllied Polymer Science, 2016, 133 (1): 42855.

[511] Hu J, Wang G L, Zhao W G, et al. Site-specific in situ growth of an interferon-polymer conjugate that outperforms PEGASYS in cancer therapy [J]. Biomaterials, 2016, 96: 84–92.

[512] Hu J, Wang G L, Zhao W G, et al. *In situ* growth of a C-terminal interferon-alpha conjugate of a phospholipid polymer that outperforms PEGASYS in cancer therapy [J]. Journal of Controlled Release, 2016, 237: 71–77.

[513] Hu Q L, Wu M, Fang C, et al. Engineering nanoparticle-coated bacteria as oral DNA vaccines for cancer immunotherapy [J]. Nano Letters, 2015, 15 (4): 2732–2739.

[514] Liu J Y, Huang W, Pang Y, et al. Hyperbranched polyphosphates: Synthesis, functionalization and biomedical applications [J]. Chemical Society Reviews, 2015, 44 (12): 3942–3953.

[515] Jiang W F, Zhou Y F, Yan D Y. Hyperbranched polymer vesicles: From self-assembly, characterization, mechanisms, and properties to applications [J]. Chemical Society Reviews, 2015, 44 (12): 3874–3889.

[516] Guan X W, Guo Z P, Lin L, et al. Ultrasensitive pH triggered charge/size dual-rebound gene delivery system [J]. Nano Letters, 2016, 16 (11): 6823–6831.

[517] Wang Y, Lin Y-X, Qiao Z-Y, et al. Self-assembled autophagy-inducing polymeric nanoparticles for breast cancer interference in-vivo [J]. Advanced Materials, 2015, 27 (16): 2627–2634.

[518] Liu X, Xiang J, Zhu D C, et al. Fusogenic reactive oxygen species triggered charge-reversal vector for effective gene delivery [J]. Advanced Materials, 2016, 28 (9): 1743–1752.

[519] Qiu N S, Liu X R, Zhong Y, et al. Esterase-activated charge-reversal polymer for fibroblast-exempt cancer gene therapy [J]. Advanced Materials, 2016, 28 (48): 10613–10622.

[520] Ai X Z, Ho C J H, Aw J, et al. *In vivo* covalent cross-linking of photon-converted rare-earth nanostructures for tumour localization and theranostics [J]. Nature Communications, 2016, 7: 10432.

[521] Liao C Z, Zhang M, Yao M Y, et al. Flexible organic electronics in biology: Materials and devices [J]. Advanced Materials, 2015, 27 (46): 7493–7527.

[522] Liu F Y, He X X, Chen H D, et al. Gram-scale synthesis of coordination polymer nanodots with renal clearance properties for cancer theranostic applications [J]. Nature Communications, 2015, 6: 8003.

[523] Liao C Z, Mak C, Zhang M, et al. Flexible organic electrochemical transistors for highly selective enzyme biosensors and used for saliva testing [J]. Advanced Materials, 2015, 27 (4): 676–681.

[524] Li S Y, Wang X, Cao B, et al. Effects of nanoscale spatial arrangement of arginine-glycine-aspartate peptides on dedifferentiation of chondrocytes [J]. Nano Letters, 2015, 15 (11): 7755–7765.

[525] Ye K, Wang X, Cao L P, et al. Matrix stiffness and nanoscale spatial organization of cell-adhesive ligands direct stem cell fate [J]. Nano Letters, 2015, 15 (7): 4720–4729.

[526] Yang H R, Liu D, Ju J Z, et al. Chain deformation on the formation of shish nuclei under extension flow: An *in situ* SANS and SAXS study [J]. Macromolecules, 2016, 49 (23): 9080–9088.

[527] Zhang Q L, Zhang R, Meng L P, et al. Biaxial stretch-induced crystallization of poly (ethylene terephthalate) above glass transition temperature: The necessary of chain mobility [J]. Polymer, 2016, 101: 15-23.

[528] Zhou Y, Deng H, Yu F L, et al. Processing condition induced structural evolution in the alternating multi-layer structure during high speed thin-wall injection molding [J]. Polymer, 2016, 99: 49-58.

[529] Ru J F, Yang S G, Zhou D, et al. Dominant β -form of poly (l-lactic acid) obtained directly from melt under shear and pressure fields [J]. Macromolecules, 2016, 49 (10): 3826-3837.

[530] Wang M, Lin B P, Yang H. A plant tendril mimic soft actuator with phototunable bending and chiral twisting motion modes [J]. Nature Communications, 2016, 7: 13981.

[531] Zhou F, Gu K H, Zhang Z Y, et al. Exploiting host-guest interactions for the synthesis of a rod-rod block copolymer with crystalline and liquid-crystalline blocks [J]. Angewandte Chemie International Edition, 2016, 55 (48): 15007-15011.

[532] Wang J, Li X Q, Ren X K, et al. Helical polyacetylene-based switchable chiral columnar phases: Frustrated chain packing and two-way shape actuator [J]. Chemistry-an Asian Journal, 2016, 11 (17): 2387-2391.

[533] Lv J A, Liu Y Y, Wei J, et al. Photocontrol of fluid slugs in liquid crystal polymer microactuators [J]. Nature, 2016, 537 (7619): 179-184.

[534] Peng X, Liu H L, Yin Q, et al. A zwitterionic gel electrolyte for efficient solid-state supercapacitors [J]. Nature Communications, 2016, 7: 11782.

[535] Fan X L, Liu H L, Gao Y T, et al. Forward-osmosis desalination with poly (ionic liquid) hydrogels as smart draw agents [J]. Advanced Materials, 2016, 28 (21): 4156-4161.

[536] Wu M, Zhu Y T, Jiang W. Release behavior of polymeric vesicles in solution controlled by external electrostatic field [J]. ACS Macro Letters, 2016, 5 (11): 1212-1216.

[537] Yin Y D, Niu L, Zhu X C, et al. Non-equilibrium behaviour in coacervate-based protocells under electric-field-induced excitation [J]. Nature Communications, 2016, 7: 10658.

[538] Gan S C, Wu Z L, Xu H L, et al. Viscoelastic behaviors of carbon black gel extracted from highly filled natural rubber compounds: Insights into the payne effect [J]. Macromolecules, 2016, 49 (4): 1454-1463.

[539] Sun S Q, Wang D, Russell T P, et al. Nanomechanical mapping of a deformed elastomer: Visualizing a self-reinforcement mechanism [J]. ACS Macro Letters, 2016, 5 (7): 839-843.

[540] Li J F, Wang Z L, Wen L G, et al. Highly elastic fibers made from hydrogen-bonded polymer complex [J]. ACS Macro Letters, 2016, 5 (7): 814-818.

[541] Zong C Y, Zhao Y, Ji H P, et al. Tuning and erasing surface wrinkles by reversible visible-light-induced photoisomerization [J]. Angewandte Chemie International Edition, 2016, 55 (12): 3931-3935.

[542] Gao N W, Zhang X Y, Liao S L, et al. Polymer swelling induced conductive wrinkles for an ultrasensitive pressure sensor [J]. ACS Macro Letters, 2016, 5 (7): 823-827.

[543] Xie N, Liu M J, Deng H L, et al. Macromolecular metallurgy of binary mesocrystals via designed multiblock terpolymers [J]. Journal of the American Chemical Society, 2014, 136 (8): 2974-2977.

[544] Liu M J, Qiang Y C, Li W H, et al. Stabilizing the Frank-Kasper phases via binary blends of AB diblock copolymers [J]. ACS Macro Letters, 2016, 5 (10): 1167-1171.

[545] Gao Y, Deng H L, Li W H, et al. Formation of nonclassical ordered phases of AB-type multiarm block copolymers [J]. Physical Review Letters, 2016, 116 (6): 068304.

[546] He Z Y, Xie W J, Liu Z Q, et al. Tuning ice nucleation with counterions on polyelectrolyte brush surfaces [J]. Science Advances, 2016, 2 (6): e1600345

[547] Luo Z L, Zhang A F, Chen Y M, et al. How big is big enough? Effect of length and shape of side chains on the single-chain enthalpic elasticity of a macromolecule [J]. Macromolecules, 2016, 49 (9): 3559-3565.

［548］ Luo Z L, Zhang B, Qian H J, et al. Effect of size of solvent moleculeS on the single-chain mechanics of poly（ethylene glycol）: Implications on a novel design of molecular motor［J］. Nanoscale, 2016, 8（41）: 17820–17827.

［549］ Zheng N, Fang Z Z, Zou W K, et al. Thermoset shape-memory polyurethane with intrinsic plasticity enabled by transcarbamoylation［J］. Angewandte Chemie International Edition, 2016, 55（38）: 11421–11425.

［550］ Zhao Q, Zou W K, Luo Y W, et al. Shape memory polymer network with thermally distinct elasticity and plasticity［J］. Science Advances, 2016, 2（1）: e1501297.

［551］ Kou F, Yang S L, Qian H J, et al. A fluorescence study on the complexation of Sm（III）, Eu（III）and Tb（III）with tetraalkyldiglycolamides（TMDGA）in aqueous solution, in solid state, and in solvent extraction［J］. Dalton Transactions, 2016, 45（46）: 18484–18493.

［552］ Kou F, Yang S L, Zhang L H, et al. Complexation of Ho（III）with tetraalkyl-diglycolamide in aqueous solutions and a solid state compared in organic solutions of solvent extraction［J］. Inorganic Chemistry Communications, 2016, 71: 41–44.

［553］ 张燕. N,N-二（2-乙基己基）二甘酰胺酸与 Nd（Ⅲ）/Eu（Ⅲ）、U（Ⅳ）、U（Ⅵ）萃合物配位化学［D］. 北京：中国原子能科学研究院，2017.

［554］ 杨贺，张虎，李丽，等. 硝酸肼还原反萃取分离镎 / 钚的研究［J］. 核技术，2016, 39（9）: 090301.

［555］ 耿俊霞，窦强，王子豪，等. 钍基熔盐堆核能系统中熔盐的蒸馏纯化与分离［J］. 核化学与放射化学，2017, 39（1）: 36–42.

［556］ 曹石巍，谈存敏，张鑫，等. 四乙基双三嗪吡啶与二酰胺荚醚在硝酸介质中对 Am（III）/E（III）的协同萃取［J］. 核化学与放射化学，2016, 38（5）: 274–281.

［557］ 王志鹏，李诗萌，胡晓阳，等. 水溶性 BisDGA 对 NTAamide（C8）萃取 Am^{3+}、Eu^{3+} 的影响［C］// 第十四届全国核化学与放射化学学术研讨会. 长春，2016: 26.

［558］ 张曦，杨素亮，丁有钱，等. 大孔硅基材料 $CyMe_4$-BTBP/SiO_2-P 的制备及其对 Am（III）、Eu（III）的分离性能研究［J］. 原子能科学技术，2016, 50（5）: 774–781.

［559］ 李荣，庞利娟，张明星，等. 辐射接枝制备聚丙烯纤维基海水提铀吸附剂［J］. 核技术，2017, 40（5）: 050301.

［560］ 熊亮萍，古梅，吕开. 钨掺杂的提铯离子筛的制备及离子交换性能研究［J］. 原子能科学技术，2016, 50（6）: 966–972.

［561］ 史冬峰，唐振平，黄华勇，等. 磁性氧化石墨烯 / β – 环糊精复合材料对 U（VI）的吸附性能及机理［J］. 原子能科学技术，2016, 50（9）: 1556–1564.

［562］ 王关全，郭欣，宋虎，等. ^{131}I 标记卟啉化合物结合光动力治疗对肿瘤生长的抑制作用［J］. 原子能科学技术，2016, 50（9）: 1555–1569.

［563］ 孙钰林，刘陆，梁积新，等. 基于 ^{64}Cu 和超顺磁性氧化铁纳米粒子的 PET/MRI 双模态淋巴显像探针的研究［J］. 原子能科学技术，2016, 50（12）: 2130–2137.

［564］ a）孙钰林，沈浪涛，梁积新，等. PET/MRI 双模态显像探针 ^{64}Cu-DOTA-SPIONs-PEG-FA 的合成及生物评价［C］//2015 年全国核医学学术年会. 上海，2015: 110; b）孙钰林，申一鸣，梁积新，等. ^{64}Cu-DOTA-SPIONs-PEG-FA: 靶向叶酸受体阳性肿瘤的 PET/MRI 双模态显像探针［J］. 核化学与放射化学，2017, 39（4）: 298–308.

［565］ 霍焱，唐磊，闫平，等. microRNA 靶向的 ^{18}F 标记探针的制备及条件优化［C］// 第十四届全国核化学与放射化学学术研讨会. 长春，2016: 137.

［566］ 王欢，汪航，王淑霞，等.［^{18}F］标记的 α7 烟碱型乙酰胆碱受体放射性配体的设计合成及生物评价［C］// 第十四届全国核化学与放射化学学术研讨会. 长春：中国化学会，2016: 147.

［567］ 何建刚，乔雪玲，杨小雨，等. 硒 –79 在北山花岗岩中的迁移扩散行为—毛细管内扩散法研究［C］// 第十四届全国核化学与放射化学学术研讨会. 长春，2016: 107.

［568］ 杨小雨，王春丽，郑仲，等．温度对硒 -75 在北山花岗岩中扩散的影响［C］// 第十四届全国核化学与放射化学学术研讨会．长春，2016：110.［19］　刘东旭，司高华，李哲，等．非均质条件下锶迁移的反向随机模拟［J］. 原子能科学技术，2017, 51（4）：609–616.

［569］ 刘东旭，司高华，李哲，等．非均质条件下锶迁移的反向随机模拟，原子能科学技术［J］, 2017, 51（4）：609–616.

［570］ 王晓丽，李士成，黄召亚，等．Th（IV）在高岭土上的吸附行为研究［J］. 化学研究与应用，2016, 28（5）：738–742.

［571］ 邬琴琴，代群威，韩林宝，等．脱氮硫杆菌的筛选及其对锶离子的矿化作用［J］. 核化学与放射化学，2017, 39（2）：187–192.

［572］ Yin X, Wang Y, Bai X, Wang, Y, et al. Rare earth separations by selective borate crystallization［J］. Nature Communications 2017, 8：14438.

［573］ Zheng T,Yang Z, Gui D, et al. Overcoming the crystallization and designability issues in the ultrastable zirconium phosphonate framework system［J］. Nature Communications 2017, 8：15369.

［574］ Wang Y, Liu Z, Li Y, et al.Umbellate Distortions of the Uranyl Coordination Environment Result in a Stable and Porous Polycatenated Framework That Can Effectively Remove Cesium from Aqueous Solutions［J］. Journal of the American Chemical Society 2015, 137（19）：6144–6147.

［575］ Xie J,Wang Y, Liu W, et al. Highly Sensitive Detection of Ionizing Radiations by a Photoluminescent Uranyl Organic Framework［J］. Angewandte Chemie International Edition 2017, 56（26）：7500–7504.

［576］ Sheng D, Zhu L, Xu C, et al. Efficient and Selective Uptake of TcO4– by a Cationic Metal–Organic Framework Material with Open Ag+ Sites［J］. Environ. Sci. Technol. 2017, 51（6）：3471–3479.

［577］ Zhu L, Xiao C, Dai X, et al, Exceptional Perrhenate/Pertechnetate Uptake and Subsequent Immobilization by a Low–Dimensional Cationic Coordination Polymer：Overcoming the Hofmeister Bias Selectivity［J］. Environmental Science & Technology Letters. 2017,4（7）：316–322.

［578］ Xu L, Zheng T,Yang S, et al. Uptake Mechanisms of Eu（III）on Hydroxyapatite：A Potential Permeable Reactive Barrier Backfill Material for Trapping Trivalent Minor Actinides［J］. Environmental Science & Technology. 2016, 50,（7）：3852–3859.

［579］ Liu W, Dai X, Bai Z,et al. Highly Sensitive and Selective Uranium Detection in Natural Water Systems Using a Luminescent Mesoporous Metal–Organic Framework Equipped with Abundant Lewis Basic Sites：A Combined Batch, X–ray Absorption Spectroscopy, and First Principles Simulation Investigation［J］. Environmental Science & Technology.2017, 51,（7）：3911–3921.

［580］ 张怡，郑佐西，朱欣研，等．放射性固体废物水泥砂浆固定配方研究［J］. 核化学与放射化学，2017, 39（1）：63–68.

［581］ 陈长伦，高阳，孙玉兵，等．Pb（Ⅱ）、Ni（Ⅱ）和 Sr（Ⅱ）在氧化石墨烯上竞争吸附机理与微观结构［C］// 第十四届全国核化学与放射化学学术研讨会．长春，2016：156.

［582］ 高阳，谭小丽，陈长伦．二氧化锆修饰还原氧化石墨烯用于铼的吸附［C］// 第十四届全国核化学与放射化学学术研讨会．长春，2016：157.

［583］ 王驹．高水平放射性废物地质处置：关键科学问题和相关进展［J］. 科技导报，2016, 34（15）：51–55.

［584］ 蒋京呈，王晓丽，蒋美玲，等．利用 CHEMSPEC 模拟计算 Np 和 Pu 在北山地下水中的种态分布及其在水合氧化铁上的吸附［J］. 中国科学：化学，2016, 46（8）：816–822.

［585］ 洪永侠，程瑛，漆明森，等．气载氚监测仪现场校准技术［J］. 核化学与放射化学，2016, 38（1）：38–42.

［586］ 丁有钱，梁小虎，毛国淑，等．核设施排放水中氚的快速自动取样装置的研制［C］// 第十四届全国核化学与放射化学学术研讨会．长春，2016：71.

［587］ 马鹏，张生栋，杨金玲，等．用于溶液中 α 核素在线检测的玻璃闪烁探测器研制［C］// 第十四届全国核

化学与放射化学学术研讨会 . 长春，2016：72.

［588］ 宋志君，张生栋，丁有钱，等 . ^{90}Sr 放化传感器的研制［C］// 第十四届全国核化学与放射化学学术研讨会 . 长春，2016：77.

［589］ Wang C，Wang S，Cai W，et al. Silver mirror for enhancing the detection ability of near-infrared diffuse reflectance spectroscopy. Talanta, 2017,162：123–129.

［590］ Cui X Y, Liu X W, Yu X M, et al. Water can be a probe for sensing glucose in aqueous solutions by temperature dependent near infrared spectra［J］. Analytica Chimica Acta, 2017, 957：47–54.

［591］ Yin X L, Wu H L, Gu H W, et al. Chemometrics-assisted high performance liquid chromatography-diode array detection strategy to solve varying interfering patterns from different chromatographic columns and sample matrices for beverage analysis［J］. Journal of Chromatography A, 2016, 1435：75–84.

［592］ Kang C, Wu H-L, Zhou C, et al. Quantitative fluorescence kinetic analysis of NADH and FAD in human plasma using three- and four-way calibration methods capable of providing the second-order advantage［J］. Analytica Chimica Acta, 2016, 910：36–44.

［593］ Hu L L, Lu J Y, Cheng J D, et al. Structural insight into substrate preference for TET-mediated oxidation［J］. Nature, 2015, 527（7576）：118–122.

［594］ Fu H H, Shao X G, Chipot C, et al. Extended adaptive biasing force algorithm. An on-the-fly implementation for accurate free-energy calculations［J］. Journal of Chemical Theory and Computation, 2016, 12（8）：3506–3513.

［595］ Liu P, Shao X G, Chipot C, et al. The true nature of rotary movements in rotaxanes［J］. Chemical Science, 2016, 7（1）：457–462.

［596］ Sun H Y, Chen P C, Li D, et al. Directly binding rather than induced-fit dominated binding affinity difference in（*S*）- and（*R*）-crizotinib bound MTH1［J］. Journal of Chemical Theory and Computation, 2016, 12（2）：851–860.

［597］ Liu H, Hou T J. CaFE：A tool for binding affinity prediction using end-point free energy methods［J］. Bioinformatics, 2016, 32（14）：2216–2218.

［598］ Rao L, Chi B, Ren Y L, et al. DOX：A new computational protocol for accurate prediction of the protein-ligand binding structures［J］. Journal of Computational Chemistry, 2016, 37（3）：336–344.

［599］ Zhang T T, Li X X, Qiao X J, et al. Initial mechanisms for an overall behavior of lignin pyrolysis through large-scale reaxFF molecular dynamics simulations［J］. Energy & Fuels, 2016, 30（4）：3140–3150.

［600］ Chen K X, Yao J, Chen Z R, et al. Structure-reactivity landscape of N-hydroxyphthalimides with ionic-pair substituents as organocatalysts in aerobic oxidation［J］. Journal of Catalysis, 2015, 331：76–85.

［601］ Zhang H N, Qiao A N, Yang D H, et al. Structure of the full-length glucagon class B G-protein-coupled receptor［J］. Nature, 2017, 546（7657）：259–264.

［602］ Wang J, Luo C, Shan C L, et al. Inhibition of human copper trafficking by a small molecule significantly attenuates cancer cell proliferation［J］. Nature Chemistry, 2015, 7（12）：968–979.

［603］ Jiang H M, Deng R, Yang X Y, et al. Peptidomimetic inhibitors of APC-Asef interaction block colorectal cancer migration［J］. Nature Chemical Biology, 2017, 13（9）：994–1001.

［604］ Wang J J, Zou J X, Xue X Q, et al. ROR-γ drives androgen receptor expression and represents a therapeutic target in castration-resistant prostate cancer［J］. Nature Medicine, 2016, 22（5）：488–496.

［605］ Liu Z H, Su M Y, Han L, et al. Forging the basis for developing protein-ligand interaction scoring functions［J］. Accounts of Chemical Research, 2017, 50（2）：302–309.

［606］ Du J, Yan X, Liu Z, et al. cBinderDB：A covalent binding agent database［J］. Bioinformatics, 2017, 33（8）：1258–1260.

［607］ Shen Q C, Wang G Q, Li S, et al. ASD v3.0：Unraveling allosteric regulation with structural mechanisms and biological networks［J］. Nucleic Acids Research, 2016, 44（D1）：D527–D535.

[608] Wu Z, Cheng F, Li J, et al. SDTNBI: An integrated network and chemoinformatics tool for systematic prediction of drug-target interactions and drug repositioning [J]. Briefings in Bioinformatics, 2016, 18 (2): 333–347.

[609] Zhang Q, Xu Z J, Shi J Y, et al. Underestimated halogen bonds forming with protein backbone in protein data bank [J]. Journal of Chemical Information and Modeling, 2017, 57 (7): 1529–1534.

[610] Zhang Q, Xu Z J, Zhu W L. The underestimated halogen bonds forming with protein side chains in drug discovery and design [J]. Journal of Chemical Information and Modeling, 2017, 57 (1): 22–26.

[611] Zhu L L, Lin H P, Li Y Y, et al. A rhodium/silicon co-electrocatalyst design concept to surpass platinum hydrogen evolution activity at high overpotentials [J]. Nature Communications, 2016, 7: 12272.

[612] Ohmann R, Ono L K, Kim H-S, et al. Real-space imaging of the atomic structure of organic-inorganic perovskite [J]. Journal of the American Chemical Society, 2015, 137 (51): 16049–16054.

[613] Jiang X M, Reiter G, Hu W B. How chain-folding crystal growth determines the thermodynamic stability of polymer crystals [J]. The Journal of Physical Chemistry B, 2016, 120 (3): 566–571.

[614] Zhang R, Zha L Y, Hu W B. Intramolecular crystal nucleation favored by polymer crystallization: Monte carlo simulation evidence [J]. The Journal of Physical Chemistry B, 2016, 120 (27): 6754–6760.

[615] Song Y H, Zheng Q. Concepts and conflicts in nanoparticles reinforcement to polymers beyond hydrodynamics [J]. Progress in Materials Science, 2016, 84: 1–58.

[616] Song Y H, Zheng Q. A Guide for hydrodynamic reinforcement effect in nanoparticle-filled polymers [J]. Critical Reviews in Solid State and Materials Sciences, 2016, 41 (4): 318–346.

[617] Gan S C, Wu Z L, Song Y L, et al. Viscoelastic behaviors of carbon black gel extracted from highly filled natural rubber compounds: Insights into the Payne effect [J]. Macromolecules, 2016, 49 (4): 1454–1463.

[618] Song Y H, Zeng L B, Guan A Z, et al. Time-concentration superpositioning principle accounting for reinforcement and dissipation of multi-walled carbon nanotubes filled polystyrene melts [J]. Polymer, 2017, 121: 106–110.

[619] Song Y, Zeng L, Zheng Q. Reconsideration of the rheology of silica filled natural rubber compounds [J]. Journal of Physics Chemistry B, 2017, 121 (23): 5867–5875.

[620] Song Y, Zeng L, Zheng Q. Unique liquid-to-solid transition of carbon filler filled polystyrene melts [J]. Composites Science and Technology, 2017, 147: 39–44.

[621] Song Y, Tan Y, Zheng Q. Linear rheology of carbon black filled polystyrene [J]. Polymer, 2017, 112: 35–42.

[622] Ma T, Yang R, Zheng Z, et al. Rheology of fumed silica/polydimethylsiloxane suspensions [J]. Journal of Rheology, 2017, 61 (2): 205–215.

[623] Yang R Q, Song Y H, Zheng Q. Payne effect of silica-filled styrene-butadiene rubber [J]. Polymer, 2017, 116: 304–313.

[624] Wang J, Guo Y, Yu W, et al. Linear and nonlinear viscoelasticity of polymer/silica nanocomposites: an understanding from modulus decomposition [J]. Rheology Acta, 2016, 55: 37–50.

[625] Yu W, Wang J, You W. Structure and linear viscoelasticity of polymer nanocomposites with agglomerated particles [J]. Polymer, 2016, 98: 190–200.

[626] You W, Yu W, Zhou C X. Cluster size distribution of spherical nanoparticles in polymer nanocomposites: rheological quantification and evidence of phase separation [J]. Soft Matter, 2017, 13: 4088–4098.

[627] Nie Z J, Yu W, Zhou C X. Nonlinear rheological behavior of multiblock copolymers under large amplitude oscillatory shear [J]. Journal of Rheology, 2016, 60 (6): 1161–1179.

[628] Yang K, Yu W. Dynamic wall slip behavior of yield stress fluids under large amplitude oscillatory shear [J]. Journal of Rheology, 2017, 61 (4): 627–641.

[629] Chen Q, Bao N Q, Wang J H H, et al. Linear viscoelasticity and dielectric spectroscopy of ionomer/plasticizer mixtures: a transition from ionomer to polyelectrolyte [J]. Macromolecules, 2015, 48: 8240–8252.

[630] Zhang Z, Liu C, Cao X, et al. Morphological evolution of ionomer/plasticizer mixtures during a transition from ionomer to polyelectrolyte [J]. Macromolecules, 2017, 50: 963–971.

[631] Zhang Z, Liu C, Cao X, et al. Linear viscoelastic and dielectric properties of strongly hydrogen-bonded polymers near the sol-gel transition [J]. Macromolecules, 2016, 49: 9192–9202.

[632] Chen Q, Zhang Z, Colby R H. Viscoelasticity of entangled random polystyrene ionomers [J]. Journal of Rheology, 2016, 60: 1031–1040.

[633] Li C W, Qiu L, Zhang B Q, et al. Robust vacuum-/air-dried graphene aerogels and fast recoverable shape-memory hybrid foams [J]. Advanced Materials, 2016, 28: 1510–1516.

[634] Liu F Y, Lv Y X, Liu J J, et al. Crystallization and rheology of poly (ethylene oxide) in imidazolium ionic liquids [J]. Macromolecules, 2016, 49: 6106–6115.

[635] Liu J J, Zhang J M, Zhang B Q, et al. Determination of intrinsic viscosity-molecular weight relationship for cellulose in BmimAc/DMSO solutions [J]. Cellulose, 2016, 23: 2341–2348.

[636] Chen W J, Chen W, Zhang B Q, et al. Thermal imidization process of polyimide film: Interplay between solvent evaporation and imidization [J]. Polymer, 2017, 109: 205–215.

[637] Zhang D D, Ruan Y B, Zhang B Q, et al. A self-healing PDMS elastomer based on acylhydrazone groups and the role of hydrogen bonds [J]. Polymer, 2017, 120: 189–196.

[638] Yin J B, Wang X X, Zhao X P. Silicone-grafted carbonaceous nanotubes with enhanced dispersion stability and electrorheological efficiency [J]. Nanotechnology, 2015, 26 (6): 065704.

[639] Li Y Z, Guan Y Q, Liu Y, et al. Highly stable nanofluid based on polyhedral oligomeric silsesquioxane-decorated graphene oxide nanosheets and its enhanced electro-responsive behavior [J]. Nanotechnology, 2016, 27 (19): 195702.

[640] Su Z X, Yin J B, Zhao X P. Terahertz dual-band metamaterial absorber based on graphene/MgF_2 multilayer structures [J]. Optics Express, 2015, 23 (2): 1679–1690.

[641] Wang S, Xuan S H, Liu M, et al. Smart wearable Kevlar-based safeguarding electronic textile with excellent sensing performance [J]. Soft Matter, 2017, 13 (13): 2483–2491.

[642] Wang S, Xuan S H, Jiang W Q, et al. Rate-dependent and self-healing conductive shear stiffening nanocomposite: A novel safe-guarding material with force sensitivity [J]. Journal of Materials Chemistry A, 2015, 3 (39): 19790–19799.

[643] Gong X L, Wang Y, Hu T, et al. Mechanical property and conductivity of a flax fibre weave strengthened magnetorheological elastomer [J]. Smart Materials and Structures, 2017, 26 (7): 075013.

[644] Pang H M, Xuan S H, Liu T X, et al. Magnetic field dependent electro-conductivity of the graphite doped magnetorheological plastomers [J]. Soft Matter, 2015, 11 (34): 6893–6902.

[645] Gong X L, Chen Q, Liu M, et al. Squeeze flow behavior of shear thickening fluid under constant volume [J]. Smart Materials and Structures, 2017, 26 (6): 065017.

[646] Yu M, Qi S, Fu J, et al. Understanding the reinforcing behaviors of polyaniline-modified carbonyl iron particles in magnetorheological elastomer based on polyurethane/epoxy resin IPNs matrix [J]. Composites Science and Technology, 2017, 139: 36–46.

[647] Yang P G, Yu M, Luo H P, et al. Improved rheological properties of dimorphic magnetorheological gels based on flower-like carbonyl iron particles [J]. Applied Surface Science, 2017, 416: 772–780.

[648] Yang P G, Yu M, Fu J, et al. Synthesis and microwave absorption properties of hierarchical Fe micro-sphere assembly by nano-plates [J]. Journal of Alloys and Compounds, 2017, 721: 449–455.

[649] Ma M, Fang B, Lu Y J, et al. Intrinsic rheo-kinetics on gelation process of hydrophobic amphoteric cellulose [J]. Journal of Dispersion Science and Technology, 2016, 37 (8): 1076–1082.

[650] Chen J, Fang B, Yu L C, et al. Interfacial rheological property and rheokinetics of a novel photoreversible micellar system [J]. Journal of Dispersion Science and Technology, 2016, 37 (2): 183–189.

[651] Chen J, Fang B, Jin H, et al. Photorheologically reversible micelle composed of polymerizable cationic surfactant and 4–phenylazo benzoic acid [J]. Chinese Journal of Chemical Engineering, 2016, 24 (2): 289–292.

[652] Yang M, Fang B, Jin H, et al. Rheology and Rheokinetics of photosensitive micelle composed of 3–chloro–2–hydroxypropyl oleyl dimethyl ammonium acetate and trans–4–phenylazo benzoic acid [J]. Journal of Dispersion Science and Technology, 2016, 37 (11): 1655–1663.

[653] Li G D, Fang B, Lu Y J, et al. Intrinsic crosslinking and gel–breaking rheokinetics of CMHEC/CTAB systems [J]. Journal of Dispersion Science and Technology, 2016, 37 (11): 1638–1644.

[654] Li G D, Fang B, Lu Y J, et al. Rheological properties and crosslinking rheo–kinetics of CMHEC/CTAB synergistic systems [J]. Journal of Dispersion Science and Technology, 2016, 37 (12): 1826–1831.

[655] Wu J, Gao W, Liang Y, et al. Spatiotemporal distribution and alpine behavior of short chain chlorinated paraffins in air at Shergyla Mountain and Lhasa on the Tibetan Plateau of China [J]. Environmental Science & Technology, 2017, 51 (19): 11136–11144.

[656] Zhou Z, Shi Y L, Vestergren R, et al. Highly elevated serum concentrations of perfluoroalkyl substances in fishery employees from Tangxun Lake, China [J]. Environmental Science & Technology, 2014, 48 (7): 3864–3874.

[657] Zhang H Y, Wang Y W, Sun C, et al. Levels and distributions of hexachlorobutadiene and three chlorobenzenes in biosolids from wastewater treatment plants and in soils within and surrounding a chemical plant in China [J]. Environmental Science & Technology, 2014, 48 (3): 1525–1531.

[658] Wu Q R, Wang S X, Li G L, et al. Temporal trend and spatial distribution of speciated atmospheric mercury emissions in China during 1978–2014 [J]. Environmental Science & Technology, 2016, 50 (24): 13428–13435.

[659] Hui M L, Wu Q R, Wang S X, et al. Mercury flows in China and global drivers [J]. Environmental Science & Technology, 2017, 51 (1): 222–231.

[660] Guo S, Hu M, Zamora M L, et al. Elucidating severe urban haze formation in China [J]. Proceedings of the National Academy of Sciences of the United States of America, 2014, 111 (49): 17373–17378.

[661] He H, Wang Y S, Ma Q X, et al. Mineral dust and NOx promote the conversion of SO_2 to sulfate in heavy pollution days [J]. Scientific Reports, 2014, 4: 4172.

[662] Wang G H, Zhang R Y, Gomez M E, et al. Persistent sulfate formation from London fog to Chinese haze [J]. Proceedings of the National Academy of Sciences of the United States of America, 2016, 113 (48): 13630–13635.

[663] Ji Y M, Zhao J, Terazono H, et al. Reassessing the atmospheric oxidation mechanism of toluene [J]. Proceedings of the National Academy of Sciences of the United States of America, 2017, 114 (31): 8169–8174.

[664] Yin Y G, Li Y B, Tai C, et al. Fumigant methyl iodide can methylate inorganic mercury species in natural waters [J] . Nature Communications, 2014, 5: 4633.

[665] Wang X, Bao Z D, Lin C J, et al. Assessment of global mercury deposition through litterfall [J]. Environmental Science & Technology, 2016, 50 (16): 8548–8557.

[666] Yu B, Fu X W, Yin R S, et al. Isotopic composition of atmospheric mercury in China: New evidence for sources and transformation processes in air and in vegetation [J]. Environmental Science & Technology, 2016, 50 (17): 9262–9269.

[667] Wang X, Luo J, Yin R S, et al. Using mercury isotopes to understand mercury accumulation in the montane forest floor of the eastern Tibetan Plateau [J]. Environmental Science & Technology, 2017, 51 (2): 801–809.

[668] Lu D W, Liu Q, Zhang T Y, et al. Stable silver isotope fractionation in the natural transformation process of silver nanoparticles [J]. Nature Nanotechnology, 2016, 11 (8): 682–687.

[669] Yao S Y, Zhang X, Zhou W, et al. Atomic–layered Au clusters on α –MoC as catalysts for the low–temperature

water-gas shift reaction [J]. Science, 2017, 357 (6349): 389–393.

[670] Gao P, Li S G, Bu X N, et al. Direct conversion of CO_2 into liquid fuels with high selectivity over a bifunctional catalyst [J]. Nature Chemistry, 2017, 9 (10): 1019–1024.

[671] Qian Q L, Zhang J J, Cui M, et al. Synthesis of acetic acid via methanol hydrocarboxylation with CO_2 and H_2 [J]. Nature Communications, 2016, 7: 11481.

[672] He Z H, Qian Q L, Ma J, et al. Water-enhanced synthesis of higher alcohols from CO_2 hydrogenation over a Pt/Co_3O_4 catalyst under milder conditions [J]. Angewandte Chemie International Edition, 2016, 55 (2): 737–741.

[673] Sun X F, Zhu Q G, Kang X C, et al. Molybdenum–bismuth bimetallic chalcogenide nanosheets for highly efficient electrocatalytic reduction of carbon dioxide to methanol [J]. Angewandte Chemie International Edition, 2016, 55 (23): 6771–6775.

[674] Liu X–F, Li X–Y, Qiao C, et al. Betaine catalysis for hierarchical reduction of CO_2 with amines and hydrosilane to form formamides, aminals, and methylamines [J]. Angewandte Chemie International Edition, 2017, 56 (26): 7425–7429.

[675] Xiong W F, Qi C R, He H T, et al. Base–promoted coupling of carbon dioxide, amines, and N–tosylhydrazones: Anovel and versatile approach to carbamates [J]. Angewandte Chemie International Edition, 2015, 54 (10): 3084–3087.

[676] Xiong W F, Qi C R, Peng Y B, et al. Base–promoted coupling of carbon dioxide, amines, and diaryliodonium salts: Aphosgene– and metal–free route to O–aryl carbamates [J]. Chemistry–A European Journal, 2015, 21 (41): 14314–14318.

[677] Su X, Lin W W, Cheng H Y, et al. Metal–free catalytic conversion of CO_2 and glycerol to glycerol carbonate [J]. Green Chemistry, 2017, 19 (7): 1775–1781.

[678] Ying Z, Dong Y, Wang J, et al. Carbon dioxide as a sustainable resource for macrocyclic oligourea [J]. Green Chemistry, 2016, 18 (8): 2528–2533.

[679] Ying Z, Wu C Y, Jiang S, et al. Synthesis of polyurethane–urea from double CO_2–route oligomers [J]. Green Chemistry, 2016, 18 (12): 3614–3619.

[680] Zhong L S, Yu F, An Y L, et al. Cobalt carbide nanoprisms for direct production of lower olefins from syngas [J]. Nature, 2016, 538 (7623): 84–87.

[681] Zhang K, Ren S H, Hou Y C, et al. Efficient absorption of SO_2 with low–partial pressures by environmentally benign functional deep eutectic solvents [J]. Journal of Hazardous Materials, 2017, 324: 457–463.

[682] Zhang K, Ren S H, Meng L Y, et al. Efficient and reversible absorption of sulfur dioxide of flue gas by environmentally benign and stable quaternary ammonium inner salts in aqueous solutions [J]. Energy & Fuels, 2017, 31 (2): 1786–1792.

[683] Zhang K, Ren S H, Yang X, et al. Efficient absorption of low–concentration SO_2 in simulated flue gas by functional deep eutectic solvents based on imidazole and its derivatives [J]. Chemical Engineering Journal, 2017, 327: 128–134.

[684] Xia Q N, Chen Z J, Shao Y, et al. Direct hydrodeoxygenation of raw woody biomass into liquid alkanes [J]. Nature Communications, 2016, 7: 11162.

[685] Jiang Z C, Yi J, Li J M, et al. Promoting effect of sodium chloride on the solubilization and depolymerization of cellulose from raw biomass materials in water [J]. ChemSusChem, 2015, 8 (11): 1901–1907.

[686] Luo Y P, Yi J, Tong D M, et al. Production of γ–valerolactone via selective catalytic conversion of hemicellulose in pubescens without addition of external hydrogen [J]. Green Chemistry, 2016, 18 (3): 848–857.

[687] He T, Jiang Z C, Wu P, et al. Fractionation for further conversion: From raw corn stover to lactic acid [J]. Scientific Reports, 2016, 6: 38623.

[688] Jiang Z C, Zhang H, He T, et al. Understanding the cleavage of inter– and intramolecular linkages in corncob residue for utilization of lignin to produce monophenols [J]. Green Chemistry, 2016, 18 (14) : 4109–4115.

[689] Jiang Z C, Hu C W. Selective extraction and conversion of lignin in actual biomass to monophenols: A review [J]. Journal of Energy Chemistry, 2016, 25 (6) : 947–956.

[690] Yi J, Luo Y P, He T, et al. High efficient hydrogenation of lignin–Derived monophenols to cyclohexanols over Pd/ γ –Al_2O_3 under mild conditions [J]. Catalysts, 2016, 6 (1) : 12.

[691] Meng Q L, Hou M Q, Liu H Z, et al. Synthesis of ketones from biomass–derived feedstock [J]. Nature Communications, 2017, 8: 14190.

[692] Song J J, Zhou B W, Zhou H C, et al. Porous zirconium–phytic acid hybrid: A highly efficient catalyst for meerwein–ponndorf–verley reductions [J] , Angewandte Chemie International Edition, 2015, 54 (32) : 9399–9403.

[693] Wang K W, Jia Z F, Yang X K, et al. Acid and base coexisted heterogeneous catalysts supported on hypercrosslinked polymers for one–pot cascade reactions [J]. Journal of Catalysis, 2017, 348: 168–176.

[694] Jia Z F, Wang K W, Tan B E, et al. Hollow hyper–cross–linked nanospheres with acid and base sites as efficient and water–stable catalysts for one–pot tandem reactions [J]. ACS Catalysis, 2017, 7 (5) : 3693–3702.

[695] Liu Z–J, Lu X, Wang G, et al. Directing group in decarboxylative cross–coupling: Copper–catalyzed site–selective C–N bond formation from nonactivated aliphatic carboxylic acids [J]. Journal of the American Chemical Society, 2016, 138 (30) : 9714–9719.

[696] Xu H, Hu J–L, Wang L J, et al. Asymmetric annulation of donor–acceptor cyclopropanes with dienes [J]. Journal of the American Chemical Society, 2015, 137 (25) : 8006–8009.

[697] Kang Q–K, Wang L J, Liu Q–J, et al. Asymmetric H_2O–nucleophilic ring opening of D–A cyclopropanes: Catalyst serves as a source of water [J]. Journal of the American Chemical Society, 2015, 137 (46) : 14594–14597.

[698] Liu Q–J, Yan W–G, Wang L J, et al. One–pot catalytic asymmetric synthesis of tetrahydrocarbazoles [J]. Organic Letters, 2015, 17 (16) : 4014–4017.

[699] Zheng M F, Chen P Q, Wu W Q, et al. Palladium–catalyzed Heck–type reaction of oximes with allylic alcohols: Synthesis of pyridines and azafluorenones [J]. Chemical Communications, 2016, 52 (1) : 84–87.

[700] Liu W J, Lindgren I. Going beyond "no–pair relativistic quantum chemistry" [J]. The Journal of Chemical Physics, 2013, 139 (1) : 014108; Liu W J, Lindgren I. Erratum: "Going beyond 'no–pair relativistic quantum chemistry'" [J]. Chem. Phys. 139, 014108(2013)] [J]. The Journal of Chemical Physics, 2016, 144 (4) : 049901.

[701] Liu W J. Advances in relativistic molecular quantum mechanics [J]. Physics Reports, 2014, 537 (2) : 59–89.

[702] Xu E H, Zhao D B, Li S H. Multireference second order perturbation theory with a simplified treatment of dynamical correlation [J]. Journal of Chemical Theory and Computation, 2015, 11 (10) : 4634–4643.

[703] Liu W J, Hoffmann M R. SDS: The 'static–dynamic–static' framework for strongly correlated electrons [J]. Theoretical Chemistry Accounts, 2014, 133: 1481.

[704] Liu W J, Hoffmann M R. iCI: Iterative CI toward full CI [J]. Journal of Chemical Theory and Computation, 2016, 12 (3) : 1169–1178.

[705] Ying F M, Su P F, Chen Z H, et al. DFVB: A density–functional–based valence bond method [J]. Journal of Chemical Theory and Computation, 2012, 8 (5) : 1608–1615.

[706] Su P F, Wu W. *Ab initio* nonorthogonal valence bond methods [J]. Wiley Interdisciplinary Reviews: Computational Molecular Science, 2013, 3 (1) : 56–68.

[707] Li S H, Li W, Ma J. Generalized energy–based fragmentation approach and its applications to macromolecules and molecular aggregates [J]. Accounts of Chemical Research, 2014, 47 (9) : 2712–2720.

[708] Zhang C Y, Yuan D D, Guo Y, et al. Efficient implementation of local excitation approximation for treating excited states of molecules in condensed phase [J]. Journal of Chemical Theory and Computation, 2014, 10 (12) : 5308–

5317.
[709] Li Z D, Li H Y, Suo B B, et al. Localization of molecular orbitals: From fragments to molecule [J]. Accounts of Chemical Research, 2014, 47 (9) : 2758–2767.
[710] Liu J, Zhang Y, Liu W J. Photoexcitation of light-harvesting C–P–C_{60} triads: A FLMO–TD–DFT study [J]. Journal of Chemical Theory and Computation, 2014, 10 (6) : 2436–2448.
[711] Zhang I Y, Xu X, Jung Y, et al. A fast doubly hybrid density functional method close to chemical accuracy: XYGJ–OS [J]. Proceedings of the National Academy of Sciences of the United States of America, 2011, 108 (50) : 19896–19900.
[712] Su N Q, Xu X. Beyond energies: Geometry predictions with the XYG3 type of doubly hybrid density functionals[J]. Chemical Communications, 2016, 52 (96) : 13840–13860.
[713] Su N Q, Xu X. The XYG3 type of doubly hybrid density functionals [J]. Wiley Interdisciplinary Reviews: Computational Molecular Science, 2016, 6 (6) : 721–747.
[714] Liu J, Liang W Z. Analytical approach for the excited-state Hessian in time-dependent density functional theory: Formalism, implementation, and performance [J]. The Journal of Chemical Physics, 2011, 135 (18) : 184111.
[715] Zeng Q, Liu J, Liang W Z. Molecular properties of excited electronic state: Formalism, implementation, and applications of analytical second energy derivatives within the framework of the time-dependent density functional theory/molecular mechanics [J]. The Journal of Chemical Physics, 2014, 140 (18) : 18A506.
[716] Ma H L, Liu J, Liang W Z. Time-dependent approach to resonance Raman spectra including duschinsky rotation and herzberg-teller effects: Formalism and its realistic applications [J]. Journal of Chemical Theory and Computation, 2012, 8 (11) : 4474–4482.
[717] Xiao C L, Xu X, Liu S, et al. Experimental and theoretical differential cross sections for a four-atom reaction: HD+OH → H_2O+D [J]. Science, 2011, 333 (6041) : 440–442.
[718] Zhang Z J, Liu T H, Fu B N, et al. First-principles quantum dynamical theory for the dissociative chemisorption of H_2O on rigid Cu (111) [J]. Nature Communications, 2016, 7: 11953.
[719] Wang T, Chen J, Yang T G, et al. Dynamical resonances accessible only by reagent vibrational excitation in the F+HD → HF+D reaction [J]. Science, 2013, 342 (6165) : 1499–1502.
[720] Yang T G, Chen J, Huang L, et al. Extremely short-lived reaction resonances in Cl + HD (v=1) → DCl + H due to chemical bond softening [J]. Science, 2015, 347 (6217) : 60–63.
[721] Wang Y T, Liu X Y, Cui G L, et al. Photoisomerization of arylazopyrazole photoswitches: Stereospecific excited-state relaxation [J]. Angewandte Chemie International Edition, 2016, 55 (45) : 14009–14013.
[722] Xia S H, Cui G L, Fang W H, et al. How photoisomerization drives peptide folding and unfolding: Insights from QM/MM and MM dynamics simulations [J]. Angewandte Chemie International Edition, 2016, 55 (6) : 2067–2072.
[723] Cui G L, Cao X Y, Fang W H, et al. Photoinduced Gold (I) –Gold (I) chemical bonding in dicyanoaurate oligomers [J]. Angewandte Chemie International Edition, 2013, 52 (39) : 10281–10285.
[724] Ye L Z, Wang X L, Hou D, et al. HEOM–QUICK: A program for accurate, efficient and universal characterization of strongly correlated quantum impurity systems [J]. Wiley Interdisciplinary Reviews: Computational Molecular Science, 2016, 6 (6) : 608–638.
[725] Yan Y J. Theory of open quantum systems with bath of electrons and phonons and spins: Many-dissipaton density matrixes approach [J]. The Journal of Chemical Physics, 2014, 140 (5) : 054105.
[726] Yan Y J, Jin J S, Xu R X, et al. Dissipation equation of motion approach to open quantum systems [J]. Frontiers of Physics, 2016, 11: 110306.
[727] Liu H, Zhu L L, Bai S M, et al. Reduced quantum dynamics with arbitrary bath spectral densities: Hierarchical equations of motion based on several different bath decomposition schemes [J]. The Journal of Chemical Physics,

2014, 140 (13): 134106.
[728] Chen L P, Zheng R H, Jing Y Y, et al. Simulation of the two-dimensional electronic spectra of the Fenna-Matthews-Olson complex using the hierarchical equations of motion method [J]. The Journal of Chemical Physics, 2011, 134 (19): 194508.
[729] Yang L J, Liu C W, Shao Q, et al. From thermodynamics to kinetics: Enhanced sampling of rare events [J]. Accounts of Chemical Research, 2015, 48 (4): 947-955.
[730] Zhang J, Yang Y I, Yang L J, et al. Dynamics and kinetics study of "in-water" chemical reactions by enhanced sampling of reactive trajectories [J]. The Journal of Physical Chemistry B, 2015, 119 (45): 14505-14514.
[731] Xie W J, Gao Y Q. A simple theory for the hofmeister series[J]. The Journal of Physical Chemistry Letters, 2013, 4: 4247-4252.
[732] Yue L, Liu Y J, Fang W H. Mechanistic insight into the chemiluminescent decomposition of firefly dioxetanone [J]. Journal of the American Chemical Society, 2012, 134 (28): 11632-11639.
[733] Yue L, Lan Z G, Liu Y J. The theoretical estimation of the bioluminescent efficiency of the firefly via a nonadiabatic molecular dynamics simulation [J]. The Journal of Physical Chemistry Letters, 2015, 6 (3): 540-548.
[734] Chen S F, Ferré N, Liu Y J. QM/MM study on the light emitters of aequorin chemiluminescence, bioluminescence, and fluorescence: A general understanding of the bioluminescence of several marine organisms [J]. Chemistry - A European Journal, 2013, 19 (26): 8466-8472.
[735] Luo Y L, Liu Y J. Bioluminophore and Flavin Mononucleotide fluorescence quenching of bacterial bioluminescence - A theoretical study [J]. Chemistry - A European Journal, 2016, 22 (45): 16243-16249.
[736] Wu P, Jiang H J, Zhang W H, et al. Lattice mismatch induced nonlinear growth of graphene [J]. Journal of the American Chemical Society, 2012, 134 (13): 6045-6051.
[737] Wu P, Zhang Y, Cui P, et al. Carbon dimers as the dominant feeding species in epitaxial growth and morphological phase transition of graphene on different Cu substrates [J]. Physical Review Letters, 2015, 114 (21): 216102.
[738] Qiu Z Y, Song L, Zhao J, et al. The nanoparticle size effect in graphene cutting: A "pac-man" mechanism [J]. Angewandte Chemie International Edition, 2016, 55 (34): 9918-9921.
[739] Shuai Z G, Peng Q. Excited states structure and processes: Understanding organic light-emitting diodes at the molecular level [J]. Physics Reports, 2014, 537 (4): 123-156.
[740] Geng H, Peng Q, Wang L J, et al. Toward quantitative prediction of charge mobility in organic semiconductors: Tunneling enabled hopping model [J]. Advanced Materials, 2012, 24 (26): 3568-3572.
[741] Shuai Z G, Wang D, Peng Q, et al. Computational evaluation of optoelectronic properties for organic/carbon materials [J]. Accounts of Chemical Research, 2014, 47 (11): 3301-3309.
[742] Long M Q, Tang L, Wang D, et al. Electronic structure and carrier mobility in graphdiyne sheet and nanoribbons: Theoretical predictions [J]. ACS Nano, 2011, 5 (4): 2593-2600.
[743] Gao Q, Xu J, Cao D P, et al. A rigid nested metal-organic framework featuring a thermoresponsive gating effect dominated by counterions [J]. Angewandet Chemie International Edition, 2016, 55 (48): 15027-15030.
[744] Zhang Y J, Chen C, Tan B, et al. A dual-stimuli responsive small molecule organic material with tunable multi-state response showing turn-on luminescence and photocoloration [J]. Chemical Communications, 2016, 52 (13): 2835-2838.
[745] Li H Y, Xu H, Zang S Q, et al. A viologen-functionalized chiral Eu-MOF as a platform for multifunctional switchable material [J]. Chemical Communications, 2016, 52 (3): 525-528.
[746] Sun Y Q, Wan F, Li X X, et al. A lanthanide complex for metal encapsulations and anion exchanges [J]. Chemical Communications, 2016, 52 (66): 10125-10128.
[747] Sui Q, Ren X T, Dai Y X, et al. Piezochromism and hydrochromism through electron transfer: New stories for

viologen materials [J]. Chemical Science, 2017, 8 (4): 2758–2768.

[748] Zhang H, Wu X T. Calix-like metal-organic complex for high-sensitivity X-Ray-induced photochromism [J]. Advanced Science, 2016, 3 (4): 1500224.

[749] Tu B B, Pang Q Q, Ning E L, et al. Heterogeneity within a mesoporous metal-organic framework with three distinct metal-containing building units [J]. Journal of the American Chemical Society, 2015, 137 (42): 13456–13459.

[750] Xu H Q, Wang K C, Ding M L, et al. Seed-mediated synthesis of metal-organic frameworks [J]. Journal of the American Chemical Society, 2016, 138 (16): 5316–5320.

[751] Zhang L, Lin L, Liu D, et al. Stacking interactions induced selective conformation of discrete aromatic arrays and borromean rings [J]. Journal of the American Chemical Society, 2017, 139 (4): 1653–1660.

[752] Dong J Q, Tan C X, Zhang K, et al. Chiral NH-controlled supramolecular metallacycles [J]. Journal of the American Chemical Society, 2017, 139 (4): 1554–1564.

[753] Deng Y K, Su H F, Xu J H, et al. Hierarchical assembly of a $\{Mn^{II}_{15}Mn^{III}_{4}\}$ brucite disc: Step-by-step formation and ferrimagnetism [J]. Journal of the American Chemical Society, 2016, 138 (4): 1328–1334.

[754] Xia Q C, Liu Y, Li Z J, et al. A Cr (salen) -based metal-organic framework as a versatile catalyst for efficient asymmetric transformations [J]. Chemical Communications, 2016, 52 (89): 13167–13170.

[755] Dong Y, Li Y, Wei Y L, et al. A N-heterocyclic tetracarbene Pd (II) moiety containing a Pd (II) -Pb (II) bimetallic MOF for three-component cyclotrimerization via benzyne [J]. Chemical Communications, 2016, 52 (69): 10505–10508.

[756] Xu H, Liu X F, Cao C S, et al. A porous metal organic framework assembled by [Cu_{30}] nanocages: Serving as recyclable catalysts for CO_2 fixation with aziridines [J]. Advanced Science, 2016, 3: 1600048.

[757] He H J, Ma E, Cui Y J, et al. Polarized three-photon-pumped laser in a single MOF microcrystal [J]. Nature Communications, 2016, 7: 11087.

[758] Cui X, Chen K J, Xing H B, et al. Pore chemistry and size control in hybrid porous materials for acetylene capture from ethylene [J]. Science, 2016, 353 (6295): 141–144.

[759] Gu Z G, Li D J, Zheng C, et al. MOF-templated synthesis of ultrasmall photoluminescent carbon-nanodot arrays for optical applications [J]. Angewandet Chemie International Edition, 2017, 56 (24): 6853–6858.

[760] Pang Z F, Xu S Q, Zhou T Y, et al. Construction of covalent organic frameworks bearing three different kinds of pores through the heterostructural mixed linker strategy [J]. Journal of the American Chemical Society, 2016, 138 (14): 4710–4713.

[761] Hu F L, Liu C P, Wu M Y, et al. An ultrastable and easily regenerated hydrogen - bonded organic molecular framework with permanent porosity [J]. Angewandet Chemie International Edition, 2017, 56 (8): 2101–2104.

[762] Popov I A, Pan F X, You X R, et al. Peculiar all-metal σ-aromaticity of the [Au_2Sb_{16}]$^{4-}$ anion in the solid [J]. Angewandet Chemie International Edition, 2016, 55 (49): 15344–15346.

[763] Min X, Popov I A, Pan F X, et al. All-metal antiaromaticity in Sb_4-type lanthanocene anions [J]. Angewandet Chemie International Edition, 2016, 55 (18): 5531–5535.

[764] Pan F X, Li L J, Wang Y J, et al. An all-metal aromatic sandwich complex [$Sb_3Au_3Sb_3$]$^{3-}$ [J]. Journal of the American Chemical Society, 2015, 137 (34): 10954–10957.

[765] Feng G D, Cheng P, Yan W F, et al. Accelerated crystallization of zeolites via hydroxyl free radicals [J]. Science, 2016, 351 (6278): 1188–1191.

[766] Kong B, Sun X T, Selomulya C, et al. Sub-5 nm porous nanocrystals: Interfacial site-directed growth on graphene for efficient biocatalysis [J]. Chemical Science, 2015, 6 (7): 4029–4034.

[767] Luo W, Zhao T, Li Y H, et al. A Micelle fusion-aggregation assembly approach to mesoporous carbon materials with rich active sites for ultrasensitive ammonia sensing [J]. Journal of the American Chemical Society, 2016, 138 (38):

12586-12595.

[768] Liao P-Q, Huang N-Y, Zhang W-X, et al. Controlling guest conformation for efficient purification of butadiene [J]. Science, 2017, 356 (6343): 1193-1196.

[769] Xue H, Chen Q H, Jiang F L, et al. A regenerative metal-organic framework for reversible uptake of Cd (II): From effective adsorption to in situ detection [J]. Chemical Science, 2016, 7 (9): 5983-5988.

[770] Lu C J, Ben T, Xu S X, et al. Electrochemical synthesis of a microporous conductive polymer based on a metal-organic framework thin film [J]. Angewandte Chemie International Edition, 2014, 53 (25): 6454-6458.

[771] Fu J R, Das S, Xing G L, et al. Fabrication of COF-MOF composite membranes and their highly selective separation of H_2/CO_2 [J]. Journal of the American Chemical Society, 2016, 138 (24): 7673-7680.

[772] Li H, Pan Q Y, Ma Y C, et al. Three-dimensional covalent organic frameworks with dual linkages for bifunctional cascade catalysis [J]. Journal of the American Chemical Society, 2016, 138 (44): 14783-14788.

[773] Fang Q R, Wang J H, Gu S, et al. 3D porous crystalline polyimide covalent organic frameworks for drug delivery[J]. Journal of the American Chemical Society, 2015, 137 (26): 8352-8355.

[774] Cui T L, Ke W Y, Zhang W B, et al. Encapsulating palladium nanoparticles inside mesoporous MFI zeolite nanocrystals for shape-selective catalysis [J]. Angewandte Chemie International Edition, 2016, 55 (32): 9178-9182.

[775] Gao Q, Xu J, Cao D P, et al. A rigid nested metal-organic framework featuring a thermoresponsive gating effect dominated by counterions [J]. Angewandte Chemie International Edition, 2016, 55 (48): 15027-15030.

[776] Wang H Y, Ge J Y, Hua C, et al. Photo- and electronically switchable spin-crossover iron (II) metal-organic frameworks based on a tetrathiafulvalene ligand [J]. Angewandte Chemie International Edition, 2017, 56 (20): 5465-5470.

[777] Wang S G, Chen Y, Li X, et al. Injectable 2D MoS_2-integrated drug delivering implant for highly efficient NIR-triggered synergistic tumor hyperthermia [J]. Advanced Materials, 2015, 27 (44): 7117-7122.

[778] Cui X L, Chen K J, Xing H B, et al. Pore chemistry and size control in hybrid porous materials for acetylene capture from ethylene [J]. Science, 2016, 353 (6295): 141-144.

[779] Peng Y, Li Y S, Ban Y J, et al. Metal-organic framework nanosheets as building blocks for molecular sieving membranes [J]. Science, 2014, 346 (6215): 1356-1359.

[780] Zhao M T, Yuan K, Wang Y, et al. Metal-organic frameworks as selectivity regulators for hydrogenation reactions [J]. Nature, 2016, 539 (7627): 76-80.

[781] Huang R W, Wei Y S, Dong X Y, et al. Hypersensitive dual-function luminescence switching of a silver-chalcogenolate cluster-based metal-organic framework [J]. Nature Chemistry, 2017, 9 (7): 689-697.

[782] Yuan Y, Sun F X, Li L N, et al. Porous aromatic frameworks with anion-templated pore apertures serving as polymeric sieves [J]. Nature Communications, 2014, 5: 4260.

[783] Chen C X, Wei Z W, Jiang J J, et al. Precise modulation of the breathing behavior and pore surface in ZR-MOFs by reversible post-synthetic variable-spacer installation to fine-tune the expansion magnitude and sorption properties [J]. Angewandte Chemie International Edition, 2016, 55 (34): 9932-9936.

[784] Xu D D, Ma Y H, Jing Z F, et al. $\pi-\pi$ interaction of aromatic groups in amphiphilic molecules directing for single-crystalline mesostructured zeolite nanosheets [J]. Nature Communications, 2014, 5: 4262.

[785] Chen J L, Liang T Y, Li J F, et al. Regulation of framework aluminum siting and acid distribution in H-MCM-22 by boron incorporation and its effect on the catalytic performance in methanol to hydrocarbons [J]. ACS Catalysis, 2016, 6 (4): 2299-2313.

[786] [24] He H J, Ma E, Cui Y J, et al. Polarized three-photon-pumped laser in a single MOF microcrystal [J]. Nature Communications, 2016, 7: 11087.

［787］［25］Wei J, Ge Q J, Yao R W, et al. Directly converting CO_2 into a gasoline fuel［J］. Nature Communications, 2017, 8: 15174.

［788］Gao P, Li S G, Bu X N, et al. Direct conversion of CO_2 into liquid fuels with high selectivity over a bifunctional catalyst［J］. Nature Chemistry, 2017, 9（10）: 1019–1024.

［789］Wang L, Wang G X, Zhang J, et al. Controllable cyanation of carbon–hydrogen bonds by zeolite crystals over manganese oxide catalyst［J］. Nature Communications, 2017, 8: 15240.

［790］Yang S L, Peng L, Huang P P, et al. Nitrogen, phosphorus, and sulfur Co–doped hollow carbon shell as superior metal–free catalyst for selective oxidation of aromatic alkanes［J］. Angewandte Chemie International Edition, 2016, 55（12）: 4016–4020.

［791］Wang T L, Hong T T, Huang Y, et al. Fluorescein derivatives as bifunctional molecules for the simultaneous inhibiting and labeling of FTO protein［J］. Journal of The American Chemical Society, 2015, 137（43）: 13736–13739.

［792］Chu T T, Gao N, Li Q Q, et al. Specific knockdown of endogenous tau protein by peptide–directed ubiquitin–proteasome degradation［J］. Cell Chemical Biology, 2016, 23（4）: 453–461.

［793］Chen H C, Tian J W, He W J, et al. H_2O_2–activatable and O_2–evolving nanoparticles for highly efficient and selective photodynamic therapy against hypoxic tumor cells［J］. Journal of the American Chemical Society, 2015, 137（4）: 1539–1547.

［794］Chen X P, Zhao C Y, Li X L, et al. Terazosin activates Pgk1 and Hsp90 to promote stress resistance［J］. Nature Chemical Biology, 2015, 11（1）: 19–25.

［795］Zhou Y Q, Li W C, Xiao Y L. Profiling of multiple targets of artemisinin activated by hemin in cancer cell proteome［J］. ACS Chemical Biology, 2016, 11（4）: 882–888.

［796］Li X, Tao R R, Hong L J, et al. Visualizing peroxynitrite fluxes in endothelial cells reveals the dynamic progression of brain vascular injury［J］. Journal of the American Chemical Society, 2015, 137（38）: 12296–12303.

［797］Wang S R, Zhang Q Y, Wang J Q, et al. Chemical targeting of a G–quadruplex RNA in the ebola virus L gene［J］. Cell Chemical Biology, 2016, 23(9): 1113–1122; Wang S R, Min Y Q, Wang J Q, et al. A highly conserved G–rich consensus sequence in hepatitis C virus core gene represents a new anti–hepatitis C target［J］. Science Advances, 2016, 2（4）: e1501535.

［798］Tian T, Song Y Y, Wang J Q, et al. Small–molecule–triggered and light–controlled reversible regulation of enzymatic activity［J］. Journal of the American Chemical Society, 2016, 138（3）: 955–961.

［799］Zhang Y Q, Zhou X P, Xie Y H, et al. Thiol specific and tracelessly removable bioconjugation via michael addition to 5–methylene pyrrolones［J］. Journal of the American Chemical Society, 2017, 139（17）: 6146–6151.

［800］Zhao Q F, Wang M, Xu D X, et al. Metabolic coupling of two small–molecule thiols programs the biosynthesis of lincomycin A［J］. Nature, 2015, 518（7537）: 115–119.

［801］Fage C D, Isiorho E A, Liu Y N, et al. The structure of SpnF, a standalone enzyme that catalyzes［4 + 2］cycloaddition［J］. Nature Chemical Biology, 2015, 11（4）: 256–258.

［802］Zhu Z W, Zhou Y J, Krivoruchko A, et al. Expanding the product portfolio of fungal type I fatty acid synthases［J］. Nature Chemical Biology, 2017, 13（4）: 360–362.

［803］Li J, Chen P R. Development and application of bond cleavage reactions in bioorthogonal chemistry［J］. Nature Chemical Biology, 2016, 12（3）: 129–137; Zhang G, Li J, Xie R, et al. Bioorthogonal chemical activation of kinases in living systems［J］. ACS Central Science, 2016, 2（5）: 325–331.

［804］Lin L, Liu L, Zhao B, et al. Carbon nanotube–assisted optical activation of TGF–β signalling by near–infrared light［J］. Nature Nanotechnology, 2015, 10（5）: 465–471.

［805］Xie R, Dong L, Du Y F, et al. In vivo metabolic labeling of sialoglycans in the mouse brain by using a liposome–

assisted bioorthogonal reporter strategy [J]. Proceedings of the National Academy of Sciences of the United States of America, 2016, 113 (19): 5173–5178.

[806] Gui Y, Qiu L Q, Li Y H, et al. Internal activation of peptidyl prolyl thioesters in native chemical ligation [J]. Journal of the American Chemical Society, 2016, 138 (14): 4890–4899.

[807] Chen X, Tang S, Zheng J S, et al. Chemical synthesis of a two-photon-activatable chemokine and photon-guided lymphocyte migration in vivo [J]. Nature Communications, 2015, 6: 7220.

[808] Tang S, Wan Z P, Gao Y R, et al. Total chemical synthesis of photoactivatable proteins for light-controlled manipulation of antigen-antibody interactions [J]. Chemical Science, 2016, 7 (3): 1891–1895; Liu K, Ding L G, Li Y L, et al. Neuronal necrosis is regulated by a conserved chromatin-modifying cascade [J]. Proceedings of the National Academy of Sciences of the United States of America, 2015, 111 (38): 13960–13965.

[809] Li F H, Dong J S, Hu X S, et al. A covalent approach for site-specific RNA labeling in mammalian cells [J]. Angewandte Chemie International Edition, 2015, 54 (15): 4597–4602.

[810] Zhang K, Zhang J, Xi Z, et al. A new H_2S-specific near-infrared fluorescence-enhanced probe that can visualize the H_2S level in colorectal cancer cells in mice [J]. Chemical Science, 2017, 8 (4): 2776–2781.

[811] Li J B, Huang L, Xiao X, et al. Photoclickable microRNA for the intracellular target identification of microRNAs[J]. Journal of the American Chemical Society, 2016, 138 (49): 15943–15949.

[812] Zhang W, Yuan J H, Yang Y, et al. Monomeric type I and type III transforming growth factor-β receptors and their dimerization revealed by single-molecule imaging [J]. Cell Research, 2010, 20 (11): 1216–1223; He K M, Yan X H, Li N, et al. Internalization of the TGF-β type I receptor into caveolin-1 and EEA1 double-positive early endosomes [J]. Cell Research, 2015, 25 (6): 738–752.

[813] Huang P C, Wu F Y, Mao L Q. Target-triggered switching on and off the luminescence of lanthanide coordination polymer nanoparticles for selective and sensitive sensing of copper ions in rat brain [J]. Analytical Chemistry, 2015, 87 (13): 6834–6841.

[814] Zhu Z, Guan Z C, Liu D, et al. Translating molecular recognition into a pressure signal to enable rapid, sensitive, and portable biomedical analysis [J]. Angewandte Chemie International Edition, 2015, 54 (36): 10448–10453.

[815] Fu Y S, Li C M, Lu S J, et al. Uniform and accurate single-cell sequencing based on emulsion whole-genome amplification[J]. Proceedings of the National Academy of Sciences of the United States of America, 2015, 112(38): 11923–11928.

[816] Chen Y L, Ding L, Song W Y, et al. Liberation of protein-specific glycosylation information for glycan analysis by exonuclease III-aided recycling hybridization[J]. Analytical Chemistry, 2016, 88 (5): 2923–2928. Wu N, Bao L, Ding L, et al. A single excitation-duplexed imaging strategy for profiling cell surface protein-specific glycoforms[J]. Angewandte Chemie International Edition, 2016, 55 (17): 5220–5224.

[817] Bao L, Ding L, Yang M, et al. Noninvasive imaging of sialyltransferase activity in living cells by chemoselective recognition [J]. Scientific Reports, 2015, 5: 10947.

[818] Zhang H L, Chao J, Pan D, et al. DNA origami-based shape IDs for single-molecule nanomechanical genotyping[J]. Nature Communications, 2017, 8: 14738.

[819] Chen H, Lin L, Li H F, et al. Aggregation-induced structure transition of protein-stabilized zinc/copper nanoclusters for amplified chemiluminescence [J]. ACS Nano, 2015, 9 (2): 2173–2183.

[820] Chen J L, Wang X, Yang F, et al. 3D structure determination of an unstable transient enzyme intermediate by paramagnetic NMR spectroscopy [J]. Angewandte Chemie International Edition, 2016, 55 (44): 13744–13748.

[821] Yang Y, Yang F, Gong Y J, et al. A reactive, rigid GdIII labeling tag for in-cell EPR distance measurements in proteins [J]. Angewandte Chemie International Edition, 2017, 56 (11): 2914–2918.

[822] Cao C, Wang S Q, Cui T X, et al. Ion and inhibitor binding of the double-ring ion selectivity filter of the

mitochondrial calcium uniporter [J]. Proceedings of the National Academy of Sciences of the United States of America, 2017, 114 (14): E2846–E2851.

[823] Wang J S, Zhao C Q, Zhao A D, et al. New insights in amyloid beta interactions with human telomerase [J]. Journal of the American Chemical Society, 2015, 137 (3): 1213–1219.

[824] Pu F, Wu L, Ran X, et al. G–quartet–based nanostructure for mimicking light–harvesting antenna [J]. Angewandte Chemie International Edition, 2015, 54 (3): 892–896; Sun H J, Zhao A D, Gao N, et al. Deciphering a nanocarbon–based artificial peroxidase: Chemical identification of the catalytically active and substrate–binding sites on graphene quantum dots [J]. Angewandte Chemie International Edition, 2015, 54 (24): 7176–7180; Chen Z W, Ji H W, Liu C Q, et al. A multinuclear metal complex based Dnase–mimetic artificial enzyme: Matrix cleavage for combating bacterial biofilms [J]. Angewandte Chemie International Edition, 2016, 55 (36): 10732–10736; Li W, Dong K, Ren J S, et al. A β–lactamase–imprinted responsive hydrogel for the treatment of antibiotic–resistant bacteria [J]. Angewandte Chemie International Edition, 2016, 55 (28): 8049–8053; Franz M, Januszewski J A, Wendinger D, et al. Cumulene rotaxanes: Stabilization and study of [9] cumulenes [J]. Angewandte Chemie International Edition, 2015, 54 (22): 6645–6649.

[825] Li W, Chen Z W, Zhou L, et al. Noninvasive and reversible cell adhesion and detachment via single–wavelength near–infrared laser mediated photoisomerization [J]. Journal of the American Chemical Society, 2015, 137 (25): 8199–8205; Shi P, Ju E G, Yan Z Q, et al. Spatiotemporal control of cell–cell reversible interactions using molecular engineering [J]. Nature Communications, 2016, 7: 13088.

[826] Wei W, Sun Y, Zhu M L, et al. Structural insights and the surprisingly low mechanical stability of the Au–S bond in the gold–specific protein GolB [J]. Journal of the American Chemical Society, 2015, 137 (49): 15358–15361.

[827] Chen W Z, Zhang Y F, Yeo W S, et al. Rapid and efficient genome editing in staphylococcus aureus by using an engineered CRISPR/Cas9 system [J]. Journal of the American Chemical Society, 2017, 139 (10): 3790–3795.

[828] Chu B B, Liao Y C, Qi W, et al. Cholesterol transport through lysosome–peroxisome membrane contacts [J]. Cell, 2015, 161 (2): 291–306.

[829] Mo F, Zhuang X X, Liu X, et al. Acetylation of aurora B by TIP60 ensures accurate chromosomal segregation [J]. Nature Chemical Biology, 2016, 12 (4): 226–232.

[830] Li L, Liu Y, Chen H Z, et al. Impeding the interaction between Nur77 and p38 reduces LPS–induced inflammation[J]. Nature Chemical Biology, 2015, 11 (5): 339–346.

[831] Shan B, Wang X X, Wu Y, et al; The metaolic ER stress sensor IRE1 α suppresses alternative activation of macrophages and impairs energy expenditure in obesity [J]. Nature Immunology, 2017, 18: 519–529.

[832] Yan W P, Song H, Song F H, et al. Endoperoxide formation by an α–ketoglutarate–dependent mononuclear non–haem iron enzyme [J]. Nature, 2015, 527 (7579): 539–542.

[833] Bai C X, Zhang Y, Zhao X J, et al. Exploiting a precise design of universal synthetic modular regulatory elements to unlock the microbial natural products in Streptomyces [J]. Proceedings of the National Academy of Sciences of the United States of America, 2015, 112 (39): 12181–12186.

[834] Yue J P, Zhang Y, Liang W G, et al. In vivo epidermal migration requires focal adhesion targeting of ACF7 [J]. Nature Communications, 2016, 7: 11692.

[835] Xie Z X, Li B Z, Mitchell L A, et al. "Perfect" designer chromosome V and behavior of a ring derivative [J]. Science, 2017, 355 (6329): eaaf4704; Wu Y, Li B Z, Zhao M, et al. Bug mapping and fitness testing of chemically synthesized chromosome X [J]. Science, 2017, 355 (6329): eaaf4706.

[836] Wang J, Luo C, Shan C L, et al. Inhibition of human copper trafficking by a small molecule significantly attenuates cancer cell proliferation [J]. Nature Chemistry, 2015, 7: 968–979.

[837] Guo Z Q, Zheng T, Chen B E, et al. Small–molecule targeting of E3 ligase adaptor SPOP in kidney cancer [J].

Cancer Cell, 2016, 30 (3) : 474–484.

[838] Chen F F, Di H X, Wang Y X, et al. Small-molecule targeting of a diapophytoene desaturase inhibits S. aureus virulence [J]. Nature Chemical Biology, 2016, 12 (3) : 174–179.

[839] Huang Y, Yan J L, Li Q, et al. Meclofenamic acid selectively inhibits FTO demethylation of m^6A over ALKBH5 [J]. Nucleic Acids Research, 2015, 43 (1) : 373–384.

[840] Wang Y Z, Wang J J, Huang Y, et al. Tissue acidosis induces neuronal necroptosis via ASIC1a channel independent of its ionic conduction [J]. eLife, 2015, 4: e05682.

[841] Zhou T T, Quan L L, Chen L P, et al. SP6616 as a new Kv2.1 channel inhibitor efficiently promotes β -cell survival involving both PKC/Erk1/2 and CaM/PI3K/Akt signaling pathways [J]. Cell Death & Disease, 2016, 7: e2216.

[842] Shen Q C, Cheng F X, Song H L, et al. Proteome-scale investigation of protein allosteric regulation perturbed by somatic mutations in 7,000 cancer genomes [J]. The American Journal of Human Genetics, 2017, 100 (1) : 5–20.

[843] Sessi P, Di Sante D, Szczerbakow A, et al. Robust spin-polarized midgap states at step edges of topological crystalline insulators [J]. Science, 2016, 354 (6317) : 1269–1273.

[844] Sun X M, Pan Q, Yuan C H, et al. A single ssDNA aptamer binding to mannose-capped lipoarabinomannan of bacillus calmette-guérin enhances immunoprotective effect against tuberculosis [J]. Journal of the American Chemical Society, 2016, 138 (36) : 11680–11689.

[845] Hao G F, Jiang W, Ye Y N, et al. ACFIS: A web server for fragment-based drug discovery [J]. Nucleic Acids Research, 2016, 44 (W1) : W550–W556.

[846] Gao S, Lin Y, Jiao X C, et al. Partially oxidized atomic cobalt layers for carbon dioxide electroreduction to liquid fuel [J]. Nature, 2016, 529 (7584) : 68–71.

[847] Zhao M T, Yuan K, Wang Y, et al. Metal-organic frameworks as selectivity regulators for hydrogenation reactions [J]. Nature, 2016, 539 (7627) : 76–80.

[848] Jiao F, Li J J, Pan X L, et al. Selective conversion of syngas to light olefins [J]. Science, 2016, 351 (6277) : 1065–1068.

[849] Zhong L S, Yu F, An Y L, et al. Cobalt carbide nanoprisms for direct production of lower olefins from syngas [J]. Nature,2016,538: 84–87.

[850] Liu P X, Zhao Y, Qin R X, et al. Photochemical route for synthesizing atomically dispersed palladium catalysts [J]. Science, 2016, 352 (6287) : 797–801.

[851] 中华人民共和国国务院 . 国务院关于深化考试招生制度改革的实施意见 [Z]. 2014-09-04.

[852] 刘前树 . 试论化学核心素养的结构 [J]. 化学教育 , 2016, 37 (21) : 4–8.

[853] 朱鹏飞 , 徐惠 . 核心素养的研究进展及对化学核心素养构建的启示 [J]. 化学教学 , 2016,(7) : 3–7.

[854] 吴俊明 . 关于核心素养及化学学科核心素养的思考与疑问 [J]. 化学教学 , 2016,(11) : 3–8, 23.

[855] 周业虹 . 基于发展化学学科核心素养的教学设计案例分析 [J]. 化学教学 , 2016,(8) : 36–39.

[856] 王云生 . 探索课堂学习活动设计 落实核心素养培养要求 [J]. 化学教学 , 2016,(9) : 3–6.

[857] Wan Y L, Bi H L. Representation and analysis of chemistry core ideas in science education standards between china and the united states [J]. Journal of Chemical Education, 2016, 93 (1) : 70–78.

[858] Lu S S, Bi H L. Development of a measurement instrument to assess students' electrolyte conceptual understanding [J]. Chemistry Education Research and Practice, 2016, 17 (4) : 1030–1040.

[859] He P, Liu X F, Zheng C L, et al. Using rasch measurement to validate an instrument for measuring the quality of classroom teaching in secondary chemistry lessons[J]. Chemistry Education Research and Practice, 2016, 17(2) : 381–393.

[860] Zhou Q, Wang T T, Zheng Q. Probing high school students' cognitive structures and key areas of learning difficulties on ethanoic acid using the flow map method [J]. Chemistry Education Research and Practice, 2015, 16 (3) :

589-602.

[861] Wang L, Zhu Y J, Jiang Y X, et al. Science education research in mainland china [M] //Chiu M H. Science Education Research and Practice in Asia. Singapore: Springer, 2016: 17-39.

[862] Wang L, Cheung D, Chiu M H, et al. Pre-service education of high school science education [M] //Chiu M H. Science Education Research and Practice in East Asia: Trends and Perspectives. Singapore: Springer, 2016.

田伟生 郝临晓 朱玉军

专题报告

非线性光学晶体材料研究进展

一、引言

非线性光学（NLO）晶体材料涉及激光技术的各个领域，现已成为激光变频、电光调制和光折变晶体记忆和存储等技术必不可少的晶体材料，其中又以二阶 NLO 晶体材料用途居多。几十年来，我国科学家做出了 β-BaB_2O_4（BBO）[1]、LiB_3O_5（LBO）[2]、和 $KBe_2BO_3F_2$（KBBF）[3] 等一系列重大发明发现。其中,BBO 和 LBO 被国际上誉为“中国牌”晶体，并被广泛用于可见光和紫外光波段的倍频激光输出；其产品已经主导了相关市场，并催生了以福晶科技有限股份公司等高科技上市公司为代表的 NLO 晶体和相关激光技术产业链。基于 KBBF 晶体，我国科学家成功研制了深紫外激光拉曼光谱仪、深紫外激光光化学反应仪等一系列我国独有的深紫外固态激光源装备，并已在石墨烯、高温超导、拓扑绝缘体和宽禁带半导体等众多前沿科学研究中获得了重要应用[4]。KBBF 晶体也成为我国对美国禁运的高科技产品，*Nature* 期刊还以“China's crystal cache”（藏匿在中国的晶体珍宝）为专题对 KBBF 晶体及其相关全固态激光技术进行了报道[5]。近年来，我国科学家在 *J. Am. Chem. Soc.*、*Angew. Chem. Int. Ed.*、*Adv. Mater.*、*Nat. Commun.* 等化学和材料科学领域的顶尖刊物和主流刊物上，发表了大量高影响力关于非线性光学晶体材料研究的学术论文。同时，我国多家大学和研究所的学者多次应邀在 *Chem. Soc. Rev.*、*Coord. Chem. Rev. Acc. Chem. Res.* 等国际著名综述性杂志上发表综述论文，并在国际学术会议上做多次大会报告和主旨报告，担任了国际晶体学和材料科学领域杂志的编辑、编委等。可以认为，我国在 NLO 晶体材料研究领域的国际领先地位得到了进一步巩固和提高。

本文将主要对近年来我国学者在 NLO 晶体材料领域发表的一些有代表性的成果，按照应用波段划分为深紫外（小于 200nm）、紫外 – 可见 – 近红外和中远红外 NLO 晶体材料，进行简要的总结介绍。

二、深紫外 NLO 晶体材料

我国在深紫外 NLO 晶体材料研究及相关固体激光技术及其应用方面处在国际领先的位置。目前唯一能在实际中直接倍频输出深紫外激光的晶体是我国科学家陈创天等发明的 KBBF 晶体。基于该晶体，我国科学家发展了一系列我国独有的相关深紫外固体激光技术和激光源装备，并在众多前沿科学研究中获得了重要应用。但是，KBBF 晶体也存在难以克服的本征缺陷，制约了其商业化生产和应用进程。一方面，KBBF 所含的铍元素有剧毒；另一方面，KBBF 晶体结构中层状结构单元之间的连接力很弱，导致 KBBF 存在严重的层状生长习性，限制了其晶体器件倍频输出的功率和效率。针对这一问题，科学家们的解决思路主要有两个：一是对 KBBF 进行结构改造，设计合成新型深紫外 NLO 晶体材料；二是探索合成全新的深紫外 NLO 晶体材料体系。

陈创天和黄洪伟等基于材料制备的阳离子工程设计，在层状硼酸盐中引入小半径阳离子，成功获得具有目前文献报道的最小层间距的新型深紫外 NLO 晶体材料 $Na_2Be_4B_4O_{11}$ 和 $LiNa_5Be_{12}B_{12}O_{33}$[6]。由于它们具有和 KBBF 类似但是更紧密的层状晶体结构，其 NLO 效应进一步增大到了磷酸二氢钾（KDP）的 1.3 倍和 1.4 倍，并且能够实现 1064 nm 波长的相位匹配。同时，由于层状结构单元之间通过较强的 Be–O 键连接，有利于其克服层状生长习性。透过光谱表明 $Na_2Be_4B_4O_{11}$ 和 $LiNa_5Be_{12}B_{12}O_{33}$ 的紫外吸收边分别为 171nm 和 169nm，表明他们完全可以实现 200 nm 以下的深紫外光透过。

洪茂椿和罗军华等在我国科学家提出的阴离子基团理论[7]基础上进一步考虑阳离子对晶体中阴离子功能基元排列的关键作用，利用阳离子配位环境强制实现［BO_3］$^{3-}$ 功能基元的层状定向排列并调节其层间作用力，设计合成了一系列新型硼酸盐无铍深紫外 NLO 材料 $Li_4Sr(BO_3)_2$（与中科院理化技术研究所合作发现）[8]、$K_3Ba_3Li_2Al_4B_6O_{20}F$[9]、$Rb_3Al_3B_3O_{10}F$[10]。它们的层状结构单元基本继承了 KBBF 中［BO_3］$^{3-}$ 活性基元的有利排列方式，从而基本保留了 KBBF 良好的光学性能。同时，层状单元间由更强的配位键连接，使得层间连接力分别达到了 KBBF 的约 4.7 倍、不小于 4 倍和不小于 9.5 倍，从而克服了层状生长习性。其中，初步生长出的 $K_3Ba_3Li_2Al_4B_6O_{20}F$ 晶体 c 向厚度最高达 8 mm，大大超过了目前 KBBF 的最大厚度 3.7 mm[11]。同时，该系列晶体材料组成元素对环境毒性小，原料成本低，且物化性能稳定。这些结果表明该系列晶体作为深紫外 NLO 晶体材料具有较好的应用前景。

探索满足“深紫外透过 – 大倍频效应 – 较大双折射”相互矛盾性能指标的深紫外 NLO 晶体是当前该领域亟待突破关键难点。通过材料结构性能关系研究，建立功能基元数据库，探索平衡制约性能微观机理，筛选并引入新的功能基团是突破深紫外用晶体的有效手段。潘世烈等提出了通过 B–F 剪裁三维网络结构获得较大双折射并保持低紫外截止边、大倍频效应的设计策略，并基于材料模拟计算筛选出了一类［BOF］功能基团：（BO_3F）$^{4-}$、（BO_2F_2）$^{3-}$和（BOF_3）$^{2-}$。基于理论计算从已知化合物中筛选出了一种深紫外 NLO 晶体 $Li_2B_6O_9F_2$[12]，理论计算评估的紫外吸收边为 155 nm，双折射值为 0.07@1064 nm，最短相位匹配波长达到 192 nm。实验表明 $Li_2B_6O_9F_2$ 可实现 266 nm 倍频输出，其倍频效应达 0.3 倍 BBO，是一种很有潜力的深紫外非线性光学晶体。潘世烈等还将有利于紫外光透过的碱土金属阳离子、卤素氟离子结合起来，合成出了具有紫外 / 深紫外透过的 NLO 材料 $Ba_4B_{11}O_{20}F$[13]。该材料具有碱金属、碱土金属卤素硼酸盐中最大的倍频效应，其粉末倍频效应达 10 倍 KDP。理论计算发现引入氟离子使得阴、阳离子基团畸变对晶体整体的 NLO 效应都有较大的贡献，丰富了 NLO 材料倍频效应起源的理论体系。此外，潘世烈等在阴离子基团理论研究的基础上，在 B–O 框架中引入刚性基团 SiO_4，迫使 B–O 框架发生较大畸变，从而成功设计合成出富硼硅酸盐深紫外 NLO 材料 $Cs_2B_4SiO_9$[14]。该晶体紫外吸收截止边短于 190 nm、能够实现相位匹配，具有合适的倍频效应（4.6KDP），物化性能稳定，是一种潜在的深紫外 NLO 材料。该发现为新型深紫外 NLO 材料的研究提供了新体系。

长期以来，深紫外 NLO 晶体材料的探索一直集中在硼酸盐体系。陈玲等报道了以三聚磷酸根为基本结构单元构建的首例非硼酸盐深紫外 NLO 晶体材料 $Ba_3P_3O_{10}X$（X =Cl、Br）[15]。其中，$Ba_3P_3O_{10}Cl$ 单晶测试的透过截止边可低至 180 nm（～7eV），呈现了很好的非线性相位匹配性能及 0.6 倍 KDP 的响应强度。因此，该化合物有望在深紫外 NLO 激光领域得以应用。与已知的三聚磷酸钠相比较，$Ba_3P_3O_{10}X$ 由于空间堆积作用，使得三聚磷酸根基本构筑单元围绕 P–O 键旋转，发生了从中心对称到非对称中心的转变。这个发现也为 NLO 晶体的结构设计提供了重要的参考。

优良的深紫外 NLO 晶体既要具有大的非线性光学效应，又要具有短的紫外吸收边，而这两种性能在某种程度上是相互冲突的，这就需要在两者之间达到一个微妙的平衡。洪茂椿和罗军华等将合适离子半径的碱土金属和碱金属阳离子同时引入到磷酸盐中，成功构建了一种深紫外透光的新型磷酸盐 NLO 晶体材料 $RbBa_2(PO_3)_5$[16]。该材料实现了线性光学性能（紫外吸收边 =163nm）和 NLO 性能（1.4 倍 KDP）的良好平衡。同时，$RbBa_2(PO_3)_5$ 同成分熔融，易于大尺寸晶体生长，这使得其作为深紫外 NLO 材料具有潜在应用前景。他们与林哲帅等合作进行理论计算揭示了其倍频效应产生的微观机理：随着磷氧结构基元中 PO_4 单元聚合程度的提高，相应磷氧结构基元的微观 NLO 系数增大，这为设计合成高性能的此类材料提供了新的研究思路。洪茂椿和罗军华等还基于一种具有结构柔性的［P_3O_{10}］$^{5-}$基元，对已知的磷酸盐 NLO 材料 $Ba_3P_3O_{10}Cl$ 进行结构裁剪（tailored synthesis），成功获得了一种结构相近的新型深紫外 NLO 材料 $Ba_5P_6O_{20}$[17]。其倍频效应基本不变，但其紫外吸收边发生显著蓝移，从而大大拓展了其深紫外波段应用范围。与林哲帅等合作通过理论计算揭示了一种与 KBBF 刚好相反的吸收边蓝移机制，深化了对此类材料光学性能的科学认知。

潘世烈等通过在磷酸盐中引入离子半径差异较大的碱金属阳离子，成功设计合成了一种磷酸盐深紫外 NLO 晶体材料 $LiCs_2PO_4$[18]。该化合物不仅展现出短的紫外截止边（174nm），而且具有大倍频效应（2.6 倍 KDP），是迄今为止磷酸盐深紫外 NLO 材料体系中倍频效应最大的化合物。同时，$LiCs_2PO_4$ 能够实现 1064 nm 下相位匹配，晶体易于生长，有望作为新型深紫外 NLO 晶体材料。理论计算发现该材料晶体结构中 LiO_4-PO_4 基团的特殊共边连接方式有利于 O-2p 非键轨道的定向排列，导致磷氧结构基元微观非线性光学系数的有效叠加，从而使 $LiCs_2PO_4$ 显示出较大的倍频效应。这为设计合成此类材料提供了一种新颖的倍频效应增益机制。

林哲帅与合作者华中科技大学李伟教授等最近发现 KBBF 晶体材料还具备新奇的负面压缩性[19]。研究发现，在三方晶体对称性的限制之下，KBBF 的负面压缩行为在（*a*，*b*）平面内完全各向同性，这是迄今为止报道的压力范围最大的负面压缩材料。理论计算显示，KBBF 的负面压缩性质主要来源于 K^+ 离子与（$Be_2BO_3F_2$）$_\infty$层之间较弱的相互作用，以及（$Be_2BO_3F_2$）$_\infty$层中（BO_3）$^{3-}$ 和（BeO_3F）$^{5-}$ 基团交错排列的特殊结构。在静水压的作用下，层与层之间发生的严重塌缩使得（$Be_2BO_3F_2$）$_\infty$层受到强烈的挤压，导致（BeO_3F）$^{5-}$ 基团中的 O_3 平面发生了扩展，使得晶体在（*a*，*b*）平面内表现出了负面压缩性质。此外，理论计算显示，KBBF 在深紫外波段的透过性能随着压力的增大进一步改善，这可为 KBBF 在深紫外波段负压缩性质应用奠定基础。

三、紫外 - 可见 - 近红外 NLO 晶体材料

虽然紫外 - 可见 - 近红外 NLO 晶体材料的研究已经比较成熟，能应用的晶体也很多，但都存在着某些 NLO 性能制备不够好而不能得到更广泛应用的问题。同时，这类材料由于物化性能稳定，具有潜在的压电、电光、铁电性能，也有可能作为新型多功能晶体材料应用。因此，探索具有更优异 NLO 性能的新型紫外 - 可见 - 近红外 NLO 晶体依然具有重要意义，这也是国内外还有众多研究者仍在从事该波段 NLO 晶体研究的原因。

根据阴离子基团理论，在硼酸盐晶体中具有共轭 π 电子体系的平面三角形 BO_3 基团比 BO_4 四面体具有更大的极化率。而硝酸根或碳酸根具有与 BO_3 基团相同的平面三角形几何构型，也是非常重要的 NLO 活性基团。提高化合物中平面三角形结构基元的密度并使得它们在三维结构中平行排列，通过这些基团的协同极化作用可以提高化合物的二阶 NLO 系数。

毛江高等提出将硼酸根与硝酸根两种平面三角形结构基元复合到同一化合物中的设计思路。在温和水热条件下得到了首例极性结构的铅硼酸 - 硝酸盐 $Pb_2(BO_3)(NO_3)$[20]。由于其结构中铅离子与硼酸根连接成蜂窝型［$Pb_2(BO_3)$］$_\infty$层，没有参与配位的硝酸根位

于层之间，所有硼酸根与硝酸根都平行排列，从而形成了共轭 π 电子和铅离子（II）孤对电子的协同极化作用，$Pb_2(BO_3)(NO_3)$ 表现出 9 倍 KDP 的二阶 NLO 效应。理论计算表明 PbO_3、BO_3 和 NO_3 基团对 d_{32} 的贡献百分比分别为 41.4%、13.0% 和 41.2%，这为今后新型硼酸盐二阶 NLO 晶体材料的设计提供了一条新的方法。

针对 KBBF 的层状生长习性和原料 BeO 的剧毒性，潘世烈等采用与 Be^{2+} 离子具有相似配位化学键的 Zn^{2+} 离子替代 KBBF 中的 Be^{2+} 离子，并且采用强的 B-O 键连接相邻的［$Zn_2BO_3O_2$］层，成功的合成出了具有类 KBBF 的 $Cs_3Zn_6B_9O_{21}$ 化合物。[21] 该化合物不仅克服了 KBBF 族化合物生长过程中强的层状习性和 BeO 的剧毒性，同时也具有 KBBF 族中最大的倍频效应。同时，该化合物紫外截止边约为 200nm，可见光区双折射率大于 0.06（计算值）。该设计思路也将有利于发展具有类铍硼酸结构的锌硼酸盐在紫外非线性光学材料的设计应用。潘世烈等系统研究了 Pb- 碱土金属替代硼酸盐结构性能关系，设计合成出第一例非中心对称结构的仅含孤立 BO_3 的铅硼酸盐 $Pb_2Ba_3(BO_3)_3Cl$[22]。其倍频效应是 KDP 的 3.2 倍，较同构化合物 $Ba_5(BO_3)_3Cl$ 增益高达 6.4 倍，且满足相位匹配条件。首次揭示了 Pb- 碱土金属替代致使倍频效应增益不同的微观机制，发现有利于获得较大倍频效应增益的主要原因是 Pb（Ⅱ）O 多面体与 BO_3 基团独特的共边连接方式。这表明含孤立 BO_3 基团的铅硼酸盐是可获得较大倍频效应增益的潜在化合物。

叶宁等以具有平面三角形结构的碳酸盐为研究对象，将含孤对电子的 Pb 原子引入具有 π 共轭平面三角形构型的该碳酸盐体系，发现了一种倍频系数达 13.4 倍 KDP 的 NLO 晶体 $CsPbCO_3F$。[23] 该晶体大 NLO 系数的产生机理不同于常见的平面基团与孤对电子协同增强的机制，而是一种未被发现的金属与平面共轭体系的 p-π 轨道相互作用。该研究成果拓展了紫外 NLO 晶体材料研究的新体系，为设计大 NLO 系数紫外晶体提供了一个崭新的思路。

硝酸根具有和 BO_3 相似的 π 共轭平面三角形构型，但存在易溶于水的问题。洪茂椿和罗军华等基于基元协同的设计思想成功合成了一例抗水硝酸盐 NLO 材料 $Bi_3TeO_6OH(NO_3)_2$[24]。由于其晶体结构中三种基元的协同作用，其倍频效应达到了 3 倍 KDP，优于已知的大多数硝酸盐 NLO 材料。同时，该材料具有很高的抗水性。他们与林哲帅等合作通过理论计算揭示了其抗水性的原因，从而为设计合成抗水硝酸盐提供了有益的参考。

林哲帅和夏明军等基于“三合一”的设计思想，将三种 NLO 功能基元 Bi_3O_9、TeO_6 和 BO_3 复合到同一个化合物中，设计合成了一种新型 NLO 晶体材料 Bi_3TeBO_9[25]。由于三种基元微观二阶 NLO 系数的有效叠加，Bi_3TeBO_9 表现出非常大的宏观倍频效应（KDP 的 20 倍）。这种“三合一”的功能基元协同策略为设计合成高倍频效应晶体材料提供了新的手段。

金属碘酸盐晶体因具有较强的倍频效应、较宽的透过波段、较高的热稳定性和光学损伤阈值在二阶 NLO 晶体材料领域占有非常重要的地位。毛江高等提出将氟离子引入碘酸盐体系，构筑新颖结构的金属碘酸盐氟化物，利用金属氟化物三维结构来限制碘酸根离子的排列方向，使各个碘酸根离子的极化作用相互叠加而表现出大的倍频系数。基于该学术思想，以水热合成方法，得到了首例碘酸盐氟化物 NLO 晶体 $Bi(IO_3)F_2$[26]。该化合物表现出非常高的倍频效应，为 KDP 的 11.5 倍，且相位匹配，同时具有较宽的透过波段（0.3 ~ 11μm），是一种具有潜在应用价值的二阶 NLO 晶体材料。该工作为今后新型碘酸盐二阶 NLO 晶体材料的设计提供了一条新的合成策略。

亚硒酸盐因其含有活性孤对电子而在二阶非线性光学晶体材料中占有非常重要的地位，但该类化合物的倍频系数一般比相应的亚碲酸盐和碘酸盐小得多。为提高其倍频系数一般采用引入畸变八面体配位构型的 d^0 过渡金属阳离子，但这样的化合物组成与结构往往比较复杂，影响其大晶体的制备。毛江高等提出对一些已知高倍频系数的化合物进行不等价取代的方法来设计新的高性能倍频晶体的新思路。对已知的高倍频系数化合物 $BiOIO_3$（倍频效应为 KDP 的 12.5 倍）进行不等价取代，在温和水热条件下得到了新型亚硒酸盐 NLO 晶体材料 $BiFSeO_3$[27]。由于 Bi^{3+} 和亚硒酸根的孤对电子产生协同极化作用，$BiFSeO_3$ 表现出很强的倍频效应，其粉末倍频系数为 KDP 的 13.5 倍，为亚硒酸盐中最高值。理论计算表明 Bi^{3+} 和 SeO_3 基团对 d_{31} 的贡献百分比分别为 62.8% 和 37.1%。该工作为今后新型二阶 NLO 晶体材料的设计提供了新的策略和思路。

四、中远红外 NLO 晶体材料

对已有的激光光源进行频率转换是目前获得中远红外可调谐激光的重要方法，其核心器件之一是中远红外非线性光学晶体材料，具有小型化、易操纵、转换效率高、光束质量好等突出的优点。中远红外非线性光学晶体材料主要分为中红外（波长 3~20 μm）和远红外（如太赫兹，波长 30 μm ~ 3 mm）两大类。我国科学家通过不懈的努力，在这些材料的探索和应用方面取得了丰硕的成果，在世界范围内的相关研究领域中处于并跑乃至领跑的地位。

通过光学参量振荡（OPO）、光学参量放大（OPA）或差频（DFG）等非线性频率转换技术[28]，是产生中红外激光输出的主要手段。中红外波段激光以其高亮度、良好的相干性和极高的空间分辨率等特点在红外对抗、红外定向干扰、激光制导、红外测距、分子光谱、医学诊疗、地球遥感探测等军事和民用领域有着广泛的应用[29]。目前，研究最多的中红外非线性光学晶体材料有硫属化合物、磷族化合物、卤化物和过渡金属氧化物等。

硫属化合物的典型代表是黄铜矿型化合物 $AgGaS_2$ 和 $AgGaSe_2$，其已经实现了一定的商业化应用。但是这两种材料有其本征的缺点，比如它们的抗激光损伤阈值比较低，无法应用于高功率的激光输出，$AgGaSe_2$ 在 1μm 处无法实现相位匹配，晶体的热膨胀各向异性明显，导致晶体容易开裂[30]。为了克服这些缺点，科学家们对新型硫属红外非线性光学材料进行了卓有成效的探索。其中比较有代表性的晶体是 $BaGa_4S_7$ 和 $BaGa_4Se_7$。$BaGa_4S_7$ 晶体是 2009 年由福建物构所叶宁课题组首次报道，生长了 $\Phi10\times100mm^3$ 的大尺寸晶体，抗激光损伤阈值是同等条件下 $AgGaS_2$ 的 3 倍，比 $LiInS_2$ 还要高[31]。2014 年，叶宁研究组通过采用双温区摇摆炉，降低了晶体生长过程中的蒸汽压，生长出了大尺寸高质量的 $BaGa_4S_7$ 晶体，测量了热导率和抗激光损伤阈值，并给出了 2.07μm 下的 OPO 相位匹配曲线[32]。$BaGa_4Se_7$ 晶体是 2010 年由中科院理化所吴以成课题组姚吉勇等首次报道，粉末倍频效应约为 2 倍的 $AgGaS_2$[33]。近几年，通过采用双温区合成大批量多晶原料，并不断优化生长条件，目前已经可以生长出 $\Phi40\times120mm^3$ 的大尺寸高质量单晶，达到了国际领先水平[34]。2013 年，吴以成课题组姚吉勇与理化所激光中心合作，采用 1.06μm 泵浦光源，在 $BaGa_4Se_7$ 晶体中首次实现了 3 ~ 5μm 的 OPA 激光输出。当泵浦能量为 9.1 mJ 时，3.9μm 处最高的闲频光输出能量为 830 μJ，峰值功率为 27 MW[35]。2015 年，又在国际上首次实现了 6.4–11μm 的可调谐 OPA 激光输出[36]，并于同年用 Marker 条纹法测量了 $BaGa_4Se_7$ 晶体的非线性光学系数（d_{11}=24.3 pm/V and d_{13}=20.4 pm/V）[37]。2016 年，姚吉勇与哈尔滨工业大学姚宝权合作，在 $BaGa_4Se_7$ 晶体中首次实现了 2.09μm 光源泵浦的 OPO 激光输出，信号光调谐范围为 3.49 ~ 4.13 μm，闲频光调谐范围为 5.19 ~ 4.34μm，光光转换效率为 14.4%[38]。2017 年，姚吉勇与天津大学徐德刚合作，采用 1.06μm 泵浦源，在 $BaGa_4Se_7$ 晶体中实现了 OPO 输出，闲频光调谐范围为 3.12 ~ 5.16 μm。在泵浦能量为 61.6 mJ 时，4.11μm 处的闲频光输出能量为 2.56 mJ，光光转换效率为 4.16%[39]。同年，姚吉勇等还测量了 $BaGa_4Se_7$ 晶体在 0.5 ~ 2.3μm 范围，25 ~ 150℃区间内的变温折射率，拟合了变温的 Sellmeier 色散方程，为今后进一步的激光应用打下良好基础[40]。

利用高电负性的 Li^+ 替换 $AgGaS_2$ 中的 Ag^+，可以显著的增加材料的带隙，提高晶体的抗激光损伤阈值，促使科学家发现了 $LiMQ_2$（M=Ga、In；Q=S，Se、Te）等系列新型的红外非线性光学材料[41]。近几年，山东大学的陶绪堂课题组对 $LiInQ_2$（Q=S、Se）单晶的生长开展了深入研究。2013 年，他们采用自发成核布里奇曼方法生长了尺寸为 $\Phi10\times100$ mm^3 的 $LiInS_2$ 单晶，并测试了其压电系数、介电常数和弹性系数[42]。$LiInS_2$ 晶体的带隙为 3.54eV，抗激光损伤阈值为 $0.9J/cm^2$，是同等条件下 $AgGaS_2$ 的 2.2 倍[43]。2014 年，他们又对 $LiInS_2$ 晶体进行了退火实验，发现在 $LiInS_2$ 气氛中退火后的晶体激光损伤阈值提高到了 $1.2J/cm^2$，热导率也有所提高[44]。同年，他们还生长了 $\Phi12\times30$ mm^3 的 $LiInSe_2$ 晶体，其热导率是 $AgGaS_2$ 的三倍以上，比较有利于高功率的中红外激光应用[45]。与 $LiInS_2$ 相比，$LiInSe_2$ 的透过范围更宽，在 0.5 ~ 12μm 区间内的透过率均超过 70%[45]。2017 年，陶绪堂课题组在 $LiInSe_2$ 晶体研究方面取得突破性进展，生长出了高质量完整的 $\Phi16\times50mm^3$ $LiInSe_2$ 晶体，与北京理化所许祖彦课题组合作，利用一块 $5.1\times6.2\times6.76mm^3$ 和 $8\times8\times8.26mm^3$ 的 $LiInSe_2$ 晶体，在国际上首次实现了 3.6 ~ 4.8μm 和 7 ~ 12μm 的宽调谐皮秒 OPA 输出[46]。通过进一步优化晶体生长工艺、晶体退火后处理以及晶体镀膜等，有望获得更高能量的可调谐中远红外激光输出。此外，$LiInSe_2$ 晶体在低频太赫兹波段有较低的吸收系数和良好的相位匹配能力，说明它在利用差频方法产生太赫兹波方面也是一个有应用潜力的晶体[47]。

围绕着实现高的抗激光损伤阈值和大的非线性光学效应之间的平衡问题，近年来国内外在新型硫族化合物研究和探索方面进行了大量的工作。2015 年，吴以成课题组姚吉勇等发现了 $BaGa_4Se_7$ 的同系化合物 $PbGa_4S_7$，带隙为 3.08 eV，粉末倍频效应为 1.2 倍的 $AgGaS_2$[48]。同年，又发现了 $BaGa_2GeS_6$ 的同系化合物 $BaGa_2SnSe_6$，表现出了较强的非线性光学效应（5.2×AGS），在 2.09 μm 处可以实现相位匹配[49]。2016 年，姚吉勇等类比硼酸盐晶体中的 $(BO_3)^{3-}$ 基团，在 $BaHgSe_2$ 中发现了具有平面 π 共轭特征的 $(HgSe_3)^{4-}$ 基团，表现出较强的非线性光学响应（1.5×AGS）[50]。第一性原理计算证明平面 $(HgSe_3)^{4-}$ 基团对非线性光学系数的贡献超过 75%，这为新型硫族红外非线性光学材料的探索提供了新的思路。2015 年，中科院理化所林哲帅课题组利用第一性原理方法，系统的研究了磷硫体系化合物的构效关系[51]，并与姚吉勇等合作，发现了 3 个新型的磷硫体系红外非线性光学材料 $Zn_3(PS_4)_2$、$LiZnPS_4$ 和 $AgZnPS_4$，它们都具有优良的非线性光学性能。其中 $AgZnPS_4$ 化合物带隙为 2.76 eV，非线性光学效应为 1.8 倍的 $AgGaS_2$，是同成分熔融化合物，是一个具有应用潜力的中红外非线性光学材料[52]。他们又系统地研究了类金刚石体系中红外非线性光学性能的构效关系，给出了类金刚石体系的化合物族谱，为今后大规模、高通量的探索新型红外非线性光学材料提供理论支持和帮助[53]。随之，新疆理化所潘世烈研究组合成了多种新型的类金刚石材料 $Li_4HgGe_2S_7$[54]、Li_2HgMS_4（M=Si、Ge、Sn）[55]等，其优秀的非线性光学性质也得到了初步验证。除此以外，潘世烈研究组还陆续合成了 $Na_2ZnGe_2S_6$[56]、$NaBaM_2Q_4$（M=Ge、Sn；Q=S、Se）[57]、$Na_2HgM_2S_8$（M=Si、Ge、Sn）[58]、$BaHgS_2$[59]、$Na_4MgM_2Se_6$（M=Si、Ge）[60]等新型的红外非线性光学材料，其中 $Na_2ZnGe_2S_6$ 具有大的带隙（3.25 eV）和较强的倍频响应（0.9×AGS），同时可以实现相位匹配，是一种比较有潜力的红外非线性光学材料[56]。2013 年，中科院福建物构所郭国聪课题组发现了高温单斜相的 Ga_2S_3，具有 0.7 倍 $AgGaS_2$ 的倍频响应和较高的激光损伤阈值（30×$AgGaS_2$），该测试值是通过采用首次提出的粉末样品激光损伤阈值测试方法获得[61]，并进行大晶体生长，目前已生长出尺寸为 4×4×1 mm^3 的高质量样品。郭国聪课题组也发现了兼备较大非线性系数和较高激光损伤阈值的新型硫族化合物 $Ba_2Ga_8GeS_{16}$[62]，M_3GaS_6（M=Dy、Y）[63]，$Na_2In_2MS_6$（M=Si、Ge）[64]，混合阴离子化合物 $BaGeOSe_2$[65]，并采用以超四面体为结构基元和非线性光学功能基元的思路合成了 $Na_2In_4SSe_6$，$NaGaIn_2Se_5$ 和 $NaIn_3Se_5$[66]。2013 年，中科院福建物构所陈玲课题组发现了一系列的 $KCd_4Ga_5Se_{12}$ 型化合物，具有超大的非线性光学响应，其中 $RbCd_4In_5Se_{12}$ 的粉末倍频强度是同等条件下 $AgGaS_2$ 的 39 倍[67]。2015 年，他们系统地研究了 R_3MTQ_7 型化合物的非线性光学效应，揭示了系统组分对非线性光学系数的影响[68]。2017 年，陈玲课题组（调入北京师范大学）又合成出了 $RbZn_4In_5Se_{12}$ 系列化合物，并分析了晶体带隙与晶胞参数之间的关系，为下一步探索具有大带隙的非线性光学材料提供了思路[69]。他们还发现了 $CsGaSn_2Se_6$ 和 $CsInSn_2Se_6$，并指出微观偶极矩的取向是决定该类型化合物能否实现相位匹配的重要因素[70]。此外，福建物构所程文旦课题组也一直致力于新型硫族化合物的探索，并发现了一系列具有非线性光学效应的新材料，如 $Ba_8Sn_4S_{15}$[71]、$SnGa_4Q_7$（Q=S、Se）[72]、$PbGa_2MSe_6$（M=Si、Ge）[73]等。在国际上，美国西北大学的 M. G. Kanatzidis 和杜肯大学的 J. A. Aitken 合成了一系列具有高倍频效应的硫族化合物，如 Li_2MnGeS_4[74]、$Li_2ZnGeSe_4$[75]、Cu_2ZnSiS_4[76]、$K_{1-x}Cs_xPSe_6$[77]、$Ba_6Ag_{2.67+4\delta}Sn_{4.33-\delta}S_{16-x}Se_x$[78]等。

由于卤素较高的电负性，将卤素原子（F、Cl、Br）引入硫族化合物体系也是目前研究比较多的一个新方向。2014 年，吴以成课题组姚吉勇等发现了宽禁带的红外非线性光学材料 $NaBa_4Ge_3S_{10}Cl$，带隙为 3.49 eV，粉末倍频效应为 1/3 的 $AgGaS_2$[79]。2015 年，物构所陈玲课题组合成了 $Ba_4MGa_4Se_{10}Cl_2$（M=Zn、Cd、Mn、Cu/Ga、Co、Fe）系列化合物，具有较宽的带隙和较大的非线性光学响应[80]。2016 年，福建物构所的郭国聪研究组提出双“功能基元”的结构设计思路，即把“抗激光损伤功能基元”和“NLO 活性功能基元”在分子水平上组装成无心结构，并发现了同时具有宽带隙与大的非线性光学效应；同时是同成分熔融的［A_3X］［Ga_3PS_8］（A=K、Rb；X=Cl、Br）系列化合物，也是一类有潜力的新材料[81]。2017 年，陈玲课题组又巧妙地对 $NaBa_4Ge_3S_{10}Cl$ 进行了分子结构剪裁，合成出了 $Ba_4Ge_3S_9Cl_2$ 系列化合物，也表现出了较强的非线性光学响应（$2.4\times AgGaS_2$）[82]。2017 年，为了总结硫族非线性光学材料的科研成果，吴以成课题组和林哲帅课题组合作在 *Crystal Growth& Design* 上撰写了综述文章，详细分析了各种微观基团对材料非线性光学性质的影响，为未来新型中红外硫族非线性光学材料的探索提供了研究方向[83]。

磷族化合物 $ZnGeP_2$ 是一种性能优秀的红外非线性光学晶体材料，被广泛应用于各种先进的光学器件，尤其是中红外波段可调谐激光器。目前国内有多家单位在研究 $ZnGeP_2$ 晶体的生长问题，包括哈尔滨工业大学、四川大学、山东大学、中国工程物理研究院、安徽光机所等，力图生长出大尺寸、高光学质量的 $ZnGeP_2$ 单晶。2011 年，哈尔滨工业大学的杨春晖课题组通过双温区法合成出了大批量的 $ZnGeP_2$ 多晶原料，并采用垂直的布里奇曼方法生长出了大尺寸（$\Phi30\times130mm^3$）、高质量的晶体[84]。2014 年，杨春晖课题组又生长出了高质量无开裂的 $ZnGeP_2$ 晶体，尺寸达到 $\Phi50\times140mm^3$。用 2.09μm 的 Ho:YAG 激光器泵浦，实现了功率高达 20.5 W 的 OPO 红外激光输出，能量总转换效率 44.3%[85]。2012 年，安徽光机所吴海信课题组采用垂直的布里奇曼方法生长出了 $\Phi20\times90\ mm^3$ 的 $ZnGeP_2$ 晶体，并研究了不同的退火条件对晶体光学吸收系数的影响[86]。2017 年，吴海信课题组利用一块经过电子辐照和退火处理后的 $ZnGeP_2$ 晶体（$6\times6\times15mm^3$），在 2.09μm 的激光泵浦下，实现了峰值功率密度高达 24 MW/cm^2 的 OPO 红外激光输出，光光转换效率为 75.7%，是目前国际上已经报道的最高水平[87]。2016 年，四川化工材料研究所康彬课题组用自制的双温区机械震荡管式炉合成了 $ZnGeP_2$ 多晶，单次合成量超过 500 g。他们在国内首次采用水平梯度冷凝法生长了质量为 350 ~ 395g 的 $ZnGeP_2$ 单晶，单晶最大尺寸为 $25 \times 32 \times 165mm^3$。他们制备了 $ZnGeP_2$ 光学参量震荡器件，在 2 μm 激光泵浦下成功输出 3 ~ 5 μm 可调谐激光，光光转化效率为 56.2%，达到国际较高水平[88]。除了 $ZnGeP_2$ 以外，其他同构的黄铜矿型磷族化合物也受到了科研人员的关注，如 $CdSiP_2$、$CdGeAs_2$ 等。2013 年，山东大学陶绪堂课题组生长了 $\Phi 8 \times 40mm^3$ 的 $CdSiP_2$ 单晶，并测试了其折射率和吸收系数，计算了 OPO 相位匹配曲线[89]。2017 年，陶绪堂课题组利用第一性原理方法，详细讨论了本征缺陷对 $CdSiP_2$ 晶体光学性能的影响[90]。2017 年，四川大学朱世富课题组生长了 Cr 掺杂的 $CdGeAs_2$ 晶体，并测量其透过光谱[91]。2015 年，山大陶绪堂课题还发现了新型的 $Ba_4AgGa_5Pn_8$（Pn=P、As）化合物[92]，理论计算的非线性光学系数约为 3 倍的 $AgGaSe_2$，也是一类有研究价值的新材料体系。

卤族化合物材料在中红外波段也具有应用潜力。2013 年，中科院理化所的林哲帅课题组系统地总结了卤素化合物的配位类型，并阐明了微观结构与光学性能之间的关系，指出结构基元是四面体型的、三角锥状的和一维链状的氯化物或溴化物能够呈现出相对均衡的中红外非线性光学性能，即保证足够的带隙，同时又维持较大的倍频效应[93]。同时，武汉大学的秦金贵课题组在新型卤族非线性光学材料实验探索中取得了丰硕的成果。2013 年至 2016 年，秦金贵课题组陆续合成出了含有混合卤素的 β-HgBrCl[94]、Hg_2Br_3I[95]、HgBrI[96]、$Rb_2CdBr_2I_2$[97]、$K_2SbF_2Cl_3$[98] 等化合物。其中 $Rb_2CdBr_2I_2$ 和 $K_2SbF_2Cl_3$ 同时具有大的光学带隙（分别为 3.35 和 4.01 eV）和较强的非线性光学效应（~4 × KDP），在中红外波段具有潜在的应用价值。潘世烈等人结合电负性较大的 Cl^- 离子及易产生二阶姜泰勒效应的 Pb^{2+} 离子成功获得一种性能优异的红外倍频晶体 $Pb_{17}O_8Cl_{18}$[99]。该晶体除了具有大的带隙和非线性效应（3.44eV，2 × $AgGaS_2$）以外，还具有较高的激光损伤阈值（12.8 × $AgGaS_2$），并且可在开放体系生长，降低了传统红外非线性光学晶体在封闭系统中生长的难度。

山东大学陶绪堂课题组等系统地开展了含碲钼 / 钨酸盐非线性光学晶体的生长、性能评价和器件研究。由于这些化合物结构中存在畸变的 Te–O 和 Mo/W–O 多面体，极化矢量有效叠加，晶体表现出大的非线性光学系数，并能实现位相匹配，如 β–$BaTeMo_2O_9$[100]、$Na_2TeW_2O_9$[101]、$Cs_2TeW_3O_{12}$[102]，其有效非线性光学系数分别为 10.3 pm/V、6.9 pm/V、4.0 pm/V。此类材料突破了目前大多数氧化物单晶无法完全覆盖中红外波段的限制，为探索新型近 – 中红外非线性光学晶体开拓了一个新的方向。同时，他们生长了全新的 α-$BaTeMo_2O_9$ 晶体[103]，发现该晶体具有大的双折射，基于该晶体的偏光器件使用范围可以覆盖从 0.5 ~ 5.0 μm 的宽波段范围，突破了偏光器件仅限于单轴晶的限制。

介于微波和红外光之间的太赫兹波段，是电子学技术和光子学技术的过渡区域，在生物医学、空间探测、国防工业和反恐等领域有重要的应用。利用有机非线性光学晶体差频或者光整流产生太赫兹波，可以将 THz 输出范围覆盖整个 THz 波段，这在太赫兹辐射产生技术中的地位日益重要。常见的有机非线性光学晶体有 DAST、DSTMS 和 OH1。吴以成课题组张国春等在国内较早的开始了这三种晶体的生长工作。通过改善生长条件，生长出了最大尺寸为 $12\times10\times1mm^3$、$13\times8\times2\ mm^3$、$15\times7\times2mm^3$ 的 DAST、DSTMS、OH1 单晶[104]。2014 年，吴以成课题组张国春等与天津大学徐德刚合作，利用差频方法，在 DSTMS 晶体中实现了 0.9 ~ 19.7THz 波段的调谐输出，在 3.8THz 处的最高脉冲能量为 85.3nJ，峰值功率达到 17.9W，是目前国际上已报道的利用 DSTMS 晶体差频方法产生单频可调谐 THz 波的最好结果[105]。2016 年，吴以成课题组张国春等和徐德刚合作，利用差频方法，在 OH1 晶体实现了 0.02 ~ 20THz 波段的调谐输出。当泵浦能量为 4.35mJ 时，1.92 THz 处最高输出功率达到了 507nJ/ 脉冲，光子转换效率为 2.9%，其调谐范围和输出能量均是目前已报道的利用 OH1 晶体差频方法产生单频可调谐 THz 波的国际最高水平[106]。2016 年，青岛大学滕冰课题组采用溶液缓慢降温法生长出了尺寸为 $14\times12\times3mm^3$ 的 DAST 晶体，并与天津大学徐德刚合作，用差频方法实现了范围为 0.51 ~ 19.69THz 的调谐输出[107]。同年，滕冰课题组通过优化生长条件，采用高浓度溶液缓慢降温斜板成核技术，长出了表面积达到 $221mm^2$，厚度达到 11 mm 的 DAST 晶体，达到国际领先水平[108]。2017 年，徐德刚等改用高重频的 Nd:YAG（1319/1338 nm）激光，在 DAST 晶体中实现了微瓦级的差频太赫兹波输出，转换效率为 6.4×10^{-7}。尽管输出功率还有提升空间，但已经足以应用于高灵敏度的太赫兹探测器成像[109]。同年，滕冰课题组在《中国科学》杂志发表综述文章，专门讨论大尺寸 DAST 晶体生长和太赫兹辐射输出[110]。2015 年，吴以成课题组张国春等还在异噁唑酮体系中发现了一种新型的有机太赫兹晶体 MLS，其带隙为 2.63eV，粉末倍频效应为 1.5 倍的 OH1[111]。2016 年，他与天津大学徐德刚合作，首次利用差频方法在 MLS 晶体中实现了 1.23 ~ 14.09THz 波段的调谐输出[112]。

参考文献

[1] Chen C T，Wu B C，Jiang A D，et al. A new - type ultraviolet SHG crystal - β-BaB_2O_4[J]. Scientia Sinica Series B-Chemical Biological Agricultural Medical & Earth Sciences，1985，28: 235-243.

[2] Chen C T，Wu Y C，Jiang A D，et al. New nonlinear-optical crystal: LiB_3O_5[J]. Journay of Optical Society of Materical B-Optical Physics，1989，6（4）: 616-621.

[3] Xia Y N，Chen C T，Tang D Y，et al. New nonlinear optical crystals for UV and VUV harmonic[J]. generationAdvanced Materials，1995，7: 79-81.

[4] 周兴江，曹凝．深紫外固态激光源光电子能谱仪系列装备中国科学院院刊．2013，28：104-110.

[5] Cyranoski D. Materials science: China's crystal cache[J]. Nature News，2009，457: 953-955.

[6] Huang H W，Liu L J，Jin S F，et al. Deep-ultraviolet nonlinear optical materials: $Na_2Be_4B_4O_{11}$ and $LiNa_5Be_{12}B_{12}O_{33}$[J]. Journal of the American Chemical Society，2013，135: 18319-18322.

[7] Chen C T. A localized quantum theoretical treatment, based on an anionic coordination polyhedron model for the EO and SHG effects in crystals of the mixed-oxide types[J]. Scientia Sinica, 1979, 22: 756-776.

[8] Zhao S G, Gong P F, Bai L, et al. Beryllium-free $Li_4Sr(BO_3)_2$ for deep-ultraviolet nonlinear optical applications[J]. Nature Communication, 2014, 5: 4019.

[9] Zhao S G, Kang L, Shen Y G, et al. Designing a beryllium-free deep-ultraviolet nonlinear optical material without a structural instability problem[J]. Journal of the American Chemical Society, 2016, 138: 2961-2964.

[10] Zhao S G, Gong P, Luo S Y, et al. Beryllium-free $Rb_3Al_3B_3O_{10}F$ with reinforced interlayer bonding as a deep-ultraviolet nonlinear optical crystal[J]. Journal of the American Chemical Society, 2015, 137(6): 2217-2210.

[11] Wang X Y, Yan X, Luo S Y et al. Flux growth of large KBBF crystals by localized spontaneous nucleation[J]. Journal of Crystal Growth, 2011, 318: 610-612.

[12] Zhang B B, Shi G Q, Yang Z H, et al. Fluorooxoborates: beryllium - free deep - ultraviolet nonlinear optical materials without layered growth[J]. Angewandte Chemie International Edition, 2017, 56: 3916-3919.

[13] Wu H P, Yu H W, Yang Z H, et al. Designing a deep-ultraviolet nonlinear optical material with a large second harmonic generation response[J]. Journal of the American Chemical Society 2013, 135: 4215-4218.

[14] Wu H P, Yu H W, Pan S L, et al. $Cs_2B_4SiO_9$: a deep - ultraviolet nonlinear optical crystal[J]. Angewandte Chemie International Edition, 2013, 52: 3406-3410.

[15] Yu P, Wu L M, Zhou L J, et al. Deep-ultraviolet nonlinear optical crystals: $Ba_3P_3O_{10}X$ (X= Cl, Br)[J]. Journal of the American Chemical Society, 2014, 136: 480-487.

[16] Zhao S G, Gong P F, Luo S Y, et al. Deep-ultraviolet transparent phosphates $RbBa_2(PO_3)_5$ and $Rb_2Ba_3(P_2O_7)_2$ show nonlinear optical activity from condensation of $[PO_4]^{3-}$ units[J]. Journal of the American Chemical Society, 2014, 136: 8560-8563.

[17] Zhao S G, Gong P F, Luo S Y, et al. Tailored synthesis of a nonlinear optical phosphate with a short absorption edge[J]. Angewandte Chemie International Edition, 2015, 54: 4217-4221.

[18] Li L, Wang Y, Lei B H, et al. A new deep-ultraviolet transparent orthophosphate $LiCs_2PO_4$ with Large second harmonic generation response[J]. Journal of the American Chemical Society, 2016, 138: 9101-9104.

[19] Jiang X X, Luo S Y, Kang L, et al. Isotropic Negative Area Compressibility over Large Pressure Range in Potassium Beryllium Fluoroborate and its Potential Applications in Deep Ultraviolet Region[J]. Advanced Materials 2015, 27: 4851-4857.

[20] Song J L, Hu C L, Xu X, et al. A facile synthetic route to a new SHG material with two types of parallel π - conjugated planar triangular units[J]. Angewandte Chemie International Edition, 2015, 54: 3679-3682.

[21] Yu H W, Wu H P, Pan S L, et al. $Cs_3Zn_6B_9O_{21}$: a chemically benign member of the KBBF family exhibiting the largest second harmonic generation response[J]. Journal of the American Chemical Society, 2014, 136: 1264-1267.

[22] Dong X Y, Jing Q, Shi Y J, et al. $Pb_2Ba_3(BO_3)_3Cl$: a material with large SHG enhancement activated by Pb-chelated BO_3 groups[J]. Journal of the American Chemical Society, 2015, 137: 9417-9422.

[23] Zou G H, Huang L, Ye N, et al. $CsPbCO_3F$: a strong second-harmonic generation material derived from enhancement via p-π interaction[J]. Journal of the American Chemical Society, 2013, 135: 18560-18566.

[24] Zhao S G, Yang Y, Shen Y G, et al. Cooperation of three chromophores generates the water - resistant nitrate nonlinear optical material $Bi_3TeO_6OH(NO_3)_2$[J]. Angewandte Chemie International Edition, 2017, 56: 540-544.

[25] Xia M J, Jiang X X, Lin Z S, et al. "All-three-in-one": a new bismuth - tellurium - borate Bi_3TeBO_9 exhibiting strong second harmonic generation response[J]. Journal of the American Chemical Society, 2016, 138: 14190-14193.

[26] Mao F F, Hu C L, Xu X, et al. Bi (IO_3) F_2: the first metal iodate fluoride with a very strong second harmonic generation effect [J]. Angewandte Chemie International Edition, 2017, 56: 2151–2155.

[27] Liang M L, Hu C L, Kong F, et al. $BiFSeO_3$: an excellent SHG material designed by aliovalent substitution [J]. Journal of the American Chemical Society, 2016, 138: 9433–9436.

[28] Isaenko L I, Yelisseyev A P. Recent studies of nonlinear chalcogenide crystals for the mid–IR [J]. Semiconductor. Science and Technology, 2016, 31: 123001.

[29] Ebrahim–Zadeh M, Sorokina I T. Mid–Infrared coherent sources and applications [J]. Springer Netherlands, 2008.

[30] Ohmer M C, Pandey R, Bairamov B H, et al. Emergence of chalcopyrites as nonlinear optical materials [J]. MRS Bulletin, 1998, 23: 16–20.

[31] Lin X, Zhang G, Ye N, et al. Growth and characterization of $BaGa_4S_7$: a new crystal for mid–IR nonlinear optics [J]. Crystal Growth and Design, 2009, 9: 1186–1189.

[32] Guo Y, Zhou Y, Lin X, et al. Growth and characterizations of $BaGa_4S_7$ crystal [J]. Optical Materical, 2014, 36: 2007–2011.

[33] Yao J Y, Mei D J, Bai L, et al. $BaGa_4Se_7$: a new congruent–melting IR nonlinear optical material [J]. Inorganic Chemistry, 2010, 49: 9212–9216.

[34] Yao J Y, Yin W L, Feng K, et al. Growth and characterization of $BaGa_4Se_7$ crystal [J]. Journal of Crystal Growth, 2012, 346: 1–4.

[35] Yang F, Yao J Y, Xu H Y, et al. High efficiency and high peak power picosecond mid–infrared optical parametric amplifier based on $BaGa_4Se_7$ crystal [J]. Optics Letters, 2013, 38: 3903–3905.

[36] Yang F, Yao J Y, Xu H Y, et al. Midinfrared optical parametric amplifier with 6.4 – 11 μm range based on $BaGa_4Se_7$ [J]. IEEE Photonics Technology Letters, 2015, 27: 1100–1103.

[37] Zhang X, Yao J Y, Yin W L, et al. Determination of the nonlinear optical coefficients of the $BaGa_4Se_7$ crystal [J]. Optical Express, 2015, 23: 552–558.

[38] Yuan J H, Li C, Yao B Q, et al. High power, tunable mid–infrared $BaGa_4Se_7$ optical parametric oscillator pumped by a 2.1 μm Ho: YAG laser [J]. Optical Express, 2016, 24: 6083–6087.

[39] Xu W T, Wang Y, Xu D G, et al. High–pulse–energy mid–infrared optical parametric oscillator based on $BaGa_4Se_7$ crystal pumped at 1.064 μm [J]. Applied Physics B–Lasers and Optics, 2017, 123: 80.

[40] Zhai N, Li C, Xu B, et al. Temperature–dependent sellmeier equations of IR nonlinear optical crystal $BaGa_4Se_7$ [J]. Crystals, 2017, 7 (3): 62.

[41] Isaenko L I, Vasilyeva I G. Nonlinear $LiB^{III}C^{YI}{}_2$ crystals for mid–IR and far–IR: novel aspects in crystal growth [J]. Journal of Crystal Growth, 2008, 310: 1954–1960.

[42] Wang S, Gao Z, Yin X, et al. Crystal growth and piezoelectric, elastic and dielectric properties of novel $LiInS_2$ crystal [J]. Journal of Crystal Growth, 2013, 362: 308–311.

[43] 王善朋，陶绪堂，刘贯东，等. 大尺寸红外非线性光学晶体 $LiInS_2$ 的生长与性能研究 人工晶体学报，2009，38：851–855.

[44] Wang S, Gao Z, Zhang X, et al. Crystal growth and effects of annealing on optical and electrical properties of mid-infrared single crystal $LiInS_2$ [J]. Crystal Growth and Design, 2014, 14: 5957–5961.

[45] Wang S, Zhang X, Zhang X, et al. Modified bridgman growth and properties of mid–infrared $LiInSe_2$ crystal [J]. Journal of Crystal Growth, 2014, 401: 150–155.

[46] (a) Dai S B, Jia N, Chen J K, et al. Picosecond mid–infrared optical parametric amplifier based on $LiInSe_2$ with tenability extending from 3.6 to 4.8 μm [J]. Optical Express, 2017, 25: 12860–12866; (b) Wang S, Dai S, Jia N, et al. Tunable 7 – 12 μm picosecond optical parametric amplifier based on a $LiInSe_2$ mid–infrared crystal [J].

Optical Letters 2017，42，2098–2101.

[47] (a) Liang Q，Wang S，Tao X，et al. Erratum: tunable semimetallic state in compressive–strained $SrIrO_3$ films revealed by transport behavior [J]. Physical Review B，2015，91: 035110; (b) Liang Q，Wang S，Tao X，et al. Optical properties of $LiInSe_2$ in the THz frequency regime [J]. Optical Materials Express，2014，4: 1336–1344.

[48] Li X S，Kang L，Li C，et al. $PbGa_4S_7$: a wide–gap nonlinear optical material [J]. Journal of Materials Chemistry C，2015，3: 3060–3067.

[49] Li X，Li C，Gong P，et al. $BaGa_2SnSe_6$: a new phase–matchable IR nonlinear optical material with strong second harmonic generation response [J]. Journal of Materials Chemistry C，2015，3: 10998–11004.

[50] Li C，Yin W L，Gong P F，et al. Trigonal planar $[HgSe_3]^{4-}$ unit: a new kind of basic functional group in IR nonlinear optical materials with large susceptibility and physicochemical stability [J]. Journal of the American Chemical Society，2016，138: 6135–6138.

[51] Kang L，Zhou M，Yao J，et al. Metal thiophosphates with good mid–infrared nonlinear optical performances: A first–principles prediction and analysis [J]. Journal of the American Chemical Society，2015，137: 13049–13059.

[52] Zhou M，Kang L，Yao J，et al. Midinfrared nonlinear optical thiophosphates from $LiZnPS_4$ to $AgZnPS_4$: A combined experimental and theoretical study [J]. Inorganic Chemistry，2016，55: 3724–3726.

[53] Liang F，Kang L，Lin Z，et al. Analysis and prediction of mid–IR nonlinear optical metal sulfideswith diamond–like structures [J]. Coordination Chemistry Reviews，2017，333: 57–70.

[54] Wu K，Yang Z，Pan S，et al. The first quaternary diamond–like semiconductor with 10–membered LiS_4 rings exhibiting excellent nonlinear optical performances [J]. Chemistry Communation，2017，53: 3010–3013.

[55] Wu K，Pan S. Li_2HgMS_4 (M= Si，Ge，Sn): new quaternary diamond–like semiconductors for infrared laser frequency conversion [J]. Crystals，2017，7: 107.

[56] Li G，Wu K，Liu Q，et al. $Na_2ZnGe_2S_6$: A new infrared nonlinear optical material with good balance between large second–harmonic generation response and high laser damage threshold [J]. Journal of the American Chemical Society，2016，138: 7422–7428.

[57] Wu K，Yang Z，Pan S. Na_2BaMQ_4 (M=Ge，Sn; Q=S，Se): infrared nonlinear optical materials with excellent performances and that undergo structural transformations [J]. Angewandte Chemie International Edition，2016，55: 6712–6714.

[58] Wu K，Yang Z，Pan S. $Na_2Hg_3M_2S_8$ (M= Si，Ge，and Sn): new infrared nonlinear optical materials with strong second harmonic generation effects and high laser–damage thresholds [J]. Chemistry of Materials，2016，28: 2795–2801.

[59] Wu K，Su X，Pan S，et al. Synthesis and characterization of mid–infrared transparency compounds: acentric $BaHgS_2$ and centric $Ba_8Hg_4S_5Se_7$ [J]. Inorganic Chemistry，2015，54: 2772–2779.

[60] Wu K，Yang Z，Pan S. $Na_4MgM_2Se_6$ (M= Si，Ge): the first noncentrosymmetric compounds with special ethane–like $[M_2Se_6]^{6-}$ units exhibiting large laser–damage thresholds [J]. Inorganic Chemistry，2015，54: 10108–10110.

[61] Zhang M J，Jiang X M，Zhou L J，et al. Two phases of Ga_2S_3: promising infrared second–order nonlinear optical materials with very high laser induced damage thresholds [J]. Journal of Materials Chemistry C，2013，1: 4754–4760.

[62] Liu B–W，Zeng H–Y，Zhang M–J，et al. Syntheses，structures，and nonlinear–optical properties of met al sulfides $Ba_2Ga_8MS_{16}$ (M= Si，Ge) [J]. Inorganic Chemistry，2015，54: 976–981.

[63] Zhang M–J，Li B–X，Liu B–W，et al. Ln_3GaS_6 (Ln=Dy，Y): new infrared nonlinear optical materials with high laser induced damage thresholds [J]. Dalton Transactions，2013，42: 14223–14229.

[64] Li S F，Liu B W，Zhang M J，et al. Syntheses，structures，and nonlinear optical properties of two sulfides

$Na_2In_2MS_6$ (M=Si, Ge) [J] . Inorganic Chemistry, 2016, 55: 1480–1485.

[65] Liu B W, Jiang X M, Wang G E, et al. Oxychalcogenide $BaGeOSe_2$: highly distorted mixed–anion building units leading to a large second–harmonic generation response [J] . Chemistry of Materials, 2015, 27: 8189–8192.

[66] Li S F, Jiang X M, Liu B W, et al. Superpolyhedron–built second harmonic generation materials exhibit large mid–Infrared conversion efficiencies and high laser–induced damage thresholds [J] . Chemistry of Materials, 2017, 29: 1796 – 1804.

[67] Lin H, Chen L, Zhou L J, et al. Functionalization based on the substitutional flexibility: strong middle IR nonlinear optical selenides $AX^{II}_4X^{III}_5Se_{12}$ [J] . Journal of the American Chemical Society, 2013, 135: 12914–12921.

[68] Shi Y–F, Chen Y–k, Chen M–C, et al. Strongest second harmonic generation in the polar R_3MTQ_7 family: atomic distribution induced nonlinear optical cooperation [J] . Chemistry of Materials, 2015, 27: 1876–1884.

[69] Lin H, Zheng Y–J, Hu X–N, et al. Template– and additive–free electrosynthesis and characterization of spherical gold nanoparticles on hydrophobic conducting polydimethylsiloxane [J] . Chemistry – An Asian Journal, 2017, 12: 453–458.

[70] Lin H, Chen L, Yu J–S, et al. Infrared SHG materials CsM3Se6 (M= Ga/Sn, In/Sn) : phase matchability controlled by dipole moment of the asymmetric building unit [J] . Chemistry of Materials, 2017, 29: 499–503.

[71] Luo Z–Z, Lin C–S, Zhang W–L, et al. $Ba_8Sn_4S_{15}$: a strong second harmonic generation sulfide with zero–dimensional crystal structure [J] . Chemistry of Materials, 2014, 26: 1093–1099.

[72] Luo Z Z, Lin C S, Cui H H, et al. SHG materials $SnGa_4Q_7$ (Q=S, Se) appearing with large conversion efficiencies, high damage thresholds, and wide transparencies in the mid–infrared region [J] . Chemistry of Materials, 2014, 26: 2743–2749.

[73] Luo Z Z, Lin C S, Cui H H, et al. $PbGa_2MSe_6$ (M=Si, Ge) : two exceptional infrared nonlinear optical crystals [J] . Chemistry of Materials, 2015, 27: 914–922.

[74] Brant J A, Clark J, Kim Y S, et al. Outstanding laser damage threshold in Li_2MnGeS_4 and tunable optical nonlinearity in diamond–like semiconductors [J] . Inorganic Chemistry, 2015, 54: 2809–2819.

[75] Zhang J H, Clark D J, Brant J A, et al. Infrared nonlinear optical properties of lithium–containing diamond–like semiconductors $Li_2ZnGeSe_4$ and $Li_2ZnSnSe_4$ [J] . Dalton Transactions, 2015, 44: 11212–11222.

[76] Rosmus K A, Brant J A, Wisneski S D, et al. Optical nonlinearity in Cu_2CdSnS_4 and α / β –Cu_2ZnSiS_4: diamond–like semiconductors with high laser–damage thresholds [J] . Inorganic Chemistry, 2014, 53: 7809–7811.

[77] Haynes A S, Saouma F O, Otieno C O, et al. Phase–change behavior and nonlinear optical second and third harmonic generation of the one–dimensional $K_{(1-x)}Cs_xPSe_6$ and metastable β –$CsPSe_6$ [J] . Chemistry of Materials, 2015, 27: 1837–1846.

[78] Lai W H, Haynes A S, Frazer L, et al. Hsu, Second harmonic generation response optimized at various optical wavelength ranges through a series of cubic chalcogenides $Ba_6Ag_{2.67+4\delta}Sn_{4.33-\delta}S_{16-x}Se_x$ [J] . Chemistry of Materials, 2015, 27: 1316–1326.

[79] Feng K, Kang L, Lin Z. S, et al. Noncentrosymmetric chalcohalide $NaBa_4Ge_3S_{10}Cl$ with large band gap and IR NLO response [J] . Journal of Materials Chemistry C 2014, 2: 4590–4596.

[80] Li Y–Y, Liu P–F, Hu L, et al. Strong IR NLO material $Ba_4MGa_4Se_{10}Cl_2$: highly improved laser damage threshold via dual ion substitution synergy. Advanced Optical Materials, 2015, 3: 957–966.

[81] Liu B W, Zeng H Y, Jiang X M, et al. [A_3X] [Ga_3PS_8] (A=K, Rb; X=Cl, Br) : promising IR non–linear optical materials exhibiting concurrently strong second–harmonic generation and high laser induced damage thresholds [J] . Chemical Science, 2016, 7: 6273–6277.

[82] Liu P–F, Li Y–Y, Zheng Y–J, et al. Tailored synthesis of nonlinear optical quaternary chalcohalides: $Ba_4Ge_3S_9Cl_2$, $Ba_4Si_3Se_9Cl_2$ and $Ba_4Ge_3Se_9Cl_2$ [J] . Dalton Transactions, 2017, 46: 2715–2721.

[83] Liang F, Kang L, Lin Z, et al. Mid-infrared nonlinear optical materials based on metal chalcogenides: structure - property relationship [J] . Crystal Growth & Design, 2017, 17: 2254–2289.

[84] Xia S X, Wang M, Yang C H, et al. Vertical Bridgman growth and characterization of large $ZnGeP_2$ single crystals [J]. Journal of Crystal Growth, 2011, 314: 306–309.

[85] Lei Z, Zhu C, Xu C, et al. Growth of crack-free $ZnGeP_2$ large single crystals for high-power mid-infrared OPO applications [J] . Journal of Crystal Growth, 2014, 389: 23–29.

[86] Wang Z, Mao M, Wu H, et al. Study on annealing of infrared nonlinear optical crystal $ZnGeP_2$ [J] . Journal of Crystal Growth, 2012, 359: 11–14.

[87] Wang L, Xing T, Hu S, et al. Mid-infrared ZGP-OPO with a high optical-to-optical conversion efficiency of 75. 7% [J] . Optics Express, 2017, 25: 3373–3380.

[88] Kang B, Dou Y, Tang M, et al. Growth of pure $ZnGeP_2$ crystals by horizontal gradient freeze method and its properties [J] . Journal of the Chinese Ceramic Society 2016, 44: 503–507.

[89] Zhang G D, Ruan H P, Zhang X, et al. Vertical bridgman growth and optical properties of $CdSiP_2$ crystals [J] . Crystengcomm, 2013, 15: 4255–4260.

[90] Wang C, Sun J, Gou H, et al. Intrinsic defects and their effects on the optical properties in the nonlinear optical crystal $CdSiP_2$: a first-principles study [J] . Physical Chemistry Chemical Physics, 2017, 19: 9558–9565.

[91] Pu Y, Zhu S, Zhao B, et al. Growth and characterization of Cr-doped $CdGeAs_2$ crystal [J] . Journal of Crystal Growth, 2017, 467: 150–154.

[92] Pan M-y, Ma Z-j., Liu X-c, et al. Wu, $Ba_4AgGa_5Pn_8$ (Pn=P, As) : new pnictide-based compounds with nonlinear optical potential [J] . Journal of Materials Chemistry C, 2015, 3: 9695–9700.

[93] Kang L, Ramo D. M, Lin Z, et al. First principles selection and design of mid-IR nonlinear optical halide crystals [J]. Journal of Materials Chemistry C, 2013, 1: 7363–7370.

[94] Dang Y, Meng X, Jiang K, et al. A promising nonlinear optical material in the Mid-IR region: new results on synthesis, crystal structure and properties of noncentrosymmetric β-HgBrCl [J] . Dalton Transactions, 2013, 42: 9893–9897.

[95] Huang Y, Meng X, Kang L, et al. Hg_2Br_3I: a new mixed halide nonlinear optical material in the infrared region [J] . Crystengcomm, 2013, 15: 4196–4200.

[96] Wu Q, Li Y, Chen H, et al. HgBrI: A promising nonlinear optical material in IR region [J] . inorganic chemistry communications, 2013, 34: 1–3.

[97] Wu Q, Meng X, Zhong C, et al. $Rb_2CdBr_2I_2$: a new IR nonlinear optical material with a large laser damage threshold [J]. Journal of the American Chemical Society, 2014, 136: 5683–5686.

[98] Huang Y, Meng X, Gong P, et al. A study on $K_2SbF_2Cl_3$ as a new mid-IR nonlinear optical material: new synthesis and excellent properties [J] . Journal of Materials Chemistry C, 2015, 3: 9588–9593.

[99] Zhang H, Zhang M, Pan S, et al. $Pb_{17}O_8Cl_{18}$: a promising IR nonlinear optical material with large laser damage threshold synthesized in an open system [J] . Journal of the American Chemical Society, 2015, 137: 8360–8363.

[100] Zhang W, Tao X, Zhang C, et al. Bulk growth and characterization of a novel nonlinear optical crystal $BaTeMo_2O_9$ [J]. Crystal Growth & Design, 2008, 8: 304–307.

[101] Gao Z L, Tian S, Zhang J, et al. Determination and analysis of the linear and second order nonlinear optical properties of $Na_2TeW_2O_9$ [J] . Optical Materials Express, 2016, 6: 106–113.

[102] Zhao P, Cong H J, Tian X X, et al. Top-seeded solution growth, structure, morphology, and functional properties of a new polar crystalb $Cs_2TeW_3O_{12}$ [J] . Crystal Growth & Design, 2015, 15: 4484–4489.

[103] Gao Z L, Wu Q, Liu X T, et al. Biaxial crystal α-$BaTeMo_2O_9$: theory study of large birefringence and wide-band polarized prisms design [J] . Optics Express, 2015, 23: 3851–3860.

[104] Li Y, Wu Z, Zhang X, et al. Crystal growth and terahertz wave generation of organic NLO crystals: OH1 [J]. Journal of Crystal Growth, 2014, 402: 53-59.

[105] Liu P, Xu D, Li Y, et al. Widely tunable and monochromatic terahertz difference frequency generation with organic crystal DSTMS [J]. Europhysics Letters, 2014, 106: 60001.

[106] Liu P, Zhang X, Yan C, et al. Widely tunable and monochromatic terahertz difference frequency generation with organic crystal 2- (3- (4-hydroxystyryl) -5, 5-dime-thylcyclohex-2-enylidene) malononitrile [J]. Applied Physics Letters, 2016, 108: 011104.

[107] Cao L, Teng B, Xu D, et al. Growth, transmission, Raman spectrum and THz generation of DAST crystal [J]. Rsc Advances, 2016, 6: 101389-101394.

[108] Cao L, Teng B, Zhong D, et al. Growth and characterization of DAST crystal with large-thickness [J]. Journal of Crystal Growth, 2016, 451: 188-193.

[109] Zhong K, Mei J, Wang M, et al. Compact high-repetition-rate monochromatic terahertz source based on difference frequency generation from a dual-wavelength Nd: YAG laser and DAST crystal [J]. Journal of Infrared and Millimeter Waves, 2017, 38: 87-95.

[110] 马玉哲，钟德高，曹丽凤，等. 有机 DAST 晶体及其在太赫兹领域的应用研究进展 [J]. 中国科学：技术科学，2017，47.

[111] Zhang X, Jiang X, Li Y, et al. Isoxazolone-based single crystals with large second harmonic generation effect [J]. Crystengcomm, 2015, 17: 7316-7322.

[112] Zhang X, Jiang X, Liu P, et al. Molecular design on isoxazolone-based derivatives with large second-order harmonic generation effect and terahertz wave generation [J]. Crystengcomm, 2016, 18: 3667-3673.

撰稿人：罗军华　叶　宁　毛江高　洪茂椿

生物矿化与无机材料仿生合成研究进展

一、引言

生物体制造和使用矿物为自身提供支撑、保护、捕猎、咀嚼、飞行、浮力控制、光感应、磁场感应等功能已有数亿年历史。经长期自然选择和进化，无论是生物矿物本身还是其矿化过程都与非生物矿物差异巨大，具有鲜明而独特的属性。由于条件限制，生物矿物所使用的原料是环境中大量且易得的钙盐、镁盐、二氧化硅、铁氧化物等无机物，以及糖类、蛋白质等有机物，在生物大分子和其他小分子作用下，通过对材料微观结构的调控，在温和条件下就可得到性能优越的生物矿物材料[1-4]。生物矿物的上述特点也正是人工材料发展所追求的方向。以此为契机，无机材料仿生合成领域的发展也逐步进入新的阶段（图1）。

图1　生物矿物、模拟生物矿化与无机材料仿生合成的关系

近年来我国在多个领域的快速发展使国家对具有突破性性能的新型材料的需求越来越迫切（见“高性能构件材料—结构一体化设计与制造”重大项目指南），对生物矿化及生物矿物材料的深入研究将为研发新型材料提供极为重要的设计思路。首先，不同于常规矿化，生物矿化过程涉及无定形相的参与及非经典的晶体生长过程；其次，生物矿化常意味着对晶体形貌、晶体取向、暴露晶面的准确调控；最后，生物矿物具有典型的跨尺度多级有序结构，此类结构能显著增强材料的宏观性能，这是基于微米 / 纳米基元制备宏观材料的指导性原理之一（如国家重大科学研究计划项目“仿生轻质高强纳米复合结构材料的可控制备与性能研究”等）[5-9]。这些特征和优势成为无机复合材料仿生合成领域重要的灵感来源。

鉴于本领域研究对材料科学和化学科学等领域发展的重要意义，本报告将从三个方面依次阐述和分析近年来本领域的发展现状和最新成果：①生物矿化及生物矿物研究进展；②模拟生物矿化研究进展；③无机材料仿生合成发展动向。本报告还将基于这些新成果，对领域发展方向和我国面临的机遇与挑战做出总结和展望。

二、发展现状及最新进展

人们对生物矿物的研究已逾百年，主要研究对象包括贝壳、骨骼、牙齿、磁小体、硅藻等，研究内容包括矿化生长机理和生物矿物微观结构等。通过对矿化生长机理的研究，人们提出并发展了软硬模板理论、无定形前驱体效应、非经典的取向搭接原理与介观晶体概念、聚合物辅助的水热合成方法等，不仅推动了我们对晶体的认识和晶体生长理论的发展，也为各种材料的制备提供了新的策略。通过对生物矿物微观结构的观察和深入研究，人们认识到多级有序结构对材料宏观性能的提升起到至关重要的作用，从而开启了对仿生结构材料的研究。因此，长期以来生物矿化和仿生合成都是该领域研究的重点和热点。尤其是近年来，仿生合成的概念进一步扩展，成为材料科学发展的一个重要方向。

（一）生物矿化及生物矿物研究进展

对生物矿物本身的研究是矿化机理和仿生合成等工作的基础。本部分将总结该领域进展。

1. 软体动物壳

对以珍珠层结构为代表的软体动物外壳的研究已持续多年，以往工作揭示了这种生物矿物材料的组成、结构、生长过程和性能特征等[6]。对软体动物壳的研究经验也可用于探索其他种类生物矿物材料。

最新的研究首先进一步关注了其生长过程。Zlotnikov 等通过对大江珧蛤外壳棱柱层断层成像，发现其结构生长可用经典的多晶体系垂直生长和熟化过程预测，表明其棱柱层形貌控制是通过设定热力学边界条件实现的[10]。Gilbert 等证明红鲍鱼中文石层中的文石可能源自准方解石型的无定形碳酸钙（ACC）[11]。这些观察有助于我们更好地理解矿化和生物矿化的本质，指导人工材料的合成。

其次，近年来对文石片颗粒状结构的关注显著增多。Estroff 等观察了在大江珧蛤外壳的珍珠层 - 棱柱层转变区纳米尺度的组装过程，认为组装过程是由纳米颗粒在有机基质中的组装驱动的[12]。Checa 等发现多种软体动物珍珠层文石片在去除有机物后，均展现出蠕虫状条纹和沙漏型结构[13]。李晓东等观察到红鲍螺珍珠层文石片由小晶粒组成，裂纹是沿着晶粒间以锯齿型扩展，而在纯无机文石矿物中，裂纹沿着解理面扩展，这种差异使得文石片对裂纹的抗性大大提高[14]。对文石片结构的观察为人工结构材料的设计提供了理论依据，即通过颗粒型结构设计可有效提高材料对裂纹扩展的抗性。

软体动物壳中各种精细结构和能量耗散过程对力学性能的贡献也受到了广泛关注。Ortiz 等发现海月的方解石质层状外壳受到外力破坏时，会在下方的方解石层内诱发产生孪晶结构并催生一系列额外的非弹性能量耗散增韧机理[15]。冯西桥等通过有限元分析计算了矿物桥尺寸效应，发现矿物桥直径在 10~50 nm 时其力学表现最好[16]。张哲峰等通过对紫石房蛤多级结构的观察，分析了多级结构中的不均匀性能促发多重增强和增韧机理[17]。随着精细结构的模型和功能被越来越清晰的揭示，它们在仿生材料设计、制备过程中也将受到更多关注。

分析软体动物壳中有机物的组分和特征有助于了解生物矿物的成因，指导人工材料的合成。谢丽萍、张荣庆等识别并研究了与马氏珍珠贝外壳形成相关的一系列基质蛋白[18, 19]。王小祥等观察了白蝶珍珠蛤珍珠层文石片中被包覆的有机物[20]。张刚生等发现翡翠股贻贝珍珠层的文石片间有机薄膜和文石片取向有很大的相关性[21]。以上这些结果表明，可溶性蛋白对碳酸钙结晶过程有调控作用，而不溶性有机框架则可能诱导结晶过程的发生。

此外，还有一些有特色的工作，如刘克松、李群仰等通过动态有限元分析法研究了捕食者螳螂虾与被捕食者鲍鱼之间的相互竞争、共同进化关系[22]。

2. 骨骼与牙齿

骨骼与牙齿是与人类关系最密切的两种生物矿物。骨骼主要由磷酸钙和胶原蛋白组成，具有复杂的多级结构[23]。而牙釉质是人和其他哺乳动物身体中最硬的组织，对保护牙齿免受日常研磨、咀嚼及化学侵蚀等磨损起着至关重要的作用。牙釉质优异的机械强度和抗疲劳性归功于周期性排列的羟基磷灰石（HAP）纳米线 / 棒组成的多级结构。

近期工作中，研究人员更加深入的探究了骨骼的微纳结构及力学性能。Reznikov 等利用双光束电子显微镜和串行表面视图（SSV）方法，研究了大鼠胫骨周围层状骨的三维胶原结构[24]，其后又证明人体层状骨由有序和无序两种不同材料组成[25]。Schwiedrzik 等研究了单个骨单元和宏观绵羊骨试样的压缩性能，发现二者压缩屈服特性和破坏机制差别显著[26]。Tertuliano 等利用 TEM 对骨小梁进行分析，发现层状骨无序相中的孔隙结构在骨受力断裂时能有效耗散能量[27]。这些工作有助于我们深入认识骨骼优异性能的根本来源，并据此设计具有优异性能的人工材料。

尽管骨骼矿化过程的研究难度较大，但基于对软体动物壳矿化研究的经验，近年来也取得了一些出色进展。牛丽娜等对不同电性聚电解质诱导下胶原蛋白纤维矿化的机理进行研究，认为渗透压平衡是造成纤维内矿化的原因[28]。Addadi 等证实了在骨架形成边缘处存在含有 HPO_4^{2-} 的瞬态前驱体，并认为该相即是酸性无定形磷酸钙，而其与胶原纤维的作用非常紧密[29]。Azaïs 和 Nassif 等发现了脊椎动物细胞外基质的钙化组织中的水分子在构建磷灰石矿物中的重要作用[30]。人工骨骼材料在医学中有重大应用价值，此类研究对制备兼具强度、力学、组成、生物相容性等特征的人工骨有重要的启发意义。

同样，对牙釉质矿物结构的研究也取得了显著进展。Joester 等研究了啮齿动物牙釉质，发现存在稳定的镁取代无定形磷酸钙（Mg-ACP）[31]。Cairney 等使用原子探针层析成像技术观察成熟牙釉质中 HAP 纳米线之间的 Mg-ACP 晶间相，发现 HAP 纳米线中富含镁的细长沉淀物和有机物[32]。这些发现的一个有趣之处在于，牙釉质中 ACP 可与软体动物壳中的 ACC 类比，故可预期 ACP 也能在晶态磷酸钙制备中作为中间体。此外，牙齿中如有铁氧化物成分存在，则其硬度会显著增加，一些研究揭示了此类含铁牙齿的矿化机理。Joester 等研究了选择性沉淀亚稳态水和氧化铁（Fh）的过程，探究了石鳖早期牙齿发育过程中铁的化学环境及其演化的特征[33]。Kisailus 等研究了斯特勒氏隐石鳖的牙齿矿化过程，发现 α－几丁质能够潜在影响磁铁矿晶体聚集体的密度和形成棒的直径和曲率，这在局部机械性能方面发挥关键作用[34]。

较之其组分，牙釉质的性能同样依赖于其多级结构。Arola 等研究了人牙釉质断裂机理，发现其复杂多级结构能有效阻止裂缝到达牙本质[35]。Kisailus 等使用微观光谱分析结合有限元模拟来探究超结构特征，揭示了斯特勒氏隐石鳖完全矿化牙齿的结构与性能间的关系，阐述了各尺度结构的演变特征[36]。王小祥等发现耐磨损的海胆牙齿由沿中心单晶纤维高度取向的纳米晶体组成，并且纤维和基质通过许多单晶纳米管结合在一起[37]。

3. 其他生物矿物

大自然还创造了很多其他的复杂结构，比如鱼鳞、墨鱼骨及节肢动物的钳子等，这些生物矿物各自具有特殊的结构和组分，并为人工材料的制备提供借鉴。

这些复杂矿物的形成机理各不相同。Joester 等发现海胆幼体中 ACC 转变为方解石可分为三个阶段，即短程有序的水合 ACC、短程有序的方解石、长程有序的方解石[38]。Scheffel 等通过体内、外实验证明，颗石藻通过可溶性分子和不溶性框架的特异性作用诱导碳酸钙前驱体颗粒聚集于框架特定位置，从而控制产物形貌[39]。Pokroy 等对红贺复海鞘骨针中的球霰石结构进行研究，发现其六方晶系主体内至少含有一个其他共存晶体结构[40]。Gubieda 等对麦芽孢杆菌进行时间分辨研究，发现水铁矿是磁铁矿生物矿化的铁离子源[41]。Baumgartner 等发现磁性细菌体内磁铁矿是通过高度无序的富含磷酸盐的氢氧化铁相转变形成的，此过程中会在磁小体细胞器内形成瞬态纳米氧化铁中间体[42]。

这些生物矿物的独特结构与其优越性能吸引了广泛关注。齐利民、马玉荣等研究了海胆脊柱，分析了同心矿物层形成机理[43]。Miserez 等发现螳螂虾螯足的表面由呈梯度分布的矿物组成，这赋予其优越的抗冲击能力，而其结晶取向又使其刚度具有各向异性[44]。Naleway 等报道了箱鲀具有高度矿化的硬质外鳞和屈服性胶原基质构成的复杂结构，能够很好的抵御穿透和破碎[45]。Imai 等发现球龟螺的外壳结构是通过文石相晶体棒旋转堆积和沿着螺旋轴的纤维束的扭转堆叠形成的特定分层结构[46]。Checa 等研究了欧洲横纹墨鱼的内壳结构，发现它是由复杂的钙化柱和有机膜排列组成的重叠腔室构成，这为我们设计轻质高强结构材料提供了新的思路[47]。Ortiz 等研究了石鳖生物矿化感光系统，其由数百个文石镜片阵列组成，形成了兼具优化的光学和结构功能的多用途系统[48]。Meyers、Ritchie 及 Allison 等观察了巨骨舌鱼由平行胶原纤维和硬质高度矿化的形成的鳞片，并研究了其 Bouligand 结构和能量耗散机制[49–51]。

综上可见，除贝壳、骨骼和牙齿外，其他生物矿物同样可为我们提供重要的材料设计策略，其结构—性能关系和矿化机理都有重要的研究价值。

（二）模拟生物矿化研究进展

本部分将总结在人工模拟环境下，对生物矿物中常见的碳酸钙、磷酸钙、二氧化硅、铁氧化物等相关无机材料生长、控制的研究。

1. 碳酸钙体系

碳酸钙是软体动物壳的主要成分，也是被研究最广泛、深入的一种生物矿物无机组分。早期的研究更多是经验性的，即着眼于大量的实验现象。近年来随着表征手段的发展，材料生长的基础原理方面有不少新进展；同时，碳酸钙的合成已不仅是关注其形貌，而是更多的面向其实际应用。

（1）添加剂调控作用

纵观生物矿物的生成可发现，其几乎总是涉及复杂的有机－无机、可溶性－不可溶性物质的协同调控作用。近期这方面研究的重点从单因素控制转向多因素控制，且从对形貌、晶型的观察和理论性机理转变为对控制机理的实验性研究。

在可溶性软模板协同调控方面，周根陶等报道了制备形貌较为均一的六边形球霰石介观晶体的工作，其特色在于使用的添加剂是两种小分子盐，而非常用的大分子[52]。俞书宏等研究了水 / 乙醇溶剂组成和温度对碳酸钙晶型变化和结构演进的协同作用，分析了不同条件下起决定作用的机理[53]。齐利民、马玉荣等通过聚丙烯酸（PAA）稳定的纳米颗粒的取向聚集和再结晶，在方解石基底上成功制备了多种形貌的碳酸钙[54]。他们还报道了一种在方解石基底上生长钙掺杂碳酸锶 / 碳酸钙复合微柱阵列的方法[55]。Sommerdijk、

Yoreo 等利用透射电镜原位观察了在碳酸钙结晶过程中钙离子与聚苯乙烯磺酸钠连接并形成小滴的过程，证明离子连接在晶体成核中扮演了一个关键角色[56]。

而对于硬模板的作用机理方面，Dove 等定量研究了一系列多糖基底在碳酸钙矿化中的动力学过程，并发现其本质是基底 - 晶体和基底 - 溶剂之间表面能最小化的过程[57]。傅正义、谢浩等研究了碳酸钙在胶原蛋白纤维内的受限结晶过程及机理，这项研究有助于我们理解骨骼矿化的方式[58]。

碳酸钙矿化中较小无机离子的新型作用机理也被不断提出。Xu 等探究了在镁 - 钙 - 碳酸盐体系中阳离子水合效应的影响，发现无水环境中碳酸盐沉淀的结晶度仍高度依赖于镁含量，这表明镁离子阻碍碳酸钙结晶能力可能不止与其水合能力有关[59]。邵正中等的研究从另一个角度揭示了无机小离子在矿化过程中的角色：三价铁离子诱导蚕丝蛋白构象转变为 β - 折叠，从而引导 ACC 前驱体生长为具有特定形貌的稳定文石相，此过程中无机小离子的调控作用是间接发生的[60]。

（2）无定形碳酸钙

生物矿物中 ACC 不仅可作为晶态碳酸钙的中间态，也可稳定存在并发挥其他诸多功能。ACC 有两个显著特征，即因其非晶性，它较易塑形、易与其他材料复合。通过无定形前驱体制备晶态材料不仅限于碳酸钙体系，因此对 ACC 的研究也对制备其他材料有启发意义。近年来，ACC 稳定、可控转化及与其他材料的复合仍是矿化研究的重要方向之一。

作为一种高活性的碳酸钙，纯相 ACC 的稳定一直是个难题，俞书宏、Cölfen 等报道了一种制备在无任何添加剂下能稳定存在的 ACC 纳米球的方法，并研究了其结晶过程[61]。ACC 与聚合物的结合则会带来一些新颖的性质，例如，Cölfen 等利用高分子量聚丙烯酸（PAA）和 ACC 结合，制备出可塑形、可拉伸、可自愈的“橡皮泥碳酸钙”水凝胶，其干燥之后形成力学性能良好的透明复合材料，再遇水则能完全恢复到初始的水凝胶状态，易与其他材料复合[62]。ACC 的不稳定性赋予了它一些独特的应用潜力，赵彦利、俞书宏等制备了一种 ACC/ 阿霉素复合的纳米球，并在其外包覆了一层二氧化硅，在 pH 约 6.5 左右时，这种结构内的阿霉素会随着 ACC 的溶解发生最大程度的释放，而这个 pH 恰与肿瘤细胞的微酸性环境类似，因此能够高选择性而持久地杀灭肿瘤细胞[63]。俞书宏等分析了在添加剂作用下，碳酸钙在多种有机物基底上高度各向异性生长的过程，他们发现纳米线的侧壁上有一层 ACC，而尖端则存在很多多晶区域，故提出侧壁上高溶解性的带负电 ACC 是一种自发形成的模板，使得碳酸钙选择性沿长度方向生长[64]。

ACC 向晶态碳酸钙转化时的控制和机理研究对多种材料制备都有重要价值。Addadi 等先沉积 ACC 再控制其晶化，发现形成的多面体方解石单晶还维持了纳米颗粒的纹理构造，这表明颗粒聚集机理在生物矿物生长过程中可能有关键性作用[65]。俞书宏等制备了镁稳定的 ACC（Mg-ACC），并通过控制反应条件调控了产物的晶型和形貌[66]。Tobler 等观测了溶液中不同浓度的柠檬酸盐对 ACC 形成和晶化过程，发现柠檬酸盐能稳定

ACC，同时诱导 ACC 形成具有方解石结构的原子排列[67]。Meldrum 等发现 ACC 在溶液和空气中（即固态转变）结晶前会伴随一个脱水过程，而最后失去的少于 15% 的水触发了结晶过程，他们提出为了克服最终失水结晶的势垒，在室温条件下晶核需要通过溶解 / 再沉淀机理方能形成，而受限空间之所以能稳定 ACC 也与受限空间中溶解 / 再沉淀过程被抑制有关[68]。

ACC 在生物体内常常要在受限空间内塑形和晶化生长，因此它与硬膜板的相互作用也值得关注。植物叶子中的钟乳体主要由稳定的 ACC 组成，同时也是一个由纤维素和木质素组成的水凝胶系统，Addadi 等研究了水环境中钟乳体通过固态转变晶化为方解石的过程，并提出该过程是离子溶解 / 沉积及纳米颗粒解体 / 结晶等动力学竞争的过程，因此凝胶可能通过控制相关过程的速率而扮演关键性角色[69]。Kato 等利用几丁质晶须构筑了具有良好周期性液晶结构的几丁质框架，然后渗入 PAA-ACC，并通过在溶液中陈化制备了几丁质 /PAA/ 晶态碳酸钙复合材料，这种方法巧妙结合了天然有机材料的自组装性能与 ACC 的前驱体作用，为构筑其他有序复合材料提供了思路[70]。

（3）介观晶体

实验中发现，在液相中碳酸钙能够以类液态颗粒的形式存在，并通过聚集 - 晶化过程转变为晶体[71]。通过这个过程形成的晶体，尽管整体呈现为单一的晶体取向，但可能仍保持了组成单元的独立性，具有可识别的单元，这种与单晶略有区别的晶体被称为介观晶体。当然，介观晶体是根据其结构定义的，即由介观尺度（1~1000 nm）的亚单元组成的超结构，而非由其形成方式定义[72]。唐睿康、潘海华等发现，碳酸钙纳米晶可先形成无序多晶聚集体，然后通过固有的表面张力诱导的取向重排而转变为单晶，这个例子中通过聚集机理形成的直接是单晶而非介观晶体[73]。介观晶体被认为是一种非经典的单晶生长方式，因此它对于材料制备科学有着重要意义。

对介观晶体形貌的真正控制一直是个难题，目前的工作还是更多依赖于经验。徐安武和 Cölfen 等报道了一步法合成特殊的异质介观晶体结构，球霰石球状核心周围有一圈取向性方解石赤道环[74]。Ihli、Meldrum 等通过气相扩散方法实现了基于聚集机理生长的碳酸钙介观晶体的可控制备，并发现在扩散生长过程中，溶液过饱和度几乎是个常数，且远高于 ACC 的溶解度，而生长只消耗了很小一部分可用的离子[75]。

由于介观晶体与单晶一样展现出一致的取向性，因此其表征存在很多困扰。Meldrum 等分析了方解石与三种聚合物共沉淀形成的晶体，他们发现一些常用的确认介观晶体的手段需谨慎使用，例如较大的比表面积可能是由于一层吸附在单晶晶核上的疏松纳米颗粒带来的，XRD 衍射峰宽化可能是由于晶格应变[76]。因此，我们需要更先进的表征手段来识别介观尺度晶粒之间界面的情况，而介观晶体作为一个非晶 / 多晶和单晶之间的状态，这种“由亚单元组成”究竟如何准确定义，究竟添加剂 / 空穴哪种程度的存在可以被认为是界面而非吞埋（occlusion），这些问题都需要进一步考虑。

（4）碳酸钙与其他材料的复合

生物利用碳酸钙作为生物矿物中的无机相，原因主要是较易从环境中获得。人工体系中，从制备角度看，人工碳酸钙的力学性能难及天然生物矿物，需要通过与其他材料的复合优化其性能；从应用角度看，碳酸钙与其他材料复合，可制备出多种功能材料。

在碳酸钙功能化方面，李峻柏、戴陆如等观察了明胶在球霰石和方解石两种碳酸钙晶体结构中的分布，并研究了球霰石向方解石转化过程中明胶被排出的过程[77]。Meldrum 等通过在方解石中吞埋不同染料，制备了具有可控荧光的功能化方解石，并表征了晶体生长各阶段染料分子的分布变化，研究了吞埋过程和形貌转变之间的联系[78]。李寒莹等在分别含有金和 Fe_3O_4 纳米颗粒琼脂糖胶体中生长碳酸钙晶体，成功制备了功能化的方解石复合材料，且未破坏晶体结构，这为制备复合材料提供了一个很好的思路[79]。

在提升碳酸钙力学性能方面，Tremel 等报道了一个非常有趣的工作，他们利用一种硅蛋白（silicatein-α）与方解石的复合，通过颗粒自组装生长了具有极强弯曲能力的方解石单晶长针，其弯折可以超过 90 度，仅含有小于 16% 的有机物；这项工作颠覆了对脆性晶体材料的传统认知，表明柔性组分的引入可带来力学性能的飞跃[80]。Meldrum 等研究了通过与氨基酸的复合实现方解石硬度的调控，发现氨基酸以独立份子的形式存在于复合材料中，且压痕强度随着氨基酸含量提高而增加，通过位错钉扎模型解释，他们认为这种增强的根源是用于切断分子内共价键的力[81]。他们还研究了胶束和方解石复合的过程中决定纳米尺度吞埋过程的不同机理，发现胶束在被吞埋时会被压缩并产生空穴，这造成了局部晶格应变，从而提升了力学性能，这为通过有机分子吞埋能增强无机材料力学性能提供了微观证据[82]。

2. 磷酸钙体系

作为人体骨骼、牙齿的主要无机成分，对磷酸钙矿化过程的研究不仅有助于理解晶体生长机理，更对移植、修复材料的制备有实际意义。

（1）介稳态磷酸钙

类似于碳酸钙，在向稳定晶型转化过程中，也存在一些介稳态磷酸钙，如 ACP、磷酸八钙（OCP）等。对介稳态中间体的稳定、转化的研究可以指导晶态磷酸钙的制备。

孙志伟、潘海华等通过对 I 型胶原纤维在模拟体液中的矿化研究了 ACP 聚集状态对羟基磷灰石成核动力学的影响[83]。Birkedal 等发现在磷酸蛋白能诱导产生新的磷酸钙中间状态，这种新的中间体具有凝聚体或聚合物诱导的液相前驱体相（PILP）的许多特征[84]。Yoreo、Sommerdijk 等原位观察了磷酸钙结晶的过程，发现了可溶性三磷酸钙离子纳米簇实际上就是之前所发现的成核前团簇，这些团簇可通过从溶液中俘获一个钙离子并聚集形成大的 ACP 颗粒、作为形成磷酸八钙、磷灰石的基础，并且可降低成核的势垒[85]。王荔军等原位观察了羟基磷灰石的形貌演化的过程，发现其同时具有经典理论中的螺旋形生长和非经典理论中的中间体颗粒搭接过程，这表明这两种理论并非互斥[86]。

Christenson、Meldrum 等通过研究生理条件下羟基磷灰石从溶液中沉积的过程，发现受限空间能有效延长羟基磷灰石介稳前驱体 ACP 和 OCP 的存在时间，作者认为其原因是传质过程受阻[87]。

（2）磷酸钙可控合成与相关复合材料

磷酸钙人工材料的可控合成对其性能和应用有着决定性影响。李建树等合成具有三磷酸酯或双膦酸酯外围基团的树枝状聚合物，其表现出非胶原类蛋白的相似功能[88]。Guagliardi 等报道了在柠檬酸根作用下制备了缺失原本六方对称性的扁平磷灰石，柠檬酸根可诱导扁平形貌的形成，他们认为骨骼中的磷酸钙矿化可能也依赖于类似机理[89]。王荔军等发现极低浓度下牙釉蛋白 C 端与透钙磷石（010）晶面有很强的相互作用，从而调控了矿物生长[90]。

具有一定有序结构的磷酸钙基材料通常具有较好的性能。常江等通过自下而上的方法制备了一种具有多级有序结构的羟基磷灰石和明胶复合的人造生物矿物，材料沿 C 轴方向的模量和硬度与人骨相当[91]。Tiller 等通过在水凝胶中的酶促磷酸钙矿化制备了一种水凝胶 /ACP 复合材料，通过调控其组成不仅可使材料透明，而且可使其杨氏模量突破 400MPa，而水凝胶本身杨氏模量不到 10MPa，断裂能可达 1300J/m^2，远超目前多数人工水凝胶材料，这种材料还具有很强的透水能力，这项工作凸显了水凝胶矿化的重要意义[92]。Wiesner、Estroff 等报道了一种利用嵌段共聚物和有机硅酸盐修饰的无定形磷酸钙合成具有良好的周期性有序结构纳米复合物的方法，其压痕模量比非结构化的材料高一个数量级[93]。朱英杰等仿生制备牙釉质结构，通过油酸钙前驱体的溶剂热转化成功合成了由超长 HAP 微管高度有序阵列制成的毫米尺度宏观三维块材[94]。他们还报道了快速、大量制备有序、超长的 HAP 纳米线的方法，在室温下自动化生产 HAP 纳米线组装体，并利用高强度高弹性纳米线实现高度柔韧织物和三维打印材料的制备[95]。

（3）生物医学应用

磷酸钙具有良好的生物相容性，可被人体降解，故在医学领域有良好的应用前景。朱英杰等报道了超长 HAP 单晶微管的合成，其显示出良好的生物相容性及优异的药物负载、持续释放性能，具有多种生物医学应用价值[96]。朱宏伟和 Eliaz 等在柔性中孔石墨烯 / 单壁碳纳米管薄膜上合成均一、规整的仿生 HAP 棒阵列，这种复合薄膜有良好的生物相容性和促进矿化的能力，故有良好的生物医学应用前景[97]。刘海清等通过在模拟体液（SBF）中仿生矿化制备了碳纳米纤维（CNF）/ 羟基磷灰石三维复合支架材料，该材料力学性能良好，在骨组织工程领域中有潜在的应用价值[98]。基于病毒载体的疫苗通常具有良好的疗效，但是容易引起免疫反应而严重抑制其效力，唐睿康等报道了一种在疫苗表面包覆一层生物友好的磷酸钙外壳的方法，不仅避免了免疫反应，甚至磷酸钙外壳还能增强抗原特异性 T 细胞响应[99]。

胶原蛋白是骨骼中主要有机组分，因此磷酸钙 / 胶原蛋白复合材料是一种最佳的骨修复材料。然而人工体系中以胶原蛋白为硬膜板的磷酸钙矿化，通常得到在胶原纤维表面生长的磷酸钙，这与骨骼中的在胶原蛋白内部矿化形成的复合结构截然不同，且降低了其性能。甘业华、周彦恒等通过同时使用PAA和三聚磷酸钠分别作为螯合剂和框架蛋白类似物，在二者协同作用下使得碳磷灰石在胶原纤维内部生长，这种复合材料的机械性能显著高于在纤维外部矿化的复合材料，这对人工骨的制备很有启示[100]。陈吉华、Tay等报道了一种在胶原蛋白纤维内实现二氧化硅－磷灰石多相共矿化方法，通过选择合适前驱体分步进行硅化和钙化，制备了非晶二氧化硅/胶原蛋白/磷灰石三相互相穿插的复合材料，其抗疲劳能力和回弹性能均得到了增强[101]。

3. 其他体系

除碳酸钙、磷酸钙外，还有其他一些与生物矿物联系紧密的无机相，对这些物质矿化过程的研究一方面可与对碳酸钙、磷酸钙的研究相互映证借鉴，另一方面可将对矿化研究的已有成果推广到更多材料体系。

（1）铁氧化物

铁的氧化物出现在多种生物矿物中，其形式多见为磁铁矿（Fe_3O_4）[102, 103]。通过仿生手段控制铁氧化物的矿化过程，为控制和理解它们的生长行为提供了经验和启示。齐利民等通过控制乙酰丙酮铁水解，合成了均一的介观晶体赤铁矿纳米片，其展示出较高的磁性和比表面[104]。他们还发展了一种自上而下的合成形貌可调的介观晶体方法，合成了分层微锥形和微盘形赤铁矿介观晶体[105]。俞书宏等利用β－环糊精和聚乙二醇控制晶相和形貌，仿生宏量制备了硫化铁磁性纳米颗粒，该材料对阿霉素展现出良好的磁诱导输送能力[106]。Sommerdijk等报道了利用含有多种不同氨基酸的共聚多肽作为调控剂，以水合氧化铁为前驱物共沉淀合成仿生磁铁矿，并且通过调节氨基酸电性，可以调控超顺磁性和亚铁磁性[107]。他们还在聚天冬氨酸调控下部分氧化氢氧化亚铁合成磁铁矿纳米晶体，其在超顺磁性的极限尺寸下仍保有剩磁和矫顽力[108]。陈新等利用丝蛋白作为模板，可控合成了结晶度良好的赤铁矿介观晶体[109]。Estroff等发现在二氧化硅水凝胶中赤铁矿的生长是一种从原子到纳米尺度到介观尺度再到微米尺度的多级结构调节过程[110]。

铁氧化物常展现出磁性，这赋予了它们自组装的可能。江雷等通过采用不对称润湿性的微柱结构模板，合成了一维仿生Fe_3O_4纳米颗粒阵列，并通过调节一维结构纵横比来控制磁性的各向异性，该工作有助于提高我们对磁场感知基本原理的理解[111]。Gálvez、Dom í nguez-Vera等通过在益生菌外表面修饰超顺磁赤铁矿纳米颗粒的方法获得了室温活体人工磁性细菌[112]。Cölfen等通过矿化法合成了一种具有生物相容性的“铁凝胶”，其有机/无机比例高度可控，磁铁矿最高含量可达70%[113]。

（2）二氧化硅

二氧化硅在生物矿物中广泛存在，它不仅单独存在，也常常和其他矿物相共存，发挥调控矿化等多种功能[69, 114]。与磷酸钙类似，它也有良好的生物相容性，因此在生物医学上也有广泛应用。

周根陶等通过在酒石酸盐分子存在下通过疏水表活剂模板合成了二维二氧化硅筛板，其中酒石酸分子表现出两种相对的功能：通过氢键有效连接相邻二氧化硅结构和通过静电排斥适度分离相邻结构[115]。Choi 等利用（RKK）$_4D_8$ 多肽在 TiBALDH 和硅酸中合成二氧化硅 - 二氧化钛纳米仿生复合物，并且将这种轻度矿化的具有生物相容性矿物包覆在小球藻细胞上，能够极大程度提高细胞的耐热性[116]。Yang 等设计了一种短链多肽，其通过静电吸附沉积于酵母细胞表面，并能催化细胞表面发生硅化反应，从而将细胞封装在硅质外壳中[117]。唐睿康等通过在蓝藻细菌表面包覆二氧化硅，减轻光抑制效应，大大提高了高光条件下蓝藻细菌的光合作用效率[118]。唐睿康、秦成峰等还报道了一种通过将人工合成的水合二氧化硅修饰在个体病毒体上获得耐热性病毒的生物仿生方法，发现无定形二氧化硅纳米团簇通过形成限制分子移动的外层来稳定内层的病毒体[119]。

对二氧化硅矿化过程的观察对于研究其他物质的矿化机理也有启发意义。Rimer 等原位观察全硅分子筛 silicalite-1 的生长，发现其生长机理既有按照经典模型方式以二氧化硅分子作为基本生长单元，也有按照非经典方式基于介稳二氧化硅纳米颗粒前驱体的聚集[120]。

（3）钙、钡盐类

对其他一些钙、钡盐类的矿化研究也取得了重要进展。Noorduin、Aizenberg 等通过设计矿化条件按一定流程变化，可实现对碳酸钡生长模式的切换控制，并以此方法制备了碳酸钡微米级“花束”雕塑。这个工作不仅展示了其对晶体形貌的控制技巧，更开启了系统性控制晶体生长、通过多步法制备具有指定形貌晶体的新方向[121]。他们又与 Mahadevan 合作，在碳酸钡、二氧化硅共沉淀体系中建立了不同条件对应的不同生长模式下生长的理论模型，并依据这些模型制备出具有球对称布拉格共振器功能的指定形貌的晶体[122]。这类研究不应仅限于这一个体系，而是要在更多体系内探索系统性控制晶体生长模式的可能性。

矿物包覆不仅能保护细胞，也能杀伤细胞，唐睿康等报道了一种癌细胞靶向钙化（CCTC）的无载药癌症治疗手段，通过 Hela 细胞选择性吸附叶酸分子再局部聚集钙离子进行特殊的细胞矿化，并且通过体内实验发现 CCTC 能够限制癌细胞生长和继发性扩散[123]。

限域环境对其他物质矿化的影响也有报道。Meldrum 等研究了限域对硫酸钙矿化的影响，发现限域阻止了无定形前驱体颗粒的扩散，使其不能聚集转化为半水合物或者石膏[124]。

（三）无机材料仿生合成发展动向

仿生合成是一个非常广泛的研究领域：在有机合成方面，仿生合成理念为我们提供了合成方法和合成目标的各种启示[125, 126]；对无机材料/无机复合材料，仿生思想带来了新的制备策略、新的材料结构和新的理化性能。本节主要关注无机材料相关的仿生合成工作。根据仿生出发点不同，我们将无机材料仿生合成大致分为结构仿生、方法仿生、材料仿生等相互联系的三类。本节将选择性介绍近年的一些重要工作，并分析该领域的全新发展动向。

1. 结构仿生

生物材料经历了长期自然选择优化，其独特结构在提升材料性能方面有极大的优越性。结构仿生即模拟生物材料所具有的结构（通常是有序或多级有序结构），从而由这些结构赋予人工材料特殊的性能[8]。

（1）软体动物珍珠层结构

珍珠层具有文石—有机物层层交叠的微观结构，这种结构能显著提升力学性能，同时也易于人工制备，因此是仿生结构的热点目标之一[6]。珍珠层结构的一般制备方法包括层层组装、基元自组装和冰晶/磁场诱导成型法。

层层组装一般通过在液相中交替组装两种材料实现。俞书宏等报道了通过层层组装法制备氧化石墨烯（GO）/层状双氢氧化物耐火涂层，并展现出结构色[127]。韩景宾、卫敏等报道了层层组装的具有超高阻气性能的层状双氢氧化物/聚丙烯酸膜[128]。Sampath等报道了一种基于喷涂法和陶瓷烧结制备的氧化铝/树脂层状复合材料[129]。Gröschel、Ikkala等首先制备粘土/聚乙烯醇复合薄膜，然后通过叠层法制备了仿珍珠层块材[130]。

基元自组装即先合成基元、再诱导自组装的方法。程群峰、唐智勇等通过抽滤自组装法制备了GO/聚合物复合薄膜，聚合物与GO之间存在共轭交联，从而增强了其力学性能[131]。他们还制备了GO/壳聚糖复合薄膜，二者间氢键/共价键发挥了协同增韧效应[132]。俞书宏等通过抽滤法制备了蒙脱土片（MTM）/壳聚糖/细菌纤维素（CNF）三元混合薄膜，其中CNF发挥了锁定MTM基元的作用[133]。Walther等利用蒸发组装法制备了一系列黏土/聚合物复合膜，并系统研究了不同边厚比基元对结构、性能的影响[134]。

冰晶/磁场诱导成型法即利用水冻结和磁场等诱导陶瓷浆料形成一定结构，再接合烧结、聚合物渗透制备仿珍珠层材料。Deville等通过控制冷冻浆料的组成和形貌，得到了完全由三种陶瓷组成，但却有极佳强度和断裂韧性的新型陶瓷[135]。柏浩、Ritchie等通过对冰模板法冻结装置的改进，得到了大范围有序的层状结构[136, 137]。郭林等通过在冷冻法中引入牺牲材料制备了层间有锁定结构的层状复合材料[138]。Studart等利用磁场诱导法处理磁性陶瓷前驱体浆料制备了仿珍珠层复合块体材料，该方法突出优势在于可以适应各种形状的模具一步成型[139]。

（2）其他结构

结构仿生的对象还包括骨骼、牙齿、荷叶、猪笼草、海星、壁虎足、足弓等各种生物结构。Andersson等报道了利用一种自组装介观有序结构框架，诱导ACP沉积并转化为磷灰石短棒，最终得到具有类骨结构的块材，其具有与松质骨相当的压缩强度[140]。Kotov等报道了通过晶体逐层生长法制备基于氧化锌的人工牙釉质，其具有类似或高于天然牙釉质的黏弹性系数，展现出对冲击和振动的抗性，同时又具有较高的硬度和较低的密度[141]。刘克松等报道了在骨架上修饰仿荷叶表面结构的海绵材料用于油水分离[142]。陈华伟、张鹏飞、江雷等研究了猪笼草口缘区多级结构和其液体单向传输性能的关系，并制备了同样具有单向传输性能的仿猪笼草口缘区人工材料[143，144]。夏振海等制备了与壁虎足类似的自清洁黏性材料[145]。齐利民等首先构筑了方解石 / 聚苯乙烯球模板，然后通过气相扩散法在模板上外延生长了方解石透镜阵列，这种透镜依据其所在方解石晶面不同而展示出不同的光学性能[146]。苏宝连、李昱等基于 Murray 定律，制备了具有大孔 – 中孔 – 微孔等多级孔道结构的仿生氧化锌材料，能极大增强物质在其中的移动力[147]。俞书宏等通过取向冷冻和烧结，制备了具有层状微拱形（仿足弓结构）的碳纳米组装体材料，具有超弹性、耐疲劳性和耐高低温能力[148]。

2. 方法仿生

生物材料不仅为我们提供了值得学习的结构模型，也为我们合成和制备人工材料提供了很多灵感。尤其是对于某些制备较为困难的材料，例如，宏观尺度的具有多级有序结构的材料，学习生物的方法和策略不失为一种明智之举。

（1）无定形相

无定形相对非生物矿物体系的合成同样有重要意义。尽管对某些合成过程的研究表明其不涉及无定形相[149]，但在很多情况中，尤其在有修饰剂或阻碍结晶因素存在条件下，我们根据奥斯特瓦尔德规则可推测，无定形相的角色可能被严重忽略了，这可能是由于人们注意力仅局限在关注反应条件和最终产物上。

傅正义等报道了一种结合无定形相和生物合成的方法，他们将无定形二氧化钛前驱体植入蚌壳培养，得到了高结晶度氮掺杂金红石相二氧化钛[150]。Hutchings、Kondrat 等先合成稳定的无定形水羟碳铜石（georgeite，通常不稳定），再通过煅烧制备出高活性甲醇合成和水煤气转化催化剂[151]。Carriere 等研究了发光材料铕掺杂矾酸钇纳米颗粒的合成过程并发现存在过渡态无定形二级网络结构，这种结构限制了结晶后的晶粒尺寸及聚集状态。他们指出对无定形中间体的研究是解决纳米合成材料尺寸、形貌等问题的关键[152]。Rimer 等报道了一种利用无定形相作为前驱体在无有机添加剂条件下可控制备高纯相钠沸石的方法，作者认为其机理是无定形介稳态先溶解再重结晶[153]。他们还研究了 SSZ-13 菱沸石分子筛的结晶过程，并发现其生长包括两个协调的机理，即非经典的经由无定形硅铝酸盐颗粒在晶面上搭接生长机理和经典的经由分子在晶面上层层生长机理[154]。Tremel 等通过观察磷锌矿的结晶过程，展示了热力学稳定的无定形磷酸锌中间体会先于磷锌矿沉积，在无水条件下，这种无定形相在加热到 400℃时仍不会结晶，这暗示了水在无定形相晶化过程中有重要意义[155]。

（2）大分子调控合成

生物大分子和人工高分子能极大影响无机材料的生长过程及最终晶型、尺寸、形貌。

可溶性软模板的不仅能调控材料生长，也往往能与材料相复合，这种复合会给材料性能带来重要改变。俞书宏等制备了由不同晶型、形貌 FeOOH 组成的阵列膜，且不同形貌的阵列膜具有不同的亲疏水特性[156]。Liang、Falcaro 等研究发现，各种生物大分子可以有效地诱导金属有机框架（MOF）的形成，并通过仿生矿化过程控制 MOF 的形态，利用壳层 MOF 保护内层大分子，使之在极端条件下仍能保持生物活性[157]。Lin 等利用星形多臂嵌段共聚物作为软模板，在高温油相环境中制备了一系列尺寸、形貌接近单分散的纳米晶[158]。Berger 等利用重组胱硫醚 γ－裂解酶进行单酶促矿化反应并以之作为晶体生长的软模板，制备了具有可控光学性能的 CdS 纳米晶[159]，他们还用类似的方法可控制备了 CdSe 和 CdSe-CdS 核壳纳米晶[160]。这些结果表明单酶促反应是一种通过矿化法制备功能化材料的有力手段。杨东等基于对氨基酸功能的理解合成所需多肽，并利用这些多肽合成金属氧化物－金属纳米复合材料，该策略可制备多种无机纳米颗粒[161]。周钰明等通过一步法合成变性牛血清白蛋白壳包覆的发光 CdSe 半导体量子点，具有优异的稳定性[162]。

硬膜板能很好的诱导晶体形貌。牛丽娜、陈吉华、Tay 等通过将乙酰丙酮稳定的氧化锆前驱体纳米液滴引入聚电解质包覆的胶原基质中，制备了胶原－氧化钇稳定的无定形氧化锆混合支架，聚电解质涂层引发前体的内部凝结缩合成无定形氧化锆，煅烧后转变成氧化钇稳定的氧化锆[163]。王生杰、徐海等报道了利用多肽纳米纤维模板，在温和条件下合成均一的高孔隙率、高比表面积的枝状二氧化锰/多肽复合纳米线[164]。此外，俞书宏等用高表面能的带电荷碲纳米线模拟可溶性大分子，同时发挥软、硬模板的作用，矿化生长得到 ACC 壳层，并诱导了 ACC 结晶，这项工作为我们提供了一种新的模板法思路[165]。

（3）生物模板法

多级有序结构模板的人工合成通常都较困难，而生物材料本身就为我们提供了这样一些优异的模板材料。俞书宏等基于细菌纤维素模板制备了一系列复合材料，展现出优越的催化活性和电化学稳定性[166, 167]。崔屹、姚宏斌等利用螃蟹壳中高度矿化的纤维作为生物模板制备了中空的碳纳米纤维，具有优异的锂离子电池电极材料性能[168]。郭新等利用莲花花粉作为模板，制备了具有优异一氧化氮检测性能的多级多孔结构氧化钨微球材料[169]。Kong、Belcher 利用 M13 噬菌体制备了力学性能优异的气凝胶，且能通过结合多种无机材料实现功能化[170]。Paris 等利用云杉木模板和正硅酸乙酯前驱体制备了木材的二氧化硅复制品，并发现其在低至几纳米尺度的结构复制上都非常完美[171]。胡良斌等通过向木材的单向微孔中填充还原氧化石墨烯，发展了一种理想的锂硫电池阴极材料[172]。Nam、Lee 等利用 M13 噬菌体模板合成了具有高结晶性和压电性能的各向异性 $BaTiO_3$ 纳米晶材料，并利用其制备出柔性纳米发电机[173]。朱申敏、张荻等以蓝闪蝶翅膀为模板制备了可通过 pH 调控颜色的光子晶体[174]。Opdenbosch 等以松果为模板，成功制备了湿度驱动的陶瓷双层驱动器[175]。这些由生物模板制备的材料的优越性能均与其来自生物模板的结构或组分有密不可分的关系。

（4）仿生材料制备 / 合成的新策略

从生物进化的策略，到生物制造宏观尺度且具有多级微纳结构的矿物的策略，包括前文中已介绍的软、硬膜板调控、无定形前驱体等策略，对制备具有良好结构的无机材料都有重要的启示。本部分仅选择一些代表性进展介绍和分析。

遗传算法是基于进化论和遗传学原理搜索最优解的方法。Bawazer、Meldrum 等利用遗传算法优化了小有机分子诱导生长的光致发光量子点矿化，通过选择、重组、突变策略快速实现了可溶性添加剂的优化组合[176]。他们与 Sommerdijk 等合作，利用遗传算法从数百种矿化条件中快速筛选出制备磁铁矿纳米颗粒的条件，并研究了添加物共聚多肽在何种条件下能促进磁铁矿的形成[177]。可以看出，这种仿生遗传算法能够用于优化合成多种固体材料，今后有望将仿生矿化的方法推广之其他材料体系。

生命体系如软体动物创造了诸如珍珠母等性能卓越的矿物材料，如何制备仿生珍珠母等结构材料是公认的难题，而一些热不稳定性或脆性材料（如 MOF 和天然珍珠层中的有机物、碳酸钙等）更加难以制备。腹足纲软体动物在构筑珍珠层时，首先形成数层不溶性有机框架（几丁质等），然后再在可溶性调控组分作用下在框架里矿化完成生长[6]。在此启发下，俞书宏等首次提出了一种介观尺度“组装与矿化”相结合的合成方法，以取向冷冻干燥法预先构筑一个具有微观层状结构的几丁质框架，在层间预留充分空间，然后泵入矿化溶液使碳酸钙在框架内矿化，最后通过热压，成功制备出结构（包括一些精细结构）、性能和组成均与天然珍珠层高度类似的宏观块材[178]，实现了人工珍珠母的矿化合成。这种从介观尺度切入，结合从上至下和自下而上两种方法构筑多级结构材料的合成新策略，可应用于制备一系列宏观尺度仿生工程材料，具有广泛应用前景[179]。

3. 材料仿生

材料仿生即模拟生物矿物的物质组成，其目的一般是为制备生物相容性较好的组织工程材料，易于被机体同化、易于诱导组织再生。

一类工作是通过植入仿生的有机支架、并在机体内诱导矿化。张志愿等借助微流体技术成功制备出具有模仿细胞外基质三维纳米纤维网络结构的壳聚糖微球，以其作为细胞载体，通过在体外与软骨细胞共培养，成功获得了具有类似软骨组织的体外软骨组织复合块[180]。Mooney 课题组发展了一种可注射水凝胶，并研究表明通过调节其弹性模量和化学特性，可以实现间充质干细胞的体外成骨以及体外和体内细胞布散[181]。毛传斌、杨明英等提出了一种利用病毒活化生物框架的策略，有效促进了体内骨生成和血管的生成[182]。

另一类工作则引入了矿化的或含有矿物的支架。陈宗刚等制备了可降解的基于硫酸钙 / 矿化胶原的骨修复材料[183]。刘文广、冯学泉等发展了利用原位沉淀矿化的策略制备出高强度、高韧性的磷酸钙矿物水凝胶材料，其能有效促进头盖骨缺损的再生[184]。Shih

等利用磷酸钙矿化基质模仿富磷酸钙的骨微环境，用于培养人体间充质干细胞，结果表明细胞外基质中的磷酸盐新陈代谢过程对骨的更新起到重要的作用[185]。王铁、周彦恒等利用一种新型仿生矿化法成功制备出具有模仿天然骨的交错纳米拓扑结构的骨组织工程支架材料[186]。陈吉华、Tay 等利用在胶原纤维上同时矿化硅和磷灰石的杂化生物矿化方法，制备了一种共矿化胶原支架材料，并发现该材料具有很好的促进成骨效果和抑制破骨细胞形成特性[187]。

三、总结与展望

对生物矿化和无机材料仿生合成的研究都具有重要的理论和应用价值：2005 年《科学》杂志以“Design for Living”专刊的形式，强调了化学、材料科学与生命科学的多学科交叉对探索仿生先进材料制备与构筑的重要性[188]；《自然》杂志于 2015 年以增刊形式再次强调了仿生概念对新材料的设计制备起到重要的指导作用，展望了仿生材料研究未来的发展趋势[189]。而对生物矿物、生物矿化的研究，则为仿生材料领域提供了源源不断的灵感。通过对探索大量材料组分和制造新结构的可行方法的研究，有望发现未知的、反直觉的材料行为，这些发现有望突破材料科学的前沿[190]。

由于国家的持续投入和广大研究人员的努力，我国目前在生物矿化和无机材料的仿生合成领域已占有一席之地，做出了不少原创性工作，在一些方向上已处于领跑位置，但仍有很多不足：对生物矿物本身的研究，我国仍显落后，研究目光局限在已发现的几种生物矿物，这可能与相关人才培养不足以及此类研究缺乏直接应用有关；对模拟生物矿化和相关矿化机理的研究，我国近年来提出的新见解不多，创新性仍显不足；对于无机材料的仿生合成，我国已处于第一梯队，然而拥有的原创性、突破性的技术尚不足，占主导地位的方向也不多。

随着研究进一步深入，我们可以预期：生物矿物的结构 - 性能关系以及形成方式的研究将不断扩散到新种类生物矿物中；模拟生物矿化和矿化机理的研究前沿将集中在对矿化过程的原位观察、对矿化基本过程的分析、晶型和形貌的系统性定量性控制及其他多种材料的复合这几个方面；无机材料仿生合成的研究前沿主要包括发展新的宏量制备技术、制备新型结构的仿生材料、解决块材制备的效率问题、人工 / 天然高分子等添加剂辅助的纳米材料可控制备、生物医学应用等。我国未来一段时间应充分把握以上发展机遇并争取取得若干重要突破，真正把先进仿生材料应用于能源、航空航天、生物医药、催化转化、光电器件等重要领域。我国目前已具备一定的科研软、硬件条件，通过不断的努力以及一些新技术、新手段、新思路的发现和使用，必能在不久的将来在国际上引领该领域的研究。

参考文献

[1] Cusack M, Freer A. Biomineralization: Elemental and organic influence in carbonate systems [J] . Chemical Reviews, 2008, 108 (11) : 4433–4454.

[2] Faivre D, Schüler D. Magnetotactic bacteria and magnetosomes [J] . Chemical Reviews, 2008, 108 (11) : 4875–4898.

[3] Hildebrand M. Diatoms, biomineralization processes, and genomics [J] . Chemical Reviews, 2008, 108 (11) : 4855–4874.

[4] Meyers M A, McKittrick J, Chen P Y. Structural biological materials: Critical mechanics–materials connections [J] . Science, 2013, 339 (6121) : 773–779.

[5] Meldrum F C, Cölfen H. Controlling mineral morphologies and structures in biological and synthetic systems [J] . Chemical Reviews, 2008, 108 (11) : 4332–4432.

[6] Yao H B, Ge J, Mao L B, et al. Artificial carbonate nanocrystals and layered structural nanocomposites inspired by nacre: Synthesis, fabrication and applications [J] . Advanced Materials, 2014, 26 (1) : 163–188.

[7] Bae W G, Kim H N, Kim D, et al. Scalable multiscale patterned structures inspired by nature: The role of hierarchy [J] . Advanced Materials, 2014, 26 (5) : 675–700.

[8] Wegst U G K, Bai H, Saiz E, et al. Bioinspired structural materials [J] . Nature Materials, 2015, 14 (1) : 23–36.

[9] Zhang Y, Gong S, Zhang Q, et al. Graphene–based artificial nacre nanocomposites [J] . Chemical Society Reviews, 2016, 45 (9) : 2378–2395.

[10] Bayerlein B, Zaslansky P, Dauphin Y, et al. Self–similar mesostructure evolution of the growing mollusc shell reminiscent of thermodynamically driven grain growth [J] . Nature Materials, 2014, 13 (12) : 1102–1107.

[11] DeVol R T, Sun C Y, Marcus M A, et al. Nanoscale transforming mineral phases in fresh nacre [J] . Journal of the American Chemical Society, 2015, 137 (41) : 13325–13333.

[12] Hovden R, Wolf S E, Holtz M E, et al. Nanoscale assembly processes revealed in the nacroprismatic transition zone of *Pinna nobilis* mollusc shells [J] . Nature Communications, 2015, 6: 10097.

[13] Checa A G, Mutvei H, Osuna–Mascaro A J, et al. Crystallographic control on the substructure of nacre tablets [J] . Journal of Structural Biology, 2013, 183 (3) : 368–376.

[14] Huang Z, Li X. Origin of flaw–tolerance in nacre [J] . Scientific Reports, 2013, 3: 1693.

[15] Li L, Ortiz C. Pervasive nanoscale deformation twinning as a catalyst for efficient energy dissipation in a bioceramic armour [J] . Nature Materials, 2014, 13 (5) : 501–507.

[16] Shao Y, Zhao H P, Feng X Q. Optimal characteristic nanosizes of mineral bridges in mollusk nacre [J] . RSC Advances, 2014, 4 (61) : 32451.

[17] Jiao D, Liu Z, Zhang Z, et al. Intrinsic hierarchical structural imperfections in a natural ceramic of bivalve shell with distinctly graded properties [J] . Scientific Reports, 2015, 5: 12418.

[18] Liang J, Xu G, Xie J, et al. Dual roles of the lysine–rich matrix protein (KRMP) –3 in shell formation of pearl oyster, *Pinctada fucata* [J] . PLoS One, 2015, 10 (7) : e0131868.

[19] Liu C, Li S, Kong J, et al. In–depth proteomic analysis of shell matrix proteins of pinctada fucata [J] . Scientific Reports, 2015, 5: 17269.

[20] Wang S N, Zhu X Q, Yan X H, et al. Nanostructured individual nacre tablet: A subtle designed organic–inorganic

composite [J] . CrystEngComm, 2015, 17 (15) : 2964–2968.

[21] Xu J, Zhang G. Direct observation of the crystallographic relationship between interlamellar membranes and aragonite tablets in bivalve nacre [J] . Journal of Structural Biology, 2017, 197 (3) : 308–311.

[22] Li X, Wang J, Du J, et al. Spear and shield: Survival war between mantis shrimps and abalones [J] . Advanced Materials Interfaces, 2015, 2 (14) : 1500250.

[23] Zimmermann E A, Ritchie R O. Bone as a structural material [J] . Advanced Healthcare Materials, 2015, 4 (9) : 1287–1304.

[24] Reznikov N, Almany–Magal R, Shahar R, et al. Three–dimensional imaging of collagen fibril organization in rat circumferential lamellar bone using a dual beam electron microscope reveals ordered and disordered sub–lamellar structures [J] . Bone, 2013, 52 (2) : 676–683.

[25] Reznikov N, Shahar R, Weiner S. Three–dimensional structure of human lamellar bone: The presence of two different materials and new insights into the hierarchical organization [J] . Bone, 2014, 59: 93–104.

[26] Schwiedrzik J, Raghavan R, Burki A, et al. In situ micropillar compression reveals superior strength and ductility but an absence of damage in lamellar bone [J] . Nature Materials, 2014, 13 (7) : 740–747.

[27] Tertuliano O A, Greer J R. The nanocomposite nature of bone drives its strength and damage resistance [J] . Nature Materials, 2016, 15 (11) : 1195–1202.

[28] Niu L N, Jee S E, Jiao K, et al. Collagen intrafibrillar mineralization as a result of the balance between osmotic equilibrium and electroneutrality [J] . Nature Materials, 2017, 16 (3) : 370–378.

[29] Akiva A, Kerschnitzki M, Pinkas I, et al. Mineral formation in the larval zebrafish tail bone occurs via an acidic disordered calcium phosphate phase [J] . Journal of the American Chemical Society, 2016, 138 (43) : 14481–14487.

[30] Wang Y, Von Euw S, Fernandes F M, et al. Water–mediated structuring of bone apatite [J] . Nature Materials, 2013, 12 (12) : 1144–1153.

[31] Gordon L M, Cohen M J, MacRenaris K W, et al. Amorphous intergranular phases control the properties of rodent tooth enamel [J] . Science, 2015, 347 (6223) : 746–750.

[32] La Fontaine A, Zavgorodniy A, Liu H, et al. Atomic–scale compositional mapping reveals mg–rich amorphous calcium phosphate in human dental enamel [J] . Science Advances, 2016, 2 (9) : e1601145.

[33] Gordon L M, Rom á n J K, Everly R M, et al. Selective formation of metastable ferrihydrite in the chiton tooth [J] . Angewandte Chemie International Edition, 2014, 53 (43) : 11506–11509.

[34] Wang Q, Nemoto M, Li D, et al. Phase transformations and structural developments in the radular teeth of *Cryptochiton stelleri* [J] . Advanced Functional Materials, 2013, 23 (23) : 2908–2917.

[35] Yahyazadehfar M, Bajaj D, Arola D D. Hidden contributions of the enamel rods on the fracture resistance of human teeth [J] . Acta Biomaterialia, 2013, 9 (1) : 4806–4814.

[36] Grunenfelder L K, De Obaldia E E, Wang Q, et al. Stress and damage mitigation from oriented nanostructures within the radular teeth of *Cryptochiton stelleri* [J] . Advanced Functional Materials, 2014, 24 (39) : 6093–6104.

[37] Zhu X, Wang S, Deng J, et al. Sophisticated nanostructure in stone part of sea urchin tooth: Enlightenment for artificial composites [J] . Crystal Growth & Design, 2015, 15 (8) : 3842–3846.

[38] Tester C C, Wu C H, Krejci M R, et al. Time–resolved evolution of short– and long–range order during the transformation of amorphous calcium carbonate to calcite in the sea urchin embryo [J] . Advanced Functional Materials, 2013, 23 (34) : 4185–4194.

[39] Gal A, Wirth R, Kopka J, et al. Macromolecular recognition directs calcium ions to coccolith mineralization sites [J] . Science, 2016, 353 (6299) : 590–593.

[40] Kabalah-Amitai L, Mayzel B, Kauffmann Y, et al. Vaterite crystals contain two interspersed crystal structures [J]. Science, 2013, 340 (6131): 454–457.

[41] Fdez-Gubieda M L, Muela A, Alonso J, et al. Magnetite biomineralization in *Magnetospirillum gryphiswaldense*: Time-resolved magnetic and structural studies [J]. ACS Nano, 2013, 7 (4): 3297–3305.

[42] Baumgartner J, Morin G, Menguy N, et al. Magnetotactic bacteria form magnetite from a phosphate-rich ferric hydroxide via nanometric ferric (oxyhydr) oxide intermediates [J]. Proceedings of the National Academy of Sciences of the United States of America, 2013, 110 (37): 14883–14888.

[43] Zhang Y, Chai S, Ma Y, et al. Investigations on the microstructures of sea urchin spines via selective dissolution [J]. CrystEngComm, 2016, 18 (48): 9374–9381.

[44] Amini S, Masic A, Bertinetti L, et al. Textured fluorapatite bonded to calcium sulphate strengthen stomatopod raptorial appendages [J]. Nature Communications, 2014, 5: 3187.

[45] Yang W, Naleway S E, Porter M M, et al. The armored carapace of the boxfish[J]. Acta Biomaterialia, 2015, 23: 1–10.

[46] Suzuki M, Sasaki T, Oaki Y, et al. Stepwise rotation of nanometric building blocks in the aragonite helix of a pteropod shell [J]. Crystal Growth & Design, 2017, 17 (1): 191–196.

[47] Checa A G, Cartwright J H, Sanchez-Almazo I, et al. The cuttlefish *Sepia officinalis* (Sepiidae, Cephalopoda) constructs cuttlebone from a liquid-crystal precursor [J]. Scientific Reports, 2015, 5: 11513.

[48] Li L, Connors M J, Kolle M, et al. Multifunctionality of chiton biomineralized armor with an integrated visual system[J]. Science, 2015, 350 (6263): 952–956.

[49] Yang W, Sherman V R, Gludovatz B, et al. Protective role of *Arapaima gigas* fish scales: Structure and mechanical behavior [J]. Acta Biomaterialia, 2014, 10 (8): 3599–3614.

[50] Zimmermann E A, Gludovatz B, Schaible E, et al. Mechanical adaptability of the bouligand-type structure in natural dermal armour [J]. Nature Communications, 2013, 4: 2634.

[51] Allison P G, Chandler M Q, Rodriguez R I, et al. Mechanical properties and structure of the biological multilayered material system, *Atractosteus spatula* scales [J]. Acta Biomaterialia, 2013, 9 (2): 5289–5296.

[52] Wang Y Y, Yao Q Z, Li H, et al. Formation of vaterite mesocrystals in biomineral-like structures and implication for biomineralization [J]. Crystal Growth & Design, 2015, 15 (4): 1714–1725.

[53] Liu L, Jiang J, Yu S H. Polymorph selection and structure evolution of $CaCO_3$ mesocrystals under control of poly (sodium 4-styrenesulfonate): Synergetic effect of temperature and mixed solvent [J]. Crystal Growth & Design, 2014, 14 (11): 6048–6056.

[54] Long X, Ma Y, Cho K R, et al. Oriented calcite micropillars and prisms formed through aggregation and recrystallization of poly (acrylic acid) stabilized nanoparticles [J]. Crystal Growth & Design, 2013, 13 (9): 3856–3863.

[55] Wu W, Ma Y, Xing Y, et al. Ca-doped strontianite-calcite hybrid micropillar arrays formed via oriented dissolution and heteroepitaxial growth on calcite [J]. Crystal Growth & Design, 2015, 15 (5): 2156–2164.

[56] Smeets P J M, Cho K R, Kempen R G E, et al. Calcium carbonate nucleation driven by ion binding in a biomimetic matrix revealed by in situ electron microscopy [J]. Nature Materials, 2015, 14 (4): 394–399.

[57] Giuffre A J, Hamm L M, Han N, et al. Polysaccharide chemistry regulates kinetics of calcite nucleation through competition of interfacial energies [J]. Proceedings of the National Academy of Sciences of the United States of America, 2013, 110 (23): 9261–9266.

[58] Ping H, Xie H, Wan Y, et al. Confinement controlled mineralization of calcium carbonate within collagen fibrils [J]. Journal of Materials Chemistry B, 2016, 4 (5): 880–886.

[59] Xu J, Yan C, Zhang F, et al. Testing the cation-hydration effect on the crystallization of Ca-Mg-CO_3 systems [J]. Proceedings of the National Academy of Sciences of the United States of America, 2013, 110 (44): 17750–

17755.

[60] Lin G, Zhong Y, Zhong J, et al. Effect of Fe^{3+}on the silk fibroin regulated direct growth of nacre-like aragonite hybrids [J]. Crystal Growth & Design, 2015, 15 (12): 5774-5780.

[61] Chen S F, Cölfen H, Antonietti M, et al. Ethanol assisted synthesis of pure and stable amorphous calcium carbonate nanoparticles [J]. Chemical Communications, 2013, 49 (83): 9564-9566.

[62] Sun S, Mao L B, Lei Z, et al. Hydrogels from amorphous calcium carbonate and polyacrylic acid: Bio-inspired materials for "mineral plastics" [J]. Angewandte Chemie International Edition, 2016, 55 (39): 11765-11769.

[63] Zhao Y, Luo Z, Li M, et al. A preloaded amorphous calcium carbonate/doxorubicin@silica nanoreactor for pH-responsive delivery of an anticancer drug [J]. Angewandte Chemie International Edition, 2015, 54 (3): 919-922.

[64] Mao L B, Xue L, Gebauer D, et al. Anisotropic nanowire growth via a self-confined amorphous template process: A reconsideration on the role of amorphous calcium carbonate [J]. Nano Research, 2016, 9 (5): 1334-1345.

[65] Gal A, Kahil K, Vidavsky N, et al. Particle accretion mechanism underlies biological crystal growth from an amorphous precursor phase [J]. Advanced Functional Materials, 2014, 24 (34): 5420-5426.

[66] Liu Y Y, Jiang J, Gao M R, et al. Phase transformation of magnesium amorphous calcium carbonate (Mg-ACC) in a binary solution of ethanol and water [J]. Crystal Growth & Design, 2013, 13 (1): 59-65.

[67] Tobler D J, Rodriguez-Blanco J D, Dideriksen K, et al. Citrate effects on amorphous calcium carbonate (ACC) structure, stability, and crystallization [J]. Advanced Functional Materials, 2015, 25 (20): 3081-3090.

[68] Ihli J, Wong W C, Noel E H, et al. Dehydration and crystallization of amorphous calcium carbonate in solution and in air. [J]. Nature communications, 2014, 5: 3169.

[69] Gal A, Habraken W, Gur D, et al. Calcite crystal growth by a solid-state transformation of stabilized amorphous calcium carbonate nanospheres in a hydrogel [J]. Angewandte Chemie International Edition, 2013, 52 (18): 4867-4870.

[70] Matsumura S, Kajiyama S, Nishimura T, et al. Formation of helically structured chitin/$CaCO_3$ hybrids through an approach inspired by the biomineralization processes of crustacean cuticles [J]. Small, 2015, 11 (38): 5127-5133.

[71] Gower L B. Biomimetic model systems for investigating the amorphous precursor pathway and its role in biomineralization [J]. Chemical Reviews, 2008, 108 (11): 4551-4627.

[72] Song R Q, Cölfen H. Mesocrystals-ordered nanoparticle superstructures [J]. Advanced Materials, 2010, 22 (12): 1301-1330.

[73] Liu Z, Pan H, Zhu G, et al. Realignment of nanocrystal aggregates into single crystals as a result of inherent surface stress [J]. Angewandte Chemie International Edition, 2016, 55 (41): 12836-12840.

[74] Wang S S, Picker A, Cölfen H, et al. Heterostructured calcium carbonate microspheres with calcite equatorial loops and vaterite spherical cores [J]. Angewandte Chemie International Edition, 2013, 52 (24): 6317-6321.

[75] Ihli J, Bots P, Kulak A, et al. Elucidating mechanisms of diffusion-based calcium carbonate synthesis leads to controlled mesocrystal formation [J]. Advanced Functional Materials, 2013, 23 (15): 1965-1973.

[76] Kim Y Y, Schenk A S, Ihli J, et al. A critical analysis of calcium carbonate mesocrystals [J]. Nature Communications, 2014, 5: 4341.

[77] Fu M, Wang A, Zhang X, et al. Direct observation of the distribution of gelatin in calcium carbonate crystals by super-resolution fluorescence microscopy [J]. Angewandte Chemie International Edition, 2016, 55 (3): 908-911.

[78] Green D C, Ihli J, Thornton P D, et al. 3d visualization of additive occlusion and tunable full-spectrum fluorescence in calcite [J]. Nature Communications, 2016, 7: 13524.

[79] Liu Y，Yuan W，Shi Y，et al. Functionalizing single crystals: Incorporation of nanoparticles inside gel-grown calcite crystals [J]. Angewandte Chemie International Edition，2014，53 (16) : 4127–4131.

[80] Natalio F，Corrales T P，Panthofer M，et al. Flexible minerals: Self-assembled calcite spicules with extreme bending strength [J]. Science，2013，339 (6125) : 1298–1302.

[81] Kim Y Y，Carloni J D，Demarchi B，et al. Tuning hardness in calcite by incorporation of amino acids [J]. Nature Materials，2016，15 (8) : 903–910.

[82] Cho K R，Kim Y Y，Yang P，et al. Direct observation of mineral-organic composite formation reveals occlusion mechanism [J]. Nature Communications，2016，7: 10187.

[83] Jiang S，Jin W，Wang Y N，et al. Effect of the aggregation state of amorphous calcium phosphate on hydroxyapatite nucleation kinetics [J]. RSC Adv.，2017，7 (41) : 25497–25503.

[84] Ibsen C J S，Gebauer D，Birkedal H. Osteopontin stabilizes metastable states prior to nucleation during apatite formation [J]. Chemistry of Materials，2016，28 (23) : 8550–8555.

[85] Habraken W J E M，Tao J，Brylka L J，et al. Ion-association complexes unite classical and non-classical theories for the biomimetic nucleation of calcium phosphate [J]. Nature Communications，2013，4: 1507.

[86] Li M，Wang L，Zhang W，et al. Direct observation of spiral growth，particle attachment，and morphology evolution of hydroxyapatite [J]. Crystal Growth & Design，2016，16 (8) : 4509–4518.

[87] Wang Y W，Christenson H K，Meldrum F C. Confinement increases the lifetimes of hydroxyapatite precursors [J]. Chemistry of Materials，2014，26 (20) : 5830–5838.

[88] Xin J，Chen T，Lin Z，et al. Phosphorylated dendronized poly (amido amine) s as protein analogues for directing hydroxylapatite biomineralization [J]. Chemical Communications，2014，50 (49) : 6491–6493.

[89] Delgado-L ó pez J M，Frison R，Cervellino A，et al. Crystal size，morphology，and growth mechanism in bio-inspired apatite nanocrystals [J]. Advanced Functional Materials，2014，24 (8) : 1090–1099.

[90] Wu S，Zhai H，Zhang W，et al. Monomeric amelogenin's C-terminus modulates biomineralization dynamics of calcium phosphate [J]. Crystal Growth and Design，2015，15 (9) : 4490–4497.

[91] Liu X，Lin K，Wu C，et al. Multilevel hierarchically ordered artificial biomineral [J]. Small，2014，10 (1) : 152–159.

[92] Rauner N，Meuris M，Zoric M，et al. Enzymatic mineralization generates ultrastiff and tough hydrogels with tunable mechanics [J]. Nature，2017，543 (7645) : 407–410.

[93] Song R Q，Hoheisel T N，Sai H，et al. Formation of periodically-ordered calcium phosphate nanostructures by block copolymer-directed self-assembly [J]. Chemistry of Materials，2016，28 (3) : 838–847.

[94] Lu B Q，Zhu Y J，Chen F，et al. Solvothermal transformation of a calcium oleate precursor into large-sized highly ordered arrays of ultralong hydroxyapatite microtubes [J]. Chemistry – A European Journal，2014，20 (23) : 7116–7121.

[95] Chen F，Zhu Y J. Large-scale automated production of highly ordered ultralong hydroxyapatite nanowires and construction of various fire-resistant flexible ordered architectures [J]. ACS Nano，2016，10 (12) : 11483–11495.

[96] Zhang Y G，Zhu Y J，Chen F，et al. Ultralong hydroxyapatite microtubes: Solvothermal synthesis and application in drug loading and sustained drug release [J]. CrystEngComm，2017，19 (14) : 1965–1973.

[97] Zhang R，Metoki N，Sharabani-Yosef O，et al. Hydroxyapatite/mesoporous graphene/single-walled carbon nanotubes freestanding flexible hybrid membranes for regenerative medicine [J]. Advanced Functional Materials，2016，26 (44) : 7965–7974.

[98] Wu M，Wang Q，Liu X，et al. Biomimetic synthesis and characterization of carbon nanofiber/hydroxyapatite composite scaffolds [J]. Carbon，2013，51: 335–345.

[99] Wang X, Sun C, Li P, et al. Vaccine engineering with dual-functional mineral shell: A promising strategy to overcome preexisting immunity [J]. Advanced Materials, 2016, 28 (4): 694-700.

[100] Liu Y, Luo D, Kou X X, et al. Hierarchical intrafibrillar nanocarbonated apatite assembly improves the nanomechanics and cytocompatibility of mineralized collagen [J]. Advanced Functional Materials, 2013, 23 (11): 1404-1411.

[101] Niu L N, Jiao K, Ryou H, et al. Multiphase intrafibrillar mineralization of collagen [J]. Angewandte Chemie International Edition, 2013, 52 (22): 5762-5766.

[102] Gordon L M, Joester D. Nanoscale chemical tomography of buried organic - inorganic interfaces in the chiton tooth [J]. Nature, 2011, 469 (7329): 194-197.

[103] Mirabello G, Lenders J J M, Sommerdijk N A J M. Bioinspired synthesis of magnetite nanoparticles [J]. Chemical Society Reviews, 2016, 45 (18): 5085-5106.

[104] Cai J, Chen S, Ji M, et al. Organic additive-free synthesis of mesocrystalline hematite nanoplates via two-dimensional oriented attachment [J]. CrystEngComm, 2014, 16 (8): 1553-1559.

[105] Cai J, Chen S, Hu J, et al. Top-down fabrication of hematite mesocrystals with tunable morphologies [J]. CrystEngComm, 2013, 15 (32): 6284-6288.

[106] Feng M, Lu Y, Yang Y, et al. Bioinspired greigite magnetic nanocrystals: Chemical synthesis and biomedicine applications [J]. Scientific Reports, 2013, 3 (1): 2994.

[107] Lenders J J M, Zope H R, Yamagishi A, et al. Bioinspired magnetite crystallization directed by random copolypeptides [J]. Advanced Functional Materials, 2015, 25 (5): 711-719.

[108] Altan C L, Lenders J J M, Bomans P H H, et al. Partial oxidation as a rational approach to kinetic control in bioinspired magnetite synthesis [J]. Chemistry - A European Journal, 2015, 21 (16): 6150-6156.

[109] Fei X, Li W, Shao Z, et al. Protein biomineralized nanoporous inorganic mesocrystals with tunable hierarchical nanostructures [J]. Journal of the American Chemical Society, 2014, 136 (44): 15781-15786.

[110] Asenath-Smith E, Hovden R, Kourkoutis L F, et al. Hierarchically structured hematite architectures achieved by growth in a silica hydrogel [J]. Journal of the American Chemical Society, 2015, 137 (15): 5184-5192.

[111] Jiang X, Feng J, Huang L, et al. Bioinspired 1d superparamagnetic magnetite arrays with magnetic field perception [J]. Advanced Materials, 2016, 28 (32): 6952-6958.

[112] Mart í n M, Carmona F, Cuesta R, et al. Artificial magnetic bacteria: Living magnets at room temperature [J]. Advanced Functional Materials, 2014, 24 (23): 3489-3493.

[113] Helminger M, Wu B, Kollmann T, et al. Synthesis and characterization of gelatin-based magnetic hydrogels [J]. Advanced Functional Materials, 2014, 24 (21): 3187-3196.

[114] Ehrlich H, Deutzmann R, Brunner E, et al. Mineralization of the metre-long biosilica structures of glass sponges is templated on hydroxylated collagen [J]. Nature Chemistry, 2010, 2 (12): 1084-1088.

[115] Shi J Y, Yao Q Z, Zhou G T, et al. Two-dimensional silica sieve plates mimicking the diatom valve [J]. Chemistry - A European Journal, 2013, 19 (25): 8073-8077.

[116] Ko E H, Yoon Y, Park J H, et al. Bioinspired, cytocompatible mineralization of silica-titania composites: Thermoprotective nanoshell formation for individual *Chlorella*cells [J]. Angewandte Chemie International Edition, 2013, 52 (47): 12279-12282.

[117] Park J H, Choi I S, Yang S H. Peptide-catalyzed, bioinspired silicification for single-cell encapsulation in the imidazole-buffered system [J]. Chemical Communications, 2015, 51 (25): 5523-5525.

[118] Xiong W, Yang Z, Zhai H, et al. Alleviation of high light-induced photoinhibition in cyanobacteria by artificially conferred biosilica shells [J]. Chemical Communications, 2013, 49 (68): 7525-7525.

[119] Wang G, Wang H J, Zhou H, et al. Hydrated silica exterior produced by biomimetic silicification confers viral

vaccine heat-resistance [J] . ACS Nano，2015，9 (1) : 799–808.

[120] Lupulescu A I, Rimer J D. In situ imaging of silicalite-1 surface growth reveals the mechanism of crystallization[J]. Science，2014，344 (6185) : 729–732.

[121] Noorduin W L, Grinthal A, Mahadevan L, et al. Rationally designed complex, hierarchical microarchitectures [J] . Science，2013，340 (6134) : 832–837.

[122] Kaplan C N，Noorduin W L，Li L，et al. Controlled growth and form of precipitating microsculptures [J] . Science，2017，355 (6332) : 1395–1399.

[123] Zhao R，Wang B，Yang X，et al. A drug-free tumor therapy strategy: Cancer-cell-targeting calcification [J] . Angewandte Chemie International Edition，2016，55 (17) : 5225–5229.

[124] Wang Y W，Christenson H K，Meldrum F C. Confinement leads to control over calcium sulfate polymorph [J] . Advanced Functional Materials，2013，23 (45) : 5615–5623.

[125] Li X H，Zhu M，Wang Z X，et al. Synthesis of atisine，ajaconine，denudatine，and hetidine diterpenoid alkaloids by a bioinspired approach [J] . Angewandte Chemie International Edition，2016，55 (50) : 15667–15671.

[126] Li F，Tu Q，Chen S，et al. Bioinspired asymmetric synthesis of hispidanin A [J] . Angewandte Chemie International Edition，2017，56 (21) : 5844–5848.

[127] Yan Y X，Yao H B，Mao L B，et al. Micrometer-thick graphene oxide-layered double hydroxide nacre-inspired coatings and their properties [J] . Small，2016，12 (6) : 745–755.

[128] Dou Y，Pan T，Xu S，et al. Transparent，ultrahigh-gas-barrier films with a brick-mortar-sand structure [J] . Angewandte Chemie International Edition，2015，54 (33) : 9673–9678.

[129] Dwivedi G，Flynn K，Resnick M，et al. Bioinspired hybrid materials from spray-formed ceramic templates [J] . Advanced Materials，2015，27 (19) : 3073–3078.

[130] Morits M，Verho T，Sorvari J，et al. Toughness and fracture properties in nacre-mimetic clay/polymer nanocomposites [J] . Advanced Functional Materials，2017，27 (10) : 1605378.

[131] Cheng Q，Wu M，Li M，et al. Ultratough artificial nacre based on conjugated cross-linked graphene oxide [J] . Angewandte Chemie International Edition，2013，52 (13) : 3750–3755.

[132] Wan S，Peng J，Li Y，et al. Use of synergistic interactions to fabricate strong，tough，and conductive artificial nacre based on graphene oxide and chitosan [J] . ACS Nano，2015，9 (10) : 9830–9836.

[133] Yan Y X，Yao H B，Yu S H. Nacre-like ternary hybrid films with enhanced mechanical properties by interlocked nanofiber design [J] . Advanced Materials Interfaces，2016，3 (17) : 1600296.

[134] Das P，Malho J M，Rahimi K，et al. Nacre-mimetics with synthetic nanoclays up to ultrahigh aspect ratios [J] . Nature Communications，2015，6: 5967.

[135] Bouville F，Maire E，Meille S，et al. Strong，tough and stiff bioinspired ceramics from brittle constituents [J] . Nature Materials，2014，13 (5) : 508–514.

[136] Bai H，Chen Y，Delattre B，et al. Bioinspired large-scale aligned porous materials assembled with dual temperature gradients [J] . Science Advances，2015，1 (11) : e1500849.

[137] Bai H，Walsh F，Gludovatz B，et al. Bioinspired hydroxyapatite/poly (methyl methacrylate) composite with a nacre-mimetic architecture by a bidirectional freezing method [J] . Advanced Materials，2016，28 (1) : 50–56.

[138] Zhao H，Yue Y，Guo L，et al. Cloning nacre's 3D interlocking skeleton in engineering composites to achieve exceptional mechanical properties [J] . Advanced Materials，2016，28 (25) : 5099–5105.

[139] Le Ferrand H，Bouville F，Niebel T P，et al. Magnetically assisted slip casting of bioinspired heterogeneous composites [J] . Nature Materials，2015，14 (11) : 1172–1179.

[140] He W X, Rajasekharan A K, Tehrani-Bagha A R, et al. Mesoscopically ordered bone-mimetic nanocomposites [J]. Advanced Materials, 2015, 27 (13): 2260-2264.

[141] Yeom B, Sain T, Lacevic N, et al. Abiotic tooth enamel [J]. Nature, 2017, 543 (7643): 95-98.

[142] Zhang X, Li Z, Liu K, et al. Bioinspired multifunctional foam with self-cleaning and oil/water separation [J]. Advanced Functional Materials, 2013, 23 (22): 2881-2886.

[143] Chen H, Zhang P, Zhang L, et al. Continuous directional water transport on the peristome surface of nepenthes alata [J]. Nature, 2016, 532 (7597): 85-89.

[144] Chen H, Zhang L, Zhang P, et al. A novel bioinspired continuous unidirectional liquid spreading surface structure from the peristome surface of *Nepenthes alata* [J]. Small, 2017, 13 (4): 1601676-1601676.

[145] Xu Q, Wan Y, Hu T S, et al. Robust self-cleaning and micromanipulation capabilities of gecko spatulae and their bio-mimics [J]. Nature Communications, 2015, 6: 8949.

[146] Ye X, Zhang F, Ma Y, et al. Brittlestar-inspired microlens arrays made of calcite single crystals [J]. Small, 2015, 11 (14): 1677-1682.

[147] Zheng X, Shen G, Wang C, et al. Bio-inspired murray materials for mass transfer and activity [J]. Nature Communications, 2017, 8: 14921.

[148] Gao H L, Zhu Y B, Mao L B, et al. Super-elastic and fatigue resistant carbon material with lamellar multi-arch microstructure [J]. Nature Communications, 2016, 7: 12920.

[149] Baumgartner J, Dey A, Bomans P H H, et al. Nucleation and growth of magnetite from solution [J]. Nature Materials, 2013, 12 (4): 310-314.

[150] Xie J, Xie H, Su B L, et al. Mussel-directed synthesis of nitrogen-doped anatase TiO_2 [J]. Angewandte Chemie International Edition, 2016, 55 (9): 3031-3035.

[151] Kondrat S A, Smith P J, Wells P P, et al. Stable amorphous georgeite as a precursor to a high-activity catalyst [J]. Nature, 2016, 531 (7592): 83-87.

[152] Fleury B, Neouze M A, Guigner J M, et al. Amorphous to crystal conversion as a mechanism governing the structure of luminescent YVO_4:Eu nanoparticles [J]. ACS Nano, 2014, 8 (3): 2602-2608.

[153] Maldonado M, Oleksiak M D, Chinta S, et al. Controlling crystal polymorphism in organic-free synthesis of Na-zeolites [J]. Journal of the American Chemical Society, 2013, 135 (7): 2641-2652.

[154] Kumar M, Luo H, Román-Leshkov Y, et al. SSZ-13 crystallization by particle attachment and deterministic pathways to crystal size control [J]. Journal of the American Chemical Society, 2015, 137 (40): 13007-13017.

[155] Bach S, Celinski V R, Dietzsch M, et al. Thermally highly stable amorphous zinc phosphate intermediates during the formation of zinc phosphate hydrate [J]. Journal of the American Chemical Society, 2015, 137 (6): 2285-2294.

[156] Liu L, Yang L Q, Liang H W, et al. Bio-inspired fabrication of hierarchical FeOOHnanostructure array films at the air-water interface, their hydrophobicity and application for water treatment [J]. ACS Nano, 2013, 7 (2): 1368-1378.

[157] Liang K, Ricco R, Doherty C M, et al. Biomimetic mineralization of metal-organic frameworks as protective coatings for biomacromolecules [J]. Nature Communications, 2015, 6: 7240.

[158] Pang X, Zhao L, Han W, et al. A general and robust strategy for the synthesis of nearly monodisperse colloidal nanocrystals [J]. Nature Nanotechnology, 2013, 8 (6): 426-431.

[159] Dunleavy R, Lu L, Kiely C J, et al. Single-enzyme biomineralization of cadmium sulfide nanocrystals with controlled optical properties [J]. Proceedings of the National Academy of Sciences of the United States of America, 2016, 113 (19): 5275-5280.

[160] Yang Z, Lu L, Kiely C J, et al. Single enzyme direct biomineralization of cdse and CdSe–CdS core–shell quantum dots [J]. ACS Applied Materials & Interfaces, 2017, 9 (15): 13430–13439.

[161] Liu C, Jiang Z, Tong Z, et al. Biomimetic synthesis of inorganic nanocomposites by a *de novo* designed peptide [J]. RSC Adv, 2014, 4 (1): 434–441.

[162] Bu X, Zhou Y, He M, et al. Bioinspired, direct synthesis of aqueous CdSe quantum dots for high–sensitive copper(Ⅱ) ion detection [J]. Dalton Transactions, 2013, 42 (43): 15411–15420.

[163] Zhou B, Niu L N, Shi W, et al. Adopting the principles of collagen biomineralization for intrafibrillar infiltration of yttria–stabilized zirconia into three–dimensional collagen scaffolds [J]. Advanced Functional Materials, 2014, 24 (13): 1895–1903.

[164] Du M, Bu Y, Zhou Y, et al. Peptide–templated synthesis of branched MnO_2 nanowires with improved electrochemical performances [J]. RSC Adv, 2017, 7 (21): 12711–12718.

[165] Liu Y Y, Liu L, Xue L, et al. Charged inorganic nanowire–directed mineralization of amorphous calcium carbonate [J]. ChemNanoMat, 2016, 2 (4): 259–263.

[166] Liang H W, Wu Z Y, Chen L F, et al. Bacterial cellulose derived nitrogen–doped carbon nanofiber aerogel: An efficient metal–free oxygen reduction electrocatalyst for zinc–air battery [J]. Nano Energy, 2015, 11: 366–376.

[167] Wu Z Y, Hu B C, Wu P, et al. Mo_2 Cnanoparticles embedded within bacterial cellulose–derived 3D N–doped carbon nanofiber networks for efficient hydrogen evolution [J]. NPG Asia Materials, 2016, 8 (7): e288.

[168] Yao H, Zheng G, Li W, et al. Crab shells as sustainable templates from nature for nanostructured battery electrodes [J]. Nano Letters, 2013, 13 (7): 3385–3390.

[169] Wang X X, Tian K, Li H Y, et al. Bio–templated fabrication of hierarchically porous WO_3 microspheres from lotus pollens for no gas sensing at low temperatures [J]. RSC Adv, 2015, 5 (37): 29428–29432.

[170] Jung S M, Qi J, Oh D, et al. M_{13} virus aerogels as a scaffold for functional inorganic materials [J]. Advanced Functional Materials, 2017, 27 (4): 1603203.

[171] Fritz–Popovski G, Van Opdenbosch D, Zollfrank C, et al. Development of the fibrillar and microfibrillar structure during biomimetic mineralization of wood [J]. Advanced Functional Materials, 2013, 23 (10): 1265–1272.

[172] Li Y, Fu K K, Chen C, et al. Enabling high–areal–capacity lithium–sulfur batteries: Designing anisotropic and low–tortuosity porous architectures [J]. ACS Nano, 2017, 11 (5): 4801–4807.

[173] Jeong C K, Kim I, Park K I, et al. Virus–directed design of a flexible $BaTiO_3$ nanogenerator [J]. ACS Nano, 2013, 7 (12): 11016–11025.

[174] Yang Q, Zhu S, Peng W, et al. Bioinspired fabrication of hierarchically structured, pH–tunable photonic crystals with unique transition [J]. ACS nano, 2013, 7 (6): 4911–4918.

[175] Van Opdenbosch D, Fritz–Popovski G, Wagermaier W, et al. Moisture–driven ceramic bilayer actuators from a biotemplating approach [J]. Advanced Materials, 2016, 28 (26): 5235–5240.

[176] Bawazer L A, Ihli J, Comyn T P, et al. Genetic algorithm–guided discovery of additive combinations that direct quantum dot assembly [J]. Advanced Materials, 2015, 27 (2): 223–227.

[177] Lenders J J M, Bawazer L A, Green D C, et al. Combinatorial evolution of biomimetic magnetite nanoparticles [J]. Advanced Functional Materials, 2017, 27 (10): 1604863.

[178] Mao L B, Gao H L, Yao H B, et al. Synthetic nacre by predesigned matrix–directed mineralization [J]. Science, 2016, 354 (6308): 107–110.

[179] Barthelat F. Growing a synthetic mollusk shell [J]. Science, 2016, 354 (6308): 32–33.

[180] Zhou Y, Gao H L, Shen L L, et al. Chitosan microspheres with an extracellular matrix–mimicking nanofibrous structure as cell–carrier building blocks for bottom–up cartilage tissue engineering [J]. Nanoscale, 2016, 8 (1): 309–317.

[181] Huebsch N, Lippens E, Lee K, et al. Matrix elasticity of void-forming hydrogels controls transplanted-stem-cell-mediated bone formation [J]. Nature Materials, 2015, 14 (12): 1269-1277.

[182] Wang J, Yang M, Zhu Y, et al. Phage nanofibers induce vascularized osteogenesis in 3D printed bone scaffolds [J]. Advanced Materials, 2014, 26 (29): 4961-4966.

[183] Chen Z, Kang L, Meng Q Y, et al. Degradability of injectable calcium sulfate/mineralized collagen-based bone repair material and its effect on bone tissue regeneration [J]. Materials Science and Engineering: C, 2014, 45: 94-102.

[184] Xu B, Zheng P, Gao F, et al. A mineralized high strength and tough hydrogel for skull bone regeneration [J]. Advanced Functional Materials, 2017, 27 (4): 1604327.

[185] Shih Y R V, Hwang Y, Phadke A, et al. Calcium phosphate-bearing matrices induce osteogenic differentiation of stem cells through adenosine signaling [J]. Proceedings of the National Academy of Sciences of the United States of America, 2014, 111 (3): 990-995.

[186] Liu Y, Liu S, Luo D, et al. Hierarchically staggered nanostructure of mineralized collagen as a bone-grafting scaffold [J]. Advanced Materials, 2016, 28 (39): 8740-8748.

[187] Jiao K, Niu L N, Li Q H, et al. Biphasic silica/apatite co-mineralized collagen scaffolds stimulate osteogenesis and inhibit RANKL-mediated osteoclastogenesis [J]. Acta Biomaterialia, 2015, 19: 23-32.

[188] Lavine M. Design for living [J]. Science, 2005, 310 (5751): 1131.

[189] Brody H. Biomaterials [J]. Nature, 2015, 519 (7544): S1.

[190] Espinosa H D, Soler-Crespo R. Lesson from tooth enamel [J]. Nature, 2017, 543 (7544): 42-43.

撰稿人：茅琳波　孟玉峰　高怀岭　俞书宏

有机光伏研究进展

一、引言

作为绿色能源的重要组成部分，光伏技术引起了世界各国的广泛关注。与传统的无机太阳能电池相比，本体异质结（简称 BHJ）型有机太阳能电池（简称 OSCs）可通过廉价的印刷工艺制备大面积柔性器件，受到世界范围内学术界和工业界的广泛关注[1-4]。经过多年的发展，OSCs 已经成为涉及有机化学、材料科学、半导体物理学等多学科交叉的前沿科学领域，连续多年成为研究热点。OSCs 的基本器件结构主要由光活性层、透明导电电极和金属电极三部分所组成（图 1a）。功率转换效率（简称 PCE）是 OSCs 的主要性能指标，它正比于电池器件的开路电压（简称 V_{OC}）、短路电流（简称 JSC）和填充因子（简称 FF）的三个参数（图 1b）。

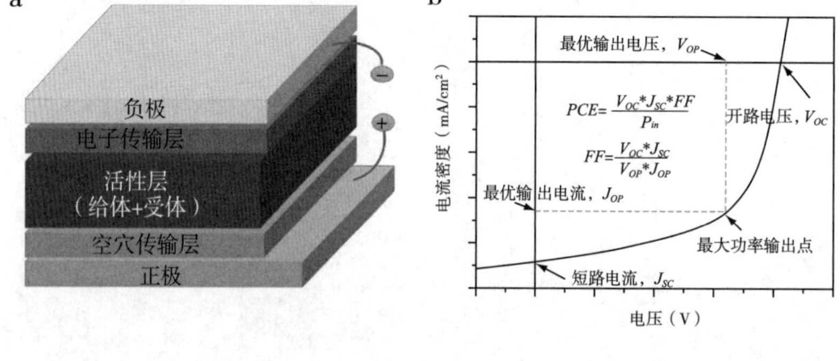

图 1 （a）BHJ 型 OSCs 的器件结构示意图；（b）OSCs 的 *J-V* 曲线以及效率与三个关键参数的关系

OSCs 器件结构优化设计和新型光伏材料的应用是推动其性能不断提升的两个关键因素[5-7]。OSCs 的光伏活性材料主要包括电子给体和电子受体两种，富勒烯衍生物如 C60

或 C70- 苯基丁酸甲酯（简称 $PC_{61}MB$、$PC_{71}MB$）、C60 或 C70- 茚双加成物（简称 $IC_{61}BA$ 或 $IC_{71}BA$）[8, 9] 是应用最为广泛的电子受体材料，在研究初期，研究者主要通过聚合物或小分子电子给体材料的设计和优化来提升 OSCs 的 PCE。但是研究发现，富勒烯类受体材料在吸收光谱、分子能级调制上有着较大的困难，另外相应的电池器件中存在较大的能量损失等诸多问题。因此，研究人员将研究的中心转移到非富勒烯型 OSCs 中，开发应用了多种非富勒烯电子受体，逐渐成为 OSCs 领域新的研究热点[10, 11]。目前，非富勒烯型 OSCs 的 PCE 已经赶上并超过了富勒烯型 OSCs，文献报道的单层器件效率已经超过了 13%[12]。OSCs 电池器件方面，从单层、双层，发展到应用广泛的本体异质结，每一次结构的创新都明显地推动着电池性能的提高。此外，界面层材料的优化[13] 或采用三元[14]、叠层[15] 的结构等，也对 OSCs 性能的提升有十分重要的帮助。

近年来，我国在 OSCs 领域的研究水平位于世界前列，涌现了许多先进成果，为该领域的发展做出了突出贡献，本部分内容从 OSCs 中给受体材料设计、高性能电池器件制备以及工作机理探索几个方面，对近几年该领域的一些代表性工作做了简要的介绍，按照具体内容可以分为电子给体材料、电子受体材料、界面修饰层材料、高性能器件制备和工作机理五个部分。

二、有机太阳能电池研究进展

（一）电子给体材料的发展

有机光伏材料的发展对于推动 OSCs 领域不断取得突破有着十分重要的作用，在早期的研究中，富勒烯衍生物如 PCBM 和 ICBA 等是使用最为广泛的电子受体材料，研究人员把研究的重心放在给体材料的开发和应用上。从材料的分子结构上来分，给体材料可以分为聚合物材料和小分子给体材料两种，并且基于两种给体材料的 OSCs 都可以取得超过 11% 的 PCE[16, 17]。

1. 聚合物给体材料

D-A 交替共聚的方法广泛地应用在光伏材料的设计中，通过调控分子内的推拉电子作用可以实现对材料吸收光谱、分子能级以及电荷迁移性能的综合调制。在种类众多的 D-A 共聚物中，基于苯并二噻吩（简称 BDT）和噻吩并噻吩（简称 TT）的交替共聚物（图 2）因具有优异的性能而得到了广泛的关注和深入的研究。PTB7 是基于烷氧基 BDT 和氟代 TT 的共聚物，其吸收范围在 300~750 nm，基于 PTB7:$PC_{71}BM$ 的 OSCs 在报道之初就取得了 7.4% 的 PCE[18]。吴宏斌和曹镛等在电池器件中采用 PFN 和 MoO_3 分别作为阴极和阳极界面层材料，制备了具有反向结构的电池器件，将电池的效率提高到 9.2%[19]。葛子义等采用非共轭的小分子作为阴极界面层制备的正向器件进一步该类型 OSCs 的效率提高到 10% 以上[20]。Chen 等将共轭支链的方法运用到高效率的 PTB7 聚合物中，设计制

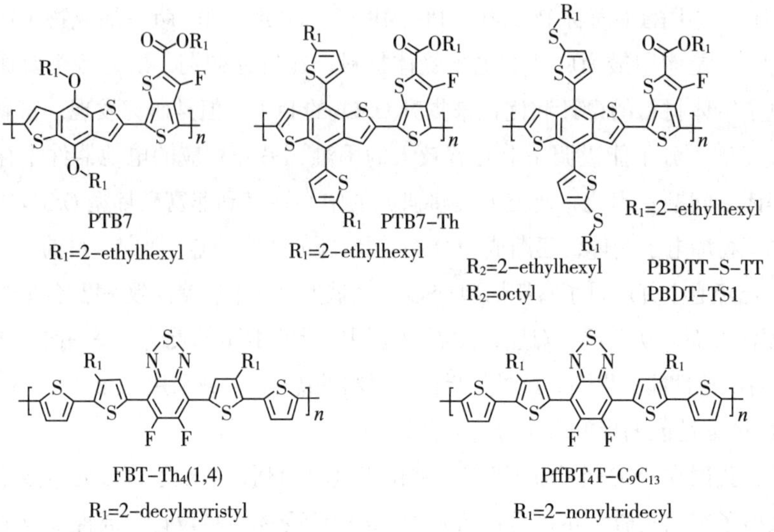

图 2　高效率聚合物给体材料举例

备了基于噻吩共轭支链 BDT 单元与氟代 TT 单元的聚合物材料 PTB7-Th，在器件中获得了 9.35% 的 PCE[21]，通过电池器件的精细优化，基于 PTB7-Th:$PC_{71}BM$ 的 OSCs 可以获得超过 10% 的 PCE[22-24]。2014 年，李永舫等和侯剑辉等分别引入烷硫基基团修饰 BDT，调制了该类材料的分子能级，提高了相应电池器件的开路电压[25, 26]，基于 PBDT-TS1 和 $PC_{71}BM$ 的 OSCs 经过优化可以获得 10.2% 的 PCE[27]。

苯并噻二唑（简称 BT）是一种常用的受体单元，BT 与很多给体单元共聚构建了种类繁多的高性能聚合物给体材料[28, 29]，其中基于 BT 和四联噻吩的聚合物体系性能最为优异。2014 年，陈军武等[30]报道了聚合物 FBT-Th_4（1，4），基于该聚合物和 $PC_{71}BM$ 的 OSCs 器件可以获得 0.76 V 的 V_{OC}，最初效率为 7.64%。随后颜河等[16, 31]进一步优化该体系的材料与器件制备方法，将基于该体系 OSCs 的 PCE 提高 11.7%，是目前简单单层富勒烯型 OSCs 的最高结果。值得注意的是，该聚合物体系具有很强的分子间聚集作用，其溶液在常温下是蓝色的，但是加热到较高温度后逐渐变为红色，表明聚合物分子在溶液中就具有较强的分子间相互作用，有较高的电荷迁移率。

除了以上介绍的几种聚合物给体之外，还有种类繁多、各有特点的聚合物给体材料广泛地应用在 OSCs 的研究中，但是需要注意的是，聚合物材料本身具有分子量的多分散性，对于其光伏特性和研究的结果都有着重要的影响。

2. 小分子给体材料

小分子是相对于聚合物的另外一类电子给体材料，它们具有明确的分子结构，无合成批次的差别，因而有机小分子太阳能电池的研究也是有机光伏电池领域一个重要课题。目前，基于富勒烯受体的小分子太阳能电池的能量转换效率已经超过了 11%[32]。近几年来，在溶液可加工给体小分子材料研究工作中，我国科研人员开发了一系列高效率的小分子给体材料体系（图 3），研究成果水平始终处于国际领先的地位，对该领域的发展有巨大的推动作用。下面对高效率小分子给体材料体系做简要的介绍。高效率有机小分子给体材料按照其结构特点可以分为以下几类：第一类是基于齐聚物噻吩为骨架的材料体系；第二类是围绕 BDT 单元为核心的小分子体系以及基于卟啉基团或其他单元的小分子体系。

陈永胜课题组对齐聚噻吩和 BDT 单元为骨架的小分子给体体系做了系统的工作，该类材料以齐聚噻吩为骨架，末端连接受体基团，具有典型的受体 – 给体 – 受体（简称 A–D–A）结构。他们通过调节齐聚噻吩的数量以及末端受体基团的拉电子能力，设计合成了一系列给体材料，其中，基于五联噻吩为核心，氰基罗丹宁为端基的小分子材料 DRCN5T，与 PCBM 共混的 OSCs 可以获得 10.1% 的 PCE[33]。基于 BDT 核心，中间桥链噻吩单元，末端由强拉电子基团构成的小分子体系，同样也具有非常优异的光伏性能。例如，他们采用烷硫基修饰 BDT 单元，以罗丹宁为端基，设计合成了 DR3TSBDT，相应的 OSCs 器件通过热退火和溶剂退火二元协同处理可以获得 9.95% 的 PCE[34]。李永舫等以噻吩烷硫基取代共轭支链修饰 BDT 单元，在其他结构保持不变的情况下，设计合成了具有共轭支链的小分子给体材料 BDTT–S–TR，也可以在电池器件中获得优异的性能[35]。2016 年，魏志祥等[32]以噻吩基 BDT 为核心、氟代茚酮为端基构建了小分子给体材料 BDIT–2F，优化了其与富勒烯型受体共混的微观形貌，将小分子给体的 OSCs 效率提高到 11.3%，是目前文献报道的最高结果。

除了以上几个例子外，研究人员也制备了化学结构各异的小分子电子给体材料用于构建 OSCs，取得了很多重要的研究成果。例如，王金亮、华南理等以给电子的二噻吩并吲哒省（简称 IDT）为核，苯并噻二唑单元为受体基团，设计合成了一系列的电子给体材料，其中以 BIT–4F–T 作为给体的 OSCs 与 $PC_{71}BM$ 制备的 OSCs 可以获得 9.1% 的 PCE[36]。2015 年，彭小彬等以卟啉为核心，通过炔键与吡咯并吡咯二酮（简称 DPP）基团相连，合成了给体材料 DPPEZnP–THE[37]，该材料在红外光区具有较好的吸收特性，基于此的电池器件在可以获得 8.08% 的 PCE。因为该类材料在红外光区具有较高的外量子效率，陈永胜等基于此制备了全小分子的叠层电池，可以获得 12.5% 的 PCE[38]。

小分子电子给体材料是 OSCs 的一个重要研究方面，相比于聚合物给体材料拥有特殊的优势，目前高效率的小分子 OSCs 虽已经取得了很多重要的进展，从材料的设计上以及器件制备工艺上，仍然有提升的空间。

（二）电子受体材料的研究前沿

相较于传统的富勒烯衍生物受体材料，非富勒烯型受体更容易通过分子结构的化学修饰实现对其光学吸收、分子能级以及分子堆积等特性的有效调制[39, 40]，自 2013 年以来，非富勒烯型 OSCs 的研究得到了迅速的发展，其 PCE 已经逐步赶上甚至超过了富勒

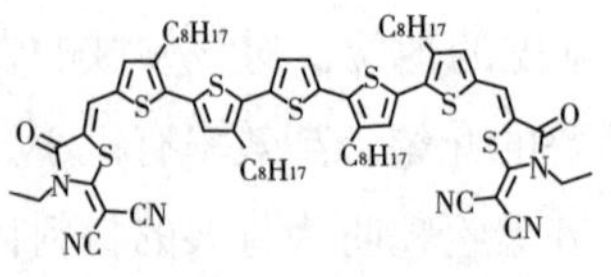

DRCN5T

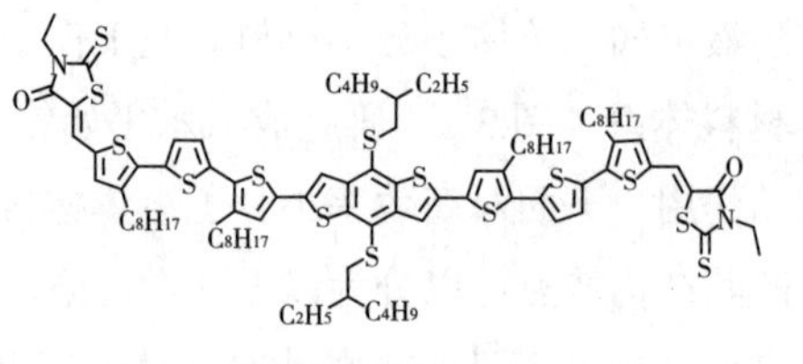

DR3TSBDT

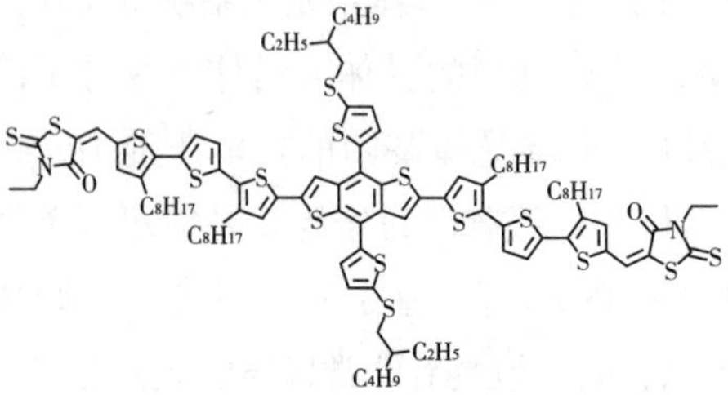

BDTT–S–TR

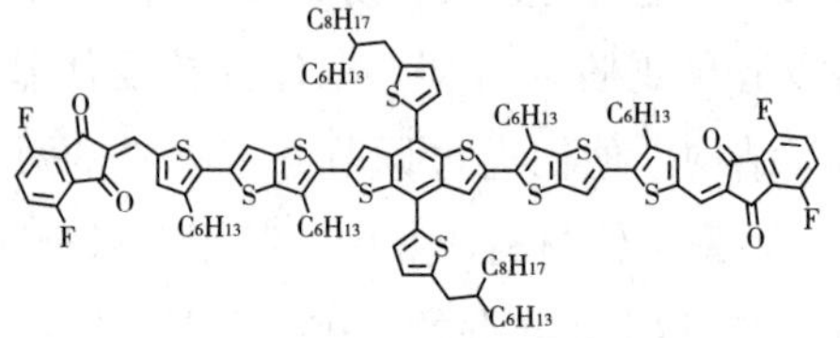

BTID–2F

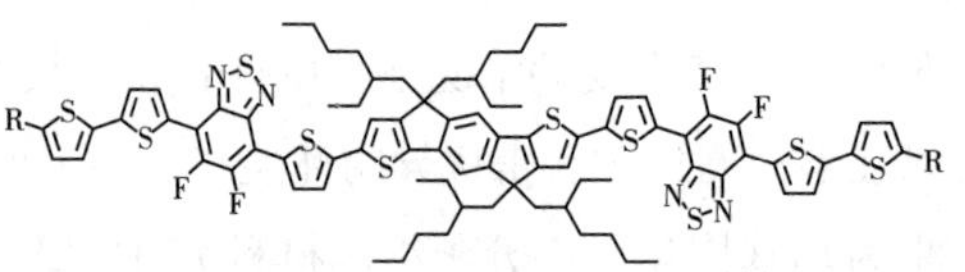

BIT4F

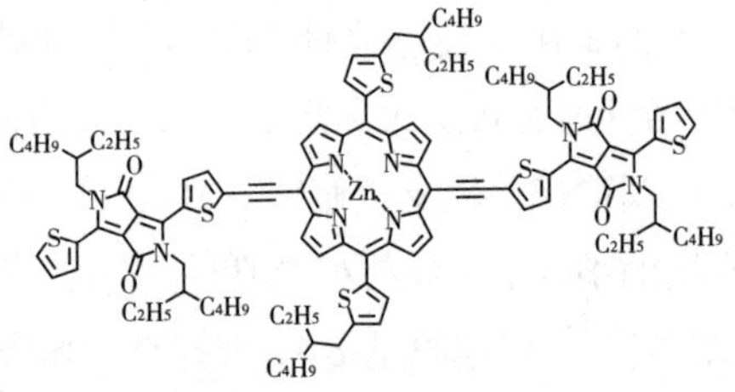

DPPEZnP–TEH

图 3　高效率的小分子给体材料举例

烯型 OSCs。国内的研究人员应用并开发了多种性能优异的受体材料，主要有萘二酰亚胺（简称 NDI）/ 苝二酰亚胺（简称 PDI）的聚合物或小分子、茚达省并二噻吩的稠环小分子，另外还有基于吡咯并吡咯二酮、苯并噻二唑以及 B–N 化学物等。

1. 酰胺类电子受体材料

基于 PDI 或 NDI 的分子具有优良的电子传输性能，且在可见光区有较强的吸收特性，使此类材料在有机光伏领域得到了广泛的关注和应用[41，42]。PDI 单元由于具有较大的 π 体系而导致该类材料较强的聚集行为，研究人员通过多种化学修饰的方法来减弱 PDI 分子的团聚趋势，主要有以下几种方式：①将两个或多个 PDI 单元通过酰亚胺位（imide）进行连接；②采用碳碳单键、乙烯键以及芳香类连接单元通过湾位（bay-position）连接 PDI 单元。2014 年，王朝晖等分别将两个和三个 PDI 分子通过胺位进行连接，合成了 H-di-PDI[43] 和 H-tri-PDI[44]（图 4），有效抑制了 PDI 分子间的聚集作用，基于此的 OSCs 可以获得 7.25% 的 PCE。赵达慧等采用螺芴基团连接两个 PDI 单元，设计了 SF-PDI2 的小分子电子受体材料，并采用 P3HT 作为给体材料制备了 OSCs，最初获得 2.35% 的 PCE[45]。2016 年，颜河等采用窄带隙聚合物 P3TEA 作为电子给体材料，制备 P3TEA:SF-PDI2 的电池器件，获得了高达 1.11 V 的 V_{OC} 和 9.5% 的 PCE[46]。另外，王朝晖课题组在联 PDI 分子的湾位分别引入杂原子 S 或 Se，设计合成了 SdiPDI-S 和 SdiPDI-

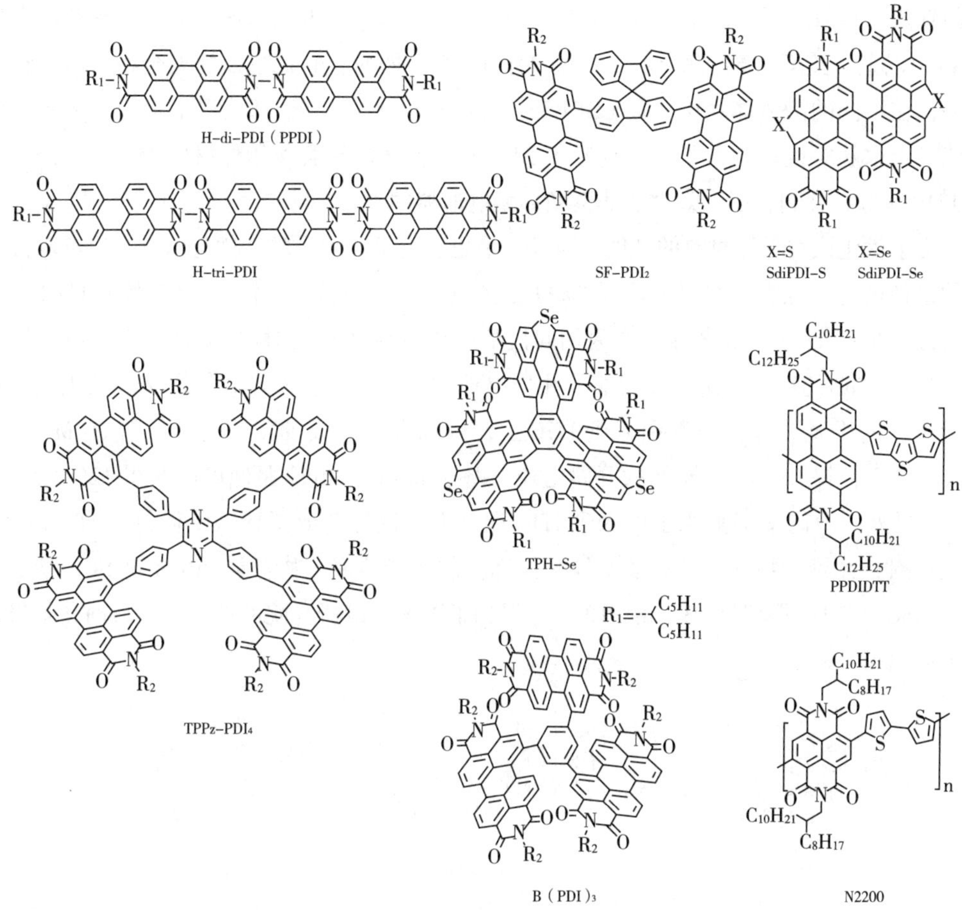

图 4　基于 PDI/NDI 的聚合物或小分子电子受体材料举例

Se，在电池器件中 PCE 达到 7.16% 和 8.4%[47, 48]。此外，陈红征等和颜河等分别采用不同的核心单元将多个 PDI 单元连在一起，构建了高度扭转的 PDI 分子 B（PDI）$_3$[49]、TPPz-PDI4[50] 和 TPH-Se[51]，同样抑制了 PDI 单元间的聚集作用，其中基于 TPH-Se 的 OSCs 可以取得高达 9.28% 的 PCE。基于 PDI 或 NDI 的聚合物受体材料也在全聚合物 OSCs 的研究中取得了十分重要的进展，例如，2014 年，占肖卫、侯剑辉等制备了基于 PPDIDTT 的全聚合物 OSCs，获得了 3.45% 的 PCE[52]。李永舫等采用 NDI 的聚合物 N2200 作为受体材料，BDT 和三氮唑的聚合物 J51 作为给体材料，进一步将全聚合物 OSCs 的效率提高到 8.27%[53]。

2. 引达省并二噻吩并噻吩类电子受体材料

2014 年，占肖卫等设计合成了一系列基于引达省并二噻吩（IDT）[54] 和引达省并二噻吩并噻吩（DTIDT）[55] 单元为核的稠环电子受体材料，基于此，非富勒烯型 OSCs 的研究取得了突破性的进展，我国科学家在该方向的研究中做出了众多开创性的工作。例如，占肖卫等设计合成了以 ITIC（图 5）为代表的新型受体材料，并采用 PTB7-Th 作为给体材料制备了 OSCs，取得了 7.52% 的 PCE，拉开了高效率非富勒烯型 OSCs 研究的序幕[55]。

侯剑辉课题组采用自主制备的给体材料 PBDB-T，将基于 ITIC 的 OSCs 光伏效率首次提高到 11% 以上，接近富勒烯型 OSCs 的最高水平[56]。同时，它们也发挥聚合物给体材料的优势，在不同的给体体系中，如 PDCBT[57] 和 PBQ-4F[58] 等，都获得了十分突出的光伏性能。李永舫等设计制备了一系列基于 BDT 和三氮唑的聚合物给体材料 J51[59]、J61[60] 和 J71[61] 等，并制备了高效率的非富勒烯型电池器件。

为了优化该类受体材料的性能，研究人员也从支链工程、端基优化以及杂原子取代等多种途径进行了探索，取得了重要的进展。例如，占肖卫等采用噻吩支链替换 ITIC 的苯基支链（ITIC-Th）[62]，李永舫等采用间位苯烷基替换 ITIC 的对位苯烷基（ITIC-*m*）[63]，都优化了 ITIC 的性能，进一步提高了相应电池的光伏性能。侯剑辉等通过端基优化，将弱给电子的甲基或噻吩甲基引入到 ITIC 上，分别设计合成了新的受体材料 IT-M[64] 和 ITCC-*m*[65]，提升了相应器件的开路电压，基于 IT-M 的电池器件可以获得 12% 以上的 PCE。另外，他们又将氟原子引入到 ITIC 端基上，设计合成了新型受体材料 IT-4F[12]，并通过对给体聚合物材料的同步氟原子修饰，制备新的聚合物材料 PBDB-T-SF，基于两者的 OSCs 可以取得超过 13% 的 PCE，并得到了国家计量研究院的认证，是目前文献报道的最高结果。

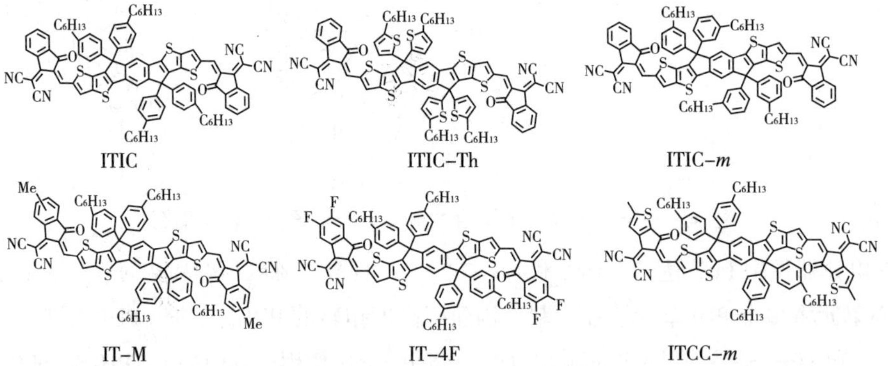

图 5　基于 DTIDT 的高效率小分子电子受体材料举例

3. B-N 类电子受体材料

B-N 化学物类电子受体材料在最近几年也得到了快速发展。相较于 C-C 共价键而言，B-N 配位键虽然等电子，但是具有更大的偶极，在 OLEDs 和 OFETs 领域有广泛的应用[66-68]。2015 年，王利祥和刘俊等首次将 B-N 配位键引入到聚合物材料中，并开发了一系列基于 B-N 单元的聚合物电子受体[69-74]（图 6）。他们将常见的二噻吩并环戊二烯给电子单元中的 C-C 替换成 B-N 键后，材料的 HOMO 和 LUMO 能级均显著的降低，具有电子受体的特性[69]。其中，基于 B-N 单元和 IID 单元的共聚物 P-BN-IID 与给体材料 PTB7-Th 共混制备的电池器件可以获得 5.04% 的能量转化效率[70]。2016 年，刘俊等又开发了 B-N 桥连双吡啶（BNBP）吸电子单元，并与硒吩或联噻吩单元共聚制备了 P-BNBP-Se 和

P-BNBP-fBT 的聚合物受体材料[71-74]。由于 BNBP 单元具有较好的平面性，基于 BNBP 单元的聚合物均具有较高的电子迁移率，基于此的电池器件可以获得 6.26% 的 PCE[73]。

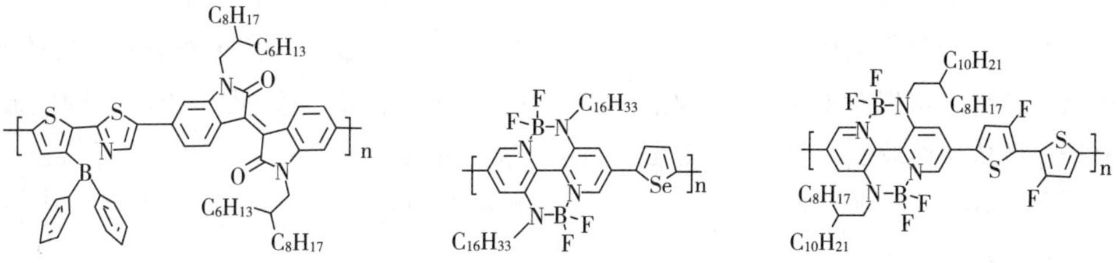

图 6　基于 B-N 配位键的电子受体材料举例

此外，还有许多研究组基于 PDI/NDI、IDT/DTIDT 或其他具有 A-D-A 型的非富勒烯受体材料，展开了十分重要的研究，在这里就不一一赘述。这一系列的工作表明，非富勒烯型 OSCs 在效率的提升上还有很大的空间，在未来的实际应用中具有较大的潜力。

（三）界面层材料

界面层是连接 OSC 电极和光伏活性层的中间层，可以有效提高电极对载流子的收集能力，对于提高 OSCs 光伏性能具有重要的意义，按照其修饰电极的不同，可以分为阴极界面层和阳极界面层两种。界面层对电极的修饰作用主要有两点：一是调节 OSC 正极和负极的功函数，增强正、负极之间的内建电场；二是使电极和光伏活性层之间形成良好的欧姆接触，降低电极与活性层之间的能量势垒，从而减少载流子收集过程中的能量损耗。在过去的几年中，我国的科研人员研发出了多种性能优越的界面层材料，取得了许多重要的研究进展，接下来简要介绍几个有代表性的工作。

1. 阴极界面层材料

共轭聚合物聚［9，9- 二辛基芴 -9，9- 双（*N*，*N*- 二甲基胺丙基）芴］（PFN）是一种应用十分广泛的阴极界面层材料，能有效降低阴极的功函数，并且具有成膜性好和适用范围广的特点，曹镛等在 2011 年报道了其在 OSCs 中的应用[75]。近几年来，国内的科研人员在 PFN 的基础上进行了各种结构上的优化不断改善 PFN 界面层的修饰能力。例如，黄飞等制备了含 Hg 原子的聚合物 PFEN-Hg，通过引入 Hg-Hg 相互作用，促进聚合物主链的聚集，从而提高界面层对电子的传输作用[76]。此外，他们通过层层自组装的方法制备了平整、均一的 PFN 薄膜[77]，并基于此制备了性能优异的 OSCs。

金属氧化物也是常见的阴极界面修饰材料，在正向、反向及叠层器件中都有着良好的应用。例如 ZnO 具有性能优异的性能，通过表面修饰减少其薄膜表面的缺陷，从而改善其光电子性能。解增旗等制备了小分子染料 PBI-H[78]、PBI-Py[79]来修饰 ZnO，使其功函数得到显著降低，并且修饰后的界面层具有较强的光导电性，有效降低了载流子在电极处的复合，该界面层在膜厚较厚的条件下依然能使电池保持良好的性能。蔡植豪等将 $CsCO_3$ 与 MoO_3 或 V_2O_5 共混，制备了新型的杂化阴极界面材料[80]。

李永舫等基于苝酰亚胺合成了两种醇溶性阴极界面层材料 PDIN 和 PDINO[81]。其中，PDINO 可以有效降低阴极的功函数，并具有较高的载流子迁移率，用于修饰 PTB7:$PC_{71}BM$ 活性层的器件，可以获得 8.2% 的能量转换效率。另外，值得注意的是，PDINO 的修饰性能对膜厚的依赖性较低，薄膜厚度达到 30 nm，依然可以达到 7.2% 的器件效率。

另外，2015 年，葛子义等采用一种非共轭的电解质 MSAPBS 修饰阴极，以 PTB7:$PC_{71}BM$ 为活性层，可以获得 10.02% 的 PCE[20]。

2. 阳极界面层材料

PEDOT:PSS 是应用最广的阳极界面层材料，目前效率最高的 OSC 器件基本都是采用 PEDOT:PSS 作界面层。但是，PEDOT:PSS 具有较强的吸水性和酸性，会腐蚀 ITO 电极，从而降低了器件的稳定性，这些缺点使 PEDOT:PSS 很难在未来有机光伏的产业化中得到真正的应用。面对这些问题，很多国内的科研工作者采用修饰、改性的办法，优化 PEDOT:PSS 的化学、电子学性质，努力改善材料的修饰性能和应用范围。谭占鳌等向 PEDOT:PSS 中掺杂一种全氟磺酸聚合物 Nafion[82]，提高了 PEDOT:PSS 的电导率，使 P3HT/$PC_{61}BM$ 活性层器件的短路电流和填充因子均获得了提高，效率也从原来的 4.0% 提高到 4.6%。马於光等利用电化学聚合的方法制备了聚咔唑阳极界面层材料 CMP[83]，相比只用 PEDOT:PSS 修饰的器件，在 ITO/PEDOT:PSS 基底上覆盖一层 10 nm 厚的 CMP，器件的短路电流可以得到明显的提升，能量转化效率也从 7.1% 提高到 8.4%。

侯剑辉等设计合成了具有自掺杂特性的阳极界面材料 PCP-Na[84]。由于良好的导电性和载流子传输特性，PCP-Na 的修饰性能受膜厚影响较小。当薄膜厚度增加到 50 nm 时，器件的 PCE 仍然保持在 7% 以上。他们还制备了一种新型的双功能界面修饰材料 PFS，不仅可以用于修饰 OSCs 的阳极，也可以用于修饰 OSCs 的阴极[85]，在同一器件中，阳极和阴极界面层全部采用 PFS，器件效率依然获得较高的效率。

得益于良好的透光性、高功函和化学稳定性，金属氧化物在阳极修饰方面获得广泛的关注。近年来，低温、溶液加工技术越来越多用于金属氧化物阳极界面层的制备。三氧化钼（MoO_3）是应用最为广泛的金属氧化物阳极界面层材料。谭占鳌等通过加热一种含钼配合物 $MoO_2(acac)_2$ 制备出高透光性、表面平整的 MoO_3 薄膜，避免了高温操作，从而简化了界面层的加工工艺[86]。他们还将金属有机配合物（乙酰丙酮配合物）配制成一定浓度的醇溶液，旋涂成膜后，再进行简单的加热退火或者紫外光辐照，即可获得均一、致密的金属氧化物薄膜。该方法不仅操作简单、原料成本低，还具有良好的普适性，适用于制备多种金属氧化物阳极界面材料，包括 CuO_x、RuO_2、ReO_x 和 CrO_x 等[87-90]。蔡植豪等通过在乙醇 / 双氧水中氧化钼粉的方法制备了阳极界面层 MoO_x[91]，这种溶液加工的方法，减少了高温退火过程带来的种种缺陷，有利于在未来的产业化中进行应用。

氧化石墨烯（GO）由于透光性好、带隙宽、电导率高、可溶液加工等优点，经常被用作 OSCs 的阳极界面层。李灿等通过紫外光辐照的方法使 GO 氯化，制备了阳极界面层材料 Cl-GO[92]，研究表明，随着 GO 氯化程度增加，材料的功函数也随之在提高，逐渐接近活性层中给体的 HOMO 能级，从而降低了空穴传输过程中的能量势垒。他们还通过层层自组装的方法制备了表面均匀、致密的 GO 阳极界面层，该方法很容易控制 GO 的薄膜厚度，在 P3HT:ICBA 活性层的器件中得到 6.0% 的能量转换效率[93]。

（四）高效率叠层、三元和大面积太阳能电池器件的研究

1. 叠层电池器件

随着光电活性材料的不断创新以及器件工艺的日益优化，单节有机 OSCs 的研究工作在近年来已经取得了突破性的进展。但是，对于单节活性层的有机光伏器件来说，其光电活性层只能利用其单节吸收范围内的光子，有较大的能量浪费。因此，为了克服单一活性层器件 OSCs 无法充分有效地利用太阳辐射的缺点，国内外的研究工作者制备了叠层 OSCs，并在近年来取得了重要的研究进展。

叠层 OSCs 通常是由两个或者两个以上的光谱互补的光电活性层通过前后串联的连接方式所构成，其中各光电活性层彼此互补的吸收光谱能显著弥补单节太阳能电池的缺陷，拓宽有机太阳电池的光谱响应，进而实现对太阳辐照更有效的利用。以双节叠层有机太阳电池为例，其前电池通常由在短波区间有较强响应的宽带隙材料组成，后电池则由在长波区间有较强响应的窄带隙材料组成。前电池的负极与后电池的正极（抑或前电池的正极与后电池的负极）通过一个中间界面电极层连接在一起，而整个叠层太阳电池则通过底部电极和顶部电极与外电路相连接。

对于叠层太阳电池而言，合理地选用吸收光谱互补的活性层材料来制作前后子电池是决定器件效率的一项关键因素。2015 年，侯剑辉等[94]分别使用 PDCBT:$PC_{71}BM$ 及 PBDT-TS1:$PC_{61}BM$ 作为前后电池的活性层，使吸收光谱得到了有效的互补，并通过调节前后电池活性层的膜厚，使前后电池的活性层的电流得到了较好的匹配，从而得到了 10.16% 的 PCE。在叠层器件中，通常使用的窄带隙材料的后电池具有较低的开路电压，限制了叠层有机太阳电池的整体电压和效率。为了改善这一情况，侯剑辉等制备了一类基于 DPP 结构单元的新型窄带隙共轭聚合物材料 PDPP4T-2F[15]，该材料在单节器件中能表现出 0.78 V 的开路电压以及 7.58% 的能量转换效率。然后，他们以该聚合物为后电池的活性层材料，并在前电池中采用了他们前期报道的高性能宽带隙聚合物 PBDD4T-2F，最终使得器件的能量转换效率达到了 11.62%。基于共轭小分子给体材料的电池器件也在叠层有机电池中有广泛的应用。陈永胜等[95]应用 DR3TSBDT:$PC_{71}BM$ 作为叠层电池的前电池，当后电池活性层分别使用 PTB7-Th:$PC_{71}BM$ 和 DPPEZnP-TBO:$PC_{61}BM$，中间层为 ZnO/n-PEDOT:PSS 时，叠层太阳电池的 PCE 可以达到 11.47% 和 12.5%[96]。

除了优化前后电池活性层外，中间连接层的设计与制备对叠层 OSCs 的发展同样起着至关重要的作用。中间层须有效地收集前后电池所产生的载流子，并为其提供一个良好的复合通道，因此中间层的设计和制备需要满足以下四个基本条件：首先，中间层必须在各子电池间形成良好的欧姆接触；其次，要确保前后子电池产生的电子和空穴在中间层中能发生有效的复合；再次，中间层要能够对加工溶剂有足够对耐侵蚀性，以阻止后电池对加工过程中溶剂对前电池对破坏；最后，中间层须有良好的光学透明度。陈红征等通过设计新型的中间层 MoO_3/ 超薄 Ag/PFN，使电荷得到有效的抽取及复合[97]，从而使叠层电池的 EQE 总和超过了 90%，最终取得了 10.98% 的能量转换效率。曹镛等设计并合成了一种新型的阴极界面层 PF3N-2TNDI，将其与 PEDOT:PSS 组合，并蒸镀了薄层 Ag 以更好地促使空穴与电子在中间层发生复合[98]。该 PF3N-2TNDI/Ag/PEDOT:PSS 中间层不但能促使前后电池形成有效的欧姆接触，并且能确保后电池的加工几乎不会对前电池造成影响。他们以 PThBDTP:PC_{71}BM 为前电池、DPPEZnP-TEH:PC_{61}BM 为后电池制备了叠层有机太阳电池，得到了 11.35% 的能量转换效率。

富勒烯型 OSCs 通常具有相对较高的能量损失[99]，从而使得叠层太阳电池的开路电压受到了限制。近年来，对非富勒烯型 OSCs 的研究结果表明，其可以在能量损失较低的情况下，实现高效的电荷生成和收集，从而获得较高的光伏性能，这对于叠层 OSCs 性能的提高带来了有利的契机。陈红征等采用 IEIC 非富勒烯小分子作为受体材料、聚合物 PTB7-Th 作为给体材料，所制备的单节电池作为后电池，其 V_{OC} 为 0.97 V；同时他们在前电池中采用 P3HT:SF（DPPB）$_4$ 作为活性层，该单节电池的 V_{OC} 高达 1.14 V。基于此制备叠层 OSCs 的 V_{OC} 达到了 1.97 V，光电转换效率达到了 8.48%[100]。颜河等基于 P3TEA:SF-PDI_2 的非富勒烯太阳电池制备了前后电池均为此活性层的叠层电池，并采用 PEDOT:PSS/ZnO 的中间层，获得了 10.8% 的 PCE[101]。值得注意的是，其开路电压达到了 2.1 V，是双节性叠层电池中的最高电压。侯剑辉等设计合成了窄带隙的非富勒烯受体 IEICO，将吸收拓宽至 900 nm 以外，并且基于此的电池器件在长波处具有较高的外量子效应[102]。它们采用宽带隙聚合物 PBDD4T-2F 与富勒烯衍生物共混的电池器件作为叠层电池的电池时，可以获得 1.71V 的 V_{OC}，11.51mA · cm^{-2} 的 J_{SC} 和 65% 的 FF，最终的 PCE 为 12.8%[103]。然后，他们进一步将前电池也替换为宽带隙的非富勒烯型 OSCs，将 V_{OC} 提高到 1.79 V，获得了最高 13.8% 的 PCE，是目前文献报道的叠层电池最高结果[65]。

2. 三元电池器件

在过去的几十年，由一种给体材料和一种受体材料所构建的二元 OSCs 的性能已经取得了显著的提高。但是，有机半导体材料通常只有一个较窄的吸收带，二元 OSCs 对太阳光的吸收利用并不充分，除了以上介绍的叠层电池之外，在二元体系中加入吸收光谱互补的第三种组分，可以构建一种三元的 OSCs。这种简单的方法既有效地扩展了 OSCs 对太阳光的吸收利用，又具有制备工艺简单的特点，得到了国内外研究人员的广泛关注。设计高效的 OSCs 除了要求共混材料具有互补的吸收光谱之外，还要求三者具有良好的相容性和平衡的电荷迁移率。

在高效的二元富勒烯型 OSCs 中添加另外一种聚合物或小分子给体是研究比较多的三元体系。魏志祥、马伟等在聚合物作为给体、富勒烯为受体的体系中添加了具有光谱互补的小分子材料作为第三组分，并研究了小分子的添加比例对器件光伏性能的影响[104]。结果表明，小分子的加入能够显著提高器件填充因子。在添加 50% 小分子的比例下，器件的效率相较二元电池有明显的提高。然后，魏志祥等继续运用结晶性较强的小分子给体作为第三组分，制备了性能优异的三元 OSCs，最高效率可达 10.5%[105]，相对于二元电池体系有很大的提高。最近，谢增旗、彭小彬等通过向传统的二元电池中加入具有近红外吸收的小分子材料 DPPEP-TEH，拓宽了 OSCs 的光电响应范围，明显提升了电池的短路电流密度，可以获得高达 11.03% 的 PCE[106]。张福俊等基于 DRCN5T:DR3TSBDT:PC71BM 制备了三元的全小分子 OSCs，获得了 10.16% 的光电转化效率，表明三元共混的策略在有机小分子 OSCs 中也具有较好的适用性[107]。黄飞等通过在 PTB7-Th:PC_{71}BM 体系中引入具有高空穴迁移率的液晶小分子（BTR）作为第三组分，改善了活性层的形貌并提高空穴迁移率，最终获得了高达 11.40% 的 PCE[108]。另外需要注意的是，他们制备的三元 OSCs 可以在较宽的膜厚范围内获得较高的性能，这对于 OSCs 的大面积制备是十分有利的。

近来，随着多种高性能非富勒烯受体材料的不断涌现，基于聚合物给体与非富勒烯受体 OSCs 的效率迅速提升，已经超过了富勒烯型 OSCs。如何在高效率非富勒烯电池中引入第三种组分，进一步提升其光伏性能是一项具有挑战性的工作。薄志山、马伟等采用富勒烯衍生物受体作为第三组分，引入到非富勒烯型 OSCs（PPBDTBT:ITIC）中，获得了 10.4% 的能量转换效率，与二元对照器件相比，三元器件的效率提高了约 35%[109]。侯剑辉等采用 Bis-PC_{71}BM 材料作为三元组分引入到 PBDB-T:IT-M 的高效率体系中，补充电池在 380~550 nm 范围内的光电响应，基于此制备的三元 OSCs 可以获得高达 12.20% 的 PCE[110]。最近，孙艳明等采用两种光谱互补的聚合物给体材料 PDBT-T1 和 PTB7-Th 与非富勒烯受体共混制备了三元 OSCs，通过调节 PDBT-T1 在三元体系中的比例，优化了电池器件的各个参数，最终该三元 OSCs 取得了 10.2% 的效率[111]。

3. 大面积电池器件制备

在 OSCs 的研究过程中大都采用小于 1 cm^2 的活性面积，而 OSCs 的产业化应用必须依赖于大面积的制备技术。因此，基于大面积高效率 OSCs 的研究是有机光伏电池领域一个重要课题。目前在该研究方向，国内科研人员已陆续开展了一些相关工作，主要可以分为以下几个方面。

在柔性基底大面积器件研究方面，魏志祥等用印刷的方式在柔性基底 PET 上沉积银纳米线（AgNWs）制备了透明导电电极 PET-AgNWs，该电极在可见光波长范围具有 95% 的光透过率和较低的表面电阻[112]。基于此，他们制备了 PPDT2FBT:PC_{71}BM 的柔性大面积电池（7 cm^2），获得了 3.04% 的 PCE。曲胜春等采用纳米压印技术制备了银栅格 PET 的透明导电极，基于 PTB7–Th:$PC_{71}BM$ 活性层制备了 2.25 cm^2 的电池器件，获得 6.4% 的 PCE[113]。在大面积印刷技术制备过程中，由于存在获得均一厚度薄膜的困难，因此通过大面积印刷的方式制备光伏器件时，就需要电池器件对活性层厚度具有一定的容忍性。魏志祥等在 PET 薄膜中内嵌银栅格作为透明导电电极，制备了 1.25cm^2 的柔性电池器件，在膜厚 300 nm 左右的条件下，可以获得 8.28% 的 PCE[114]。更为重要的是，他们用印刷的方式制备了面积为 20cm^2 的电池器件，仍可以获得 5.18% 的 PCE。陈立桅、李耀文等采用高导电性 PEDOT:PSS（PH1000）内嵌银栅格制备了杂化透明电极，并基于 PTB7:$PC_{71}BM$ 的活性层制备了面积为 1.21cm^2 的柔性器件，获得了 5.85% 的 PCE[115]。

开发可溶液加工、高透过率且具有低表面电阻的透明导电电极，是实现高效率大面积柔性器件急需突破的关键性技术之一。杨小牛等采用 AgNWs 网络作为透明导电电极，并利用"毛细管 – 辅助流体组装"技术一步法实现了 AgNWs 的有序化，使其具有较高的透过率和较低的表面电阻，相比于传统的 ITO 电极有众多的优势[116]。另外，他们也采用喷涂的方法，采用 PBTI3T:$PC_{71}BM$ 作为活性材料，在 ITO/PEDOT 基底上制备了面积为 38.5 cm^2 的电池器件，最终获得了 5.27% 的 PCE[117]。

（五）有机太阳能电池机理的研究

OSCs 的光电转换物理过程主要分为五个步骤，分别是吸收光子产生激子、激子在活性层中的扩散、激子解离产生载流子、载流子传输和收集。其中，载流子的传输过程中还有可能发生复合，从而导致光电转换效率的下降。深刻地理解光电转换过程与材料的化学结构、本体异质结形貌以及器件光伏性能之间的关系对于从根本上提高 OSCs 的性能非常重要。并且，这对于光伏材料的分子设计、活性层形貌的优化和新颖的器件结构的构造也具有指导性的意义。近几年，随着非富勒烯型 OSCs 的快速发展，很多研究结果表明非富勒烯型 OSCs 不仅仅展现了优于富勒烯受体的光伏性能，其光电转换过程也展现了许多新特点。因此，很多研究者从光物理方面出发，试图从源头上将非富勒烯受体材料的优异性能同光电转化的特点联系起来，以期能够更好的进行分子设计和形貌优化。

侯剑辉和高峰等使用傅里叶转换光电流谱 – 外量子效率技术（FTPS–EQE）测量了非富勒烯材料体系和富勒烯材料体系下电荷转移态的能量（E_{CT}）[56]。他们发现对于同样的给体材料 PBDB–T，使用非富勒烯受体 ITIC 时体系的 E_{CT} 要高于使用富勒烯受体 PCBM 时的 E_{CT}，同时电荷转移态吸收带宽也会收窄，从而降低了能量损失。因此可以在更高的开路电压，仍然能保证较高的短路电流密度和填充因子。此外，他们还使用飞秒瞬态吸收光谱研究了给受体之间分子轨道能级（HOMO，LUMO）差异对于电荷转移的影响，他们发现 ΔHOMO 和 ΔLUMO 的缩小并没有对空穴转移产生负面影响[58]。ΔHOMO 从 0.35 eV 降低到 0.04 eV，电荷转移依然可以高效进行（一般认为，在富勒烯体系中，ΔHOMO 要大于 0.3 eV 才能保证有效的给受体间电荷转移）。因此，他们提出在这种非富勒烯材料体系中，是给受体间的偶极作用，而非能级差对给受体间的电荷转移提供了主要驱动作用。

颜河和高峰等针对苝酰亚胺类小分子受体 OSCs 中电荷转移与复合以及光电转换中能量损失的问题展开了研究，他们发现，在这种材料体系中，电荷转移态的能量等于带隙能量（E_{CT}=E_g），电荷转移的驱动力（E_g-E_{CT}）几乎可以忽略。给受体共混层没有明显的电荷转移态吸收峰，表明该类电池器件的辐射复合能量损失大大降低，与无机太阳能电池处于同一水平，远小于富勒烯型 OSCs。同时，荧光发射光谱的结果表明共混层具有很高的发光效率，非辐射能量损失也明显小于传统的富勒烯电池[118]。尽管如此，瞬态吸收光谱却表明在给受体之前的电荷转移仍然能高效进行，而且电荷复合也没有因此而显著提升，内量子效率可以保持在 90% 以上。他们认为在非富勒烯受体体系中，可以通过降低驱动力损失来提高开路电压，这为改善有机光伏体系能量转换效率提供了新途径。此外，国内的许多科研人员使用瞬态吸收光谱、傅里叶转换光电流谱 - 外量子效率等技术表征了共混体系的电荷转移、能量损失等问题[48，119，120]。

三、总结与展望

在过去的几年，OSCs 的研究取得了众多的突破性的进展，我国的科研团队在该领域的研究中走在最前列，对推动该领域的发展中起着十分重要的作用，本部分的内容只对其中少量的工作做了简要的介绍。新型光伏材料、界面层材料的设计与应用以及电池器件的优化都为 OSCs 性能的提升起着至关重要的作用，目前文献报道的单层和叠层电池都获得了超过 13% 的能量转换效率。研究人员对材料分子结构与器件参数之间的构效关系以及电池器件工作机理都也逐渐清晰。大量的研究发现，以富勒烯衍生物作为电子受体材料制备的 OSCs，普遍具有较大的能量损失，例如，目前文献报道效率最高的电池器件，能量损失约为 0.8 eV，其效率为 11.7%，已经逐步接近于其理论值，进一步的提升面临着较大的挑战。最近几年，非富勒烯型 OSCs 的研究有了重大突破，可以在较低能量损失下（0.5~0.6 eV）工作，因而其效率具有很大提升空间。另外，在提升 OSCs 效率之外，还要求较好的稳定性以及友好的制备条件，所以从光伏材料和器件制备两个方面开展相关的研究也具有十分重要的意义。

参考文献

[1] 李永舫 . 聚合物太阳能电池高效共轭聚合物给体和富勒烯受体光伏材料［J］. 高分子通报，2011，10: 33–49.
[2] 姚惠峰；侯剑辉 . 高性能聚合物光伏材料的设计与应用［J］. 高分子学报，2016，11: 1468–1481.

[3] Li G, Zhu R, Yang Y. Polymer solar cells[J]. Nature Photonics, 2012, 6 (3): 153–161.

[4] Polman A, Knight M, Garnett E C, et al. Photovoltaic materials: Present efficiencies and future challenges[J]. Science, 2016, 352 (6283).

[5] Yao H, Ye L, Zhang H, et al. Molecular design of benzodithiophene–based organic photovoltaic materials[J]. Chemical Reviews, 2016, 116 (12): 7397–7457.

[6] Lu L, Zheng T, Wu Q, et al. Recent advances in bulk heterojunction polymer solar cells[J]. Chemical Reviews, 2015, 115 (23): 12666–12731.

[7] Li Y. Molecular design of photovoltaic materials for polymer solar cells: toward suitable electronic energy levels and broad absorption[J]. Accounts of Chemical Research, 2012, 45 (5): 723–733.

[8] He Y, Chen H Y, Hou J, et al. Indene–C (60) bisadduct: a new acceptor for high–performance polymer solar cells [J]. Journal of the American Chemical Society, 2010, 132 (4): 1377–1382.

[9] He Y, Zhao G, Peng B, et al. High–Yield Synthesis and Electrochemical and Photovoltaic Properties of Indene–C70 Bisadduct[J]. Advanced Functional Materials, 2010, 20 (19): 3383–3389.

[10] Li S, Zhang Z, Shi M, et al. Molecular electron acceptors for efficient fullerene–free organic solar cells[J]. Physical Chemistry Chemical Physics, 2017, 19 (5): 3440–3458.

[11] Fan H, Zhu X. Development of small–molecule materials for high–performance organic solar cells[J]. Science China Chemistry, 2015, 58 (6): 922–936.

[12] Zhao W, Li S, Yao H, et al. Molecular optimization enables over 13% efficiency in organic solar cells[J]. Journal of the American Chemical Society, 2017, 139 (21): 7148–7151.

[13] Yip H L, Jen A K Y. Recent advances in solution–processed interfacial materials for efficient and stable polymer solar cells[J]. Energy & Environmental Science, 2012, 5 (3): 5994–6011.

[14] Zhao W, Li S, Zhang S, et al. Ternary polymer solar cells based on two acceptors and one donor for achieving 12.2% efficiency[J]. Advanced Materials 2017, 29 (2): 1604059.

[15] Zheng Z, Zhang S, Zhang J, et al. Over 11% efficiency in tandem polymer solar cells featured by a low–band–gap polymer with fine–tuned properties[J]. Advanced Materials, 2016, 28 (25): 5133–5138.

[16] Liu Y, Zhao J, Li Z, et al. Aggregation and morphology control enables multiple cases of high–efficiency polymer solar cells[J]. Nature Communications, 2014, 5: 5293.

[17] Deng D, Zhang Y, Zhang J, et al. Fluorination–enabled optimal morphology leads to over 11% efficiency for inverted small–molecule organic solar cells[J]. Nature Communications, 2016, 7: 13740.

[18] Liang Y, Xu Z, Xia J, et al. For the bright future—bulk heterojunction polymer solar cells with power conversion efficiency of 7.4%[J]. Advanced Materials, 2010, 22 (20): E135–E138.

[19] He Z, Zhong C, Su S, et al. Enhanced power–conversion efficiency in polymer solar cells using an inverted device structure[J]. Nature Photonics, 2012, 6 (9): 591–595.

[20] Ouyang X, Peng R, Ai L, et al. Efficient polymer solar cells employing a non–conjugated small–molecule electrolyte[J]. Nature Photonics, 2015, 9 (8): 520–524.

[21] Liao S H, Jhuo H J, Cheng Y S, et al. fullerene derivative–doped zinc oxide nanofilm as the cathode of inverted polymer solar cells with low–band gap polymer (PTB7–Th) for high performance[J]. Advanced Materials, 2013, 25 (34): 4766–4771.

[22] Chen J D, Cui C, Li Y Q, et al. Single–junction polymer solar cells exceeding 10% power conversion efficiency[J]. Advanced Materials, 2015, 27 (6): 1035–1041.

[23] He Z, Xiao B, Liu F, et al. Single–junction polymer solar cells with high efficiency and photovoltage[J]. Nature Photonics, 2015, 9 (3): 174–179.

[24] Liao S H, Jhuo H J, Yeh P N, et al. Single junction inverted polymer solar cell reaching power conversion

efficiency 10.31% by employing dual-doped zinc oxide nano-film as cathode interlayer [J] . Scientific Reports, 2014, 4: 6813.

[25] Cui C, Wong W Y, Li Y. Improvement of open-circuit voltage and photovoltaic properties of 2D-conjugated polymers by alkylthio substitution [J] . Energy & Environmental Science, 2014, 7 (7) : 2276-2284.

[26] Ye L, Zhang S, Zhao W, et al. Highly efficient 2D-conjugated benzodithiophene-based photovoltaic polymer with linear alkylthio side chain [J] . Chemistry of Materials, 2014, 26 (12) : 3603-3605.

[27] Zhang S Q, Ye L, Zhao W C, et al. Realizing over 10% efficiency in polymer solar cell by device optimization [J] . Science China-Chemistry, 2015, 58 (2) : 248-256.

[28] Gao C, Wang L, Li X, et al. Rational design on D-A conjugated P(BDT-DTBT)polymers for polymer solar cells [J]. Polymer Chemistry, 2014, 5 (18) : 5200-5210.

[29] 李婉宁，姚惠峰，侯剑辉 . 苯并噻二唑单元在聚合物光伏材料设计中的应用 [J] . 高分子通报，2016，9: 71-86.

[30] Chen Z, Cai P, Chen J, et al. Low band-gap conjugated polymers with strong interchain aggregation and very high hole mobility towards highly efficient thick-film polymer solar cells [J] . Advanced Materials, 2014, 26 (16) : 2586-2591.

[31] Zhao J, Li Y, Yang G, et al. Efficient organic solar cells processed from hydrocarbon solvents [J] . Nature Energy, 2016, 1: 15027.

[32] Deng D, Zhang Y, Zhang J, et al. Fluorination-enabled optimal morphology leads to over 11% efficiency for inverted small-molecule organic solar cells [J] . Nature Communications, 2016, 7: 13740.

[33] Kan B, Li M, Zhang Q, et al. A series of simple oligomer-like small molecules based on oligothiophenes for solution-processed solar cells with high efficiency [J] . Journal of the American Chemical Society, 2015, 137 (11) : 3886-3893.

[34] Kan B, Zhang Q, Li M, et al. Solution-processed organic solar cells based on dialkylthiol-substituted benzodithiophene unit with efficiency near 10% [J] . Journal of the American Chemical Society, 2014, 136 (44) : 15529-15532.

[35] Cui C, Guo X, Min J, et al. High-performance organic solar cells based on a small molecule with alkylthio-thienyl-conjugated side chains without extra treatments [J] . Advanced Materials, 2015, 27 (45) : 7469-7475.

[36] Wang J L, Liu K K, Yan J, et al. Series of multifluorine substituted oligomers for organic solar cells with efficiency over 9% and fill factor of 0.77 by combination thermal and solvent vapor annealing [J] . Journal of the American Chemical Society, 2016, 138 (24) : 7687-7697.

[37] Gao K, Li L, Lai T, et al. Deep absorbing porphyrin small molecule for high-performance organic solar cells with very low energy losses [J] . Journal of the American Chemical Society, 2015, 137 (23) : 7282-7285.

[38] Li M, Gao K, Wan X, et al. Solution-processed organic tandem solar cells with power conversion efficiencies >12% [J] . Nature Photonics, 2016, 11 (2) : 85-90.

[39] Lin Y Z, Zhan X W. Non-fullerene acceptors for organic photovoltaics: an emerging horizon [J] . Materials Horizons, 2014, 1 (5) : 470-488.

[40] Nielsen C B, Holliday S, Chen H Y, et al. Non-Fullerene Electron Acceptors for Use in Organic Solar Cells [J] . Accounts of Chemical Research, 2015, 48 (11) : 2803-2812.

[41] Chen Z J, Lohr A, Saha-Moller C R, et al. Self-assembled pi-stacks of functional dyes in solution: structural and thermodynamic features [J] . Chemical Society Reviews, 2009, 38 (2) : 564-584.

[42] Jiang W, Li Y, Wang Z. Tailor-made rylene arrays for high performance n-channel semiconductors [J] . Accounts of Chemical Research, 2014, 47 (10) : 3135-3147.

[43] Ye L, Sun K, Jiang W, et al. Enhanced efficiency in fullerene-free polymer solar cell by incorporating fine-

designed donor and acceptor materials [J] . ACS Applied Materials & Interfaces, 2015, 7 (17) : 9274–9280.

[44] Liang N, Sun K, Zheng Z, et al. Perylene diimide trimers based bulk heterojunction organic solar cells with efficiency over 7% [J] . Advanced Energy Materials, 2016, 6 (11) : 1600060.

[45] Yan Q, Zhou Y, Zheng Y Q, et al. Towards rational design of organic electron acceptors for photovoltaics: a study based on perylenediimide derivatives [J] . Chemical Science, 2013, 4 (12) : 4389–4394.

[46] Liu J, Chen S, Qian D, et al. Fast charge separation in a non–fullerene organic solar cell with a small driving force [J] . Nature Energy, 2016, 1 (7) : 16089.

[47] Sun D, Meng D, Cai Y H, et al. Non–fullerene–acceptor–based bulk–heterojunction organic solar cells with efficiency over 7% [J] . Journal of the American Chemical Society, 2015, 137 (34) : 11156–11162.

[48] Meng D, Sun D, Zhong C, et al. High–performance solution–processed non–fullerene organic solar cells based on selenophene–containing perylene bisimide acceptor [J] . Journal of the American Chemical Society, 2016, 138 (1) : 375–380.

[49] Koster L J A, Shaheen S E, Hummelen J C. Pathways to a new efficiency regime for organic solar cells [J] . Advanced Energy Materials, 2012, 2 (10) : 1246–1253.

[50] Lin H, Chen S, Hu H, et al. Reduced intramolecular twisting improves the performance of 3D molecular acceptors in non–fullerene organic solar cells [J] . Advanced Materials, 2016, 28 (38) : 8546–8551.

[51] Meng D, Fu H, Xiao C, et al. Three–bladed rylene propellers with three–dimensional network assembly for organic electronics [J] . Journal of the American Chemical Society, 2016, 138 (32) : 10184–10190.

[52] Cheng P, Ye L, Zhao X, et al. Binary additives synergistically boost the efficiency of all–polymer solar cells up to 3.45% [J] . Energy & Environmental Science, 2014, 7 (4) : 1351–1356.

[53] Gao L, Zhang Z G, Xue L, et al. All–Polymer solar cells based on absorption–complementary polymer donor and acceptor with high power conversion efficiency of 8.27% [J] . Advanced Materials, 2016, 28 (9) : 1884–1890.

[54] Lin Y, Zhang Z G, Bai H, et al. High–performance fullerene–free polymer solar cells with 6.31% efficiency [J] . Energy & Environmental Science, 2015, 8 (2) : 610–616.

[55] Lin Y, Wang J, Zhang Z G, et al. An electron acceptor challenging fullerenes for efficient polymer solar cells [J] . Advanced Materials, 2015, 27 (7) : 1170–1174.

[56] Zhao W, Qian D, Zhang S, et al. Fullerene–free polymer solar cells with over 11% efficiency and excellent thermal stability [J] . Advanced Materials, 2016, 28 (23) : 4734–4739.

[57] Qin Y, Uddin M A, Chen Y, et al. Highly efficient fullerene–free polymer solar cells fabricated with polythiophene derivative [J] . Advanced Materials, 2016, 28 (42) : 9416–9422.

[58] Zheng Z, Awartani O M, Gautam B, et al. Efficient charge transfer and fine–tuned energy level alignment in a thf–processed fullerene–free organic solar cell with 11.3% efficiency [J] . Advanced Materials 2017, 29 (5) : 1604241.

[59] Gao L, Zhang Z G, Bin H, et al. High–efficiency nonfullerene polymer solar cells with medium bandgap polymer donor and narrow bandgap organic semiconductor acceptor [J] . Advanced Materials, 2016, 28 (37) : 8288–8295.

[60] Bin H, Zhang Z G, Gao L, et al. Non–fullerene polymer solar cells based on alkylthio and fluorine substituted 2D–conjugated polymers reach 9.5% efficiency [J] . Journal of the American Chemical Society, 2016, 138 (13) : 4657–4664.

[61] Bin H, Gao L, Zhang Z G, et al. 11.4% Efficiency non–fullerene polymer solar cells with trialkylsilyl substituted 2D–conjugated polymer as donor [J] . Nature Communications, 2016, 7: 13651.

[62] Lin Y, Zhao F, He Q, et al. High–performance electron acceptor with thienyl side chains for organic photovoltaics [J] . Journal of the American Chemical Society, 2016, 138 (14) : 4955–4961.

[63] Yang Y, Zhang Z G, Bin H, et al. Side–chain isomerization on an n–type organic semiconductor ITIC acceptor

makes 11.77% high efficiency polymer solar cells [J] . Journal of the American Chemical Society, 2016, 138 (45) : 15011–15018.

[64] Li S, Ye L, Zhao W, et al. Energy–level modulation of small–molecule electron acceptors to achieve over 12% efficiency in polymer solar cells [J] . Advanced Materials, 2016, 28 (42) : 9423–9429.

[65] Cui Y, Yao H, Gao B, et al. Fine tuned photoactive and interconnection layers for achieving over 13% efficiency in a fullerene–free tandem organic solar cell [J] . Journal of the American Chemical Society, 2017, 139 (21) : 7302–7309.

[66] Li D, Zhang H Y, Wang Y. Four–coordinate organoboron compounds for organic light–emitting diodes(OLEDs) [J]. Chemical Society Reviews, 2013, 42 (21) : 8416–8433.

[67] Rao Y L, Wang S N. Four–coordinate organoboron compounds with a pi–conjugated chelate ligand for optoelectronic applications [J] . Inorganic Chemistry, 2011, 50 (24) : 12263–12274.

[68] Wakamiya, A, Taniguchi, T, Yamaguchi, S. Intramolecular B–N coordination as a scaffold for electron–transporting materials: Synthesis and properties of boryl–substituted thienylthiazoles [J] . Angewandte Chemie International Edition 2006, 45 (19) : 3170–3173.

[69] Dou C D, Ding Z C, Zhang Z J, et al. Developing conjugated polymers with high electron affinity by replacing a c–c unit with a B<– N Unit [J] . Angewandte Chemie International Edition, 2015, 54 (12) : 3648–3652.

[70] Zhao R Y, Dou C D, Xie Z Y, et al. Polymer acceptor based on BN units with enhanced electron mobility for efficient all–polymer solar cells [J] . Angewandte Chemie International Edition, 2016, 55 (17) : 5313–5317.

[71] Dou C D, Long X J, Ding Z C, et al. An electron–deficient building block based on the B<– N unit: an electron acceptor for all–polymer solar cells [J] . Angewandte Chemie International Edition, 2016, 55 (4) : 1436–1440.

[72] Ding Z C, Long X J, Dou C D, et al. A polymer acceptor with an optimal LUMO energy level for all–polymer solar cells [J] . Chemical Science, 2016, 7 (9) : 6197–6202.

[73] Long X J, Ding Z C, Dou C D, et al. Polymer acceptor based on double B <– N bridged bipyridine (BNBP) unit for high–efficiency all–polymer solar cells [J] . Advanced Materials, 2016, 28 (30) : 6504–6508.

[74] Long X J, Wang N, Ding Z C, et al. Low–bandgap polymer electron acceptors based on double B <– N bridged bipyridine (BNBP) and diketopyrrolopyrrole (DPP) units for all–polymer solar cells [J] . Journal of Materials Chemistry C, 2016, 4 (42) : 9961–9967.

[75] He Z C, Zhong C M, Huang X, et al. Simultaneous enhancement of open–circuit voltage, short–circuit current density, and fill factor in polymer solar cells [J] . Advanced Materials, 2011, 23 (40) : 4636–4643.

[76] Liu S J, Zhang K, Lu J M, et al. High–efficiency polymer solar cells via the incorporation of an amino–functionalized conjugated metallopolymer as a cathode interlayer [J] . Journal of the American Chemical Society, 2013, 135 (41) : 15326–15329.

[77] Zhang K, Hu Z C, Xu R G, et al. High–performance polymer solar cells with electrostatic layer–by–layer self–assembled conjugated polyelectrolytes as the cathode interlayer [J] . Advanced Materials 2015, 27 (24) : 3607–3613.

[78] Nian L, Zhang W Q, Zhu N, et al. Photoconductive Cathode Inter layer for Highly Efficient Inverted Polymer Solar Cells [J] . Journal of the American Chemical Society, 2015, 137 (22) : 6995–6998.

[79] Nian L, Chen Z H, Herbst S, et al. Aqueous solution processed photoconductive cathode interlayer for high performance polymer solar cells with thick interlayer and thick active layer [J] . Advanced Materials, 2016, 28 (34) : 7521–7526.

[80] Li X C, Xie F X, Zhang S Q, et al. Over 1.1 eV workfunction tuning of cesium intercalated metal oxides for functioning as both electron and hole transport layers in organic optoelectronic devices [J] . Advanced Functional Materials, 2014, 24 (46) : 7348–7356.

[81] Zhang Z G, Qi B Y, Jin Z W, et al. Perylene diimides: a thickness–insensitive cathode interlayer for high performance polymer solar cells [J] . Energy & Environmental Science, 2014, 7 (6) : 1966–1973.

[82] Hou X L, Li Q X, Cheng T, et al. Improvement of the power conversion efficiency and long term stability of polymer solar cells by incorporation of amphiphilic Nafion doped PEDOT–PSS as a hole extraction layer [J] . Journal of Materials Chemistry A, 2015. 3, 18727–18734.

[83] Gu C, Chen Y C, Zhang Z B, et al. Achieving high efficiency of PTB7–based polymer solar cells via integrated optimization of both anode and cathode interlayers [J] . Advanced Energy Materials, 2014, 4 (8) .1301771.

[84] Cui Y, Xu B W, Yang B, et al. A novel pH neutral self–doped polymer for anode interfacial layer in efficient polymer solar cells [J] . Macromolecules, 2016, 49 (21) : 8126–8133.

[85] Xu B W, Zheng Z, Zhao K, et al. A bifunctional interlayer material for modifying both the anode and cathode in highly efficient polymer solar cells [J] . Advanced Materials, 2016, 28 (3) : 434–439.

[86] Tan Z A, Qian D P, Zhang W Q, et al. Efficient and stable polymer solar cells with solution–processed molybdenum oxide interfacial layer [J] . Journal of Materials Chemistry A, 2013, 1 (3) : 657–664.

[87] Xu Q, Wang F Z, Tan Z A, et al. High–Performance Polymer Solar Cells with Solution–Processed and Environmentally Friendly CuOx Anode Buffer Layer [J] . ACS Applied Materials & Interfaces, 2013, 5 (21) : 10658–10664.

[88] Wang F Z, Xu Q, Tan Z A, et al. Efficient polymer solar cells with a solution–processed and thermal annealing–free RuO_2 anode buffer layer [J] . Journal of Materials Chemistry A, 2014, 2 (5) : 1318–1324.

[89] Tan Z A, Li L J, Wang F Z, et al. Solution–processed rhenium oxide: a versatile anode buffer layer for high performance polymer solar cells with enhanced light harvest [J] . Advanced Energy Materials, 2014, 4 (1) : 1–7.

[90] Tu X H, Wang F Z, Li C, et al. Solution–processed and low–temperature annealed CrOx as anode buffer layer for efficient polymer solar cells [J] . Journal of Physical Chemistry C, 2014, 118 (18) : 9309–9317.

[91] Xie F X, Choy W C H, Wang C D, et al. Low–temperature solution–processed hydrogen molybdenum and vanadium bronzes for an efficient hole–transport layer in organic electronics [J]. Advanced Materials, 2013, 25(14): 2051–2055.

[92] Yang D, Zhou L Y, Yu W, et al. Work–function–tunable chlorinated graphene oxide as an anode interface layer in high–efficiency polymer solar cells [J] . Advanced Energy Materials, 2014, 4 (15) .

[93] Zhou L Y, Yang D, Yu W, et al. An efficient polymer solar cell using graphene oxide interface assembled via layer–by–layer deposition [J] . Organic Electronics, 2015, 23: 110–115.

[94] Zheng Z, Zhang S, Zhang M, et al. Highly efficient tandem polymer solar cells with a photovoltaic response in the visible light range [J] . Advanced Materials, 2015, 27 (7) : 1189–1194.

[95] Zhang Q, Wan X, Liu F, et al. Evaluation of small molecules as front cell donor materials for high–efficiency tandem solar cells [J] . Advanced Materials, 2016, 28 (32) : 7008–7012.

[96] Li M, Gao K, Wan X, et al. Solution–processed organic tandem solar cells with power conversion efficiencies >12% [J] . Nature Photonics, 2016, 11 (2) : 85–90.

[97] Zuo L, Chang C Y, Chueh C C, et al. Design of a versatile interconnecting layer for highly efficient series–connected polymer tandem solar cells [J] . Energy & Environmental Science, 2015, 8 (6) : 1712–1718.

[98] Zhang K, Gao K, Xia R, et al. High–performance polymer tandem solar cells employing a new n–type conjugated polymer as an interconnecting layer [J] . Advanced Materials, 2016, 28 (24) : 4817–4823.

[99] Bin H, Zhang Z G, Gao L, et al. Non–fullerene polymer solar cells based on alkylthio and fluorine substituted 2D–conjugated polymers reach 9.5% efficiency [J] . Journal of the American Chemical Society, 2016, 138 (13) : 4657–4664.

[100] Liu W, Li S, Huang J, et al. Nonfullerene Tandem Organic Solar Cells with High Open–Circuit Voltage of 1.97 V [J] . Advanced Materials, 2016, 28 (44) : 9729–9734.

[101] Chen S, Zhang G, Liu J, et al. An all–solution processed recombination layer with mild post–treatment enabling

efficient homo-tandem non-fullerene organic solar cells [J] . Advanced Materials 2017, 29 (6) .

[102] Yao H, Chen Y, Qin Y, et al. Design and synthesis of a low bandgap small molecule acceptor for efficient polymer solar cells [J] . Advanced Materials, 2016, 28 (37) : 8283-8287.

[103] Qin Y, Chen Y, Cui Y, et al. Achieving 12.8% Efficiency by Simultaneously Improving Open-Circuit Voltage and Short-Circuit Current Density in Tandem Organic Solar Cells [J] . Advanced Materials 2017, 29 (24) : 1606340.

[104] Fang J, Wang Z Y, Zhang J Q, et al. Understanding the impact of hierarchical nanostructure in ternary organic solar cells [J] . Advanced Science, 2015, 2 (10) : 1500250.

[105] Zhang J Q, Zhang Y J, Fang J, et al. Conjugated polymer-small molecule alloy leads to high efficient ternary organic solar cells [J] . Journal of the American Chemical Society, 2015, 137 (25) : 8176-8183.

[106] Nian L, Gao K, Liu F, et al. 11% Efficient ternary organic solar cells with high composition tolerance via integrated near-IR sensitization and interface engineering [J] . Advanced Materials, 2016, 28 (37) : 8184-8190.

[107] An Q S, Zhang F J, Yin X X, et al. High-performance alloy model-based ternary small molecule solar cells [J] . Nano Energy, 2016, 30: 276-282.

[108] Zhang G, Zhang K, Yin Q, et al. High-performance ternary organic solar cell enabled by a thick active layer containing a liquid crystalline small molecule donor [J]. Journal of the American Chemical Society, 2017, 139 (6): 2387-2395.

[109] Lu H, Zhang J C, Chen J Y, et al. Ternary-blend polymer solar cells combining fullerene and nonfullerene acceptors to synergistically boost the photovoltaic performance [J] . Advanced Materials, 2016, 28 (43) : 9559-9566.

[110] Zhao W, Li S, Zhang S, et al. Ternary polymer solar cells based on two acceptors and one donor for achieving 12.2% efficiency [J] . Advanced Materials, 2016, 29 (2) : 1604059.

[111] Liu T, Huo L J, Sun X B, et al. Ternary organic solar cells based on two highly efficient polymer donors with enhanced power conversion efficiency [J] . Advanced Energy Materials, 2016, 6 (6) :1502109.

[112] Zhao Y F, Zou W J, Li H, et al. Large-area, flexible polymer solar cell based on silver nanowires as transparent electrode by roll-to-roll printing [J] . Chinese Journal of Polymer Science, 2016, 35 (2) : 261-268.

[113] Lu S, Lin J, Liu K, et al. Large area flexible polymer solar cells with high efficiency enabled by imprinted Ag grid and modified buffer layer [J] . Acta Materialia, 2017, 130: 208-214.

[114] Zhang J, Zhao Y, Fang J, et al. Enhancing Performance of Large-Area Organic Solar Cells with Thick Film via Ternary Strategy [J] . Small, 2017, 13 (21) : 1700388.

[115] Mao L, Chen Q, Li Y, et al. Flexible silver grid/PEDOT:PSS hybrid electrodes for large area inverted polymer solar cells [J] . Nano Energy, 2014, 10: 259-267.

[116] Wu F, Li Z, Ye F, et al. Aligned silver nanowires as transparent conductive electrodes for flexible optoelectronic devices [J] . Journal of Materials Chemistry C, 2016, 4 (47) : 11074-11080.

[117] Zhang T, Chen Z, Yang D, et al. Fabricating high performance polymer photovoltaic modules by creating large-scale uniform films [J] . Organic Electronics, 2016, 32: 126-133.

[118] Liu J, Chen S, Qian D, et al. Fast charge separation in a non-fullerene organic solar cell with a small driving force [J]. Nature Energy, 2016, 1: 16089.

[119] Bin H, Gao L, Zhang Z G, et al. 11.4% Efficiency non-fullerene polymer solar cells with trialkylsilyl substituted 2D-conjugated polymer as donor [J] . Nature Communications, 2016, 7: 13651.

[120] Lin Y, Zhao F, Wu Y, et al. Mapping polymer donors toward high-efficiency fullerene free organic solar cells [J] . Advanced Materials, 2017, 29 (3) :1604155.

撰稿人：侯剑辉　姚惠峰

碳氢键活化研究进展

一、引言

合成科学是创造新物质的科学，它使人类生活更加丰富多彩，对人类物质生活品质的提高起到了至关重要的作用。传统的有机合成科学是基于有机化合物活性官能团之间的相互转化。但除了少数生物发酵和提纯得到的合成工业的原料之外，绝大部分合成工业原料是从石油工业中直接获得的碳氢化合物通过相对苛刻、费时、耗能的化学转化而获得的，同时这些转化也是合成工业污染的主要来源，如硝化等。因此，避免冗长、耗能、高污染的合成路径，从广普、廉价、容易获得的化学品直接制备高附加值的有用产物，不仅可以大大提高合成科学的步骤经济性和原子经济性，同时将改变传统合成路径，实现合成工业的革命。而这些化学品一般均含有丰富的碳氢键结构单元，因此此类物质转化的最关键基本科学问题是组成此类物质中最基本、最丰富的碳氢键的高选择性活化和转化问题。

由于一般情况下碳氢键只有 σ－轨道，解离能高，是一类非极性键，同时在原料和产物的分子中都存在着多种不同碳氢键，使碳氢键的选择性活化和转化面临活性和选择性双重挑战。但是，碳氢键直接选择性活化和转化可以避免传统有机合成中冗长步骤，简单、高效、经济、绿色，是化学家们追求的目标。鉴于此领域研究的挑战性和在基础研究中的重要性及其广泛的应用前景，碳氢键活化被誉为化学的“圣杯”[1]。

严格意义上的碳氢键活化的概念是基于金属有机化学的基元反应的角度提出来的，特指在金属试剂的作用下实现碳氢键的选择性切断形成碳金属键的过程[2]。在提出此概念伊始，从某种意义上而言，碳氢键活化特指碳氢键对低价过渡金属氧化加成实现碳氢键断裂和活化的过程[3]。自从 20 世纪 60 年代 Bergman 和 Graham 提出碳氢键活化的概念以来，碳氢键活化逐渐引起了化学家们的广泛关注，直到 1990 年开始，此领域的研究开始爆炸式增长，而且碳氢键活化的概念被进一步延伸，从碳氢键断裂过程中金属催化剂与底物的作用模式的不同被划分为四种不同的类型，包括碳氢键对低价金属的氧化加成、高价过渡金属对富电子体系的亲电取代、σ－键复分解、金属参与的碳氢键均裂等[4]。根据严格的碳氢键活化的定义，这些过程实现了碳氢键的断裂同时伴随着新的碳金属键的形成。近年来，随着此领域研究的深入，一种通过六元环状过渡态的碳氢键活化的新模式被提出：在分子内碱的作用下，碳金属键的形成和碳氢键的断裂协同发生，此过程被定义为“协同脱质子金属化”（concerted metalation–deprotonation，CMD）[5]。

另一方面，随着此领域研究的发展，碳氢键活化的概念进一步被扩展，在很多情况下伴随着碳氢键的选择性断裂并没有形成含有碳金属键的稳定的中间体，但化学家们依然把这些过程归纳到碳氢键活化的范畴。有时候为了区分二者的不同，很多时候化学家相对应的也将这些化学过程称之为“碳氢键转化”（C–H transformation）或者“碳氢键官能团化”（C–H functionalization）[5]。最近几年来在“碳氢键转化”的研究领域已经取得了一系列重要进展。对于脂肪族碳氢键的官能团化中最有效的两种策略，卡宾（乃春）插入[6]和自由基氧化[7]，一般认为都没有碳金属键的形成，目前也被认为属于碳氢键活化领域的重要进展；高效非均相催化剂，特别是具有独特结构的分子筛和金属氧簇合物，由于其独特的催化活性，在惰性碳氢键的催化转化，特别是甲烷的碳氢活化和定向转化领域（例如转化为苯和乙烯等[8]）也取得了重要突破。另一方面，新型仪器的开发进一步促进了电化学和光化学的二次繁荣，利用光或电化学促进的碳氢键高效高选择性转化在最近几年也获得了迅速发展，成为碳氢键转化的一个重要分支[9]。同时，一些新的碳氢键转化的途径被发展，例如利用 FLP 的新体系促进碳氢键的选择性转化等，也拓展了在此领域进一步发展的新思路[10]。由于篇幅有限，本报告将主要集中于阐述近三年来以含碳金属键的关键中间体的传统的碳氢键活化的研究进展。为了进一步阐述碳氢键转化研究领域的新趋势，在展望部分我们也稍微涉及目前碳氢键转化领域的新策略和新思路。

二、近三年来碳氢键活化的研究进展

传统的碳氢键活化经历了近 70 年的发展，取得了一系列重要的研究成果，碳氢键活化的各种模型及后续转化的反应历程相对比较清楚，在进入 21 世纪以来此领域更是蓬勃发展[11]。但随着更多的人力物力的投入和此领域研究的深入开展，过渡金属催化的碳氢键活化领域的研究也逐渐进入了瓶颈阶段，遇到的前所未有的挑战，其主要包括以下几个方面：①由于碳氢键广泛存在于有机分子中，其活化过程中的反应性和选择性是面临的最核心科学问题和挑战。目前解决此关键科学问题的策略是利用不同碳氢键之间微弱的化学环境的差别（如，电子效应、位阻效应和立体电子效应等）以及导向基团的配位导向策略，因此必然面临着底物局限性的问题；②为了解决反应性和催化效率的问题，过渡金属催化的碳氢键活化和转化的反应条件相对于传统有机化学转化而言更加剧烈，反应体系非常复杂，例如通常需要较高的温度、较长的反应时间、需要不同的添加剂和氧化剂等；③基于碳氢键活化的机理和不同过渡金属催化剂的催化性能不同，在传统的碳氢键活化中一般选用的是较为昂贵的后过渡金属催化剂，如较常用的 Pd、Rh、Ir 等，而且催化剂的用量也较大。这不仅大大增加了合成成本，同时还面临合成过程中催化剂重金属残留的问题；④相对于传统的有机合成官能之间的化学转化，从碳氢键出发的化学转化的反应类型较为单一；⑤由于碳氢键活化中一般反应条件相对较为苛刻，同时配体在很多体系中难以发挥作用，因此不对称催化转化面临更大挑战；⑥尽管此领域的有了很大发展，但各方面的局限在一定程度上限制了各类碳氢键活化反应在合成中的应用。

概括说来，自 2014 年以来，尽管从过渡金属促进的碳氢键断裂生成碳金属键的基元反应本质的角度来讲碳氢键活化领域并没有质的突破，但面对过渡金属催化的碳氢键活化所面临的各方面的挑战，经过化学家们的共同努力都取得了重要进展。我们将根据碳氢键活化面临的不同挑战及取得的重要突破，分别来阐述近三年来碳氢键活化领域的最新进展。

（一）导向基主流策略的发展

在过去的三年中，导向基策略仍然是碳氢键活化领域研究的主流。尽管早在 2005 年就有双齿螯合导向基团被报道[12]，近几年此类导向基团在碳氢键活化研究领域的应用才得到了迅速发展，特别是以导向 sp^3 碳氢键的定向活化和转化[13]。一些新的导向基团被设计、应用乃至商品化（如 PIP），并发现他们在导向 sp^3 碳氢键的活化和转化中体现独特的反应性[14]。一般情况下，利用双导向策略实现碳氢键活化都是通过 5–5 钯杂环中间体实现 β–位选择性。最近的研究表明，形成 5–6 钯杂环中间体可以进一步实现 γ–位碳氢键的烯基化，实现了导向基团控制的脂肪族碳氢键活化位置选择性研究领域的一个突破[15]。尽管通过底物和时间作用现场生成导向基的原理早在本世纪初就已经报道，但利用氨基酸作为催化的导向基实现碳氢键活化过程中的选择性控制依然是此领域的一个亮点[16]。相对于双齿导向策略，利用底物本身自有官能团实现惰性 sp^3 碳氢键的活化是此领域未来重点发展方向。目前，除了利用较为常见的弱羧酸及其盐的导向之外[17]，最近发展的利用胺及其衍生物（包括酰胺及盐等）再结合底物本身的结构特征，实现其特定位置的碳氢键选择性官能团化也是此领域的重要进展[18]。对于普通的脂肪族化合物，如烷烃、醚等，目前除了利用相对较为成熟的位阻控制实现其活化的选择性外还没有更好的突破[19]。

一般而言，芳香体系的碳氢键活化导向基团的模板效应都控制其邻位选择性[20]。研究表明，利用不同的过渡金属催化剂，通过反应机制的调控可以实现普通导向基团（如酰胺等）的间位选择性[21]。最近发展的模板控制的长程导向也可以实现间位甚至对位活化，是近些年此领域的重要的突破性进展，成为实现取代苯间（对）位官能团化的利器之一[22]。最近，利用支持金属催化剂的配体和底物官能团之间的相互作用实现底物碳氢键活化的选择性控制成为碳氢键活化的一个新的方向[23]。当然，在底物中引入具有不同导向能力的多个官能团，通过对反应体系的调节实现底物不同位置的碳氢键的官能团化，可以为进一步实现产物分子复杂结构的建立提供快速有效地渠道[24]。对于不同杂环体系的碳氢键活化，早在2008年就有报道[25]，但最近的研究实现了不同杂环体系不同位置碳氢键活化的控制，并实现了化学转化的多样性[26]，同时对同一杂环体系不同碳氢键的程序活化将是此领域的未来研究一个重要方向[27]。但是，目前对普通非官能团化的芳烃（如甲苯等）的高选择性（特别是单一选择性）活化呼唤新的策略和手段。

目前发展的碳氢键活化的反应体系一般耗时、耗能，而且大多需要添加剂，相对复杂，这些都阻碍了碳氢键活化在工业合成中的应用[28]。有机化学家一直致力于发展切实有效的操作过程来开拓碳氢键活化的应用前景。在过去几年里，一些温和、高效的反应体系逐渐被开发，使碳氢键活化甚至在室温下就可以顺利进行[29]。为了避免大量无机金属氧化剂在碳氢键活化，特别是在碳氢键氧化偶联中的使用，化学家一方面发展了具有氧化性能的导向基团，在碳氢键活化中同时扮演导向基和氧化剂的角色[29]。无疑，洁净的氧化剂的使用是大势所趋。因此，利用氧气或空气作为氧化剂和氧合试剂的碳氢键活化在过去三年也获得了重要进展[30]，尽管此领域的研究已有较长的历史[31]。

（二）配体调控碳氢键活化

配体对过渡金属催化的碳氢键活化的反应性和选择性的调控是最近三年碳氢键活化领域最重要的进展。基于过渡金属对碳氢键的活化过程机理的分析，一般认为目前碳氢键活化中相对比较成熟且具有应用前景的是碳氢键对低价过渡金属的氧化加成和高价过渡金属对碳氢键的亲电两种途径。通过氧化加成实现碳氢键活化的研究相对比较成熟，主要应用于铑或者铱配合物催化的对非极性C–H键，特别是烷烃和普通芳烃碳氢键的硼化和硅化反应[32]。此类反应中一般需要电子云密度高、位阻较大的Cp*[33]或者双氮类配体[34]促进反应的顺利进行。该领域的一个重要突破是，通过对反应历程的了解，一类新型的含硼配体被设计并在铱催化碳氢键硼化中展现了其威力[35]。需要强调的是，尽管此领域在反应类型上并没有太大的突破，但在底物拓展方面最近取得了突破性进展：分子内反应实现醇硅醚的特定位置碳氢键的硅化反应，为醇向二醇及其衍生物转化提供了一条最直接的途径[36]；醚、胺及吡啶导向的烷烃碳氢键硼化，突破原本只有简单烷烃才能实现末端硅、硼化的瓶颈[37]；甲烷的直接硼化反应，为天然气的直接官能团化应用提供重要途径[38]。该领域另一个重要进展是利用不同种类手性配体的调控，实现了芳烃底物碳氢键的不对称去对称化硼化或硅化以及相应的动力学拆分，为光学活性的二芳基甲烷及其衍生物的制备提供了有效的途径[39]。据此推测，利用此策略实现脂肪族碳氢键的不对称硅化和硼化在不久的将来就会实现。另外，利用对Cp*结构的修饰实现惰性碳氢键的不对称转化是此领域近几年的另一大亮点[40]，而利用非手性Cp*Rh配合物和酶的手性空腔的协同作用实现碳氢键的不对称转化为此领域开辟了一条崭新的途径[41]。

相对于氧化加成实现碳氢键活化，亲电取代实现碳氢键活化的过程需要高价过渡金属的参与，金属中心低电子云密度有利于亲电金属化的发生，因此一般认为在此过程中配体的参与不利于反应的顺利进行[42]。特别是在利用导向基促进的碳氢键活化体系，尽管导向基团的引入提高了碳氢键活化的效率和选择性[20-43]，但在反应过程中，由于导向杂原子与中心金属的配位作用一方面增加了中心金属的电子云密度，同时占用了中心金属的配位位点，配体的加入在原则上对此类反应更加不利[44]。但随着碳氢键活化领域的进展，特别是以后过渡金属钯络合物为代表的催化剂催化碳氢键活化的研究逐渐深入，传统意义上认为是 Pd（II）/Pd（0）的催化循环被进一步拓展到 Pd（II）/Pd（IV）、甚至 Pd（III）参与的催化过程[45]。另一方面，新的 CMD 机理的提出为过渡金属催化的碳氢键活化打开了一条全新的通道[5]。这些实验上的发现和新的理论的提出促使化学家重新审视配体在钯催化的碳氢键活化领域中的角色，进一步探索不同配体在实现碳氢键活化过程中反应性和选择性的调控。在最近的几年里，一系列含氮杂环和氨基酸及其衍生的配体开始逐渐被应用于钯催化的碳氢键活化，起到了加速反应、调控反应立体选择性等作用[46]。即使是被认为对后过渡金属催化剂有毒害作用的含硫配体，如具有一定位阻效应的硫醚配体，在钯催化的碳氢键活化中也具有良好的促进和调控效应[47]。此外，配体在碳氢键活化中的作用和不对称控制获得了计算化学和实验的强力支撑[48]。目前此领域的研究渐有燎原之势，必将掀起配体调控的碳氢键活化的一翻浪潮。

（三）廉价金属催化碳氢活化的发展

较大用量的后过渡贵金属催化剂的使用大大限制了碳氢键活化在合成中的应用，因此发展高效的催化体系也是此领域面临的重要挑战之一[49]。如上所述，利用配体的调控可以加速反应、提高效率，从而降低催化剂的用量。另一个解决方案是利用地球上丰度较大的廉价金属，特别是毒性较小的短周期过渡金属，来代替后过渡贵金属实现同样的化学转化，在降低合成成本的同时解决重金属残留的问题。近期研究发现，对于相对较早研究的芳烃与卤化物直接偶联的反应[50]，基本上所有具有单电子转移能力的过渡金属都可以促进此类反应的进行，实现了此领域从贵金属催化到普通金属催化的拓展[51]。尽管不同金属催化的反应结果一致，但一般认为，前过渡金属催化的历程是单电子转移的历程，与后过渡金属催化的反应历程有本质差别，并不是传统意义上的碳氢键活化。在此基础上，化学家进一步发展了无过渡金属促进的普通芳烃（烯烃）和卤代芳烃直接偶联，已成为一类非常重要的“均裂芳香烃取代反应”催化体系（HAS，homolytic aromatic substitution）[52]。近几年，此领域继续迅速发展，不同种类的催化剂逐渐被开发[53]，而利用类似的策略实现芳烃碳氢键的直接硅基化反应成为此领域一道亮丽的风景线[54]，相应催化历程也被详细探究[55]。

近几年的研究表明，事实上所有后过渡金属催化的碳氢键活化的反应基本上都可以利用不同的过渡金属催化剂来代替。相对于广泛应用的 Pd[56]、Rh[57]、Ir[58]等，Ru 和 Co

成为了两类明星金属催化剂，展现了其独特的催化性能。相比较而言，过渡金属 Ru 的催化性能和后过渡金属的性能更加接近，且其具有价格优势，相对毒性也较小，在过渡金属催化反应，特别是烯烃复分解反应中，体现了其独特性[59]。最近，Ru 催化的碳氢键活化得到了迅速发展，特别是在导向的芳香烃碳氢键活化的研究领域获得了巨大成功，实现了芳香烃碳氢键的不同的转化[60]。而最为廉价的金属钴不仅在催化碳氢键活化中体现了其独特的反应性（例如，目前只有钴催化剂可以催化碳氢键和活泼金属试剂格式试剂的偶联反应）[61]，同时也展现了在碳氢键和不同的偶联对象之间的催化活性[62]。而且，钴的价电子结构特点也决定了其催化历程的特异性，例如，在钴催化的碳氢键活化过程中往往需要特殊活性金属试剂，如格式试剂等的参与[63]。因此，无论是从应用的角度，发展廉价金属催化的碳氢键活化体系，还是从学术研究的角度，探索短周期过渡金属的独特反应性和催化性能而言，此领域的研究都非常重要。

铁和铜作为生命体系中绝大多数金属酶的催化活性中心的核心金属，被称之为“金属之王”[64]。不同配体支撑的铁和铜配合物体现了独特的催化活性，催化过程一般是以单电子转移的方式进行[65]。利用铁催化的单电子转移过程实现碳氢键的氧化也是碳氢键转化中的经典过程，包括“fenton chemistry”[66]和“giff chemistry”[67]。近几年来，利用新型配体的设计和调控，铁催化的碳氢键选择性氧化也取得了重大进展[68]。但此领域不属于严格意义上的碳氢键活化的范畴。最近的研究表明，铁催化剂也可以实现碳氢键的活化，以碳铁键作为关键中间体来实现催化循环，此领域的研究无疑为铁催化剂在碳氢键活化和转换中的角色重新定位，开辟了一片广袤的空间[69]。铜催化的末端炔碳氢键活化具有悠久历史，而且在合成化学中已经获得了广泛应用[70]。铜催化脂肪族碳氢键转化较早用于交叉脱氢偶联反应（CDC，cross dehydrogenative coupling）[71]，此类反应也被认为是经过单电子转移的过程[71]。近期的研究发现，通过碳铜中间体，高价铜物种可以催化惰性碳氢键的直接活化，大大拓展了此研究领域[72]。另外，Mn 催化的碳氢键活化的过程也体现其独特性，实现了和后过渡金属催化历程完全不同的新反应[73]。

前过渡金属，包括稀土金属配合物，一般没有或很少含有 d 电子，同时配位数较大而且离子半径多变，并且此类金属一遍较为亲氧，使它们在碳氢键活化的过程中体现了完全不同的催化活性。近年来，这些前过渡金属配合物也逐渐被应用于碳氢键活化[72]。例如，一般作为导向基团的苯甲醚中的甲氧基，在利用稀土配合物作为催化剂的碳氢键活化中展现出优秀的邻位定位效应，实现苯甲醚邻位的碳氢键对烯烃的加成反应[74]。此类过渡金属催化剂还在催化聚合领域展现了优秀而独特的催化活性[75]，因此，目前它们在碳氢键活化领域的发展将为这两类重要反应的组合提供广阔的空间[76]。

（四）主族金属在碳氢活化中的发展

除了过渡金属催化剂在碳氢键活化领域的迅速发展之外，主族金属最近几年在这个重要的研究领域也开始崭露头角。由于主族金属配合物中心金属本身没有变价，因此主族金属催化剂从原理上来说并不能实现碳氢键参与的氧化还原转化过程。因此，在过去的研究中，主族金属在其参与的碳氢键的转化中往往扮演强碱的抗衡阳离子或 Lewis 酸的角色，例如经典的傅克反应等[77]。最近的研究表明，碱金属、碱土金属等配合物也可以直接参与碳氢的选择性断裂[78]；利用具有氧化还原能力的配体支撑的配合物还可以促进碳氢键参与的氧化还原转化[79]。但目前此领域的报道只有零星的几例，而且都是利用当量金属试剂，但此领域是一个重要的方向，相信在未来会有突破性进展。

（五）碳氢键反应性的突破

制约碳氢键活化发展的另一个重要方面是反应类型的局限性。目前，在碳氢键活化领域发展的最为成熟的反应是在过渡金属催化下实现的碳氢键和卤化物之间的偶联反应[80]。无论是相对较为活泼的杂环碳氢键还是普通芳烃的碳氢键，以及最近迅速发展的脂肪族碳氢键，都可以与不同的芳基卤化物、烷基卤化物以及炔基卤化物等进行偶联，实现不同种类的碳碳键的构建[81]。随着此领域的发展，芳基卤化物的替代物，如对甲苯磺酸苯酯等，也可以成功应用于此类偶联[82]，并且一些重要进展已经在天然产物和药物分子的合成中获得了较为广泛的应用[83]。但目前相对较为惰性的卤化物，如烷基溴化物、卤化物等基本上不能参与反应，因此还有很大的发展空间。利用类似的原理，醇类化合物及其衍生物[84]氧、氮杂张力环等也可以代替卤化物作为亲电试剂参与反应[85]，从而进一步拓展了此类偶联反应。

碳氢键的氧化是另一类重要的反应，可以实现碳氢键直接到碳杂原子间的化学转化。此类化学转化中扮演重要角色的是自由基反应（包括 fenton chemistry 和 giff chemistry 等）[86]以及金属乃春及其类似物对碳氢键的插入反应[87]，由于没有碳金属中间体的存在，在此不做赘述。近年来，通过导向的策略实现了过渡金属参与的碳氢键活化形成金属杂环状中间体后，在强氧化剂的存在下实现中心金属的进一步氧化或基团转移，进而实现碳杂原子键的构建方面获得了蓬勃发展，成为直接构建碳杂原子的有效手段[88]。利用氧气作为洁净氧化剂实现碳氢键到酚或醇的转化更体现了其经济性和重要性，也为此领域的发展提出了重要的方向[89]。另一方面，除了碳氢键还原硼化在底物和配体领域的突破[90]，利用钯催化的氧化偶联也取得了重要进展，可以实现天然和非天然氨基酸及其衍生物的脂肪族碳氢键的直接硼化[91]。由于条件相对温和，此类硼化反应体现了比还原硼化更好的官能团兼容性。随后，sp^3 碳氢键的直接氧化硅化方面也取得了重要进展[92]。

有机合成化学的核心任务之一是碳碳键的构建，而从碳氢键出发实现碳碳键的构建将大大缩短合成路径，提高合成效率。除了碳氢键和卤化物及其衍生物的直接偶联，从碳氢键出发与不同的金属试剂、甚至另一分子的碳氢键的氧化偶联是直接有效的手段[93]。自 2000 年以来，利用不同金属试剂和碳氢键的直接氧化偶联构建联芳基化合物的方法获得了长足发展[94]，但在过去三年里此领域的研究并没有突破性进展。目前，此类氧化偶联还主要局限于在导向基作用下或杂环体系中 sp^2 碳氢键与不同金属试剂的氧化偶联，对于脂肪族碳氢键只有一些零星的报道[95]，这也是此领域有待突破的一个方向。

碳氢键与烯烃的氧化偶联反应是碳氢键活化最早涉及的反应之一，由 Fujiwara 于 20 世纪 60 年代首先发展起来，因此也叫作 Fujiwura 偶联反应[96]。这是碳氢键活化领域发展相对成熟的反应，但也主要集中于芳烃碳氢键的烯基化[97]。近几年，通过对不同烯烃底物的结构调控，实现两种不同烯烃直接的氧化偶联获得了很大进展，成为制备多取代共轭二烯的最有效手段[98]。另一方面，从官能团化芳烃出发，利用过渡金属催化的碳氢键活化启动反应实现烯烃或炔烃串联的氧化环化在近几年获得了一定的发展[99]，通过此过程实现苯并饱和杂环体系的对映选择性和非对映选择性的控制也是碳氢键活化领域的亮点[100]，在手性配体支撑下的单一催化剂催化通过碳氢键活化和不对称去芳构化相结合构建光学活性的螺环体系是此领域的一个重要突破[101]。此外，利用导向策略的 sp^3 碳氢键和烯烃的直接偶联也有一些零星的进展，但都不能停留在烯基化的产物，一般都是获得进一步加成的产物，这也将是此领域未来研究的重点[102]。

从碳氢键出发及炔烃衍生物出发制备内炔的反应这几年才发展起来，但主要是利用炔基溴和碳氢键在导向基团的协助下实现偶联[103]。对于贫电子芳烃和杂环，通过双金属协同催化可以实现其碳氢键和末端炔的氧化偶联[104]。尽管化学家们一直在探索脂肪族碳氢键与末端炔的偶联，但一直没有获得进展。在此过程中，张玉红等首次报道了导向基团协助的脂肪族碳氢键与末端炔氧化偶联后进一步加成的串联反应[105]。直到近两年，此领域才获得了重要突破：利用 Ni/Cu 作为协同催化剂，导向策略在脂肪族碳氢键和末端炔的偶联中获得了成功[106]；而对于普通的脂肪族碳氢键，利用双金属协同催化，通过自由基活化的策略，也可以实现氧化偶联，但目前无论是选择性的控制还是对反应的机理探究都仍需进一步努力[107]。

不同种类的惰性碳氢键之间氧化偶联构建碳碳键无疑是最理想的合成碳骨架的方法，但也面临巨大的挑战。因为此类转化一方面要面对每个碳氢键活化所面临的所有挑战，同时还需要解决两个碳氢键活化过程中的选择性和反应性匹配以及如何在相同条件下避免单一底物的自身偶联等问题，特别是对于两类结构相近的底物将是更大的挑战。通过分子内的两个芳烃碳氢键的氧化偶联构建稠环方向体系相对较为简单，此领域的发展也较为成熟[108]，但不同芳烃之间的分子间氧化偶联发展就更为缓慢。为了实现在不同种类的芳烃之间的氧化偶联领域的突破，科学家们发展了不同的策略：①利用两种不同芳烃的反应性差别，通过底物比例的调控实现交叉偶联的控制[109]；②利用芳烃，特别是杂环芳烃和氧化偶联对象不同位点立体和电子效应的控制实现交叉氧化偶联[110]；③利用导向基团和立体 / 电子效应的协同，实现不同种类芳烃的直接交叉氧化偶联[111]。这三种策略在几年前都已经建立，在近三年都有了一定发展，但在绝大部分情况下至少其中有一个芳烃面临着选择性问题，特别是对于单取代芳烃底物[112]。为了彻底解决两个芳烃反应过程中的选择性问题，2015 年双导向策略被发展起来。从两类容易获得的具有不同导向能力的取代苯出发，在单一催化剂的作用下可以实现二者之间的单一选择性氧化偶联，进一步实现分子内环化[113]。双官能团的存在一方面解决了氧化偶联的选择性，同时也为构建复杂和多向性的联芳基化合物提供了最直接的合成方法。最近，利用双金属协同催化，此领域的研究获得了进一步拓展[114]。

由于普通 sp^3 碳氢键活化的挑战性和活化手段的匮乏，除了相对活性相对较高的 sp^3 碳氢键，如羰基 α－位、苄位、烯丙位、杂原子邻位的碳氢键等，可以通过氧化形成的自由基或碳正离子中间体，再和另外一分子碳氢键实现氧化偶联[115]。此领域的研究可以进一步拓展到普通烷烃，如环己烷等的碳氢键[116]。但是，在此过程中至少有一种碳氢键是相对活泼的碳氢键。而无论是分子内还是分子间的反应，发生在惰性芳烃和普通烷基碳氢键之间的氧化偶联都鲜有进展。利用两种碳氢键直接完全不同的反应性，通过各种促进不同的碳氢键断裂的方法的组合是实现分子间芳烃碳氢键和 sp^3 碳氢键氧化偶联的一个重要策略。利用此策略，结合芳烃碳氢键活化中的导向策略和苄位碳氢键容易发生自由基氧化的特点，化学家最终实现了芳烃和甲苯之间的氧化偶联[117]。尽管之前利用自由基氧化后形成的中间体直接对富电子芳环发生亲电进攻也可以得到类似的结果，但后者仅局限于富电子芳环和较为活泼的双芳基甲烷类底物[118]。

利用吡咯类化合物 2 位独特的反应性，同时结合导向基团的导向作用，从原理上实现吡咯 2 位的碳氢键活化，再进一步与分子内通过酰胺键桥连的新戊酰基的甲基上的碳氢键活化，再实现二者之间的氧化偶联，是首例实现杂环芳烃上碳氢键和甲基碳氢键之间的氧化偶联的报道[119]。但是，由于普通苯环上碳氢键的反应性更低，迄今为止还没有芳烃和脂肪族甲基碳氢键直接的氧化偶联的报道。最近此领域有了重要突破，研究者实现了从芳基烷基醚出发，通过氧化偶联直接构建多取代苯并二氢呋喃的新方法[120]。在此过程中，配体的调控具有关键作用。这也是首次实现没有导向基存在下的醚 β－位伯碳和仲碳的碳氢键的选择性活化，这为进一步拓展芳烃和烷烃碳氢键以及两个不同的烷烃碳氢键之间通过碳氢键活化实现氧化偶联构建碳碳键提供了一种新思路。

（六）其他方面的发展

碳氢键从电性简单分析可以看成是碳基金属试剂（如格式试剂）的替代物，因此，催化的碳氢键代替金属试剂，实现对碳杂原子不饱和键的加成将是最理想的合成醇和胺的手段，其可以避免卤化物和金属试剂制备过程中的冗长步骤、有毒危险品的使用以及无水无氧等复杂实验操作。2011 年此领域获得了重要突破，在导向基存在下利用 Rh 催化可以实现芳基碳氢键对亚胺的加成[121]，随后碳氢键对醛[122]、酮[123]的加成也相继被实现。最近，利用廉价金属，如 Mn 等也可以实现碳氢键对醛的加成[124]。由于 Mn 催化剂独特的反应性，反应的底物得到了更好的拓展。此领域的突破从原理上来说表明了碳氢键对碳杂原子双键加成的可能性。但是，无论是从反应效率还是从底物的类型来看，目前发展的方法都具有很大的局限性，而且至今也没有普通 sp^3 碳氢键对醛、酮、亚胺的加成的报道。此类反应的不对称催化也将是下一个吸引人的挑战性课题。

最近，碳氢键活化领域的重要发展还包括将碳氢键活化反应与其他不同的领域相结合，实现不同反应的串联或接力。首先，在基于碳氢键的转化方面，利用碳氢键活化生成的产物不同的淬灭方式或进一步的化学转化，可以实现不同种类杂环化合物的构建，为碳氢键活化在材料化学或药物化学中的应用提供了坚实的实验基础，也为碳氢键活化反应中的不对称催化历程提供重要反应模型[125]。过去三年里，利用烯烃或炔烃作为反应试剂，不同种类的苯并环系乃至螺环骨架被高效构建，由此产生的不对称催化反应也引领了碳氢键活化相关的不对称催化反应的潮流，成为碳氢键活化发展最为迅速的领域之一[126]；利用多种反应物参与的串联反应不仅增加生成产物的多样性和复杂性，同时更进一步提高了合成效率，此领域的研究在最近逐渐发展起来[127]；利用同一分子不同位置碳氢键的反应性的差别，通过不同类型的碳氢键活化组合，发展程序化碳氢键活化，实现同一分子多官能团化过程[128]。其次，在碳氢键活化和转化领域不同催化体系的组合更体现了合成化学的效率和美妙。利用过渡金属催化实现碳氢键的选择性活化和有机小分子催化调控转化的立体选择性，可以直接完成烯烃烯丙位的不对称烷基化[129]。同时，结合金属有机化学或其他方法学领域的最新进展，可以大大提高碳氢键活化在有机合成中的应用潜力。结合发展相对成熟的烷烃脱氢、烯烃异构化以及烯烃的硅氢化反应，可以间接实现烷烃的高选择性末端硅化，为廉价碳氢化合物向高附加值产品的转化提供重要途径[130]。通过碳氢键活化的脱氢过程结合借氢还氢的原理，可以实现烷烃的交叉复分解，此方法在烃类化合物的规整化和石油工业中具有重要意义。最近，化学家利用此原理实现聚烯烃类高聚物到短链烷烃的转化，为将高聚物降解成有用有机化合物，实现白色污染“变废为宝”提供了重要手段[131]。

随着其他实验手段和实验方法的日新月异，碳氢键活化与其他新手段的结合也使此领域的发展变得更加绚丽多彩，同时也为碳氢键活化领域的发展提供了更多的契机。光化学和电化学一直伴随着化学科学的发展，具有悠久的历史。最近光化学和电化学又二次繁荣，在碳氢键的转化，特别是直接氧化的领域展现了优势[132]，同时在过渡金属催化的碳氢键活化的领域也发挥了重要作用，成为碳氢键活化领域的近几年的亮点之一。光化学在碳氢键活化转化中的参与主要贡献在两个方面：①催化循环中实现催化剂的光氧化或还原，协助催化循环的进行[133]；②通过光化学对底物或者试剂的活化促进碳氢键活化和转化的顺利进行[134]。与光化学的作用类似，利用电化学可以产生活性中间体与碳氢键断裂后的金属中间体作用，或者通过对金属中间体的直接电子转移完成碳氢键活化循环。例如，最近的研究表明，在导向策略下，利用电化学可以代替外加氧化剂，实现温和洁净条件下钯催化的芳烃碳氢键的卤化反应[135]。

不对称催化是合成科学中最重要的研究领域，基于碳氢键活化的不对称催化方法学也一直是合成科学家们最关注、最具挑战性的研究领域之一。由于碳氢键活化的反应条件相对苛刻，通常在温度较高的条件下进行，为此类反应的不对称催化增加了困难[136]。一般情况下，芳烃碳氢键的活化过程不涉及手性中心的建立，因此手性中心的产生一般在活化后转化的过程。目前在此领域利用手性配体[137]或酶的手性环境[138]下实现不对称催化的过程都已经取得了重要进展。当然，在手性配体作用下对对称底物的芳香碳氢键的不对称识别，实现不对称去对称化反应也是此领域的一个亮点[139]。对于 sp^3 碳氢键活化的不对称反应的研究，近几年也取得了突破性进展。最近研究表明，通过配体的调控，可以通过不对称去对称化反应实现异丙基上两个甲基的碳氢键的手性区分[140]。更重要的是，根据近几年不同配体在钯催化碳氢键活化中的作用的研究积累，化学家从氨基酸出发设计合成了一类新型配体，可以实现潜手性二级碳中心两个不同碳氢键的手性识别和转化，此研究是目前不对称催化碳氢键活化领域的重要突破[141]。目前此领域方兴未艾，需要更多的合成化学家共同努力。

随着实验科学的深入，近些年，在碳氢键活化领域的理论研究也获得了重要进展。目前，相关的理论研究所用的方法主要是通过动力学实验和中间体的分离[142]，同时计算化学成为不可或缺的最有力的手段，在碳氢键活化中研究配体的作用[143]、反应位置选择性[144]、以及对映选择性[145]控制等方面都起到了关键作用。最重要的是，近年来我国化学家通过对碳氢键键能数据的积累，进一步总结和阐述了碳氢键键能和其断裂过程中的构效关系，此领域的研究对碳氢键活化的理论发展具有重要的指导作用[146]。

随着碳氢键活化领域的方法学研究如火如荼的开展，这些新方法在各个领域的应用也逐渐成为化学家们关注的目标。过去三年，此领域也已经取得了可喜进展。将碳氢键活化方法拓展到更加简单易得的原料，将为它们在合成中的应用铺平道路。例如，利用乙烯作为乙基化试剂，通过过渡金属催化的碳氢键对乙烯的加成实现烷基化，将是最高效、最洁净、最经济的乙基化方法[147]。类似地，从苯和乙烯直接合成苯乙烯也体现了化学的美和魅力[148]。芳烃碳氢键活化及其进一步转化在大共轭体系的建立以及构建稠环体系中具有很好的优势。因此，利用此方法在材料化学中的应用已经获得了一系列重要进展，是碳氢键活化的应用领域最重要的发展之一[149]。碳氢键活化可以实现杂环等化合物的快速官能团化，对不复杂分子在合成后期的官能团化也体现出的独特优势，因此利用碳氢键活化实现分子多样化和复杂化，快速构建具有优势结构的化合物库具有重要意义，此类碳氢键活化在药物化学领域的应用已经崭露头角[150]。另一方面，利用碳氢键活化作为核心和关键步骤实现复杂天然产物的全合成在过去三年也有了重要进展[151]，而碳氢键活化在特殊结构化合物的制备方面也体现出了独特的优势[152]。事实上，碳氢键活化化学还可以实现传统合成化学难以实现的目标，例如对于高聚物的直接结构改造实现官能团化[152]以及聚烯烃类高聚物的降解[131]等。

碳氢键活化的目标是实现碳氢键的选择性转化，没有碳金属键参与的碳氢键的高效高选择性转化与碳氢键活化的目标一致，都是以实现从碳氢键出发的更加洁净、经济、绿色合成为目标，二者在一定程度上相互促进，相辅相成。在过去三年里，碳氢键转化的领域也取得了飞速发展，特别是在脂肪族碳氢键的高效高选择性转化领域。脂肪族碳氢键的选择性氧化是自然界中有机物质转化的重要过程，利用底物立体电子效应的控制，在配体支撑的铁催化可以实现惰性脂肪族碳氢键的选择性氧化[153]，甚至在没有过渡金属作用下的自由基氧化方面也展现了其魅力[154]。结合快速发展的光化学和电化学合成方法，惰性碳氢键的直接选择性氧化也展现了其巨大优势[155]。在不对称催化的碳氢键转化领域，利用碳氢键活化的手段不能实现的过程也可以通过自由基化学或卡宾插入的手段来实现[156]。例如，利用铜催化的自由基反应，在手性恶唑啉配体的支撑下，可以在没有导向基团的作用下实现苄位碳氢键的立体选择性氰基化[157]；而对于普通方法根本无法实现的直链烷烃二级碳上的两个不同碳氢键的位置选择性和立体选择性区分，目前只有通过配体调控的卡宾插入来实现[156b]。这些工作都是碳氢键转化领域的重要突破，也代表着在脂肪族碳氢键转化的重要方向。

三、总结与展望

总而言之，在过去几年里，碳氢键活化及其转化的化学得到了长足发展，也取得了令人瞩目的成绩，在底物的拓展、催化体系的更新、新催化体系的建立、不对称催化以及在合成中的应用等领域都取得了丰硕的成果。令人欣慰的是，此领域的每一项重要进展都可以看到中国或华裔科学家的身影，而更难能可贵的是，华人科学家在碳氢键活化领域的研究工作逐渐从幕后走向台前，从主要对前人工作的修饰和改进，开始了更具创新性的、独特的工作，相信我国科学家在此领域将会做出更重要的贡献。但是，尽管此领域发展如火如荼，在过去几年中碳氢键活化的研究工作从原理上并没有质的突破，很多挑战性的研究课题，例如同一分子中不同化学环境的亚甲基碳氢键的选择性活化等，还没有质的突破，有待化学家们共同努力。绿色和可持续是化学转化的永恒主题，碳氢键活化及高效转化将一直伴随着合成化学中这一主题的发展，我们期待着在此领域更丰富多彩的化学将引领合成科学的未来，甚至推动合成科学的革命。

参考文献

[1] Yu J Q, Shi Z J. C–H activation [M]. Berlin: Springer, 2010.
[2] Bergman R G. Organometallic chemistry: C – H activation [J]. Nature, 2007, 446 (7134): 391–393.

[3] Janowicz A H, Bergman R G. Carbon-hydrogen activation in completely saturated hydrocarbons: Direct observation of M+R-H. Fwdarw. M (R) (H) [J] . Journal of the American Chemical Society, 1982, 104 (1) : 352-354.

[4] Dyker G. Handbook of C-H Transformations [M] . Weinheim, Germany: Wiley-VCH Velag GmbH & Co. KGaA, 2005.

[5] Lapointe D, Fagnou K. Overview of the mechanistic work on the concerted metallation - deprotonation pathway [J] . Chemistry Letters, 2010, 39 (11) : 1118-1126.

[6] Walsh P J, Hollander F J, Bergman R G. Generation, alkyne cycloaddition, arene carbon-hydrogen activation, nitrogen-hydrogen activation and dative ligand trapping reactions of the first monomeric imidozirconocene(Cp2Zr:NR) complexes [J] . Journal of the American Chemical Society, 1988, 110 (26) : 8729-8731.

[7] Sherry A E, Wayland B B. Metalloradical activation of methane [J] . Journal of the American Chemical Society, 1990, 112 (3) : 1259-1261.

[8] a) Solymosi F, Erdöhelyi A, Szöke A. Dehydrogenation of methane on supported molybdenum oxides. Formation of benzene from methane [J] . Catalysis Letters, 1995, 32 (1-2) : 43-53; b) Saroğlan A, Erdem-Şenatalar A, Tunç Savaşçı Ö, et al. The effect of dealumination on the apparent and actual rates of aromatization of methane over MFI-supported molybdenum catalysts [J] . Journal of Catalysis, 2004, 226 (1) : 210-214; c) Lacheen H S, Iglesia E. Isothermal activation of Mo2O52+-ZSM-5 precursors during methane reactions: Effects of reaction products on structural evolution and catalytic properties [J]. Physical Chemistry Chemical Physics, 2005, 7 (3): 538-547; d) Natesakhawat S, Means N C, Howard B H, et al. Improved benzene production from methane dehydroaromatization over Mo/HZSM-5 catalysts via hydrogen-permselective palladium membrane reactors [J] . Catalysis Science & Technology, 2015, 5 (11) : 5023-5036.

[9] a) Kakiuchi F, Kochi T, Mutsutani H, et al. Palladium-catalyzed aromatic C-H halogenation with hydrogen halides by means of electrochemical oxidation [J] . Journal of the American Chemical Society, 2009, 131 (32) : 11310-11311; b) Yang Q L, Li Y Q, Ma C, et al. Palladium-catalyzed C (sp^3) —H oxygenation via electrochemical oxidation [J] . Journal of the American Chemical Society, 2017, 139 (8) : 3293-3298; c) Qvortrup K, Rankic D A, MacMillan D W C. A general strategy for organocatalytic activation of C - H bonds via photoredox catalysis: Direct arylation of benzylic ethers [J] . Journal of the American Chemical Society, 2014, 136 (2) : 626-629; e) Romero N A, Margrey K A, Tay N E, et al. Site-selective arene C-H amination via photoredox catalysis [J] . Science, 2015, 349 (6254) : 1326-1330.

[10] Han Y X, Zhang S T, He J H, et al. B (C_6F_5) $_3$-catalyzed (convergent) disproportionation reaction of indoles [J] . Journal of the American Chemical Society, 2017, 139 (21) : 7399-7407.

[11] a) Kakiuchi F, Murai S. Catalytic C-H/olefin coupling [J] . Accounts of Chemical Research, 2002, 35 (10) : 826-834; b) Davies H M L, Beckwith R E J. Catalytic enantioselective C-H activation by means of metal-carbenoid-induced C-H insertion [J] . Chemical Reviews, 2003, 103 (8) : 2861-2904; c) Seregin I V, Gevorgyan V. Direct transition metal-catalyzed functionalization of heteroaromatic compounds [J] . Chemical Society Reviews, 2007, 36 (7) : 1173-1193; d) Alberico D, Scott M E, Lautens M. Aryl-aryl bond formation by transition-metal-catalyzed direct arylation [J] . Chemical Reviews, 2007, 107 (1) : 174-238; e) Liu C, Zhang H, Shi W, et al. Bond formations between two nucleophiles: Transition metal catalyzed oxidative cross-coupling reactions [J] . Chemical Reviews, 2011, 111 (3) : 1780-1824; f) Song G Y, Wang F, Li X W. C - C, C - O and C - N bond formation via rhodium (III) -catalyzed oxidative C - H activation [J] . Chemical Society Reviews, 2012, 41 (9) : 3651-3678.

[12] Zaitsev V G, Shabashov D, Daugulis O. Highly regioselective arylation of sp^3 C-H bonds catalyzed by palladium acetate [J] . Journal of the American Chemical Society, 2005, 127 (38) : 13154-13155.

[13] a) Reddy B V S, Reddy L R, Corey E J. Novel acetoxylation and C-C coupling reactions at unactivated positions in

α-amino acid derivatives [J]. Organic Letters, 2006, 8 (15): 3391–3394; b) Shang R, Ilies L, Matsumoto A, et al. β-arylation of carboxamides via iron-catalyzed C (sp^3) - H bond activation [J]. Journal of the American Chemical Society, 2013, 135 (16): 6030–6032; c) Zhang S Y, Li Q, He G, et al. Stereoselective synthesis of β-alkylated α-amino acids via palladium-catalyzed alkylation of unactivated methylene C (sp^3) - H bonds with primary alkyl halides [J]. Journal of the American Chemical Society, 2013, 135 (32): 12135–12141; d) Pan F, Shen P X, Zhang L S, et al. Direct arylation of primary and secondary sp^3 C - H bonds with diarylhyperiodonium salts via Pd catalysis [J]. Organic Letters, 2013, 15 (18): 4758–4761; e) Ju L, Yao J Z, Wu Z H, et al. Palladium-catalyzed oxidative acetoxylation of benzylic C - H bond using bidentate auxiliary [J]. The Journal of Organic Chemistry, 2013, 78 (21): 10821–10831; f) Wu X S, Zhao Y, Ge H B. Nickel-catalyzed site-selective alkylation of unactivated C (sp^3) -H bonds [J]. Journal of the American Chemical Society, 2014, 136 (5): 1789–1792.

[14] a) Chen K, Hu F, Zhang S Q, et al. Pd (II) -catalyzed alkylation of unactivated C (sp^3) - H bonds: Efficient synthesis of optically active unnatural α-amino acids [J]. Chemical Science, 2013, 4 (10): 3906–3911; b) Zhang Q, Chen K, Rao W H, et al. Stereoselective synthesis of chiral α-amino-β-lactams through palladium (II) -catalyzed sequential monoarylation/amidation of C (sp^3) H bonds [J]. Angewandte Chemie International Edition, 2013, 52 (51): 13588–13592.

[15] Xu J W, Zhang Z Z, Rao W H, et al. Site-selective alkenylation of δ-C (sp^3) - H bonds with alkynes via a six-membered palladacycle [J]. Journal of the American Chemical Society, 2016, 138 (34): 10750–10753.

[16] Zhang F L, Hong K, Li T J, et al. Functionalization of C (sp^3) - H bonds using a transient directing group [J]. Science, 2016, 351 (6270): 252–256.

[17] Giri R, Maugel N, Li J J, et al. Palladium-catalyzed methylation and arylation of sp^2 and sp^3 C–H bonds in simple carboxylic acids [J]. Journal of the American Chemical Society, 2007, 129 (12): 3510–3511.

[18] a) Chan K S L, Wasa M, Chu L, et al. Ligand-enabled cross-coupling of C (sp^3) - H bonds with arylboron reagents via Pd (II) /Pd (0) catalysis [J]. Nature Chemistry, 2014, 6 (2): 146–150; b) McNally A, Haffemayer B, Collins B S L, et al. Palladium-catalysed C - H activation of aliphatic amines to give strained nitrogen heterocycles [J]. Nature, 2014, 510 (7503): 129–133; c) Willcox D, Chappell B G N, Hogg K F, et al. A general catalytic β-C - H carbonylation of aliphatic amines to β-lactams [J]. Science, 2016, 354 (6314): 851–857.

[19] a) Chen M S, White M C. A predictably selective aliphatic C - H Oxidation reaction for complex molecule synthesis [J]. Science, 2007, 318 (5851): 783–787; b) Litvinas N D, Brodsky B H, Du Bois J. C–H hydroxylation using a heterocyclic catalyst and aqueous H_2O_2 [J]. Angewandte Chemie International Edition, 2009, 48 (25): 4513–4516; c) Ochiai M, Miyamoto K, Kaneaki T, et al. Highly regioselective amination of unactivated alkanes by hypervalent sulfonylimino-λ^3-bromane [J]. Science, 2011, 332 (6028): 448–451.

[20] Colby D A, Bergman R G, Ellman J A. Rhodium-catalyzed C–C bond formation via heteroatom-directed C–H bond activation [J]. Chemical Reviews, 2010, 110 (2): 624–655.

[21] Phipps R J, Gaunt M J. A meta-selective copper-catalyzed C - H bond arylation [J]. Science, 2009, 323 (5921): 1593–1597.

[22] a) Leow D, Li G, Mei T S, et al. Activation of remote meta-C - H bonds assisted by an end-on template [J]. Nature, 2012, 486 (7404): 518–522; b) Dai H X, Li G, Zhang X G, et al. Pd (II) -catalyzed *ortho*- or meta-C - H olefination of phenol derivatives [J]. Journal of the American Chemical Society, 2013, 135 (20): 7567–7571; c) Yang Y F, Cheng G J, Liu P, et al. Palladium-catalyzed *meta*-selective C - H bond activation with a nitrile-containing template: Computational study on mechanism and origins of selectivity [J]. Journal of the American Chemical Society, 2014, 136 (1): 344–355; d) Bag S, Patra T, Modak A, et al. Remote

para–C – H functionalization of arenes by a D–shaped biphenyl template–based assembly [J]. Journal of the American Chemical Society, 2015, 137 (37): 11888–11891.

[23] Das S, Incarvito C D, Crabtree R H, et al. Molecular recognition in the selective oxygenation of saturated C–H bonds by a dimanganese catalyst [J]. Science, 2006, 312 (5782): 1941–1943.

[24] a) Zhang X S, Zhang Y F, Li Z W, et al. Synthesis of dibenzo [c, e] oxepin–5 (7*H*) –ones from benzyl thioethers and carboxylic acids: Rhodium–catalyzed double C – H activation controlled by different directing groups [J]. Angewandte Chemie International Edition, 2015, 54 (18): 5478–5482; b) Wang Y, Gu J Y, Shi Z J. Palladium–catalyzed direct annulation of benzoic acids with phenols to synthesize dibenzopyranones [J]. Organic Letters, 2017, 19 (6): 1326–1329.

[25] Li é gault B, Fagnou K. Palladium–catalyzed intramolecular coupling of arenes and unactivated alkanes in air [J]. Organometallics, 2008, 27 (19): 4841–4843.

[26] a) Liu X W, Shi J L, Yan J X, et al. Reigoselective arylation of thiazole derivatives at 5–position via pd catalysis under ligand–free conditions [J]. Organic Letters, 2013, 15 (22): 5774–5777; b) Zhang S, Niu Y H, Ye X S. General approach to five–membered nitrogen heteroaryl *C*–glycosides using a palladium/copper cocatalyzed C – H functionalization strategy [J]. Organic Letters, 2017, 19 (13): 3608–3611.

[27] Liu X W, Shi J L, Wei J B, et al. Diversified syntheses of multifunctionalized thiazole derivatives via regioselective and programmed C – H activation [J]. Chemical Communications, 2015, 51 (22): 4599–4602.

[28] Hartwig J F. Evolution of C – H bond functionalization from methane to methodology [J]. Journal of the American Chemical Society, 2016, 138 (1): 2–24.

[29] a) Nishikata T, Abela A R, Lipshutz B H. Room temperature C – H activation and cross–coupling of aryl ureas in water [J]. Angewandte Chemie International Edition, 2010, 49 (4): 781–784; b) Peron F, Fossey C, Santos J S D O, et al. Room–temperature *ortho*–alkoxylation and –halogenation of *N*–tosylbenzamides by using palladium (II) –catalyzed C–H activation [J]. Chemistry – A European Journal, 2014, 20 (24): 7507–7513.

[30] a) Shi Z Z, Zhang C, Tang C H, et al. Recent advances in transition–metal catalyzed reactions using molecular oxygen as the oxidant [J]. Chemical Society Reviews, 2012, 41 (8): 3381–3430; b) Gulzar N, Schweitzer-Chaput B, Klussmann M. Oxidative coupling reactions for the functionalisation of C – H bonds using oxygen [J]. Catalysis Science & Technology, 2014, 4 (9): 2778–2796.

[31] Scholl R, Seer C, Weitzenböck R. Perylen, ein hoch kondensierter aromatischer Kohlenwasserstoff $C_{20}H_{12}$ [J]. Berichte Der Deutschen Chemischen Gesellschaft, 1910, 43 (2): 2202–2209.

[32] a) Chen H Y, Schlecht S, Semple T C, et al. Thermal, catalytic, regiospecific functionalization of alkanes [J]. Science, 2000, 287 (5460): 1995–1997; b) Tsukada N, Hartwig J F. Intermolecular and intramolecular, platinum–catalyzed, acceptorless dehydrogenative coupling of hydrosilanes with aryl and aliphatic methyl C–H bonds [J]. Journal of the American Chemical Society, 2005, 127 (14): 5022–5023.

[33] Hartwig J F, Cook K S, Hapke M, et al. Rhodium boryl complexes in the catalytic, terminal functionalization of alkanes [J]. Journal of the American Chemical Society, 2005, 127 (8): 2538–2552.

[34] Boller T M, Murphy J M, Hapke M, et al. Mechanism of the mild functionalization of arenes by diboron reagents catalyzed by iridium complexes. Intermediacy and chemistry of bipyridine–ligated iridium trisboryl complexes [J]. Journal of the American Chemical Society, 2005, 127 (41): 14263–14278.

[35] Hartwig J F. Borylation and silylation of C – H bonds: A platform for diverse C–H bond functionalizations [J]. Accounts of Chemical Research, 2012, 45 (6): 864–873.

[36] Simmons E M, Hartwig J F. Iridium–catalyzed arene ortho–silylation by formal hydroxyl–directed C–H activation[J]. Journal of the American Chemical Society, 2010, 132 (48): 17092–17095.

[37] a) Li Q, Liskey C W, Hartwig J F. Regioselective borylation of the C – H bonds in alkylamines and alkyl ethers.

Observation and origin of high reactivity of primary C－H bonds beta to nitrogen and oxygen [J] . Journal of the American Chemical Society, 2014, 136 (24) : 8755–8765; b) Zhang L S, Chen G H, Wang X, et al. Direct borylation of primary C–H bonds in functionalized molecules by palladium catalysis [J] . Angewandte Chemie International Edition, 2014, 53 (15) : 3899–3903.

[38] a) Cook A K, Schimler S D, Matzger A J, et al. Catalyst–controlled selectivity in the C－H borylation of methane and ethane [J] . Science, 2016, 351 (6280) : 1421–1424; b) Smith K T, Berritt S, Gonz á lez–Moreiras M, et al. Catalytic borylation of methane [J] . Science, 2016, 351 (6280) : 1424–1427.

[39] a)Su B, Zhou T G, Xu P L, et al. Enantioselective borylation of aromatic C–H bonds with chiral dinitrogen ligands[J]. Angewandte Chemie International Edition, 2017, 56 (25) : 7205–7208; b) Su B, Zhou T G, Li X W, et al. A chiral nitrogen ligand for enantioselective, iridium–catalyzed silylation of aromatic C–H bonds [J] . Angewandte Chemie International Edition, 2017, 56 (4) : 1092–1096.

[40] a) Ye B H, Cramer N. Chiral cyclopentadienyl ligands as stereocontrolling element in asymmetric C－H functionalization [J] . Science, 2012, 338 (6106) : 504–506; b) Ye B H, Cramer N. A tunable class of chiral Cp ligands for enantioselective rhodium (III) –catalyzed C－H allylations of benzamides [J] . Journal of the American Chemical Society, 2013, 135 (2) : 636–639; c) Zheng J, Cui W J, Zheng C, et al. Synthesis and application of chiral spiro Cp ligands in rhodium–catalyzed asymmetric oxidative coupling of biaryl compounds with alkenes [J] . Journal of the American Chemical Society, 2016, 138 (16) : 5242–5245.

[41] Hyster T K, Knörr L, Ward T R, et al. Biotinylated Rh (III) complexes in engineered streptavidin for accelerated asymmetric C－H activation [J] . Science, 2012, 338 (6106) : 500–503.

[42] Albrecht M. Cyclometalation using d–block transition metals: Fundamental aspects and recent trends [J] . Chemical Reviews, 2010, 110 (2) : 576–623.

[43] a) Lyons T W, Sanford M S. Palladium–catalyzed ligand–directed C–H functionalization reactions [J] . Chemical Reviews, 2010, 110 (2) : 1147–1169.

[44] Ye B H, Cramer N. Chiral cyclopentadienyls: Enabling ligands for asymmetric Rh (III) –catalyzed C－H functionalizations [J] . Accounts of Chemical Research, 2015, 48 (5) : 1308–1318.

[45] [45]Xu L M, Li B J, Yang Z, et al. Organopalladium (IV) chemistry [J] . Chemical Society Reviews, 2010, 39 (2) : 712–733.

[46] a)Wang D H, Engle K M, Shi B F, et al. Ligand–enabled reactivity and selectivity in a synthetically versatile Aryl C－H olefination [J] . Science, 2010, 327 (5963) : 315–319; b) Engle K M, Wang D H, Yu J Q. Constructing multiply substituted arenes using sequential palladium (II) –catalyzed CH olefination [J] . Angewandte Chemie International Edition, 2010, 49 (35) : 6169–6173; c) Engle K M, Wang D H, Yu J Q. Ligand–accelerated C–H activation reactions: Evidence for a switch of mechanism [J] . Journal of the American Chemical Society, 2010, 132 (40) : 14137–14151; d) Yanagisawa S, Ueda K, Sekizawa H, et al. Programmed synthesis of tetraarylthiophenes through sequential C–H arylation[J]. Journal of the American Chemical Society, 2009, 131(41): 14622–14623; e) Zhang Y H, Shi B F, Yu J Q. Pd (II) –catalyzed olefination of electron–deficient arenes using 2, 6–dialkylpyridine ligands [J] . Journal of the American Chemical Society, 2009, 131 (14) : 5072–5074.

[47] a) Steinhoff B A, Stahl S S. Mechanism of Pd (OAc) $_2$/DMSO–catalyzed aerobic alcohol oxidation: Mass–transfer–limitation effects and catalyst decomposition pathways [J] . Journal of the American Chemical Society, 2006, 128 (13) : 4348–4355; b) He C Y, Fan S L, Zhang X G. Pd–catalyzed oxidative cross–coupling of perfluoroarenes with aromatic heterocycles [J] . Journal of the American Chemical Society, 2010, 132 (37) : 12850–12852; c) Li H, Liu J, Sun C L, et al. Palladium–catalyzed cross–coupling of polyfluoroarenes with simple arenes [J] . Organic Letters, 2011, 13 (2) : 276–279.

[48] a) Cheng G J, Yang Y F, Liu P, et al. Role of *N*–acyl amino acid ligands in Pd (II) –catalyzed remote C－

H activation of tethered arenes [J] . Journal of the American Chemical Society, 2014, 136 (3) : 894–897; b) Chen G, Gong W, Zhuang Z, et al. Ligand–accelerated enantioselective methylene C (sp^3) – H bond activation [J] . Science, 2016, 353 (6303) : 1023–1027; c) Yang Y F, Chen G, Hong X, et al. The origins of dramatic differences in five–membered vs six–membered chelation of Pd(II)on efficiency of C(sp^3) – H bond activation[J]. Journal of the American Chemical Society, 2017, 139 (25) : 8514–8521.

[49] He J, Wasa M, Chan K S L, et al. Palladium–catalyzed transformations of alkyl C – H bonds [J] . Chemical Reviews, 2017, 117 (13) : 8754–8786.

[50] Alberico D, Scott M E, Lautens M. Aryl–aryl bond formation by transition–metal–catalyzed direct arylation [J] . Chemical Reviews, 2007, 107 (1) : 174–238.

[51] Li H, Sun C L, Yu M, et al. The catalytic ability of various transition metals in the direct functionalization of aromatic C–H bonds [J] . Chemistry – A European Journal, 2011, 17 (13) : 3593–3597.

[52] Augood D R, Hey D H, Nechvatal A, et al. Homolytic aromatic substitution [J] . Nature, 1951, 167 (4253) : 725.

[53] a) Cheng Y N, Gu X Y, Li P X. Visible–light photoredox in homolytic aromatic substitution: Direct arylation of arenes with aryl halides [J] . Organic Letters, 2013, 15 (11) : 2664–2667; b) Dewanji A, Murarka S, Curran D P, et al. Phenyl hydrazine as initiator for direct arene C – H arylation via base promoted homolytic aromatic substitution [J] . Organic Letters, 2013, 15 (23) : 6102–6105; c) Leifert D, Daniliuc C G, Studer A. 6–aroylated phenanthridines via base promoted homolytic aromatic substitution (BHAS) [J] . Organic Letters, 2013, 15 (24) : 6286–6289; d) Prier C K, MacMillan D W C. Amine α –heteroarylation via photoredox catalysis: A homolytic aromatic substitution pathway [J] . Chemical Science, 2014, 5 (11) : 4173–4178.

[54] a) Shang X J, Liu Z Q. Recent developments in free–radical–promoted C – Si formation via selective C – H/Si – H functionalization [J] . Organic & Biomolecular Chemistry, 2016, 14 (33) : 7829–7831; b) Bähr S, Oestreich M. Electrophilic aromatic substitution with silicon electrophiles: Catalytic friedel – crafts C–H silylation [J] . Angewandte Chemie International Edition, 2017, 56 (1) : 52–59.

[55] Rubio–Perez L, Iglesias M, Munárriz J, et al. A well–defined NHC – Ir (III) catalyst for the silylation of aromatic C – H bonds: Substrate survey and mechanistic insights [J] . Chemical Science, 2017, 8 (7) : 4811–4822.

[56] For recent examples on Pd–catalyzed C–H bond activation, see: a) Zhao R, Lu W J. Palladium–catalyzed β –mesylation of simple amide via primary sp^3 C – H activation [J] . Organic Letters, 2017, 19 (7) : 1768–1771; b) Wang Y, Gu J Y, Shi Z J. Palladium–catalyzed direct annulation of benzoic acids with phenols to synthesize dibenzopyranones [J] . Organic Letters, 2017, 19 (6) : 1326–1329; c) Smalley A P, Cuthbertson J D, Gaunt M J. Palladium–catalyzed enantioselective C – H activation of aliphatic amines using chiral anionic BINOL–phosphoric acid ligands [J] . Journal of the American Chemical Society, 2017, 139 (4) : 1412–1415; d) Mantenuto S, Ciccolini C, Lucarini S, et al. Palladium (II) –catalyzed intramolecular oxidative C – H/C – H cross–coupling reaction of C3, N–linked biheterocycles: Rapid access to polycyclic nitrogen heterocycles [J] . Organic Letters, 2017, 19 (3) : 608–611; e) He C, Gaunt M J. Ligand–assisted palladium–catalyzed C – H alkenylation of aliphatic amines for the synthesis of functionalized pyrrolidines [J] . Chemical Science, 2017, 8 (5) : 3586–3592; f) Baudoin O. Ring construction by palladium (0) –catalyzed C (sp^3) –H activation [J] . Accounts of Chemical Research, 2017, 50 (4) : 1114–1123; g) Zakrzewski J, Smalley A P, Kabeshov M A, et al. Continuous–flow synthesis and derivatization of aziridines through palladium–catalyzed C (sp^3) –H activation [J] . Angewandte Chemie International Edition, 2016, 55 (31) : 8878–8883; h) Yin G Y, Mu X, Liu G S. Palladium (II) –catalyzed oxidative difunctionalization of alkenes: Bond forming at a high–valent palladium center [J] . Accounts of Chemical Research, 2016, 49 (11) : 2413–2423; i) Yang Y Q, Qiu X D, Zhao Y, et al. Palladium–catalyzed C – H arylation of indoles at the C7 position [J] . Journal of the American Chemical Society,

2016, 138 (2): 495–498; j) Xu Y, Su T S, Huang Z X, et al. Practical direct α-arylation of cyclopentanones by palladium/enamine cooperative catalysis [J]. Angewandte Chemie International Edition, 2016, 55 (7): 2559–2563; k) Xiao K J, Chu L, Chen G, et al. Kinetic resolution of benzylamines via palladium (II)-catalyzed C–H cross-coupling [J]. Journal of the American Chemical Society, 2016, 138 (24): 7796–7800; l) Wang H, Tong H R, He G, et al. An Enantioselective bidentate auxiliary directed palladium-catalyzed benzylic C–H arylation of amines using a BINOL Phosphate ligand [J]. Angewandte Chemie International Edition, 2016, 55 (49): 15387–15391; m) Topczewski J J, Cabrera P J, Saper N I, et al. Palladium-catalysed transannular C–H functionalization of alicyclic amines [J]. Nature, 2016, 531 (7593): 220–224.

[57] For recent examples on Rh-catalyzed C–H bond activation, see: a) Webster-Gardiner M S, Chen J Q, Vaughan B A, et al. Catalytic synthesis of "super" linear alkenyl arenes using an easily prepared Rh (I) catalyst [J]. Journal of the American Chemical Society, 2017, 139 (15): 5474–5480; b) Cruz F A, Dong V M. Stereodivergent coupling of aldehydes and alkynes via synergistic catalysis using Rh and jacobsen's amine [J]. Journal of the American Chemical Society, 2017, 139 (3): 1029–1032; c) Zhou S G, Wang J H, Wang L L, et al. Enaminones as synthons for a directed C–H functionalization: Rh^{III}-catalyzed synthesis of naphthalenes [J]. Angewandte Chemie International Edition, 2016, 55 (32): 9384–9388; d) Yu S J, Tang G D, Li Y Z, et al. Anthranil: An aminating reagent leading to bifunctionality for both C (sp^3) –H and C (sp^2) –H under rhodium (III) catalysis [J]. Angewandte Chemie International Edition, 2016, 55 (30): 8696–8700; e) Yang Y D, Li K Z, Cheng Y Y, et al. Rhodium-catalyzed annulation of arenes with alkynes through weak chelation-assisted C–H activation [J]. Chemical Communications, 2016, 52 (14): 2872–2884; f) Xu P, Wang G Q, Wu Z K, et al. Rh (III)-catalyzed double C–H activation of aldehyde hydrazones: A route for functionalized 1H-indazole synthesis [J]. Chemical Science, 2017, 8 (2): 1303–1308; g) Qi Z S, Yu S J, Li X W. Rh (III)-catalyzed synthesis of *N*-unprotected indoles from imidamides and diazo ketoesters via C–H activation and C–C/C–N bond cleavage [J]. Organic Letters, 2016, 18 (4): 700–703; h) Matsumoto K, Yoshida M, Shindo M. Heterogeneous rhodium-catalyzed aerobic oxidative dehydrogenative cross-coupling: nonsymmetrical biaryl amines [J]. Angewandte Chemie International Edition, 2016, 55 (17): 5272–5276; i) Fu L B, Guptill D M, Davies H M L. Rhodium (II)-catalyzed C–H functionalization of electron-deficient methyl groups [J]. Journal of the American Chemical Society, 2016, 138 (18): 5761–5764; j) Zhang X S, Zhang Y F, Li Z W, et al. Synthesis of dibenzo [c, e] oxepin-5 (7*H*) -ones from benzyl thioethers and carboxylic acids: Rhodium-catalyzed double C–H activation controlled by different directing groups [J]. Angewandte Chemie International Edition, 2015, 54 (18): 5478–5482; k) Li X Y, Li X W, Jiao N. Rh-catalyzed construction of quinolin-2 (1*H*) -ones via C–H bond activation of simple anilines with CO and alkynes [J]. Journal of the American Chemical Society, 2015, 137 (29): 9246–9249; l) Gong H, Zeng H Y, Zhou F, et al. Rhodium (I)-catalyzed regiospecific dimerization of aromatic acids: Two direct C–H bond activations in water [J]. Angewandte Chemie International Edition, 2015, 54 (19): 5718–5721.

[58] For recent examples on Ir-catalyzed C–H bond activation, see: a) Su B, Zhou T G, Xu P L, et al. Enantioselective borylation of aromatic C–H bonds with chiral dinitrogen ligands [J]. Angewandte Chemie International Edition, 2017, 56 (25): 7205–7208; b) Su B, Zhou T G, Li X W, et al. A chiral nitrogen ligand for enantioselective, iridium-catalyzed silylation of aromatic C–H bonds [J]. Angewandte Chemie International Edition, 2017, 56 (4): 1092–1096; c) Zhu F X, Li Y H, Wang Z C, et al. Iridium-catalyzed carbonylative synthesis of chromenones from simple phenols and internal alkynes at atmospheric pressure [J]. Angewandte Chemie International Edition, 2016, 55 (45): 14151–14154; d) Xiao X S, Hou C, Zhang Z H, et al. Iridium (III)-catalyzed regioselective intermolecular unactivated secondary Csp^3–H bond amidation [J]. Angewandte Chemie International Edition, 2016, 55 (39): 11897–11901; e) Nguyen K D, Herkommer D,

Krische M J. Enantioselective formation of all–carbon quaternary centers via C – H functionalization of methanol: Iridium–catalyzed diene hydrohydroxymethylation [J] . Journal of the American Chemical Society, 2016, 138 (43) : 14210–14213; f) Larsen M A, Cho S H, Hartwig J. Iridium–catalyzed, hydrosilyl–directed borylation of unactivated alkyl C – H bonds [J] . Journal of the American Chemical Society, 2016, 138 (3) : 762–765; g) Kim Y, Park J, Chang S. A direct access to 7–aminoindoles via iridium–catalyzed mild C – H amidation of *N*–pivaloylindoles with organic azides [J] . Organic Letters, 2016, 18 (8) : 1892–1895; h) Jia X Q, Huang Z. Conversion of alkanes to linear alkylsilanes using an iridium–iron–catalysed tandem dehydrogenation–isomerization–hydrosilylation [J] . Nature Chemistry, 2016, 8 (2) : 157–161; i) Crisenza G E M, Sokolova O O, Bower J F. Branch–selective alkene hydroarylation by cooperative destabilization: Iridium–catalyzed *ortho*–alkylation of acetanilides [J] . Angewandte Chemie International Edition, 2015, 54 (49) : 14866–14870; j) Li B J, Driess M, Hartwig J F. Iridium–catalyzed regioselective silylation of secondary alkyl C – H bonds for the synthesis of 1, 3–diols [J] . Journal of the American Chemical Society, 2014, 136 (18) : 6586–6589.

[59] a) Vougioukalakis G C, Grubbs R H. Ruthenium–based heterocyclic carbene–coordinated olefin metathesis catalysts [J] . Chemical Reviews, 2010, 110 (3) : 1746–1787; b) Trnka T M, Grubbs R H. The development of $L_2X_2RuC=HR$ olefin metathesis catalysts: An organometallic success story [J] . Accounts of Chemical Research, 2001, 34 (1) : 18–29.

[60] a) Wu Y X, Zhou B. Ruthenium–catalyzed direct hydroxymethylation of aryl C – H bonds [J] . ACS Catalysis, 2017, 7 (3) : 2213–2217; b) Ruan Z X, Zhang S L, Zhu C J, et al. Ruthenium (II) –catalyzed *meta* C–H mono– and difluoromethylations by phosphine/carboxylate cooperation [J] . Angewandte Chemie International Edition, 2017, 56 (8) : 2045–2049; c) Li J, Korvorapun K, De Sarkar S, et al. Ruthenium (II) –catalysed remote C – H alkylations as a versatile platform to meta–decorated arenes [J] . Nature Communications, 2017, 8: 15430; d) Mei R H, Zhu C J, Ackermann L. Ruthenium (II) –catalyzed C – H functionalizations on benzoic acids with aryl, alkenyl and alkynyl halides by weak–O–coordination [J] . Chemical Communications, 2016, 52 (89) : 13171–13174; e) Leitch J A, Wilson P B, McMullin C L, et al. Ruthenium (II) –catalyzed C – H functionalization using the oxazolidinone heterocycle as a weakly coordinating directing group: Experimental and computational insights [J] . ACS Catalysis, 2016, 6 (8) : 5520–5529; f) Shi X Y, Dong X F, Fan J, et al. Ru (II) –catalyzed *ortho*–amidation and decarboxylation of aromatic acids: A versatile route to meta–substituted *N*–aryl benzamides [J] . Science China Chemistry, 2015, 58 (8) : 1286–1291; g) Ackermann L. Carboxylate–assisted ruthenium–catalyzed alkyne annulations by C – H/Het – H bond functionalizations [J] . Accounts of Chemical Research, 2014, 47 (2): 281–295; h) Arockiam P B, Bruneau C, Dixneuf P H. Ruthenium(II)–catalyzed C – H bond activation and functionalization [J] . Chemical Reviews, 2012, 112 (11) : 5879–5918.

[61] a) Ackermann L. Cobalt–catalyzed C – H arylations, benzylations, and alkylations with organic electrophiles and beyond [J] . The Journal of Organic Chemistry, 2014, 79 (19) : 8948–8954; b) Chen Q, Ilies L, Yoshikai N, et al. Cobalt–catalyzed coupling of alkyl grignard reagent with benzamide and 2–phenylpyridine derivatives through directed C – H bond activation under air [J] . Organic Letters, 2011, 13 (12) : 3232–3234; c) Li B, Wu Z H, Gu Y F, et al. Direct cross–coupling of C–H bonds with grignard reagents through cobalt catalysis [J] . Angewandte Chemie International Edition, 2011, 50 (5) : 1109–1113.

[62] a) Zhang Z Z, Liu B, Xu J W, et al. Indole synthesis via cobalt (III) –catalyzed oxidative coupling of *N*–arylureas and internal alkynes [J] . Organic Letters, 2016, 18 (8) : 1776–1779; b) Zhang L B, Zhang S K, Wei D, et al. Cobalt (II) –catalyzed C – H amination of arenes with simple alkylamines [J] . Organic Letters, 2016, 18 (6) : 1318–1321; c) Tan G Y, He S, Huang X L, et al. Cobalt–catalyzed oxidative C–H/C–H cross–coupling between two heteroarenes [J] . Angewandte Chemie International Edition, 2016, 55 (35) : 10414–10418; d) Liang Y J, Jiao N. Cationic cobalt (III) catalyzed indole synthesis: The regioselective intermolecular cyclization

of N-nitrosoanilines and alkynes [J] . Angewandte Chemie International Edition, 2016, 55 (12) : 4035-4039; e) Boerth J A, Hummel J R, Ellman J A. Highly stereoselective cobalt (III) -catalyzed three-component C-H bond addition cascade [J] . Angewandte Chemie International Edition, 2016, 55 (41) : 12650-12654; f) Zhao D B, Kim J H, Stegemann L, et al. Cobalt (III) -catalyzed directed C-H coupling with diazo compounds: Straightforward access towards extended π-systems [J]. Angewandte Chemie International Edition, 2015, 54 (15): 4508-4511.

[63] Gao K, Yoshikai N. Low-Valent cobalt catalysis: New opportunities for C - H functionalization [J] . Accounts of Chemical Research, 2014, 47 (4) : 1208-1219.

[64] Sazinsky M H, Lippard S J. Methane monooxygenase: Functionalizing methane at iron and copper [M] //Kroneck P, Sosa Torres M. Sustaining Life on Planet Earth: Metalloenzymes Mastering Dioxygen and Other Chewy Gases. Cham: Springer, 2015, 15: 205-256.

[65] a) Sun C L, Li B J, Shi Z J. Direct C-H Transformation via Iron Catalysis [J] . Chemical Reviews, 2011, 111 (3) : 1293-1314; b) Shi J L, Zhang J C, Wang B Q, et al. Fe-promoted chlorobenzylation of terminal alkynes through benzylic C (sp^3) - H bond functionalization [J] . Organic Letters, 2016, 18 (6) : 1238-1241; c) Ragupathi A, Sagadevan A, Lin CC, et al. Copper (I) -catalysed oxidative C - N coupling of 2-aminopyridine with terminal alkynes featuring a C ≡ C bond cleavage promoted by visible light [J] . Chemical Communications, 2016, 52 (79) : 11756-11759.

[66] a) Fenton H J H. LXXIII.—Oxidation of tartaric acid in presence of iron [J] . Journal of the Chemical Society, Transactions, 1894, 65: 899-910; b) Sawyer D T, Sobkowiak A, Matsushita T. Metal [ML_x; M=Fe, Cu, Co, Mn] /hydroperoxide-induced activation of dioxygen for the oxygenation of hydrocarbons: Oxygenated fenton chemistry [J] . Accounts of Chemical Research, 1996, 29 (9) : 409-416; c) Walling C. Intermediates in the reactions of fenton type reagents [J] . Accounts of Chemical Research, 1998, 31 (4) : 155-157.

[67] Barton D H R, Doller D. The selective functionalisation of saturated hydrocarbons: Gif and all that [J] . Pure and Appllied Chemistry, 1991, 63 (11) : 1567-1576.

[68] a) Hennessy E T, Betley T A. Complex N-heterocycle synthesis via iron-catalyzed, direct C - H bond amination [J]. Science, 2013, 340 (6132) : 591-595; b) Narute S, Parnes R, Toste F D, et al. Enantioselective oxidative homocoupling and cross-coupling of 2-naphthols catalyzed by chiral iron phosphate complexes [J] . Journal of the American Chemical Society, 2016, 138 (50) : 16553-16560; c) Osberger T J, Rogness D C, Kohrt J T, et al. Oxidative diversification of amino acids and peptides by small-molecule iron catalysis [J] . Nature, 2016, 537 (7619) : 214-219; d) Pony Yu R, Hesk D, Rivera N, et al. Iron-catalysed tritiation of pharmaceuticals [J] . Nature, 2016, 529 (7585) : 195-199.

[69] a) Shang R, Ilies L, Nakamura E. Iron-catalyzed directed C (sp^2) - H and C (sp^3) - H functionalization with trimethylaluminum [J] . Journal of the American Chemical Society, 2015, 137 (24) : 7660-7663; b) Wong M Y, Yamakawa T, Yoshikai N. Iron-catalyzed directed C2-alkylation and alkenylation of indole with vinylarenes and alkynes [J] . Organic Letters, 2015, 17 (3) : 442-445; c) Shang R, Ilies L, Nakamura E. Iron-catalyzed *ortho* C - H methylation of aromatics bearing a simple carbonyl group with methylaluminum and tridentate phosphine ligand [J] . Journal of the American Chemical Society, 2016, 138 (32) : 10132-10135.

[70] Siemsen P, Livingston R C, Diederich F. Acetylenic coupling: A powerful tool in molecular construction [J] . Angewandte Chemie International Edition, 2000, 39 (15) : 2632-2657.

[71] a) Li C J. Cross-dehydrogenative coupling (CDC) : Exploring C-C bond formations beyond functional group transformations [J] . Accounts of Chemical Research, 2009, 42 (2) : 335-344; b) Girard S A, Knauber T, Li C J. The Cross-dehydrogenative coupling of C_{sp3}_H bonds: A versatile strategy for C_C bond formations [J] . Angewandte Chemie International Edition, 2014, 53 (1) : 74-100.

[72] Wu X S, Zhao Y, Ge H B. Pyridine-enabled copper-promoted cross dehydrogenative coupling of C (sp^2) - H and unactivated C (sp^3) - H bonds [J] . Chemical Science, 2015, 6 (10) : 5978–5983.

[73] a) Liu X G, Sun B, Xie Z Y, et al. Manganese dioxide - methanesulfonic acid promoted direct dehydrogenative alkylation of sp^3 C - H bonds adjacent to a heteroatom [J] . The Journal of Organic Chemistry, 2013, 78 (7) : 3104–3112; b) Hattori K, Ziadi A, Itami K, et al. Manganese-catalyzed intermolecular C - H/C - H coupling of carbonyls and heteroarenes [J] . Chemical Communications, 2014, 50 (31) : 4105–4107.

[74] a) Shi X C, Nishiura M, Hou Z M. Simultaneous chain-growth and step-growth polymerization of methoxystyrenes by rare-earth catalysts [J] . Angewandte Chemie International Edition, 2016, 55 (47) : 14812–14817; b) Shi X C, Nishiura M N, Hou Z M. C - H Polyaddition of dimethoxyarenes to unconjugated dienes by rare earth catalysts [J] . Journal of the American Chemical Society, 2016, 138 (19) : 6147–6150.

[75] a) Guo F, Meng R, Li Y, et al. Highly *cis*-1, 4-selective terpolymerization of 1, 3-butadiene and isoprene with styrene by a C_5H_5-ligated scandium catalyst [J] . Polymer, 2015, 76: 159–167; b) Kang X H, Yamamoto A, Nishiura M, et al. Computational analyses of the effect of lewis bases on styrene polymerization catalyzed by cationic scandium half-sandwich complexes [J] . Organometallics, 2015, 34 (23) : 5540–5548; c) Yamamoto A, Nishiura M, Oyamada J, et al. Scandium-catalyzed syndiospecific chain-transfer polymerization of styrene using anisoles as a chain transfer agent [J] . Macromolecules, 2016, 49 (7) : 2458–2466.

[76] Nishiura M, Guo F, Hou Z M. Half-sandwich rare-earth-catalyzed olefin polymerization, carbometalation, and hydroarylation [J] . Accounts of Chemical Research, 2015, 48 (8) : 2209–2220.

[77] Groves J K. The Friedel - Crafts acylation of alkenes [J] . Chemical Society Reviews, 1972, 1 (1) : 73–97.

[78] a) Liu P, Zhou C Y, Xiang S, et al. Highly efficient oxidative carbon - carbon coupling with SBA-15-support iron terpyridine catalyst [J] . Chemical Communications, 2010, 46 (16) : 2739–2741; b) Li Z P, Li C J. CuBr-catalyzed direct indolation of tetrahydroisoquinolines via cross-dehydrogenative coupling between sp^3 C–H and sp^2 C–H bonds [J] . Journal of the American Chemical Society, 2005, 127 (19) : 6968–6969.

[79] Yazaki K, Noda S, Tanaka Y, et al. An M_2L_4 molecular capsule with a redox switchable polyradical shell [J] . Angewandte Chemie International Edition, 2016, 55 (48) : 15031–15034.

[80] a) Alberico D, Scott M E, Lautens M. Aryl-aryl bond formation by transition-metal-catalyzed direct arylation [J] . Chemical Reviews, 2007, 107 (1) : 174–238; b) Ackermann L, Vicente R, Kapdi A R. Transition-metal-catalyzed direct arylation of (hetero) arenes by C_H bond cleavage [J] . Angewandte Chemie International Edition, 2009, 48 (52) : 9792–9826; c) McGlacken G P, Bateman L M. Recent advances in aryl - aryl bond formation by direct arylation [J] . Chemical Society Reviews, 2009, 38 (8) : 2447–2464; d) Liu C, Zhang H, Shi W, et al. Bond formations between two nucleophiles: Transition metal catalyzed oxidative cross-coupling reactions [J] . Chemical Reviews, 2011, 111 (3) : 1780–1824; e) Segawa Y, Maekawa T, Itami K. Synthesis of extended π -systems through C - H activation [J] . Angewandte Chemie International Edition, 2015, 54 (1) : 66–81.

[81] a) He J, Wasa M, Chan K S L, et al. Palladium (0) -catalyzed alkynylation of C (sp^3) - H bonds [J] . Journal of the American Chemical Society, 2013, 135 (9) : 3387–3390; b) Zhu R Y, He J, Wang X C, et al. Ligand-promoted alkylation of C (sp^3) - H and C (sp^2) - H bonds [J] . Journal of the American Chemical Society, 2014, 136 (38) : 13194–13197; c) Liu Y B, Ge H B. Site-selective C - H arylation of primary aliphatic amines enabled by a catalytic transient directing group [J] . Nature Chemistry, 2016, 9 (1) : 26–32.

[82] Ackermann L, Althammer A, Born R. Catalytic arylation reactions by C_H bond activation with aryl tosylates [J] . Angewandte Chemie International Edition, 2006, 45 (16) : 2619–2622.

[83] Hong B K, Li C, Wang Z, et al. Enantioselective total synthesis of (-) -incarviatone A [J] . Journal of the American Chemical Society, 2015, 137 (37) : 11946–11149.

[84] Jiang T S, Wang G W. Palladium–catalyzed ortho–alkoxylation of anilides via C – H activation [J] . The Journal of Organic Chemistry, 2012, 77 (21) : 9504–9509.

[85] Xu T, Savage N A, Dong G B. Rhodium (I) –catalyzed decarbonylative spirocyclization through C_C bond cleavage of benzocyclobutenones: An efficient approach to functionalized spirocycles [J] . Angewandte Chemie International Edition, 2014, 53 (7) : 1891–1895.

[86] Liu C, Liu D, Lei A W. Recent advances of transition–metal catalyzed radical oxidative cross–couplings [J] . Accounts of Chemical Research, 2014, 47 (12) : 3459–3470.

[87] Xiao X S, Hou C, Zhang Z H, et al. Iridium (III) –catalyzed regioselective intermolecular unactivated secondary Csp^3–H bond amidation [J] . Angewandte Chemie International Edition, 2016, 55 (39) : 11897–11901.

[88] McNally A, Haffemayer B, Collins B S L, et al. Palladium–catalysed C – H activation of aliphatic amines to give strained nitrogen heterocycles [J] . Nature, 2014, 510 (7503) : 129–133.

[89] a) Dick A R, Hull K L, Sanford M S. A highly selective catalytic method for the oxidative functionalization of C–H bonds [J] . Journal of the American Chemical Society, 2004, 126 (8) : 2300–2301; b) Xiao B, Gong T J, Liu Z K, et al. Synthesis of dibenzofurans via palladium–catalyzed phenol–directed C – H activation/C – O cyclization [J] . Journal of the American Chemical Society, 2011, 133 (24) : 9250–9253; c) Wang Y, Gu J Y, Shi Z J. Palladium–catalyzed direct annulation of benzoic acids with phenols to synthesize dibenzopyranones [J] . Organic Letters, 2017, 19 (6) : 1326–1329.

[90] a) Boebel T A, Hartwig J F. Silyl–directed, iridium–catalyzed *ortho*–borylation of arenes. A one–pot ortho-borylation of phenols, arylamines, and alkylarenes [J] . Journal of the American Chemical Society, 2008, 130 (24) : 7534–7535; b) Cho S H, Hartwig J F. Iridium–catalyzed diborylation of benzylic C – H bonds directed by a hydrosilyl group: Synthesis of 1, 1–benzyldiboronate esters [J] . Chemical Science, 2014, 5 (2) : 694–698;

[91] He J, Jiang H, Takise R, et al. Ligand–promoted borylation of C (sp^3) [BOND] H bonds with palladium (II) catalysts [J] . Angewandte Chemie International Edition, 2016, 55 (2) : 785–789;

[92] a) Kakiuchi F, Tsuchiya K, Matsumoto M, et al. Ru_3 (CO) $_{12}$–catalyzed silylation of benzylic C–H bonds in arylpyridines and arylpyrazoles with hydrosilanes via C–H bond cleavage [J] . Journal of the American Chemical Society, 2004, 126 (40) : 12792–12793; b) Kuninobu Y, Nakahara T, Takeshima H, et al. Rhodium–catalyzed intramolecular silylation of unactivated C (sp^3) – H bonds [J] . Organic Letters, 2013, 15 (2) : 426–428; c) Hua Y D, Jung S, Roh J, et al. Modular approach to reductive Csp^2 – H and Csp^3 – H silylation of carboxylic acid derivatives through single–pot, sequential transition metal catalysis [J] . The Journal of Organic Chemistry, 2015, 80 (9) : 4661–4671; d) Cheng C, Hartwig J F. Catalytic silylation of unactivated C – H bonds [J] . Chemical Reviews, 2015, 115 (17) : 8946–8975.

[93] a) Yeung C S, Dong V M. Catalytic dehydrogenative cross–coupling: forming carbon–carbon bonds by oxidizing two carbon–hydrogen bonds [J] . Chemical Reviews, 2011, 111 (3) : 1215–1292; b) Girard S A, Knauber T, Li C J. The cross–dehydrogenative coupling of C_{sp3}_H bonds: A versatile strategy for C_C bond formations [J] . Angewandte Chemie International Edition, 2014, 53 (1) : 74–100; c) Liu C, Yuan J W, Gao M, et al. Oxidative coupling between two hydrocarbons: An update of recent C – H functionalizations [J] . Chemical Reviews, 2015, 115 (22) : 12138–12204; d) Yang Y D, Lan J B, You J S. Oxidative C – H/C – H coupling reactions between two (hetero) arenes [J] . Chemical Reviews, 2017, 117 (13) : 8787–8863.

[94] a) Alberico D, Scott M E, Lautens M. Aryl–aryl bond formation by transition–metal–catalyzed direct arylation [J] . Chemical Reviews, 2007, 107 (1) : 174–238; b) Liu C, Zhang H, Shi W, et al. Bond formations between two nucleophiles: Transition metal catalyzed oxidative cross–coupling reactions [J] . Chemical Reviews, 2011, 111 (3) : 1780–1824.

[95] a) Chen X, Goodhue C E, Yu J Q. Palladium–catalyzed alkylation of sp^2 and sp^3 C–H bonds with methylboroxine

and alkylboronic acids: Two distinct C–H activation pathways [J] . Journal of the American Chemical Society, 2006, 128 (39) : 12634–12635; b) Pastine S J, Gribkov D V, Sames D. sp^3 C–H bond arylation directed by amidine protecting group: α –arylation of pyrrolidines and piperidines [J] . Journal of the American Chemical Society, 2006, 128 (44) : 14220–14221;

[96] Moritani I, Fujiwara Y. Aromatic substitution of olefins by palladium salts [J] . Synthesis, 1973, 1973 (9) : 524–533.

[97] Le Bras J, Muzart J. Intermolecular dehydrogenative heck reactions [J] . Chemical Reviews, 2011, 111 (3) : 1170–1214.

[98] Hatamoto Y, Sakaguchi S, Ishii Y. Oxidative cross–coupling of acrylates with vinyl carboxylates catalyzed by a Pd (OAc) $_2$/HPMoV/O_2 system [J] . Organic Letters, 2004, 6 (24) : 4623–4625.

[99] a) Wasa M, Engle K M, Yu J Q. Pd (II) –catalyzed olefination of sp^3 C–H bonds [J] . Journal of the American Chemical Society, 2010, 132 (11): 3680–3681; b) Stuart D R, Alsabeh P, Kuhn M, et al. Rhodium (III) –catalyzed arene and alkene C–H bond functionalization leading to indoles and pyrroles [J] . Journal of the American Chemical Society, 2010, 132 (51) : 18326–18339.

[100] Tran D N, Cramer N. Enantioselective rhodium (I) –catalyzed [3+2] annulations of aromatic ketimines induced by directed C_H activations [J] . Angewandte Chemie International Edition, 2011, 50 (47) : 11098–11102.

[101] a) Reddy Chidipudi S, Khan I, Lam H W. Functionalization of C_{sp3}_H and C_{sp2}_H bonds: Synthesis of spiroindenes by enolate–directed ruthenium–catalyzed oxidative annulation of alkynes with 2–aryl–1, 3–dicarbonyl compounds [J] . Angewandte Chemie International Edition, 2012, 51 (48) : 12115–12119; b) Dooley J D, Reddy Chidipudi S, Lam H W. Catalyst–controlled divergent C – H functionalization of unsymmetrical 2–aryl cyclic 1, 3–dicarbonyl compounds with alkynes and alkenes [J] . Journal of the American Chemical Society, 2013, 135 (29) : 10829–10836; c) Nan J, Zuo Z J, Luo L, et al. Ru^{II}–catalyzed vinylative dearomatization of naphthols via a C (sp^2) – H bond activation approach [J] . Journal of the American Chemical Society, 2013, 135 (46) : 17306–17309; d) Burns D J, Lam H W. Catalytic 1,4–rhodium (III) migration enables 1,3–enynes to function as one–carbon oxidative annulation partners in C_H functionalizations [J] . Angewandte Chemie International Edition, 2014, 53 (37) : 9931–9935.

[102] a) Lahm G, Opatz T. Unique regioselectivity in the C (sp^3) – H α –alkylation of amines: the benzoxazole moiety as a removable directing group [J] . Organic Letters, 2014, 16 (16) : 4201–4203; b) Pan S G, Endo K, Shibata T. Ir (I) –catalyzed enantioselective secondary sp^3 C – H bond activation of 2– (alkylamino) pyridines with alkenes [J] . Organic Letters, 2011, 13 (17) : 4692–4695; c) Li Q, Yu Z X. Conjugated diene–assisted allylic C–H bond activation: Cationic Rh (I) –catalyzed syntheses of polysubstituted tetrahydropyrroles, tetrahydrofurans, and cyclopentanes from ene–2–dienes [J] . Journal of the American Chemical Society, 2010, 132 (13) : 4542–4543; d) Chatani N, Asaumi T, Yorimitsu S, et al. Ru_3 (CO) $_{12}$–catalyzed coupling reaction of sp^3 C–H bonds adjacent to a nitrogen atom in alkylamines with alkenes [J] . Journal of the American Chemical Society, 2001, 123 (44) : 10935–10941.

[103] a) He J, Wasa M, Chan K S L, et al. Palladium (0) –catalyzed alkynylation of C (sp^3) – H bonds [J] . Journal of the American Chemical Society, 2013, 135 (9) : 3387–3390; b) Ano Y, Tobisu M, Chatani N. Palladium–catalyzed direct ethynylation of C (sp^3) – H bonds in aliphatic carboxylic acid derivatives [J] . Journal of the American Chemical Society, 2011, 133 (33) : 12984–12986.

[104] a) Matsuyama N, Kitahara M, Hirano K, et al. Nickel– and copper–catalyzed direct alkynylation of azoles and polyfluoroarenes with terminal alkynes under O_2 or atmospheric conditions [J] . Organic Letters, 2010, 12 (10) : 2358–2361; b) Wei Y, Zhao H Q, Kan J, et al. Copper–catalyzed direct alkynylation of electron–deficient polyfluoroarenes with terminal alkynes using O_2 as an oxidant [J] . Journal of the American Chemical Society,

2010, 132 (8): 2522–2523; c) Kim S H, Yoon J, Chang S. Palladium–catalyzed oxidative alkynylation of heterocycles with terminal alkynes under air conditions [J]. Organic Letters, 2011, 13 (6): 1474–1477.

[105] Zhang J T, Chen H, Lin C, et al. Cobalt–catalyzed cyclization of aliphatic amides and terminal alkynes with silver–cocatalyst [J]. Journal of the American Chemical Society, 2015, 137 (40): 12990–12996.

[106] Luo F X, Cao Z C, Zhao H W, et al. Nickel–catalyzed oxidative coupling of unactivated C (sp^3) – H bonds in aliphatic amides with terminal alkynes [J]. Organometallics, 2017, 36 (1): 18–21.

[107] a) Liu C, Liu D, Lei A W. Recent advances of transition–metal catalyzed radical oxidative cross–couplings [J]. Accounts of Chemical Research, 2014, 47 (12): 3459–3470; b) Girard S A, Knauber T, Li C J. The cross–dehydrogenative coupling of C_{sp3}_H bonds: A versatile strategy for C_C bond formations [J]. Angewandte Chemie International Edition, 2014, 53 (1): 74–100.

[108] a) Shiotani A, Itatani H. Dibenzofurans by Intramolecular ring closure reactions [J]. Angewandte Chemie International Edition, 1974, 13 (7): 471–472; b) Åkermark B, Eberson L, Jonsson E, et al. Palladium–promoted cyclization of diphenyl ether, diphenylamine, and related compounds [J]. The Journal of Organic Chemistry, 1975, 40 (9): 1365–1367; c) Hagelin H, Oslob J D, Åkermark B. Oxygen as oxidant in palladium–catalyzed inter and intramolecular coupling reactions [J]. Chemistry–A European Journal, 1999, 5 (8): 2413–2416; d) Li é gault B, Lee D, Huestis M P, et al. Intramolecular Pd (II) –Catalyzed Oxidative Biaryl Synthesis Under Air: Reaction Development and Scope [J]. The Journal of Organic Chemistry, 2008, 73 (13): 5022–5028; e) Okada T, Unoh Y, Satoh T, et al. Rhodium (III) –catalyzed intramolecular Ar – H/Ar – H coupling directed by carboxylic group to produce dibenzofuran carboxylic acids [J]. Chemistry Letters, 2015, 44 (11): 1598–1600.

[109] a) Li R S, Jiang L, Lu W J. Intermolecular cross–coupling of simple arenes via C–H activation by tuning concentrations of arenes and TFA [J]. Organometallics, 2006, 25 (26): 5973–5975; b) Wei Y, Su W P. Pd (OAc)$_2$–catalyzed oxidative C–H/C–H cross–coupling of electron–deficient polyfluoroarenes with simple arenes [J]. Journal of the American Chemical Society, 2010, 132 (46): 16377–16379; c) Lafrance M, Rowley C N, Woo T K, et al. Catalytic intermolecular direct arylation of perfluorobenzenes [J]. Journal of the American Chemical Society, 2006, 128 (27): 8754–8759; d) He M, Soul é J F, Doucet h. Synthesis of (poly) fluorobiphenyls through metal–catalyzed C_H bond activation/arylation of (poly) fluorobenzene derivatives [J]. ChemCatChem, 2014, 6 (7): 1824–1859; e) Hull K L, Sanford M S. Catalytic and highly regioselective cross–coupling of aromatic C–H substrates [J]. Journal of the American Chemical Society, 2007, 129 (39): 11904–11905;

[110] Kita Y, Morimoto K, Ito M, et al. Metal–free oxidative cross–coupling of unfunctionalized aromatic compounds [J]. Journal of the American Chemical Society, 2009, 131 (5): 1668–1669.

[111] a) Wencel–Delord J, Nimphius C, Wang H G, et al. Rhodium (III) and hexabromobenzene—A catalyst system for the cross–dehydrogenative coupling of simple arenes and heterocycles with arenes bearing directing groups [J]. Angewandte Chemie International Edition, 2012, 51 (52): 13001–13005; b) Hull K L, Sanford M S. Catalytic and highly regioselective cross–coupling of aromatic C–H substrates [J]. Journal of the American Chemical Society, 2007, 129 (39): 11904–11905; c) Li B J, Tian S L, Fang Z, et al. Multiple C_H activations to construct biologically active molecules in a process completely free of organohalogen and organometallic components [J]. Angewandte Chemie International Edition, 2008, 47 (6): 1115–1118; d) Brasche G, Garc í a–Fortanet J, Buchwald S L. Twofold C–H functionalization: Palladium–catalyzed *ortho* arylation of anilides [J]. Organic Letters, 2008, 10 (11): 2207–2210; e) Jiao L Y, Smirnov P, Oestreich M. Exceptionally mild palladium (II) –catalyzed dehydrogenative C – H/C – H arylation of indolines at the C–7 position under air [J]. Organic Letters, 2014, 16 (22): 6020–6023.

[112] a) Hull K L，Sanford M S. Catalytic and highly regioselective cross–coupling of aromatic C–H substrates [J] . Journal of the American Chemical Society，2007，129 (39) : 11904–11905；b) Brasche G，Garc í a–Fortanet J，Buchwald S L. Twofold C–H functionalization: Palladium–catalyzed *ortho* arylation of anilides [J] . Organic Letters，2008，10 (11) : 2207–2210；c) Li B J，Tian S L，Fang Z，et al. Multiple C_H activations to construct biologically active molecules in a process completely free of organohalogen and organometallic components [J] . Angewandte Chemie International Edition，2008，47 (6) : 1115–1118；d) Zhao X D，Yeung C S，Dong V M. Palladium–catalyzed ortho–arylation of O–phenylcarbamates with simple arenes and sodium persulfate [J] . Journal of the American Chemical Society，2010，132 (16) : 5837–5844；e) Karthikeyan J，Cheng C H. Synthesis of phenanthridinones from *N*–methoxybenzamides and arenes by multiple palladium–catalyzed C_H activation steps at room temperature [J] . Angewandte Chemie International Edition，2011，50 (42) : 9880–9883；f) Wang X S，Leow D，Yu J Q. Pd (II) –catalyzed *para*–selective C – H arylation of monosubstituted arenes [J] . Journal of the American Chemical Society，2011，133 (35) : 13864–13867；g) Wencel–Delord J，Nimphius C，Patureau F W，et al. [Rh^{III}Cp*] –catalyzed dehydrogenative aryl_aryl bond formation [J] . Angewandte Chemie International Edition，2012，51 (9) : 2247–2251；h) Ito M，Kubo H，Itani I，et al. Organocatalytic C – H/C – H ′ cross–biaryl coupling: C–selective arylation of sulfonanilides with aromatic hydrocarbons [J] . Journal of the American Chemical Society，2013，135 (38) : 14078–14081；i) Jiao L Y，Oestreich M. Oxidative palladium (II) –catalyzed dehydrogenative C_H/C_H cross–coupling of 2，3–substituted indolines with arenes at the C7 position [J] . Chemistry – A European Journal，2013，19 (33) : 10845–10848；j) Reddy V P，Qiu R H，Iwasaki T，et al. Rhodium–catalyzed intermolecular oxidative cross–coupling of (hetero) arenes with chalcogenophenes [J] . Organic Letters，2013，15 (6) : 1290–1293；k) Kitahara M，Umeda N，Hirano K，et al. Copper–mediated intermolecular direct biaryl coupling [J] . Journal of the American Chemical Society，2011，133 (7) : 2160–2162；l) Wang Z，Song F J，Zhao Y S，et al. Elements of regiocontrol in the direct heteroarylation of indoles/pyrroles: Synthesis of Bi– and fused polycyclic heteroarenes by twofold or tandem fourfold C_H activation [J] . Chemistry – A European Journal，2012，18 (52) : 16616–16620.

[113] Zhang X S，Zhang Y F，Li Z W，et al. Synthesis of dibenzo [c，e] oxepin–5 (7*H*) –ones from benzyl thioethers and carboxylic acids: Rhodium–catalyzed double C_H activation controlled by different directing groups [J] . Angewandte Chemie International Edition，2015，54 (18) : 5478–5482.

[114] Wang Y，Gu J Y，Shi Z J. Palladium–catalyzed direct annulation of benzoic acids with phenols to synthesize dibenzopyranones [J] . Organic Letters，2017，19 (6) : 1326–1329.

[115] a) Liégault B，Fagnou K. Palladium–catalyzed intramolecular coupling of arenes and unactivated alkanes in air [J] . Organometallics，2008，27 (19) : 4841–4843；b) Yang W B，Ye S Q，Fanning D，et al. Orchestrated triple C_H activation reactions using two directing groups: Rapid assembly of complex pyrazoles [J] . Angewandte Chemie International Edition，2015，54 (8) : 2501–2504；c) Fukumoto Y，Hirano M，Chatani N. Iridium–catalyzed regioselective C (sp^3) – H silylation of 4–alkylpyridines at the benzylic position with hydrosilanes leading to 4– (1–silylalkyl) pyridines [J] . ACS Catalysis，2017，7 (5) : 3152–3156.

[116] Kubo T，Aihara Y，Chatani N. Pd (II) –catalyzed chelation–assisted cross dehydrogenative coupling between unactivated C (sp^3) – H bonds in aliphatic amides and benzylic C – H bonds in toluene derivatives [J] . Chemistry Letters，2015，44 (10) : 1365–1367.

[117] Aihara Y，Tobisu M，Fukumoto Y，et al. Ni (II) –catalyzed oxidative coupling between C (sp^2) – H in benzamides and C (sp^3) – H in toluene derivatives [J] . Journal of the American Chemical Society，2014，136 (44) : 15509–15512.

[118] Schweitzer–Chaput B，Sud A，Pint é r A，et al. Synergistic effect of ketone and hydroperoxide in Brønsted acid catalyzed oxidative coupling reactions [J] . Angewandte Chemie International Edition，2013，52 (50) : 13228–

13232.

[119] Liégault B, Fagnou K. Palladium-catalyzed intramolecular coupling of arenes and unactivated alkanes in air [J]. Organometallics, 2008, 27 (19): 4841-4843.

[120] Shi J L, Wang D, Zhang X S, et al. Oxidative coupling of *sp*^{2} and *sp*^{3} carbon - hydrogen bonds to construct dihydrobenzofurans [J]. Nature Communications, 2017, 8: 238.

[121] Li Y, Li B J, Wang W H, et al. Rhodium-catalyzed direct addition of aryl C_H Bonds to *N*-sulfonyl aldimines [J]. Angewandte Chemie International Edition, 2011, 50 (9): 2115-2119.

[122] a) Yang L, Correia C A, Li C J. Grignard-type arylation of aldehydes *via* a rhodium-catalyzed C_H activation under mild conditions [J]. Advanced Synthesis & Catalysis, 2011, 353 (8): 1269-1273; b) Li Y, Zhang X S, Zhu Q L, et al. Olefinic C - H Bond addition to aryl aldehyde and its *N*-sulfonylimine *via* rh catalysis [J]. Organic Letters, 2012, 14 (17): 4498-4501; c) Li Y, Zhang X S, Chen K, et al. *N*-directing group assisted rhodium-catalyzed aryl C - H addition to aryl aldehydes [J]. Organic Letters, 2012, 14 (2): 636-639.

[123] Wang G W, Zhou A X, Wang J J, et al. Palladium-catalyzed *sp*^{2} and *sp*^{3} C - H bond activation and addition to isatin toward 3-hydroxy-2-oxindoles [J]. Organic Letters, 2013, 15 (20): 5270-5273; b) Zhang X S, Zhu Q L, Luo F X, et al. Aromatic C - H addition to ketones: The effect of directing groups [J]. European Journal of Organic Chemistry, 2013, 2013 (29): 6530-6534.

[124] Zhou B W, Hu Y Y, Wang C Y. Manganese-catalyzed direct nucleophilic C (sp^2) _H addition to aldehydes and Nitriles [J]. Angewandte Chemie International Edition, 2015, 54 (46): 13659-13663.

[125] a) Stuart D R, Fagnou K et al. The catalytic cross-coupling of unactivated arenes [J]. Science, 2007, 316 (5828): 1172-1175; b) Dong J X, Long Z, Song F J, et al. Rhodium or ruthenium-catalyzed oxidative C_H/C_H cross-coupling: Direct access to extended π-conjugated systems [J]. Angewandte Chemie International Edition, 2013, 52 (2): 580-584; c) Qin X R, Li X Y, Huang Q, et al. Rhodium (III) -catalyzed ortho C_H heteroarylation of (hetero) aromatic carboxylic acids: A rapid and concise access to π-conjugated poly-heterocycles [J]. Angewandte Chemie International Edition, 2015, 54 (24): 7167-7170; d) Zhang Y F, Zhao H Q, Zhang M, et al. Carboxylic acids as traceless directing groups for the rhodium (III) -catalyzed decarboxylative C_H arylation of thiophenes [J]. Angewandte Chemie International Edition, 2015, 54 (12): 3817-3821; e) Cheng Y Y, Li G C, Liu Y, et al. Unparalleled ease of access to a library of biheteroaryl fluorophores via oxidative cross-coupling reactions: Discovery of photostable NIR probe for mitochondria [J]. Journal of the American Chemical Society, 2016, 138 (14): 4730-4738.

[126] a) Kakiuchi F, Le Gendre P, Yamada A, et al. Atropselective alkylation of biaryl compounds by means of transition metal-catalyzed C - H/olefin coupling [J]. Tetrahedron: Asymmetry, 2000, 11 (13): 2647-2651; b) Zheng J, You S L. Construction of axial chirality by rhodium-catalyzed asymmetric dehydrogenative heck coupling of biaryl compounds with alkenes [J]. Angewandte Chemie International Edition, 2014, 53 (48): 13244-13247; c) Faber K. Non-sequential processes for the transformation of a racemate into a single stereoisomeric product: Proposal for stereochemical classification [J]. Chemistry - A European Journal, 2001, 7 (23): 5004-5010; d) Steinreiber J, Faber K, Griengl H. De-racemization of enantiomers versus de-epimerization of diastereomers—classification of dynamic kinetic asymmetric transformations (DYKAT) [J]. Chemistry - A European Journal, 2008, 14 (27): 8060-8072.

[127] a) Zhao Y S, Li S Q, Zheng X S, et al. Rh/Cu-catalyzed cascade [4+2] Vinylic C-H *O*-annulation and ring contraction of α-aryl enones with alkynes in air [J]. Angewandte Chemie International Edition, 2017, 56 (15): 4286-4289; b) Tang J B, Li S Q, Liu Z, et al. Cascade C - H annulation of aldoximes with alkynes using O_2 as the sole oxidant: One-pot access to multisubstituted protoberberine skeletons [J]. Organic Letters, 2017, 19 (3):

604–607; c) Li J, Yang Y D, Wang Z G, et al. Rhodium (III) –catalyzed annulation of pyridinones with alkynes via double C – H activation: A route to functionalized quinolizinones [J] . Organic Letters, 2017, 19 (12) : 3083–3086.

[128] Wu J, Cheng Y Y, Lan J B, et al. Molecular engineering of mechanochromic materials by programmed C – H arylation: Making a counterpoint in the chromism trend [J] . Journal of the American Chemical Society, 2016, 138 (39) : 12803–12812.

[129] Avila C M, Patel J S, Reddi Y, et al. Enantioselective heck – matsuda arylations through chiral anion phase-transfer of aryl diazonium salts [J] . Angewandte Chemie International Edition, 2017, 56 (21) : 5806–5811.

[130] a) Du X Y, Huang Z. Advances in base–metal–catalyzed alkene hydrosilylation [J] . ACS Catalysis, 2017, 7 (2) : 1227–1243;

[131] a) Haibach M C, Kundu S, Brookhart M, et al. Alkane metathesis by tandem alkane–dehydrogenation – olefin-metathesis catalysis and related chemistry [J] . Accounts of Chemical Research, 2012, 45 (6) : 947–958; b) Nawara–Hultzsch A K, Hackenberg J D, Punji B, et al. Rational design of highly active "hybrid" phosphine – phosphinite pincer iridium catalysts for alkane metathesis [J] . ACS Catalysis, 2013, 3 (11) : 2505–2514.

[132] a) Twilton J, Le C C, Zhang P, et al. The merger of transition metal and photocatalysis [J] . Nature Reviews Chemistry, 2017, 1 (7) : 0052; b) Kawamata Y, Yan M, Liu Z Q, et al. Scalable, electrochemical oxidation of unactivated C – H bonds [J] . Journal of the American Chemical Society, 2017, 139 (22) : 7448–7451; c) Horn E J, Rosen B R, Baran P S. Synthetic organic electrochemistry: An enabling and innately sustainable method [J] . ACS Central Science, 2016, 2 (5) : 302–308; d) Horn E J, Rosen B R, Chen Y, et al. Scalable and sustainable electrochemical allylic C – H oxidation [J] . Nature, 2016, 533 (7601) : 77–81; e) Das A, Stahl S S. Noncovalent immobilization of molecular electrocatalysts for chemical synthesis: Efficient electrochemical alcohol oxidation with a pyrene–TEMPO conjugate [J] . Angewandte Chemie International Edition, 2017, 56 (30) : 8892–8897.

[133] a) Prier C K, Rankic D A, MacMillan D W C. Visible light photoredox catalysis with transition metal complexes: Applications in organic synthesis [J] . Chemistry Reviews, 2013, 113 (7) : 5322–5363; b) Zuo Z W, Ahneman D T, Chu L L, et al. Merging photoredox with nickel catalysis: Coupling of α –carboxyl sp^3–carbons with aryl halides [J] . Science, 2014, 345 (6195) : 437–440.

[134] a) Pirnot M T, Rankic D A, Martin D B C, et al. Photoredox activation for the direct β –arylation of ketones and aldehydes [J] . Science, 2013, 339 (6127) : 1593–1596; b) Cecere G, König C M, Alleva J L, et al. Enantioselective direct α –amination of aldehydes via a photoredox mechanism: A strategy for asymmetric amine fragment coupling [J] . Journal of the American Chemical Society, 2013, 135 (31) : 11521–11524; c) Petronijević F R, Nappi M, MacMillan D W C. Direct β –functionalization of cyclic ketones with aryl ketones via the merger of photoredox and organocatalysis [J] . Journal of the American Chemical Society, 2013, 135 (49) : 18323–18326; d) Terrett J A, Clift M D, MacMillan D W C. Direct β –alkylation of aldehydes via photoredox organocatalysis [J] . Journal of the American Chemical Society, 2014, 136 (19) : 6858–6861; e) Le C, Liang Y F, Evans R W, et al. Selective sp^3 C – H alkylation via polarity–match–based cross–coupling [J] . Nature, 2017, 547 (7661) : 79–83.

[135] Yang Q L, Li Y Q, Ma C, et al. Palladium–catalyzed C (sp3) —H oxygenation via electrochemical oxidation [J] . Journal of the American Chemical Society, 2017, 139 (8) : 3293–3298.

[136] a) Giri R, Shi B F, Engle K M, et al. Transition metal–catalyzed C – H activation reactions: diastereoselectivity and enantioselectivity [J] . Chemical Society Reviews, 2009, 38 (11) : 3242–3272; b) Zheng C, You S L. Recent development of direct asymmetric functionalization of inert C – H bonds [J] . RSC Advances, 2014, 4 (12) : 6173–6214; c)Newton C G, Wang S G, Oliveira C C, et al. Catalytic enantioselective transformations involving C –

H bond cleavage by transition-metal complexes [J] . Chemical Reviews, 2017, 117 (13) : 8908–8976.

[137] a) Shi B F, Maugel N, Zhang Y H, et al. Pd^{II}-catalyzed enantioselective activation of C (sp^2) H and C (sp^3) H bonds using monoprotected amino acids as chiral ligands [J] . Angewandte Chemie International Edition, 2008, 47 (26) : 4882–4886; b) Zhang Q W, An K, Liu L C, et al. Rhodium-catalyzed enantioselective intramolecular C_H silylation for the syntheses of planar-chiral metallocene siloles [J] . Angewandte Chemie International Edition, 2015, 54 (23) : 6918–6921; c) Murai M, Matsumoto K, Takeuchi Y, et al. Rhodium-catalyzed synthesis of benzosilolometallocenes via the dehydrogenative silylation of C(sp^2) – H bonds[J]. Organic Letters, 2015, 17 (12) : 3102–3105; d) Shibata T, Shizuno T. Iridium-catalyzed enantioselective C_H alkylation of ferrocenes with alkenes using chiral diene ligands [J] . Angewandte Chemie International Edition, 2014, 53 (21) : 5410–5413; e) Gwon D, Park S, Chang S. Dual role of carboxylic acid additive: Mechanistic studies and implication for the asymmetric C – H amidation [J] . Tetrahedron, 2015, 71 (26–27) : 4504–4511.

[138] a) Both P, Busch H, Kelly P P, et al. Whole-cell biocatalysts for stereoselective C–H amination reactions [J] . Angewandte Chemie International Edition, 2016, 55 (4) : 1511–1513; b) Prier C K, Zhang R K, Buller A R, et al. Enantioselective, intermolecular benzylic C – H amination catalysed by an engineered iron-haem enzyme [J]. Nature Chemistry, 2017, 9 (7) : 629–634.

[139] a) Albicker M R, Cramer N. Enantioselective palladium-catalyzed direct arylations at ambient temperature: Access to indanes with quaternary stereocenters [J] . Angewandte Chemie International Edition, 2009, 48 (48) : 9139–9142; b) Saget T, Cramer N. Enantioselective C_H arylation strategy for functionalized dibenzazepinones with quaternary stereocenters [J] . Angewandte Chemie International Edition, 2013, 52 (30) : 7865–7868; c) Lin Z Q, Wang W Z, Yan S B, et al. Palladium-Catalyzed enantioselective C_H arylation for the synthesis of P-stereogenic compounds [J] . Angewandte Chemie International Edition, 2015, 54 (21) : 6265–6269; d) Liu L T, Zhang A A, Wang Y F, F et al. Asymmetric synthesis of P-stereogenic phosphinic amides via Pd (0) -catalyzed enantioselective intramolecular C – H arylation [J] . Organic Letters, 2015, 17 (9) : 2046–2049.

[140] Chen G, Gong W, Zhuang Z, et al. Ligand-accelerated enantioselective methylene C (sp^3) –H bond activation [J]. Science, 2016, 153 (6303) : 1023–1027.

[141] a) Wasa M, Engle K M, Lin D W, et al. Pd (II) -catalyzed enantioselective C – H activation of cyclopropanes [J]. Journal of the American Chemical Society, 2011, 133 (49) : 19598–19601; b) Xiao K J, Lin D W, Miura M, et al. Palladium (II) -catalyzed enantioselective C (sp^3) – H activation using a chiral hydroxamic acid ligand [J]. Journal of the American Chemical Society, 2014, 136 (22) : 8138–8142; c) Chan K S L, Fu H Y, Yu J Q. Palladium (II) -catalyzed highly enantioselective C – H arylation of cyclopropylmethylamines [J] . Journal of the American Chemical Society, 2015, 137 (5) : 2042–2046;

[142] a) Lee T, Hartwig J F. Mechanistic studies on rhodium-catalyzed enantioselective silylation of aryl C – H bonds [J] . Journal of the American Chemical Society, 2017, 139 (13) : 4879–4886; b) Lee S Y, Hartwig J F. Palladium-catalyzed, site-selective direct allylation of aryl C – H bonds by silver-mediated C – H activation: A synthetic and mechanistic investigation [J] . Journal of the American Chemical Society, 2016, 138 (46) : 15278–15284.

[143] Cheng G J, Yang Y F, Liu P, et al. Role of *N*-acyl amino acid ligands in Pd (II) -catalyzed remote C – H activation of tethered arenes [J] . Journal of the American Chemical Society, 2014, 136 (3) : 894–897.

[144] Yang Y F, Cheng G J, Liu P, et al. Palladium-catalyzed *meta*-selective C – H bond activation with a nitrile-containing template: Computational study on mechanism and origins of selectivity [J] . Journal of the American Chemical Society, 2014, 136 (1) : 344–355.

[145] Jiang J L, Yu J Q, Morokuma K. Mechanism and stereoselectivity of directed C (sp^3) – H activation and

arylation catalyzed by Pd（II）with pyridine ligand and trifluoroacetate: A computational study［J］. ACS Catalysis，2015，5（6）: 3648–3661.

［146］ Xue X S，Ji P J，Zhou B Y，et al. The essential role of bond energetics in C－H activation/functionalization［J］. Chemical Reviews，2017，117（13）: 8622–8648.

［147］ a）McKeown B A，Gonzalez H E，Friedfeld M R，et al. Mechanistic studies of ethylene hydrophenylation catalyzed by bipyridyl Pt（II）complexes［J］. Journal of the American Chemical Society，2011，133（47）: 19131–19152；b）McKeown B A，Gonzalez H E，Gunnoe T B，et al. Pt^{II}–catalyzed ethylene hydrophenylation: Influence of dipyridyl chelate ring size on catalyst activity and longevity［J］. ACS Catalysis，2013，3（6）: 1165–1171；c）McKeown B A，Prince B M，Ramiro Z，et al. Pt^{II}–catalyzed hydrophenylation of α–olefins: Variation of linear/branched products as a function of ligand donor ability［J］. ACS Catalysis，2014，4（5）: 1607–1615；d）McKeown B A，Foley N A，Lee J P，et al. Hydroarylation of unactivated olefins catalyzed by platinum（II）complexes［J］. Organometallics，2008，27（16）: 4031–4033；e）McKeown B A，Gonzalez H E，Michaelos T，et al. Control of olefin hydroarylation catalysis via a sterically and electronically flexible platinum（II）catalyst scaffold［J］. Organometallics，2013，32（14）: 3903–3913；f）McKeown B A，Gonzalez H E，Friedfeld M R，et al. Platinum（II）–catalyzed ethylene hydrophenylation: Switching selectivity between alkyl– and vinylbenzene production［J］. Organometallics，2013，32（9）: 2857–2865.

［148］ Vaughan B A，Webster–Gardiner M S，Cundari T R，et al. A rhodium catalyst for single–step styrene production from benzene and ethylene［J］. Science，2015，348（6233）: 421–424.

［149］ a）Takahashi M，Masui K，Sekiguchi H，et al. Palladium–catalyzed C–H homocoupling of bromothiophene derivatives and synthetic application to well–defined oligothiophenes［J］. Journal of the American Chemical Society，2006，128（33）: 10930–10933；b）Xia J B，Wang X Q，You S L. Synthesis of biindolizines through highly regioselective palladium–catalyzed C–H functionalization［J］. The Journal of Organic Chemistry，2009，74（1）: 456–458；c）Zhang Q，Li Y F，Lu Y，et al. Pd–catalysed oxidative C－H/C－H coupling polymerization for polythiazole–based derivatives［J］. Polymer，2015，68: 227–233.

［150］ a）Yamaguchi J，Yamaguchi A D，Itami K. C_H bond functionalization: emerging synthetic tools for natural products and pharmaceuticals［J］. Angewandte Chemie International Edition，2012，51（36）: 8960–9009；b）Shang M，Wang M M，Saint–Denis T G，et al. Copper–mediated late–stage functionalization of heterocycle–containing molecules［J］. Angewandte Chemie International Edition，2017，56（19）: 5317–5321；c）Wu J Q，Zhang S S，Gao H，et al. Experimental and theoretical studies on rhodium–catalyzed coupling of benzamides with 2，2–difluorovinyl tosylate: Diverse synthesis of fluorinated heterocycles［J］. Journal of the American Chemical Society，2017，139（9）: 3537–3545.

［151］ a）Giri R，Lam J K，Yu J Q. Synthetic applications of Pd（II）–catalyzed C–H carboxylation and mechanistic insights: Expedient routes to anthranilic acids，oxazolinones，and quinazolinones［J］. Journal of the American Chemical Society，2010，132（2）: 686–693；b）Orito K，Horibata A，Nakamura T，et al. Preparation of benzolactams by Pd（OAc）$_2$–catalyzed direct aromatic carbonylation［J］. Journal of the American Chemical Society，2004，126（44）: 14342–14343；c）Zhang H，Shi R Y，Gan P，et al. Palladium–catalyzed oxidative double C_H functionalization/carbonylation for the synthesis of xanthones［J］. Angewandte Chemie International Edition，2012，51（21）: 5204–5207;（d）Lian Z，Friis S D，Skrydstrup T. C－H activation dependent Pd–catalyzed carbonylative coupling of（hetero）aryl bromides and polyfluoroarenes［J］. Chemical Communications，2015，51（10）: 1870–1873;（e）Wu X F，Anbarasan P，Neumann H，et al. Palladium–catalyzed carbonylative C_H activation of heteroarenes［J］. Angewandte Chemie International Edition，2010，49（40）: 7316–7319;（f）Tjutrins J，Arndtsen B A. An electrophilic approach to the palladium–catalyzed carbonylative C－H functionalization of heterocycles［J］. Journal of the American Chemical Society，2015，137（37）: 12050–

12054. (g) McNally A, Haffemayer B, Collins B S L, et al. Palladium–catalysed C – H activation of aliphatic amines to give strained nitrogen heterocycles [J] . Nature, 2014, 510 (7503) : 129–133; (h) Willcox D, Chappell B G N, Hogg K F, et al. A general catalytic β –C – H carbonylation of aliphatic amines to β –lactams [J] . Science, 2016, 354 (6314) : 851–857; (i) Li W, Duan Z L, Zhang X Y, et al. From anilines to isatins: Oxidative palladium–catalyzed double carbonylation of C_H bonds [J] . Angewandte Chemie International Edition, 2015, 54 (6) : 1893–1896; j) Guan Z H, Chen M, Ren Z H. Palladium–catalyzed regioselective carbonylation of C – H bonds of *N*–alkyl anilines for synthesis of isatoic anhydrides [J] . Journal of the American Chemical Society, 2012, 134 (42) : 17490–17493.

[152] Kondo Y, Garc í a–Cuadrado D, Hartwig J F, et al. Rhodium–catalyzed, regiospecific functionalization of polyolefins in the melt [J] . Journal of the American Chemical Society, 2002, 124 (7) : 1164–1165.

[153] a) Howell J M, Feng K B, Clark J R, et al. Remote oxidation of aliphatic C – H bonds in nitrogen–containing molecules [J] . Journal of the American Chemical Society, 2015, 137 (46) : 14590–14593; b) Gormisky P E, White M C. Catalyst–controlled aliphatic C – H oxidations with a predictive model for site–selectivity [J] . Journal of the American Chemical Society, 2013, 135 (38) : 14052–14055; c) Bigi M A, Reed S A, White M C. Diverting non–haem iron catalysed aliphatic C–H hydroxylations towards desaturations [J] . Nature Chemistry, 2011, 3 (3) : 216–222.

[154] Tang S, Wang P, Li H R, et al. Multimetallic catalysed radical oxidative C (sp^3) – H/C (sp) – H cross-coupling between unactivated alkanes and terminal alkynes [J] . Nature Communications, 2016, 7: 11676.

[155] a) Yang Q L, Li Y Q, Ma C, et al. Palladium–catalyzed C (sp^3)—H oxygenation via electrochemical oxidation [J]. Journal of the American Chemical Society, 2017, 139 (8) : 3293–3298; b) Le C, Liang Y F, Evans R W, et al. Selective sp^3 C – H alkylation via polarity–match–based cross–coupling [J] . Nature, 2017, 547 (7661) : 79–83.

[156] a) Zhang W, Wang F, McCann S D, et al. Enantioselective cyanation of benzylic C – H bonds via copper–catalyzed radical relay [J] . Science, 2016, 353 (6303) : 1014–1018; b) Liao K B, Negretti S, Musaev D G, et al. Site–selective and stereoselective functionalization of unactivated C – H bonds [J]. Nature, 2016, 533(7602): 230–234.

[157] Zhang W, Wang F, McCann S D, et al. Enantioselective cyanation of benzylic C – H bonds via copper–catalyzed radical relay [J] . Science, 2016, 353 (6303) : 1014–1018.

撰稿人：房华毅　施章杰

聚集诱导发光研究进展

一、引言

人类对发光材料的系统研究可追溯到二世纪中叶，当时人们发现常见的有机荧光分子在稀溶液中发光很强，但在高浓度溶液中荧光变弱甚至消失，于是认为“浓度猝灭荧光”是发光材料的普遍现象。“浓度猝灭荧光”逐渐成为了教科书中的常识性知识。2001 年，唐本忠等基于观察到的一类在溶液中不发光或者发光微弱的分子聚集后荧光显著增强的独特现象，提出了“聚集诱导发光（Aggregation-Induced Emission，以下简称 AIE）”的新概念，并在大量的实验验证和理论模拟的基础上，提出了分子内运动受限的机理，得到了国内外同行的广泛认可和采用[1, 2]。AIE 与传统的发光截然相反，打破了“浓度猝灭发光”的经典论断，改变了人们对发光材料的传统认知，在化学和材料等学科领域开辟了新的研究领域。由于 AIE 材料在聚集态或者固态表现出强的荧光发射，接近于实际应用，因此，自 AIE 概念提出后即受到了国内外学者的广泛关注和快速跟进。AIE 领域无论是发表的论文数还是引用数均呈指数增长，已经发展成一个由中国科学家引领、60 余个国家和地区约 1100 个研究单位的科学家跟进的新兴领域。根据中国科学院文献情报中心和汤森路透联合发布的《2015 研究前沿》[3]，AIE 位列化学和材料研究领域十大前沿研究的第二位，且为重点热点前沿。AIE 研究在能源、健康、环境等领域取得了一系列有影响力的创新成果[4]，AIE 纳米材料也被 2016 年发表在《自然》上的新闻深度分析文章列为支撑和驱动“纳米光革命”的四大纳米材料之一[5]。所以，AIE 不仅在科学上推动了新发光模型或机理的建立，扩展和加深了对发光过程的理解，而且在应用上提供了原创性和自主知识产权的新材料和新技术。本报告将对近年来 AIE 领域在分子设计、理论机制、智能材料、光电器件、生物检测与成像等方面的最新研究进展进行总结分析。

二、AIE 研究现状与进展

（一）AIE 分子体系设计

AIE 材料的发展经历了 AIE 概念的提出、零星 AIE 材料的报道、AIE 机理的提出及验证、机理指导下新的 AIE 分子设计制备、AIE 机理的完善、AIE 材料高技术应用等过程。在 AIE 发展的过程中，AIE 分子设计至关重要。其大致可分为两个部分：一是对已知 AIE 的核心结构进行改造；二是全新的 AIE 核心结构的设计开发。在此过程中还发现了独特的纯有机室温磷光（RTP）的 AIE 分子体系。

1. 基于已知 AIE 核心结构的分子设计

通过以 AIE 基元（比如四苯基乙烯，TPE）来修饰 ACQ 分子，或以 ACQ 基元代替 AIE 分子的部分结构，即可得到新的 AIE 分子。AIE 基元提供可运动的基团，并使整个分子呈非平面的构象，并在溶液中旋转消耗激发态能量，使分子不发光；而在聚集态，AIE 基元分子内运动受到限制，无辐射跃迁受到抑制，而非平面的构象又限制了分子间的 $\pi-\pi$ 堆积，从而使发光增强。同理，李振等以 TPE 为核心，将其与另外的 TPE 或其他生色团键接，通过控制苯环上的键接位置，或在键接点邻位引入位阻基团，使分子处于极度扭曲的构象，控制分子的共轭长度，可得最大发射在 450nm 左右的高效蓝光 OLED 材料（图 1）。研究结果也证明了控制键接点位置是一个非常有效的控制分子共轭程度从而控制分子发光颜色的手段（如化合物 1 和 2）。唐本忠等以共轭程度较小的三苯基乙烯代替 TPE，将两个三苯基乙烯单元与二苯基菲并咪唑结合，得到了最大荧光发射为 455nm 的高效蓝光材料，以其为发光层制备的 OLED 实现了最大发射在 450nm 的蓝光发射（化合物 3）。

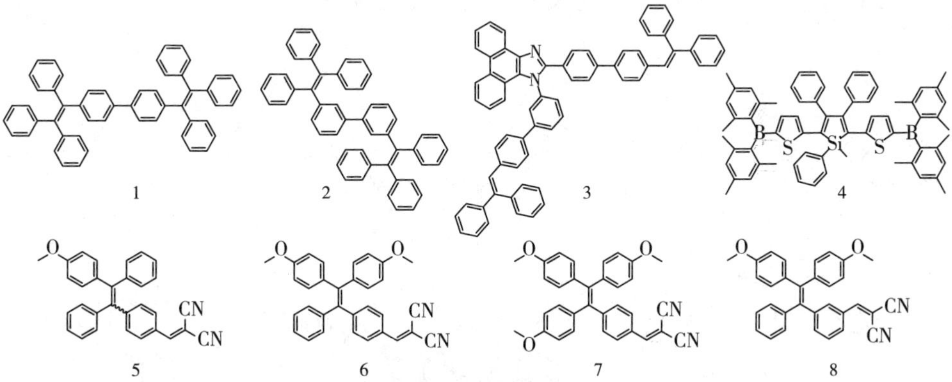

图 1　AIE 分子结构调控发光颜色示例

除蓝光外，红光 AIE 材料也受到了极大关注。研究人员将电子给体 – 受体（D–A）结构引入到 AIE 分子中，成功制备了一系列红光 AIE 材料。赵祖金、唐本忠等[6]在噻咯的 2，5– 位引入噻吩作为电子给体，而噻咯环作为电子受体，成功制备了最大发射在 598nm、效率高达 32% 的红光噻咯衍生物（化合物 4）。张德清、张关心等将二氰基乙烯基作为电子受体、甲氧基作为电子给体引入到 TPE 基元中，得到了一系列高效（48%~61%）的红光（603~645 nm）AIE 分子 5~7[7]。电子给体数量的增加，使得光谱逐渐红移。而电子受体键接到对位可比间位的光谱红移 30 nm 左右（化合物 6 与 8 比较）。因此，通过将 D–A 结构引入到 AIE 基元中制备红光 AIE 材料是一种行之有效的方法。

2. 新 AIE 核心结构的分子设计

继芳基取代的噻咯之后，在分子内转动受限（RIR）机理的指引下，已有大量的 AIE 核心分子报道。这些分子以单键将芳环和共轭核心链接起来，芳环可绕单键转动，如 TPE、四苯基丁二烯、二苯基苯并富烯、二苯乙烯基蒽等。目前，也有少量分子的可转动基团为酯基、甲基或氰基，其 AIE 现象也可用 RIR 机理来解释。比如靛蓝染料的叔丁氧酰胺衍生物（9），其在溶液和无定形态均不发光，但其晶态具有强荧光[8]。该分子可转动的部分是叔丁氧酰基团，而非芳环（图 2）。随着研究的深入，唐本忠等发现了一些不具有可旋转芳香环的化合物同样具有 AIE 性质（比如化合物 10），由此提出了分子内振动受限（RIV）的机理[9]。含环辛四烯结构和柔性 V– 型结构的一些分子也具有 AIE 的性质[10–12]。环辛四烯具有船式和椅式两种构型，且在溶液中两种构型间能相互转换，消耗激发态能量（化合物 11 和 12）。含柔性 V– 型结构的分子（13）在溶液中 V 的两臂的

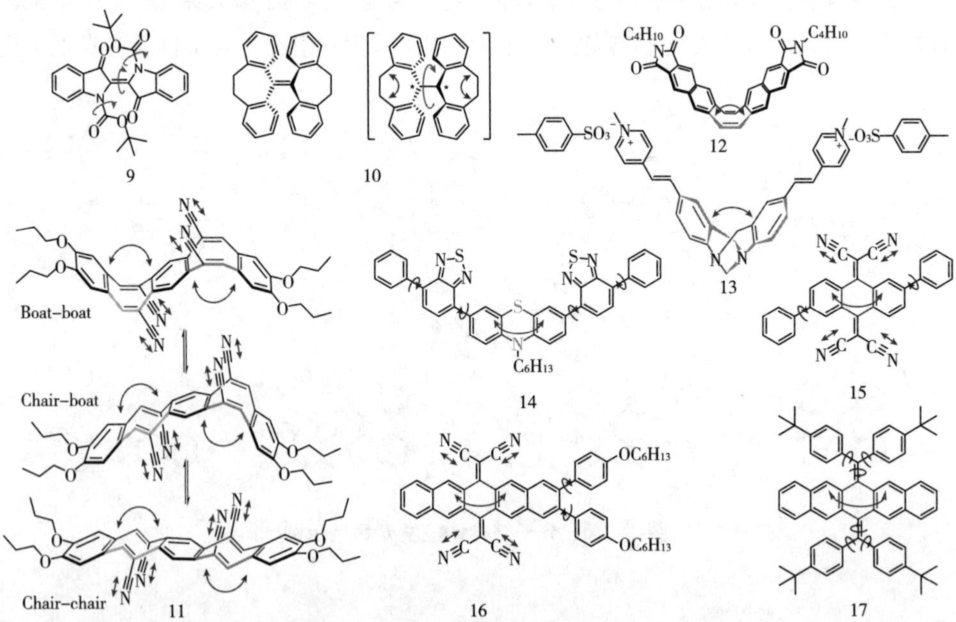

图 2　新型 AIE 分子运动模型示例

摆动也消耗激发态能量；而在聚集态，这种分子内的摆动被抑制，分子发光大大增强。但是，基于 RIV 机理的 AIE 分子还需进一步发展。研究人员还将开合摆动的中心与可绕单键旋转的基团结合，基于 RIR 和 RIV 机制制备了新型的 AIE 分子 14~17[13-16]。分子内基团的转动或摆动都是会消耗激发态能量，故可将 RIR 和 RIV 机理统一为分子内运动受限（RIM）机理。

3. 有机室温磷光材料

室温磷光在金属配合物中很常见，但纯有机体系的室温磷光则鲜有报道。2010 年唐本忠等[17]发现二苯甲酮及其衍生物在晶态具有室温下发射磷光的特性，即结晶诱导磷光（CIP）。在介质中，分子间的碰撞或分子内转动和振动导致无辐射跃迁，且三重态易被氧气猝灭，磷光不明显。而在晶体中，分子间紧密排列，隔绝了氧气；分子间的弱相互作用〔如 C－H⋯O、N－H⋯O、C－H⋯π、氢键、C－H⋯X（（X=F，Cl，Br））和 C－Br⋯Br－C〕限制了分子的运动，抑制了非辐射跃迁，与 AIE 的 RIR 机理相似。

（a）

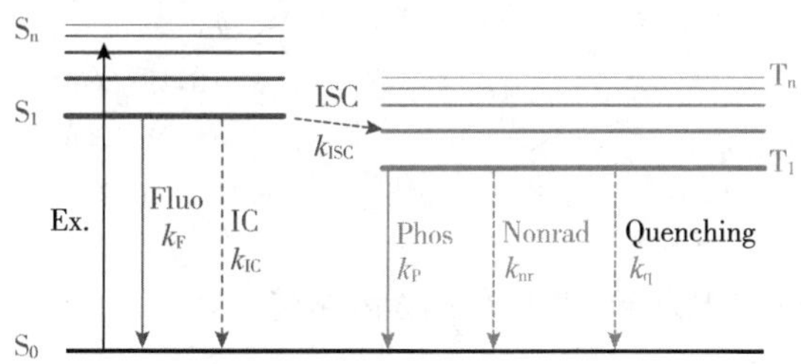

（b）

$$\tau_P = \frac{1}{k_P + k_{nr} + k_q} \quad (1)$$

$$\Phi_{ISC} = \frac{k_{ISC}}{k_{ISC} + k_{IC} + k_F} \quad (2)$$

$$\Phi_P = \Phi_{ISC} k_P \tau_P = \Phi_{ISC}(1 - k_{nr}\tau_P - k_q\tau_P) \quad (3)$$

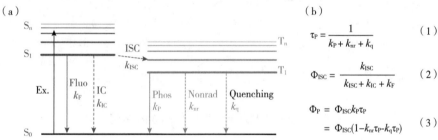

图 3 （a）有机发光分子的 Jablonski 能级图；（b）磷光寿命（τ_P），系间窜越效率（Φ_{ISC}）及磷光效率（Φ_P）公式。k_F：荧光发射速率常数；k_{IC}：内转换速率常数；k_{ISC}：系间窜越速率常数；k_P：磷光发射速率常数；k_{nr}：无辐射跃迁速率常数；k_q：磷光猝灭速率常数

一般的纯有机 RTP 材料，其磷光寿命在 10 毫秒以下，而具有长寿命（大于 100 毫秒）的纯有机 RTP 材料，其磷光衰减过程裸眼可见，因而具有更广阔的应用前景。兼具高磷光效率和长磷光寿命的材料是研究人员追求的目标。图 3 示例了有机分子发光过程中各个竞争因素[18]。研究人员通过制备晶态材料或将发光材料掺杂在聚合物或其他刚性介质中，限制分子的运动，降低 k_{nr}；同时，隔绝了氧气等磷光猝灭物质，降低 k_q，从而开发出了高效纯有机 RTP 材料。2015 年黄维、陈润锋等[19]报道一系列具有长寿命的单组份有机分子（化合物 18~20）（图 4），这些分子在固态形成 H- 聚集，寿命最长可达 1.35 秒（19b）。之后，董永强、卢然、袁望章、李振等陆续报道了一些含羰基和砜基的室温长寿命材料，且量子效率逐步提高[20-23]。池振国等将长寿命（0.28 秒）有机 RTP 材料的效率提高到了 5%[24]。唐本忠、彭谦等在仔细分析已报道的室温磷光材料结构与性能关系基础上，设计并合成了一系列兼具高效率和长寿命的有机 RTP 分子 24~28，其中化合物 24 发光效率高达 34.5%，寿命长达 230 毫秒[18]。

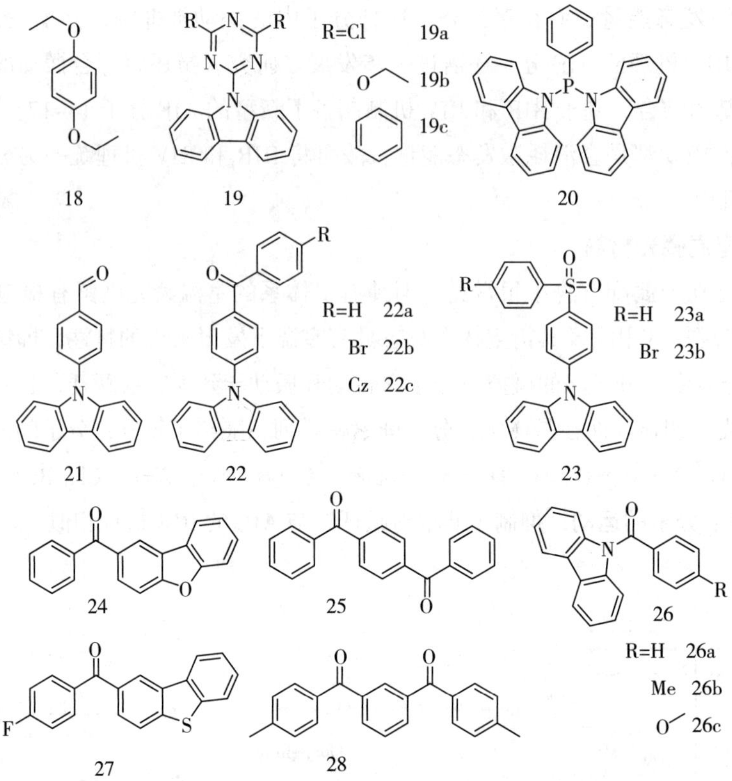

图 4　纯有机长寿命室温磷光分子结构示例

（二）AIE 机理拓展

揭示 AIE 机理、探究 AIE 的根源，是创新和扩展 AIE 体系及应用的关键。唐本忠等陆续提出了 RIM 机理，并通过系列实验证实了该机理的有效性和普遍性[4, 25]。研究结果表明，在束缚较小的环境下，AIE 分子内基团（例如芳环等）的面外运动非常自由，从而使激发态能量主要以热的形式耗散；而在受限条件（聚集态或者固态）下，这些分子内的运动往往受到限制，激发态能量以辐射，即发光的方式耗散。通过空间位阻、共轭效应、分子间配位、键连闭环等化学手段，阐明了分子结构对 AIE 效应的影响。最近，香港中文大学、剑桥大学和香港科技大学联合利用温度分辨的太赫兹时域光谱（THz–TDS）探测出在分子激发态衰减过程中低频分子内运动的贡献与温度的依赖关系，直观地印证了 RIM 机理[9]。目前，RIM 机理已得到广泛的认可和应用，但还需进一步研究，使其实现从定性到定量的转变。同时，研究人员也基于其发展的 AIE 体系，提出了诸如 *E*/*Z* 异构化、构象平面化以及 J– 聚集体生成等机制。

田禾等在二氢二苯并吩嗪类分子中发现了独特的发光现象，提出了振动诱导发光（VIE）的有发光机理[26]，并通过稳态和超快光谱等测试手段，结合反应势能面模拟，验证了该机理[27]。最近，他们进一步利用化学合成手段将二氢二苯并吩嗪类化合物进行分子内环化，限制激发态的结构振动，通过超快光谱及理论计算，再次验证了 VIE 发光机理[28]。唐本忠等在研究 D-A 型有机分子的 AIE 效应时发现，这类分子存在分子内电荷转移态（TICT）和 RIR 两个过程，后者主导时，分子则表现为 AIE 或者聚集增强发光的性能[29]。孙洪波等通过泵浦探测实验也证实了 D-A 型 AIE 体系存在的 TICT 过程[30]。马於光等通过研究 DSDPB 体系的光物理性质，得出了光异构化受限会引起 AIE 现象的结论[31]。吴长峰等利用这一机理，合成了氧杂环丁烷取代的 TPE 的 *Z*/*E* 型异构体，完成了分离、提纯及性能测试，发现 *Z* 式比 *E* 式异构体表现出更强的自修复性能[32]。这为研究 TPE 异构化光物理性质提供了分子模板。田文晶、徐斌等将苯乙烯基吡啶衍生物的 AIE 现象归因于 J- 型聚集体的形成[33]。孔林和杨家祥等将吲哚并咔唑衍生物的 AIEE 现象同样归因于 J- 型聚集体形成[34]。杨国强等研究了激发态分子内质子化转移（ESIPT）过程和 AIE/AEE 效应的关系[35]。

帅志刚、彭谦等提出了无辐射衰减过程受限的 AIE 机理，即固态堆积形成的分子间相互作用有效抑制分子低频振动模式的电声子耦合（重整能），阻塞了无辐射跃迁通道，致使辐射过程占主导，荧光增强[36-38]。最近几年，他们相继完成了化合物在气相、溶液、晶态、无定型态等多形态多尺度的光物理性质模拟计算，全面考查了不同环境下的分子本征的发光特性。建立了不可测量的重整能与实验可测的谱学信号之间的多种正向关系，从理论上提出了验证 AIE 机理的可行性实验方案。①建立了 Stokes 位移与重整能之间的近似相等关系。相较于溶液，聚集状态下重整能变小导致 AIE 体系的发射蓝移[39]。②建立了共振拉曼光谱强度与重整能之间的正比例关系。共振拉曼光谱信号从溶液到固态的变化特征等同于重整能的变化，印证了 AIE 机理，期待着实验验证[40]。③利用同位素效应直接探测无辐射跃迁变化特征[41]。AIE 体系在溶液中表现出了反常同位素效应，即无辐射跃迁速率几乎没变化；而在聚集状态下表现出正常同位素效应。这种 AIE 分子从溶液到固相同位素效应的“暴涨”行为，直接验证了 AIE 机理。④ AIE 体系从溶液到无定型薄膜到晶体的发射光谱和发光效率的渐变特征，印证了 AIE 机理[42]。结合分子动力学、量子力学 / 分子力学及光谱和激发态衰减速率理论，研究了典型 AIE 体系六苯基噻咯（HPS）在不同形态下的发光性质。从溶液到无定形态到静态，随着分子堆积的紧密分子间作用的增加，HPS 的发射光谱逐渐蓝移，发光效率依次增加现象。⑤ AIE 体系在固态下的激子相互作用很小，可以忽略不计，进一步印证了 AIE 机理[43]。采用 Frenkel 激子模型和 QM/MM 方法研究了一系列 AIE 体系和传统非 AIE 体系的晶态光谱，发现 AIE 聚集体光谱特征主要是由分子内电声子耦合所主导。进而，从溶液到固态的电声子耦合减小导致的分子内无辐射跃迁受限是诱导 AIE 发光的主要原因。

梁万珍、赵仪等采用分子动力学、量子力学 / 分子力学方法模拟了甲氧基取代的 TPE（DMO-TPE）在水溶液中的聚集过程并研究了 AIE 机理[44]。研究发现 DMO-TPE 的 AIE 效应起源于分子内转动受限和部分的 J- 聚集。李全松、Blancafort 等提出了限制锥形交叉点通路模型，从理论角度论证了噻咯分子的 AIE 机制[45]。刘峰毅等采用 QM 方法和 ONIOM 方法研究了 D–A 型苯乙烯衍生物（BIM）在甲醇溶液和晶态下的分子激发态性质[46]。结果表明，在甲醇溶液中 BIM 从发光态（EM）无势垒地弛豫到电荷转移态（CT），进而达到 CT 态附近的锥形交叉点，快速衰减到基态；在聚集状态下，这些过程被周围的空间位阻和静电相互作用所限制，致使荧光增强。

帅志刚等在紧束缚密度泛函理论框架下实现了混合量子 / 经典非绝热动力学方法，实时模拟了“打开”和“闭合”的 AIE 活性的和非 AIE 活性的环戊二烯衍生物（DPDBF）的激发态无辐射衰减过程[47]。结果表明，AIE 分子的低频扭转模式有效地与电子态耦合并接受能量，快速无辐射衰减到基态；而非 AIE 分子的这一模式被化学键所限制，无辐射通道受限，进而发强荧光。由此可知，分子发生聚集是低频模式运动受限，无辐射衰减通道受到堵塞，进而聚集体发光。

帅志刚、彭谦等针对室温下纯有机晶体的长寿命磷光提出了分子间静电相互作用诱导的机理[48]。通常情况下，三重态到基态的跃迁是偶极禁阻的，只有通过自旋轨道耦合借助于中间单重态才能发生。理论计算发现，静电相互作用能够诱导氧原子孤对电子发生变化，使得中间态 S_1 态由气相下的弱偶极跃迁的（$n\pi^*$）态转变为晶态下强偶极跃迁的（$\pi\pi^*$）态，致使 T_1 到 S_0 的跃迁偶极矩增加两个数量级。同时，C = O 振动被抑制，无辐射衰减受限，所以有机晶体在室温下发出长寿命磷光。江俊等提出了聚集诱导系间窜越机理来解释有机晶体的长寿命磷光[49]。相较于溶液，分子聚集后发生电子态能量劈裂，产生了更多的单线态和三线态。这样，最低激发单线态和邻近三线态的能量差变小，两者之间的系间窜越速率加快，导致磷光效率增加。

（三）AIE 材料智能

智能材料是能够感知外部的刺激，又可以判断并适当处理的新型功能性材料，它是在天然材料、高分子材料和人工设计材料后的下一代材料，是当今高新技术和材料发展的重要研究方向。刺激响应发光材料是指在外界刺激作用下，例如力、光、电、热、pH 等，材料的发光性能（发光强度和 / 或发光颜色等）产生明显变化的一类智能材料。它不同于一般的基于吸收光谱变化的刺激响应变色材料，它主要基于发射光谱对于外界刺激的响应性变化，具有更好的抗干扰性和检测灵敏度，因此在传感、防伪、信息存储等领域具有重要的潜在应用，受到相关领域研究工作者的高度关注[50–51]。传统荧光材料由于受到 ACQ 的限制，一般只能在溶液状态下进行基于荧光信号改变的检测，而在固体状态下的应用一直处于理论与应用研究的薄弱环节。近年来，AIE 在刺激响应材料中的相关探索与应用得到了可喜的发展，相关的理论研究和实际应用都已经初步建立。

1. 力刺激响应 AIE 材料

2008 年，唐本忠等利用机械力来改变基于 AIE 发光层的 OLED 器件发光性能以研究

AIE 现象产生的机理，发现施加的力在一定范围内时，器件的亮度呈正反馈趋势[52]。尽管当时并没有观察到力刺激发光颜色改变现象，但这是第一次明确提出把力作用于 AIE 分子考察其发光强度的变化情况。2010 年，Park 等报道了一例含乙烯腈结构的 AIE 化合物 DBDCS 具有力致发光变色效应[66]。

池振国等发现绝大多数 AIE 分子均具有力刺激发光变色性能。因此，他们在 2011 年首先指出力刺激发光变色响应是 AIE 分子的共性，并提出了力致发光变色 AIE 材料的概念[53, 54]。这一发现为进一步设计合成力刺激响应发光材料开辟了一个新的有效途径。此后，有关力刺激响应发光 AIE 材料的研究报道如雨后之春笋般大量涌现，并受到了研究人员的极大关注[55]。同时，双发射力刺激响应发光材料也逐渐被开发出来，并在诸如商标防伪、应力传感等应用领域表现出更大的优势[56–58]。

田禾等报道的 VIE 发光分子就是一类非常典型的荧光 - 荧光双发射分子，其中一个二氢吩嗪类衍生物在溶剂化过程中可以诱导出从蓝光到红光的无数种发光颜色，甚至实现白光发射[59]。池振国等发现了许多新奇的基于固态双发射的现象，如荧光 - 磷光双发射、荧光 - 热激活延迟荧光（TADF）双发射、荧光 - 长余辉双发射、力刺激荧光 - 磷光双发射直接发光等[60–62]。这些新型双发射化合物中大多数具有固态多波长力刺激发光变色性能，即可实现固态单组份发光调色。

2. 光致变色 AIE 材料

2014 年，刘浪等报道了一系列具有 AIE 和光致变色性质的吡唑啉酮衍生物[63]。吡唑啉酮结构具有特殊的激发态质子转移的特性，当分子被紫外光照激发后，由原来的醇式结构转变成酮式结构，酮式结构中氧原子的吸电子能力远远强于醇式结构，因此酮式结构的起始吸收波长要比醇式结构的长。在加热的条件下酮式结构发生了逆反应，重新回到醇式结构。2015 年，田文晶等将 TPE 和具有光致变色性能的螺吡喃基团通过双键的方式连接起来，得到具有 AIE 性质的化合物，其二氯甲烷溶液在紫外光的照射下逐渐从无色分别变成了淡红色和淡黄色，紫外 - 可见吸收光谱图上可以明显看到原来的起始吸收峰强度下降，而在可见光区则新增了一个吸收峰[64]。2015 年，向宇等合成了一系列水杨醛缩腙类化合物，发现该系列化合物在聚集态下同时具有 AIE 和光致变色性质[65]。在紫外光照射下，分子能发生醚键断裂的光化学反应，导致发光开关过程。该研究通过这种巧妙的方法得到了波长可控的聚集态荧光。

2015 年，唐本忠等合成了具有稠环结构以及能够自由旋转的苯环取代基团的 AIE 体系，该体系分子的稠环结构能够进行平面振动，从而在溶液中的荧光量子产率非常低，但由于 RIV 机理，固态下发光强烈[66]。伴随着分子的聚集，分子在紫外灯的照射下逐渐变成红色，而且固体状态下的蓝色荧光也随之消失。这种显色状态在常温下能够维持很长时间不变，而通过用 498 nm 可见光对其进行照射，红色固体在数十秒内迅速恢复到原始的无色晶体状态，蓝色的荧光也能够重新显现。其原因主要是关、开环过程所致。2016 年，池振国等报道了卤代三苯乙烯的光致变色现象[67]。该化合物在紫外光照前发射强烈的蓝光，光照后则不发光，而样品晶体本身从无色变成了红色。另外，溶液成膜过程可以被紫外光调控，薄膜表面可以从疏水性变成亲水性。上述化合物呈现出多重光刺激响应性能。同年，朱琳娜等研究了己基噻吩取代 TPE 的光致变色现象，也发现了类似的现象[68]。

3. 酸碱响应 AIE 材料

2011 年，王悦等合成了一个氰基取代二苯乙烯衍生物。通过研磨和加热或溶剂熏蒸的重复可逆过程，分子的荧光可在橙红和黄色之间转换[69]。而且其固体粉末在三氟乙酸蒸气和三乙胺蒸气的交替刺激下能够观察到荧光在蓝色和黄色之间来回变化。他们认为该响应过程的反应位点为苯胺上的氮原子。原始状态的分子中存在苯胺到氰基的分子内电荷转移（ICT）过程，苯胺属于给电子基团，而在质子化之后，苯胺上的氮原子的孤对电子被占用，原来的给电子性转变成了吸电子性，分子内的 ICT 过程不能够发生，导致分子荧光发生蓝移。

2011 年，童爱军等合成了基于水杨醛的 ESIPT 过程的 pH 响应 AIE 分子体系[70]。该体系很好地利用了两个质子的分段电离，在 pH 变化中显示出分段指示的功能，同时使分子的聚集状态在纳米颗粒和良溶液之间转换，增加了对比度。该化合物能够对不同 pH 时的 HepG2 细胞染色，具有很好的选择性、稳定性和细胞穿透性，是一种非常有前景的 pH 定量指示探针。2012 年，唐本忠等合成了具有 pH 刺激响应性的同时含有 TPE 以及花青素结构的化合物[71]。在 pH 小于 5 的溶液体系中，化合物发射出较强的红色荧光（最大发射波长 630 nm），并且荧光强度几乎不随 pH 的改变而发生变化；当 pH 在 5~7 的范围内，体系的荧光发射强度随 pH 的增大呈现线性下降的趋势，同样可以归结于该化合物的溶解度的变化而导致的聚集状态的改变（酸性条件下磺酸根质子化导致化合物的疏水性增强）；而当 pH 大于 10 时，体系开始出现蓝色的荧光，并且荧光强度随 pH 增大而增强。2013 年，唐本忠等又报道了另外一个具有 pH 刺激响应性的含吡啶结构 AIE 化合物[72]。在酸性气氛下，涂覆了化合物样品的 TLC 薄板发射微弱的红色荧光，而继续在碱性气氛下熏制一段时间，荧光颜色又恢复成蓝色，并且荧光强度增加。在轮流使用酸碱气氛熏制下，这种荧光变化过程可以无损地重复多次。2013 年，田文晶等也合成了两个含吡啶的乙烯基蒽衍生物，这两个化合物的固体粉末在酸碱蒸气的刺激下同样可以实现荧光颜色的可逆转变[73, 74]。这两个化合物样品对酸碱刺激产生荧光变色的原因同样在于吡啶氮原子的质子化 - 去质子化作用。

（四）AIE 在光电领域的应用研究

作为构建有机发光二极管（OLED）的重要组成部分，有机发光材料的发展对其性能的提升起着至关重要的作用，所以研究和开发新型高效的有机发光材料一直受到学术界和产业界的强烈关注。传统的有机发光材料由于 ACQ 问题，往往需要使用掺杂技术来制备器件，使得器件存在相分离等问题，而且制作成本较高。AIE 材料在固态下具有优异的发光效率，恰好能解决传统发光材料的 ACQ 问题，因此在高效非掺杂 OLED 的研发中具有巨大潜力。由于 AIE 材料在 OLED 中的重要作用，到现在为止，研究人员已开发出具有从蓝光到红光的多种发光颜色的 AIE 材料。

1. AIE 蓝光材料

2013 年，李振等将两个简单的 TPE 单元通过不同位置的连接成功地合成了 4 种 AIE 特性的发光波长在 435~459nm 的蓝光材料[75]。利用这些蓝光材料作为发光层制备的非掺杂 OLED 的最大发光亮度为 3266 cd/m^2，最大外量子效率达 1.9%。赵祖金、唐本忠等合成了一系列 AIE 特性的 TPE 衍生物[76]。利用这些分子制备的非掺杂 OLED 的最大发光亮度和效率分别为 49030cd/m^2 和 6.6 cd/A。唐本忠等报道了一类 AIE 特性的咔唑和三苯胺取代的乙烯衍生物，用其制备的非掺杂 OLED 发射天蓝光，其最大发光亮度和外量子效率分别为 11700 cd/m^2 和 3.3%[77]。施和平等设计合成了一类 AIE 特性的、热稳定性和化学稳定性好的含有二米基硼单元的咔唑衍生物，利用其作为发光层制备的非掺杂 OLED 发射蓝光，开启电压为 3.8 伏，最大发光亮度为 2784 cd/m^2[78]。

2015 年，唐本忠等在典型的 AIE 基团三苯基乙烯上连接不同的功能基团，设计合成了一种高效的发射深蓝光的 AIE 分子[79]。利用该分子制备的非掺杂 OLED 显示良好的电致发光性能。同年，苏忠民等合成了三种以吡啶唑为功能基团的 TPE 衍生物[80]。以该类材料作为发光层制备的非掺杂 OLED 的功率效率为 2.0lm/W。袁望章等合成了三种刚性的三苯胺衍生物，它们在溶液和聚集态均具有很强发光，以其制备的非掺杂 OLED 发射天蓝光，其最大发光亮度为 10954cd/m^2[81]。李立东等报道的包含芴基团和 TPE 基团的分子作为发光层通过溶液法制备的非掺杂 OLED 发射天蓝光，其最大电流效率和外量子产率分别为 4.55cd/A 和 2.17%[82]。

2016 年，秦安军、唐本忠等将电子供体二苯胺与电子受体四苯基吡嗪（TPP）相结合，成功地获得了三种 AIE 分子。这些分子具有优良的多功能性质[83]。用含单个二苯胺基团的分子制备的非掺杂 OLED 器件的最大亮度可达 17459 cd/m^2，并且还可用作空穴传输层，大大简化器件结构。张晶莹等将一个三苯基乙烯基团和菲并咪唑单元组合设计合成了具有优良的电子传输和空穴传输性能两种 AIE 蓝光分子。利用它们制备的非掺杂 OLED 发射深蓝光，最大外量子效率为 5.52%[84]。李振等报道了两种仅含有苯基和三苯胺基团的 AIE 分子 2TPA-CN 和 3TPA-CN[85]。其中 3TPA-CN 具有优良的热稳定性和光物理性能。利用 3TPA-CN 作为发光层和空穴传输层制备的非掺杂 OLED 发射深蓝光，其外量子效率和电流效率分别为 3.89% 和 5.21 cd/A。

2. AIE 绿光材料

赵祖金等报道了两种包含芴基团和咔唑基团的 TPE 聚合物。这两种聚合物具有良好的热稳定性和明显的 AIE 特性[86]。利用这两种聚合物制备的聚合物发光二极管（PLED）

发射绿光，其最大亮度和最大电流效率分别为 6500 cd/m^2 和 2.11cd/A。2015 年，刘飞等设计合成了两种新型的以三苯基乙烯为核心的 AIE 分子[87]。以这些分子作为发光层制备的非掺杂 OLED 发射绿光，最大亮度和最大电流效率分别为 24721.8cd/m^2 和 14.7cd/A。同年，廖良生等将具有良好的空穴传输性能的 SAF 基团与 AIE 特性的三苯基乙烯相结合制备了具有良好热稳定性的新型 AIE 分子 SAF-2-TriPE[88]，以其制备的绿光 OLED 显示出优异的电致发光性能，其最大电流效率和最大外量子效率分别为 10.5cd/A 和 4.22%。路萍等报道了两种芘 - 咪唑结构的同分异构体[89]，它们都有明显的 AIE 特性，所构建的 OLED 的电流效率高达 11.4cd/A。2016 年，唐本忠等以三苯基乙烯 - 咔唑为骨架，以间位取代的芳基硼为功能单元设计合成了新型 AIE 分子[90]，该材料具有优良的光物理性能和热稳定性，其非掺杂 OLED 发射绿光，最大亮度为 30210cd/m^2。之后，赵祖金等又将 TPE 基团与具有空穴传输性能的三苯胺基团和具有电子传输性能的苯并咪唑基团相结合[91]，合成了两种新型 AIE 绿光分子。这类分子在薄膜中的荧光量子产率可达 100%，作为发光层制备的非掺杂 OLED 的最大亮度和最大电流效率分别为 125300cd/m^2 和 16.8cd/A。

3. AIE 红光材料

2011 年，池振国等报道了两种红光 AIE 分子，它们具有良好的热稳定性，其非掺杂 OLED 的最大发光亮度为 13535 cd/m^2[92]。2015 年，唐本忠等设计合成了两种具有优异 AIE 特性和良好空穴传输性能的红光材料[93]。利用这两种材料同时作为发光层和空穴传输层制备的非掺杂 OLED 器件显示出良好的性能，最大外量子效率为 3.9%。2016 年，周箭等设计合成了一系列 TPE 衍生物[94]，其中一个衍生物发射 638 nm 的红光，而且具有明显的 AIE 特性和良好的电化学稳定性，利用该分子制备的非掺杂 OLED 的最大亮度为 16770 cd/m^2，最低开启电压为 2.9 伏。

4. AIE 材料制备白光 OLED

一般情况下，理想的白光发射可以通过三基色（红、绿和蓝）或两互补色（橙色和蓝色）以适当比例调配来实现。近几年，具有各种颜色发射 AIE 分子已有大量报道，它们在构建白光 OLED（WOLED）方面的研究也取得了巨大进步[95]。2015 年赵祖金、宁洪龙等利用深蓝光 AIE 材料（BTPEAn）制备了双色 WOLED[96]，器件性能稳定，效率达 70.2cd/A。同年，Somanathan 等制备了含芴基团的蓝光 AIE 聚合物（FCP），获得了高效的纯白色电致发光材料[97, 98]，利用其制备的 WPLED 最大亮度为 9332 cd/m^2。2016 年，他们又将黄色和蓝色 AIE 材料共聚，得到了白色发光聚合物，该聚合物的荧光量子产率高达 80.2%，其 WPLED 的最大亮度为 13455 cd/m^2。同年，Chen 等首次成功地利用蓝光 AIE 材料（CzPATPE）作为主体材料、橙色磷光材料作为客体材料制备了双色 WOLED，该器件的发光效率为 3.4cd/A[99]。

自 AIE 现象被报道后，AIE 材料在 OLED 领域的应用研究获得了广泛关注，所取得的大量研究成果不仅证明了 AIE 材料在 OLED 领域巨大的潜在应用价值，也大大推动了该领域的发展。对于 OLED 而言，进一步提高器件的效率、延长其寿命、降低生产成本以及简化操作工艺是其发展的目标。因此，为了加快 OLED 的发展，需要不断通过材料改进和器件优化等多种方法来得到高亮度、高效率和高稳定性的器件。在 AIE 材料的研究过程中，研究人员通过对分子结构与分子性能的分析，不断对分子结构加以改进和优化，如在 AIE 分子中引入一些具有空穴传输或电子传输性能的基团（例如三苯胺、咔唑、噁二唑、二米基硼、二苯基氧化物等），提高了 AIE 材料的载流子传输能力，进而提高了器件的性能。到目前为止，蓝光 AIE 材料的发展相对滞后，加强蓝光 AIE 材料的研究非常重要，是提高蓝光 OLED 和白光器件的效率及寿命的关键。

（五）AIE 在生物医学领域的应用研究

生物荧光技术只需光照即可进行检测，无需涉及化学反应以及放射性同位素，具有简便、快捷、低成本、低毒性等优点。生物荧光技术的发展取决于荧光材料的创新，但是传统探针分子也面临 ACQ 的问题，且光稳定性比较差，容易在强激发光照下被光漂白。此外，这些探针在生理缓冲液中也发光，很难区分探针分子在水相或背景中的发光和定位在目标生物大分子上的发光，从而使得它们在荧光检测中的灵敏度和信噪比较差，在生物标记与成像时，其成像对比度和分辨率也受到极大限制。AIE 材料能够在聚集态高效率发光，与传统荧光材料相比具有抗光漂白能力强、免洗、对比度高的优点，非常适合于“点亮”型检测、高分子浓度成像和长期示踪，是优异的生物荧光检测和成像材料。

1. AIE 在荧光生物传感中的应用

2012 年，唐本忠等设计了一种水溶性的含磺酸根的 TPE 衍生物，并用于研究和检测胰岛素的纤维化过程[100]。这一分子探针与胰岛素混合之后并不发光，直到胰岛素纤维化后，探针分子才发射荧光，并且还可清晰指示纤维化的各个阶段。同年，孙景志、唐本忠等设计了一种醛基功能化的噻咯分子[101]，可用于快速和高选择性地检测半胱氨酸。2012 年，Sanji 等设计了一个三聚氰酸功能化的 TPE 衍生物[102]，通过三聚氰酸与三聚氰胺之间的多价氢键的相互作用实现对三聚氰胺的高选择性的荧光检测。2013 年，唐本忠等开发了一种基于 AIE 特性的可穿透细胞膜的细胞内 pH 检测探针 TPE-Cy[103]。随着 pH 的降低，探针的颜色会由青色逐渐变为红色，通过检测红色与青色的比值即可得到 pH 值。2015 年，Smet 等开发了一种具有 AIE 特性的超支化聚合物，并基于该聚合物制备了 pH 检测探针，可实现对细胞内 pH 的探测[104]。

2013 年，童爱军等基于具有 AIE 特性的阳离子型水杨醛吖嗪衍生物，通过静电吸引作用实现了对肝素的检测[105]。肝素是一种具有很强负电性的生物分子，将其加入到水杨醛吖嗪衍生物的水溶液中后，肝素与水杨醛吖嗪衍生物由于静电吸引而发生相互作用，激活了 RIM 过程，从而使体系发射荧光。2014 年，张德清等设计了一种硼酸频哪醇酯功能化的 TPE，在葡萄糖氧化酶的辅助下可以实现对溶液中的葡萄糖的检测[106]。唐本忠等设计了一种磷酸化的TPE衍生物探针来检测碱性磷酸酶的活性[107]。在ALP存在的条件下，探针分子的磷酸基团将会被转变为羟基官能团并且引起不溶性酶残渣的聚集从而使探针分子发光。2014年李新明等设计了一种带有4个羧酸酯基团的TPE分子，通过形成超分子的微纤维聚集实现了对羧酸酯酶的检测[108]。刘斌等设计了一种简单的具有AIE特性的只标记单链DNA的探针实现了特异性的DNA杂化检测[109]。值得指出的是这是第一种利用AIE-DNA偶联物来检测特异性DNA序列的点亮型探针。2012年，刘斌等报道了首例利用AIE-多肽偶联物探针来特异性检测多种肿瘤细胞中非常重要的一种生物标志物蛋白——整合素[110]。这种探针分子通过“点击”反应将噻咯与环状RGD（cRGD）多肽偶联而得。cRGD对于整合素有非常高的亲和性，当探针与整合素相互作用时，体系发强荧光。2014年，张德清等设计了一种针对肿瘤标志物LAPTM4B检测的双响应点亮型探针分子[111]。通过在TPE上连接可以特异性靶向在肿瘤细胞上过表达的LAPTM4B的多肽实现对标志物的检测。这种探针在低pH条件下可以产生更高的荧光亮度，表明在检测肿瘤标志物的同时也可以用来检测酸性微环境。2016年，刘斌等报道了一种具有AIE特性的基于荧光共振能量转移（FRET）的分子探针，通过开启双荧光信号实现了对半胱天冬酶的实时和高灵敏度检测[112]。

2. AIE在荧光生物成像中的应用

荧光成像在生物医学领域具有举足轻重的地位，它具有简单、快速、直观等特点，在研究生物体各种生命活动中发挥了巨大的作用，帮助人们认识了从细胞到整个宏观活体的生命过程。一般情况下，荧光成像需要通过对目标物标记外源荧光染料来实现，这个过程中作为造影剂的荧光染料分子显得尤为重要。同理，AIE分子可以解决传统有机荧光分子的ACQ问题，对于生物成像来说具有重大意义，尤其是在活体荧光成像中需要荧光分子在某些病灶位置（例如肿瘤）富集时，可以充分利用AIE特性实现增强的荧光成像。

利用AIE分子在聚集态和固态下强发光的优势，对其进行功能化可以制备出各种具有特异性和高灵敏度的生物荧光成像探针。2014年，梁兴杰等设计了一种点亮型AIE探针，实现了对细胞膜的示踪成像[113]。这种探针是在TPE上连接了4个精氨酸单元和软脂酸。带正电的精氨酸基团负责靶向带有负电性的细胞膜，而软脂酸中的长链烷基则用来插入细胞膜。2013年，唐本忠等设计了一种可以靶向细胞线粒体的AIE荧光成像探针[114]。利用在TPE上修饰两个对线粒体具有特异性靶向的三苯基磷基团来实现对线粒体的示踪成像。随后他们又在TPE上修饰了阳离子型的吡啶单元，得到了一种可以靶向线粒体的发射橙光的荧光探针（TPE-Py）[115]。2015年，他们又在TPE-Py的基础上继续连接了异硫氰酸酯，得到了TPE-Py-NCS探针[116]，该探针可以通过化学连接的方式与线粒体蛋白作用，进而实现了对线粒体自噬过程的长时间监测。同年，蒋兴宇等和唐本忠等分别设计了两种含有吲哚基团的红光发射的AIE荧光探针，实现了对线粒体的靶向以及监测线粒体膜电位的变化情况[117, 118]。

在活体生物荧光成像中，拥有长波长激发和发射的荧光染料可以避免生物组织的自发背景荧光，更有利于实现高对比度和大深度的荧光成像。2012 年，刘斌等制备了一种基于 TPE-TPA-DCM 的深红色发光的 AIE 纳米颗粒[119]。这种 AIE 纳米颗粒通过尾静脉注射之后可以在小鼠肿瘤组织中富集，聚集在肿瘤组织中的纳米颗粒可以发射出明亮的深红色荧光，实现对肿瘤的定位成像。2013 年，刘斌等利用脂质体聚乙二醇包覆 TPE-TPA-FN 制备了发射红色荧光的 AIE 纳米颗粒，并利用 HIV-1 反式转录调节蛋白（Tat）对其表面进行修饰[120]。在活体 C6 细胞的示踪中，Tat-AIE 纳米颗粒可以实现最长 21 天的示踪成像，而商业化的 Qtracker665 量子点示踪时间不到 7 天。这项结果表明，有机 AIE 纳米颗粒在生物活体中的细胞示踪能力要优于商业化的量子点，在长期的细胞示踪研究中有广阔的发展前景。

2013 年，刘斌等首次利用 AIE 纳米颗粒实现了活体双光子荧光成像[121]。利用 DSPE-PEG 将具有 AIE 特性的荧光分子 BTPEBT 包覆形成 AIE 纳米颗粒，这种颗粒的双光子吸收截面要高于商业化的量子点 QD655 和 Evans Blue。它能实现小鼠大脑和耳朵以及骨髓中的血管网络的双光子荧光成像，得到了血管网络的三维结构分布图。钱骏等通过包覆 TPE-TPA-FN 制备了发射红色荧光的 AIE 纳米颗粒，并且发现在单光子和双光子激发模式下，提高 AIE 染料的包覆量可以得到更强的荧光发射[122]。通过双光子荧光成像可以揭示 AIE 纳米颗粒在血液循环中的动态变化。2015 年，钱骏等进一步研究了 AIE 荧光分子的高阶非线性光学特性，实现了基于 TPE-TPA-FN 的 AIE 纳米颗粒的细胞和活体脑血管的三光子荧光成像[123]。2015 年，他们利用基于 BODIPY-TPE 的 AIE 纳米颗粒在 1040 nm 飞秒激光的激发下实现了 700 μm 深度的小鼠脑血管网络三光子荧光成像[124]。2015 年，又利用微注射 TPE-TPP 的方法实现了对活体小鼠小神经胶质细胞的标记，并且得到了高信噪比的双光子荧光成像图片[125]。近些年来其他的具有较高的双光子吸收截面的 AIE 荧光染料分子也被开发出来，在细胞和活体双光子荧光成像中表现出优异的性能[126, 127]。

3. AIE 在药物运输与治疗中的应用

在药物的输运过程中，靶向给药以及可控释放对于提高病灶部位的药物量、减少副作用以及增强药效至关重要。一般的方法是修饰与病灶部位的细胞或者组织可以发生特异性相互作用的靶向配体来实现药物的靶向输运。另一方面，通过引入环境刺激或者相互作用响应性的机制，例如酸碱性、温度变化、酶活性变化等，实现药物的可控释放。2014 年，刘斌等提出了首例基于 AIE 的点亮型探针实现了对肿瘤治疗效果的原位监测[128]。接着他们又设计了两种不同的 AIE 探针来揭示抗肿瘤药物是通过何种方式、什么时间在细胞中被激活的[129, 130]。2015 年，他们[131]进一步设计了一种能够实现化疗与光动力治疗协同作用的 AIE 探针，以及一种利用特异性靶向细胞线粒体实现化疗和光动力协同治疗的 AIE 探针[132]。2015 年，唐本忠等也设计了一种靶向线粒体的探针 TPE-In，在与 Hela 细胞孵育之后可以对其产生一定的抗肿瘤效应[133]。

除了基于 AIE 的分子探针之外，充分利用 AIE 分子的聚集增强效应，以 AIE 纳米颗粒为依托，可以在药物运输与可控释放等应用领域发挥更大的作用。2014 年，梁兴杰等报道了一种基于 AIE 效应的可以自监测药物释放的纳米颗粒，通过荧光发射的强度、颜色等来监测纳米颗粒在细胞中的分布、药物的释放位点和时间等[134]。2015 年，他们进一步将 TPE 与 DOX 连接到一种对 pH 有响应的分子上，随后通过自组装形成纳米颗粒，实现了对 pH 响应的药物运输与释放[135]。基于类似的原理，王云兵等设计了利用 pH 响应的链接分子将 TPE 与亲水性的聚乙二醇高分子连接起来形成双亲性的偶联物，进而再将 DOX 包覆形成纳米颗粒，最终实现 pH 响应的药物释放[136]。

三、总结与展望

自 AIE 概念的提出开始，经过近 16 年的发展，AIE 这一我国科学家引领的研究在国际上已经产生了广泛的影响力。AIE 材料作为一类新兴发光材料，已被广泛应用于光电器件和生物检测与成像等各个前沿领域和交叉学科，RIM 机理也已被国内外同行广泛接受和并用于解释各种分子聚集发光现象。但 AIE 研究方兴未艾，无论是 AIE 理论体系的发展和完善，还是 AIE 新材料的开发以及应用等方面都有待进一步的深入和加强。在 AIE 理论方面，目前尚未有完善的分子聚集发光理论能够对 AIE 材料的性能进行准确的定量预测和理论模拟。要关注和理解一些未知的新型聚集态发光现象和过程，建立普适的光物理理论，例如非芳香体系和簇发光工作机制以及纯有机室温磷光内在规律的掌握等将会产生新的学科增长点，为化学和材料科学及前沿技术带来新的突破，极大地促进科技和经济的发展。在 AIE 应用方面，还需要进一步开发功能性 AIE 新材料和新技术，在新体系拓展、性能提升和功能集成等方面取得突破，在能源、健康和环境等领域真正实现产业化应用，解决当前社会的重大挑战，为推动国家的原始创新能力和社会经济的发展做出贡献。

参考文献

[1] Luo J，Xie Z，Lam J W，et al. Aggregation-induced emission of 1-methyl-1，2，3，4，5- pentaphenylsilole[J]. Chemical Communications，2001，(18)：1740-1741.

[2] Tang B Z，Zhan X，Yu G，et al. Efficient blue emission from siloles[J]. Journal of Materials Chemistry，2001，11 (12)：2974-2978.

[3] 中国科学院《2015 研究前沿》. ip-science.thomsonreuters.com.cn/media/2015_research_1030.pdf.

[4] Mei J，Leung N L，Kwok R T，et al. Aggregation-induced emission: together we shine，united we soar! [J]. Chemical Reviews，2015，115 (21)：11718-11940.

[5] Lim X. The nanolight revolution is coming[J]. Nature，2016，531 (7592)：26.

[6] Chen B, Feng G, He B, et al. Silole-based red fluorescent organic dots for bright two-photon fluorescence in vitro cell and in vivo blood vessel imaging[J]. Small, 2016, 12 (6): 782-792.

[7] Gu X, Yao J, Zhang G, et al. New electron-donor/acceptor-substituted tetraphenylethylenes: aggregation-induced emission with tunable emission color and optical-waveguide behavior[J]. Chemistry - An Asian Journal, 2013, 8 (10): 2362-2369.

[8] Yang C, Trinh Q T, Wang X, et al. Crystallization-induced red emission of a facilely synthesized biodegradable indigo derivative[J]. Chemical Communications, 2015, 51 (16): 3375-3378.

[9] Leung N L C, Xie N, Yuan W, et al. Restriction of intramolecular motions: the general mechanism behind aggregation-induced emission[J]. Chemistry-A European Journal, 2014, 20 (47): 15349-15353.

[10] Nishiuchi T, Tanaka K, Kuwatani Y, et al. Solvent-induced crystalline-state emission and multichromism of a bent π-surface system composed of dibenzocyclooctatetraene units[J]. Chemistry-A European Journal, 2013, 19 (13): 4110-4116.

[11] Yuan C X, Saito S, Camacho C, et al. Hybridization of a flexible cyclooctatetraene core and rigid aceneimide wings for multiluminescent flapping π systems[J]. Chemistry-A European Journal, 2014, 20 (8): 2193-2200.

[12] Yuan C X, Tao X T, Ren Y, et al. Synthesis, structure, and aggregation-induced emission of a novel lambda (Lambda)-shaped pyridinium salt based on Troger's base[J]. Journal of Physical Chemistry C, 2007, 111 (34): 12811-12816.

[13] Yao L, Zhang S T, Wang R, et al. Highly efficient near-infrared organic light-emitting diode based on a butterfly-shaped donor-acceptor chromophore with strong solid-state fluorescence and a large proportion of radiative excitons[J]. Angewandte Chemie International Edition, 2014, 53 (8): 2119-2123.

[14] Sharma K, Kaur S, Bhalla V, et al. Pentacenequinone derivatives for preparation of gold nanoparticles: facile synthesis and catalytic application[J]. Journal of Materials Chemistry A, 2014, 2 (22): 8369-8375.

[15] Banal J L, White J M, Ghiggino K P, et al. Concentrating aggregation-induced fluorescence in planar waveguides: a proof-of-principle[J]. Scientific Reports, 2014, 4: 4635.

[16] Liu J, Meng Q, Zhang X T, et al. Aggregation-induced emission enhancement based on 11, 11, 12, 12, -tetracyano-9, 10-anthraquinodimethane[J]. Chemical Communications, 2013, 49 (12): 1199-1201.

[17] Yuan W Z, Shen X Y, Zhao H, et al. Crystallization-induced phosphorescence of pure organic luminogens at room temperature[J]. Journal of Physical Chemistry C, 2010, 114 (13): 6090-6099.

[18] Zhao W, He Z, Lam J W Y, et al. Rational molecular design for achieving persistent and efficient pure organic room-temperature phosphorescence[J]. Chem, 2016, 1 (4): 592-602.

[19] An Z, Zheng C, Tao Y, et al. Stabilizing triplet excited states for ultralong organic phosphorescence[J]. Nature Materials, 2015, 14 (7): 685-690.

[20] Li C, Tang X, Zhang L, et al. Reversible luminescence switching of an organic solid: controllable on - off persistent room temperature phosphorescence and stimulated multiple fluorescence conversion[J]. Advanced Optical Materials, 2015, 3 (9): 1184-1190.

[21] Xue P, Sun J, Chen P, et al. Luminescence switching of a persistent room-temperature phosphorescent pure organic molecule in response to external stimuli[J]. Chemical Communications, 2015, 51 (52): 10381-10384.

[22] Gong Y, Chen G, Peng Q, et al. Achieving persistent room temperature phosphorescence and remarkable mechanochromism from pure organic luminogens[J]. Advanced Materials, 2015, 27 (40): 6195-201.

[23] Xie Y, Ge Y, Peng Q, et al. How the molecular packing affects the room temperature phosphorescence in pure organic compounds: ingenious molecular design, detailed crystal analysis, and rational theoretical calculations[J]. Advanced Materials, 2017, 29 (17): 1606829-n/a.

[24] Yang Z, Mao Z, Zhang X, et al. Intermolecular electronic coupling of organic units for efficient persistent room-

temperature phosphorescence [J] . Angewandte Chemie International Edition, 2016, 55 (6), 2181–2185.

[25] Mei J, Hong Y N, Tang B Z, et al. Aggregation-induced emission: the whole is more brilliant than the parts [J] . Advanced Materials, 2014, 26: 5429–5479.

[26] Huang W, Sun L, Tian H, et al. Colour-tunable fluorescence of single molecules based on the vibration induced emission of phenazine [J] . Chemical Communication, 2015, 51 (21): 4462–4464.

[27] Zhang Z Y, Tian H, Chou P T, et al. Excited-state conformational/electronic responses of saddle shaped N, N'isubstituted-dihydrodibenzo [a, c] phenazines: widetuning emission from red to deep blue and white light combination [J] . Journal of the American Chemical Society, 2015, 137 (26): 8509–8520.

[28] Chen W, Tian H, Chou P T, et al. Snapshotting the excited-state planarization of chemically locked N, N′ -disubstituted dihydrodibenzo [a, c] phenazines [J] . Journal of the American Chemical Society, 2017, 139 (4): 1636–1644.

[29] Hu R R, Lager E, Tang B Z, et al. Twisted intramolecular charge transfer and aggregation-induced emission of BODIPY derivatives [J] . The Journal of Physical Chemistry C, 2009, 113 (36): 15845–15853.

[30] Gao B R, Wang H Y, Sun H B, et al. Time-resolved fluorescence study of aggregation-induced emission enhancement by restriction of intramolecular charge transfer state [J] . The Journal of Physical Chemistry B, 2010, 114 (1) :128–134.

[31] Xie Z Q, Ma Y G, Liu S Y, et al. Supramolecular interactions induced fluorescence in crystal: anomalous emission of 2, 5-diphenyl-1, 4-distyrylbenzene with all cis double bonds [J] . Chemistry of Materials, 2005, 17 (6): 1287–1289.

[32] Fang X F, Liu Y F, Wu C F, et al. Facile synthesis, macroscopic separation, e/z isomerization, and distinct AIE properties of pure stereoisomers of an oxetane-substituted tetraphenylethene luminogen [J] . Chemistry of Materials, 2016, 28 (18): 6628–6636.

[33] Ma S Q, Zhang J B, Tian W J, et al. Organic fluorescent molecule with high solid state luminescent efficiency and protonation stimuli-response [J] . Chinese Journal of Chemistry, 2013, 31 (11): 1418–1422.

[34] Jia W B, Tao X T, Yang J X, et al. Synthesis of two novel indolo [3, 2-b] carbazole derivatives with aggregation-enhanced emission property [J] . Journal of Materials Chemistry C, 2013, 1 (42): 7092–7101.

[35] Yang G Q, Li S Y, Wang S Q, et al. Emission properties and aggregation-induced emission enhancement of excited-state intramolecular proton-transfer compounds [J] . Comptes Rendus Chimie, 2011, 14 (9): 789–798.

[36] Peng Q, Yi Y P, Shuai Z G, et al. Toward quantitative prediction of molecular fluorescence quantum efficiency: role of duschinsky rotation [J] . Journal of the American Chemical Society, 2007, 129 (30): 9333–9339.

[37] Shuai Z G, Peng Q. Organic light-emitting diodes: theoretical understanding of highly efficient materials and development of computational methodology [J] . National Science Review, 2017, 4 (2): 224–239.

[38] Shuai Z G, Peng Q. Excited states structure and processes: Understanding organic light-emitting diodes at the molecular level [J] . Physics Reports, 2014, 537 (4): 123–156.

[39] Wu Q Y, Peng Q, Shuai Z G, et al. Aggregation induced blue-shifted emission-the molecular picture from a QM/MM study [J] . Physical Chemistry Chemical Physics, 2014, 16 (12): 5545–5552.

[40] Zhang T, Peng Q, Shuai Z G, et al. Spectroscopic signature of the aggregation-induced emission phenomena caused by restricted nonradiative decay: a theoretical proposal [J] . The Journal of Physical Chemistry C, 2015, 119 (9): 5040–5047.

[41] Zhang T, Peng Q, Shuai Z G, et al. Using the isotope effect to probe an aggregation induced emission mechanism: theoretical prediction and experimental validation [J] . Chemical Science, 2016, 7 (8): 5573–5580.

[42] Zheng X Y, Peng Q, Shuai Z G, et al. Unraveling the aggregation effect on amorphous phase AIE luminogens: a computational study [J] . Nanoscale, 2016, 8 (33): 15173–15180.

[43] Li W Q, Peng Q, Shuai Z G, et al. Effect of intermolecular excited-state interaction on vibrationally resolved optical spectra in organic molecular aggregates[J]. Acta Chimica Sinica, 2016, 74(11): 902-909.

[44] Sun G X, Zhao Y, Liang W Z. Aggregation-induced emission mechanism of dimethoxy-tetraphenylethylene in water solution: molecular dynamics and QM/MM investigations[J]. Journal of Chemical Theory and Computation, 2015, 11(5): 2257-2267.

[45] Peng X L, Li Q S, Blancafort L, et al. Restricted access to a conical intersection to explain aggregation induced emission in dimethyl tetraphenylsilole[J]. Journal of Materials Chemistry C, 2016, 4(14): 2802.

[46] Wang B, Wang W L, Liu F Y, et al. Exploring the mechanism of fluorescence quenching and aggregation-induced emission of a phenylethylene derivative by QM (CASSCF and TDDFT) and ONIOM (QM:MM) calculations[J]. The Journal of Physical Chemistry C, 2016, 120(38): 21850-21857.

[47] Gao X, Wang D, Shuai Z G, et al. Theoretical insight into the aggregation induced emission phenomena of diphenyldibenzofulvene: a nonadiabatic molecular dynamics study[J]. Physical Chemistry Chemical Physics, 2012, 14(41): 14207-14216.

[48] Ma H L, Peng Q, Shuai Z G, et al. Electrostatic interaction-induced room-temperature phosphorescence in pure organic molecules from QM/MM calculations[J]. The Journal of Physical Chemistry Letters, 2016, 7(15): 2893-2898.

[49] Yang L, Wang X J, Jiang J, et al. Aggregation-induced intersystem crossing: a novel strategy for efficient molecular phosphorescence[J]. Nanoscale, 2016, 8(40): 17422-17426.

[50] Sagara Y, Kato T. Mechanically induced luminescence changes in molecular assemblies[J]. Nature Chemistry, 2009, 1(8): 605-610.

[51] Sagara Y, Yamane S, Mitani M, et al. Mechanoresponsive luminescent molecular assemblies: an emerging class of materials[J]. Advanced Materials, 2016, 28(6): 1073-1095.

[52] Fan X, Sun J, Wang F, et al. Photoluminescence and electroluminescence of hexaphenylsilole are enhanced by pressurization in the solid state[J]. Chemical Communications, 2008(26): 2989-2991.

[53] Li H, Zhang X, Chi Z, et al. New thermally stable piezofluorochromic aggregation-induced emission compounds[J]. Organic Letters, 2011, 13(4): 556-559.

[54] Zhang X, Chi Z, Li H, et al. Piezofluorochromism of an aggregation-induced emission compound derived from tetraphenylethylene[J]. Chemistry-An Asian Journal, 2011, 6(3): 808-811.

[55] Chi Z, Zhang X, Xu B, et al. Recent advances in organic mechanofluorochromic materials[J]. Chemical Society Reviews, 2012, 41(10): 3878-3896.

[56] Zhang Q W, Li D, Li X, et al. Multicolor photoluminescence including white-light emission by a single host-guest complex[J]. Journal of the American Chemical Society, 2016, 138(41): 13541-13550.

[57] Zhang G, Palmer G M, Dewhirst M W, et al. A dual-emissive-materials design concept enables tumour hypoxia imaging[J]. Nature Materials, 2009, 8(9): 747-751.

[58] Tanaka H, Shizu K, Nakanotani H, et al. Dual intramolecular charge-transfer fluorescence derived from a phenothiazine-triphenyltriazine derivative[J]. The Journal of Physical Chemistry C, 2014, 118(29): 15985-15994.

[59] 陈薇，苏建华，田禾．“振动诱导发光”(VIE)：二氢吩嗪类衍生物的双荧光发射机理[J]．中国科学：化学，2016，(4)：325-332.

[60] Xu S, Liu T, Mu Y, et al. An organic molecule with asymmetric structure exhibiting aggregation-induced emission, delayed fluorescence, and mechanoluminescence[J]. Angewandte Chemie International Edition, 2015, 54(3): 874-878.

[61] Cheng X, Zhang Z, Zhang H, et al. CEE-active red/near-infrared fluorophores with triple-channel solid-state “ON/

OFF" fluorescence switching [J]. Journal of Materials Chemistry C, 2014, 2 (35): 7385–7391.

[62] Xu J, Chi Z. Mechanochromic Fluorescent Materials: Phenomena, Materials and Applications [M]. Royal Society of Chemistry, 2014.

[63] Ning T, Liu L, Jia D, et al. Aggregation–induced emission, photochromism and self–assembly of pyrazolone phenlysemicarbazones [J]. Journal of Photochemistry and Photobiology A: Chemistry, 2014, 291 (1): 48–53.

[64] Qi Q, Qian J, Ma S, et al. Reversible multistimuli–response fluorescent switch based on tetraphenylethene–spiropyran molecules [J]. Chemistry–A European Journal, 2015, 21 (3): 1149–1155.

[65] Peng L, Zheng Y, Wang X, et al. Photoactivatable aggregation–induced emission fluorophores with multiple–color fluorescence and wavelength–selective activation [J]. Chemistry–A European Journal, 2015, 21 (11): 4326–4332.

[66] He Z, Shan L, Mei J, et al. Aggregation–induced emission and aggregation–promoted photochromism of bis (diphenylmethylene) dihydroacenes [J]. Chemical Science, 2015, 6 (6): 3538–3543.

[67] Ou D, Yu T, Yang Z, et al. Combined aggregation induced emission (AIE), photochromism and photoresponsive wettability in simple dichloro–substituted triphenylethylene derivatives [J]. Chemical Science, 2016, 7 (8): 5302–5306.

[68] Zhu L, Wang R, Tan L, et al. Aggregation–induced emission and aggregation–promoted photo–oxidation in thiophene–substituted tetraphenylethylene derivative [J]. Chemistry–An Asian Journal, 2016, 11 (20): 2932–2937.

[69] Dou C, Han L, Zhao S, et al. Multi–stimuli–responsive fluorescence switching of a donor– acceptor π–conjugated compound [J]. The Journal of Physical Chemistry Letters, 2011, 2 (6): 666–670.

[70] Song P, Chen X, Xiang Y, et al. A ratiometric fluorescent pH probe based on aggregation–induced emission enhancement and its application in live–cell imaging [J]. Journal of Materials Chemistry, 2011, 21 (35): 13470–13475.

[71] Chen S, Liu J, Liu Y, et al. An AIE–active hemicyanine fluorogen with stimuli–responsive red/blue emission: extending the pH sensing range by "switch+knob" effect [J]. Chemical Science, 2012, 3 (6): 1804–1809.

[72] Yang Z, Qin W, Lam J W Y, et al. Fluorescent pH sensor constructed from a heteroatom–containing luminogen with tunable AIE and ICT characteristics [J]. Chemical Science, 2013, 4 (9): 3725–3730.

[73] Zhang J, Chen J, Xu B, et al. Remarkable fluorescence change based on the protonation–deprotonation control in organic crystals [J]. Chemical Communications, 2013, 49 (37): 3878–3880.

[74] Dong Y, Zhang J, Tan X, et al. Multi–stimuli responsive fluorescence switching: the reversible piezochromism and protonation effect of a divinylanthracene derivative [J]. Journal of Materials Chemistry C, 2013, 1 (45): 7554–7559.

[75] Huang J, Sun N, Dong Y, et al. Similar or totally different: the control of conjugation degree through minor structural modifications, and deep–blue aggregation–induced emission luminogens for Non–doped OLEDs [J]. Advanced Functional Materials, 2013, 23 (18): 2329–2337.

[76] Zhang Y, He B, Luo W, et al. Aggregation–enhanced emission and through–space conjugation of tetraarylethanes and folded tetraarylethenes [J]. Journal of Materials Chemistry C, 2016, 4 (39): 9316–9324.

[77] Chan C Y K, Lam J W Y, Zhao Z, et al. Aggregation–induced emission, mechanochromism and blue electroluminescence of carbazole and triphenylamine–substituted ethenes [J]. Journal of Materials Chemistry C, 2014, 2 (21): 4320–4327.

[78] Shi H P, Zhang W X, Dong X Q, et al. A novel carbazole derivative containing dimesitylboron units: Synthesis, photophysical, aggregation induced emission and electroluminescent properties [J]. Dyes and Pigments, 2014, 104, 34–40.

[79] Qin W, Yang Z, Jiang Y, et al. Construction of efficient deep blue aggregation-induced emission luminogen from triphenylethene for nondoped organic light-emitting diodes [J]. Chemistry of Materials, 2015, 27 (11): 3892-3901.

[80] Cui Y, Yin Y M, Cao H T, et al. Efficient piezochromic luminescence from tetraphenylethene functionalized pyridine-azole derivatives exhibiting aggregation-induced emission [J]. Dyes and Pigments, 2015, 119: 62-69.

[81] Chen G, Li W, Zhou T, et al. Conjugation-induced rigidity in twisting molecules: filling the gap between aggregation-caused quenching and aggregation-induced emission [J]. Advanced Materials, 2015, 27 (30): 4496-4501.

[82] Tang F, Peng J, Liu R, et al. A sky-blue fluorescent small molecule for non-doped OLED using solution-processing [J]. RSC Advances, 2015, 5 (87): 71419-71424.

[83] Chen M, Nie H, Song B, et al. Triphenylamine-functionalized tetraphenylpyrazine: facile preparation and multifaceted functionalities [J]. Journal of Materials Chemistry C, 2016, 4 (14): 2901-2908.

[84] Li C, Wei J, Han J, et al. Efficient deep-blue OLEDs based on phenanthro [9, 10-d] imidazole-containing emitters with AIE and bipolar transporting properties [J]. Journal of Materials Chemistry C, 2016, 4 (42): 10120-10129.

[85] Zhan X, Wu Z, Lin Y, et al. Benzene-cored AIEgens for deep-blue OLEDs: high performance without hole-transporting layers, and unexpected excellent host for orange emission as a side-effect [J]. Chemical Science, 2016, 7 (7): 4355-4363.

[86] He B R, Ye S H, Guo Y J, et al. Aggregation-enhanced emission and efficient electroluminescence of conjugated polymers containing tetraphenylethene units [J]. Science China Chemistry, 2013, 56 (9): 1221-1227.

[87] Jin S, Tian Y, Liu F, et al. Intense green-light emission from 9, 10-bis (4- (1, 2, 2-triphenylvinyl) styryl) anthracene emitting electroluminescent devices [J]. Journal of Materials Chemistry C, 2015, 3 (31): 8066-8073.

[88] Xue M M, Xie Y M, Cui L S, et al. The control of conjugation lengths and steric hindrance to modulate aggregation-induced emission with high electroluminescence properties and interesting optical properties [J]. Chemistry-A European Journal, 2016, 22 (3): 916-924.

[89] Liu Y, Shan T, Yao L, et al. Isomers of pyrene-imidazole compounds: synthesis and configuration effect on optical properties [J]. Organic Letters, 2015, 17 (24): 6138-6141.

[90] Shi H, Xin D, Gu X, et al. The synthesis of novel AIE emitters with the triphenylethene-carbazole skeleton and para-/meta-substituted arylboron groups and their application in efficient non-doped OLEDs [J]. Journal of Materials Chemistry C, 2016, 4 (6): 1228-1237.

[91] Lin G, Peng H, Chen L, et al. Improving electron mobility of tetraphenylethene-based AIEgens to fabricate nondoped organic light-emitting diodes with remarkably high luminance and efficiency [J]. ACS Applied Materials & Interfaces, 2016, 8 (26): 16799-16808.

[92] Li H, Chi Z, Zhang X, et al. New thermally stable aggregation-induced emission enhancement compounds for non-doped red organic light-emitting diodes [J]. Chemical Communications, 2011, 47 (40): 11273-11275.

[93] Qin W, Lam J W Y, Yang Z, et al. Red emissive AIE luminogens with high hole-transporting properties for efficient non-doped OLEDs [J]. Chemical Communications, 2015, 51 (34): 7321-7324.

[94] Zhou J, He B, Xiang J, et al. Tuning the AIE Activities and emission wavelengths of tetraphenylethene-containing luminogens [J]. ChemistrySelect, 2016, 1 (4): 812-818.

[95] Kamtekar K T, Monkman A P, Bryce M R. Recent advances in white organic Light-Emitting materials and devices (WOLEDs) [J]. Advanced Materials, 2010, 22 (5): 572-582.

[96] Liu B, Nie H, Zhou X, et al. Manipulation of charge and exciton distribution based on blue aggregation-induced

emission fluorophors: a novel concept to achieve high-performance hybrid white organic light-emitting diodes [J] . Advanced Functional Materials, 2016, 26 (5) : 776–783.

[97] Ravindran E, Ananthakrishnan S J, Varathan E, et al. White light emitting single polymer from aggregation enhanced emission: a strategy through supramolecular assembly [J]. Journal of Materials Chemistry C, 2015, 3(17): 4359–4371.

[98] Ravindran E, Varathan E, Subramanian V, et al. Self-assembly of a white-light emitting polymer with aggregation induced emission enhancement using simplified derivatives of tetraphenylethylene [J] . Journal of Materials Chemistry C, 2016, 4 (34) : 8027–8040.

[99] Lee Y T, Chang Y T, Chen C T, et al. The first aggregation-induced emission fluorophore as a solution processed host material in hybrid white organic light-emitting diodes [J] . Journal of Materials Chemistry C, 2016, 4 (29) : 7020–7025.

[100] Hong Y, Meng L, Chen S, et al. Monitoring and inhibition of insulin fibrillation by a small organic fluorogen with aggregation-induced emission characteristics [J] . Journal of the American Chemical Society, 2012, 134 (3) : 1680–1689.

[101] Mei J, Tong J, Wang J, et al. Discriminative fluorescence detection of cysteine, homocysteine and glutathione via reaction-dependent aggregation of fluorophore-analyte adducts [J] . Journal of Materials Chemistry, 2012, 22 (33) : 17063–17070.

[102] Sanji T, Nakamura M, Kawamata S, et al. Fluorescence "turn-on" detection of melamine with aggregation-induced-emission-active tetraphenylethene [J] . Chemistry - A European Journal, 2012, 18 (48) : 15254–15257.

[103] Chen S, Hong Y, Liu Y, et al. Full-range intracellular pH sensing by an aggregation-induced emission-active two-channel ratiometric fluorogen [J] . Journal of the American Chemical Society, 2013, 135 (13) : 4926–4929.

[104] Bao Y, De Keersmaecker H, Corneillie S, et al. Tunable ratiometric fluorescence sensing of intracellular pH by aggregation-induced emission-active hyperbranched polymer nanoparticles [J] . Chemistry of Materials, 2015, 27 (9) : 3450–3455.

[105] Liu H, Song P, Wei R, et al. A facile, sensitive and selective fluorescent probe for heparin based on aggregation-induced emission [J] . Talanta, 2014, 118: 348–352.

[106] Hu F, Huang Y, Zhang G, et al. A highly selective fluorescence turn-on detection of hydrogen peroxide and D-glucose based on the aggregation/deaggregation of a modified tetraphenylethylene [J] . Tetrahedron Letters, 2014, 55 (8) : 1471–1474.

[107] Song Z, Hong Y, Kwok R. T, et al. A dual-mode fluorescence "turn-on" biosensor based on an aggregation-induced emission luminogen [J] . Journal of Materials Chemistry B, 2014, 2 (12) : 1717–1723.

[108] Wang X, Liu H, Li J, et al. A fluorogenic probe with aggregation-induced emission characteristics for carboxylesterase assay through formation of supramolecular microfibers [J] . Chemistry - An Asian Journal, 2014, 9 (3) : 784–789.

[109] Li Y, Kwok R. T, Tang B. Z, et al. Specific nucleic acid detection based on fluorescent light-up probe from fluorogens with aggregation-induced emission characteristics [J] . RSC Advances, 2013, 3 (26) : 10135–10138.

[110] Shi H, Liu J, Geng J, et al. Specific detection of integrin αvβ3 by light-up bioprobe with aggregation-induced emission characteristics [J] . Journal of the American Chemical Society, 2012, 134 (23) : 9569–9572.

[111] Huang Y, Hu F, Zhao R, et al. Tetraphenylethylene conjugated with a specific peptide as a fluorescence turn-on bioprobe for the highly specific detection and tracing of tumor markers in live cancer cells [J] . Chemistry - A European Journal, 2014, 20 (1) : 158–164.

[112] Yuan Y, Zhang R, Cheng X, et al. A FRET probe with AIEgen as the energy quencher: dual signal turn-on for

self-validated caspase detection [J] . Chemical Science，2016，7 (7) : 4245-4250.

[113] Zhang C，Jin S，Yang K，et al. Cell membrane tracker based on restriction of intramolecular rotation [J] . ACS Applied Materials & Interfaces，2014，6 (12) : 8971-8975.

[114] Leung C W T，Hong Y，Chen S，et al. A photostable AIE luminogen for specific mitochondrial imaging and tracking [J]. Journal of the American Chemical Society，2012，135 (1) : 62-65.

[115] Zhao N，Li M，Yan Y，et al. A tetraphenylethene-substituted pyridinium salt with multiple functionalities: synthesis，stimuli-responsive emission，optical waveguide and specific mitochondrion imaging [J] . Journal of Materials Chemistry C，2013，1 (31) : 4640-4646.

[116] Ryan T，Chris Y，Jacky W，et al. Real-time monitoring of the mitophagy process by a photostable fluorescent mitochondrion-specific bioprobe with AIE characteristics [J] . Chemical Communications，2015，51 (43) : 9022-9025.

[117] Zhang L，Liu W，Huang X，et al. Old is new again: a chemical probe for targeting mitochondria and monitoring mitochondrial membrane potential in cells [J] . Analyst，2015，140 (17) : 5849-5854.

[118] Zhao N，Chen S，Hong Y，et al. A red emitting mitochondria-targeted AIE probe as an indicator for membrane potential and mouse sperm activity [J] . Chemical Communications，2015，51 (71) : 13599-13602.

[119] Geng J，Li K，Ding D，et al. Lipid-PEG-folate encapsulated nanoparticles with aggregation induced emission characteristics: cellular uptake mechanism and two-photon fluorescence imaging [J] . Small，2012，8 (23) : 3655-3663.

[120] Jin G，Mao D，Cai P，et al. Conjugated polymer nanodots as ultrastable long-term trackers to understand mesenchymal stem cell therapy in skin regeneration [J] . Advanced Functional Materials，2015，25 (27) : 4263-4273.

[121] Ding D，Goh C C，Feng G，et al. Ultrabright organic dots with aggregation-induced emission characteristics for real-time two - photon intravital vasculature imaging [J] . Advanced Materials，2013，25 (42) : 6083-6088.

[122] Wang D，Qian J，Qin W，et al. Biocompatible and photostable AIE dots with red emission for in vivo two-photon bioimaging [J] . Scientific Reports，2014，4，4279.

[123] Qian J，Zhu Z，Qin A，et al. High-order non-linear optical effects in organic luminogens with aggregation-induced emission [J] . Advanced Materials，2015，27 (14) : 2332-2339.

[124] Wang Y，Hu R，Xi W，et al. Red emissive AIE nanodots with high two-photon absorption efficiency at 1040 nm for deep-tissue in vivo imaging [J] . Biomedical Optics Express，2015，6 (10) : 3783-3794.

[125] Qian J，Zhu Z，Leung C W T，et al. Long-term two-photon neuroimaging with a photostable AIE luminogen [J] . Biomedical Optics Express，2015，6 (4) : 1477-1486.

[126] Xiang J，Cai X，Lou X，et al. Biocompatible green and red fluorescent organic dots with remarkably large two-photon action cross sections for targeted cellular imaging and real-time intravital blood vascular visualization [J] . ACS Applied Materials & Interfaces，2015，7 (27) : 14965-14974.

[127] Zhao Z，Chen B，Geng J，et al. Red emissive biocompatible nanoparticles from tetraphenylethene-decorated BODIPY luminogens for two-photon excited fluorescence cellular imaging and mouse brain blood vascular visualization [J] . Particle & Particle Systems Characterization，2014，31 (4) : 481-491.

[128] Yuan Y，Kwok R T，Tang B Z，et al. Targeted theranostic platinum (IV) prodrug with a built-in aggregation-induced emission light-up apoptosis sensor for noninvasive early evaluation of its therapeutic responses in situ [J] . Journal of the American Chemical Society，2014，136 (6) : 2546-2554.

[129] Yuan Y，Chen Y，Tang B Z，et al. A targeted theranostic platinum (IV) prodrug containing a luminogen with aggregation-induced emission (AIE) characteristics for in situ monitoring of drug activation [J] . Chemical Communications，2014，50 (29) : 3868-3870.

[130] Yuan Y, Kwok R T, Zhang R, et al. Targeted theranostic prodrugs based on an aggregation–induced emission (AIE) luminogen for real–time dual–drug tracking [J] . Chemical Communications, 2014, 50 (78) : 11465–11468.

[131] Yuan Y, Zhang C J, Liu B. A platinum prodrug conjugated with a photosensitizer with aggregation–induced emission (AIE) characteristics for drug activation monitoring and combinatorial photodynamic – chemotherapy against cisplatin resistant cancer cells [J] . Chemical Communications, 2015, 51 (41) : 8626–8629.

[132] Zhang C J, Hu Q, Feng G, et al. Image–guided combination chemotherapy and photodynamic therapy using a mitochondria–targeted molecular probe with aggregation–induced emission characteristics [J] . Chemical Science, 2015, 6 (8) : 4580–4586.

[133] Zhao N, Chen S, Hong Y, et al. A red emitting mitochondria–targeted AIE probe as an indicator for membrane potential and mouse sperm activity [J] . Chemical Communications, 2015, 51 (71) : 13599–13602.

[134] Xue X, Zhao Y, Dai L, et al. Spatiotemporal Drug Release Visualized through a Drug Delivery System with Tunable Aggregation–Induced Emission [J] . Advanced Materials, 2014, 26 (5) : 712–717.

[135] Zhang C, Jin S, Li S, et al. Imaging intracellular anticancer drug delivery by self–assembly micelles with aggregation–induced emission (AIE micelles) [J] . ACS Applied Materials & Interfaces, 2014, 6 (7) : 5212–5220.

[136] Wang H, Liu G, Gao H, et al. A pH–responsive drug delivery system with an aggregation–induced emission feature for cell imaging and intracellular drug delivery [J] . Polymer Chemistry, 2015, 6 (26) : 4715–4718.

撰稿人：唐本忠　董永强　彭　谦　池振国　施和平　钱　骏　赵祖金　秦安军

单细胞分析化学研究进展

一、引言

细胞是生物体形态结构和生命活动的基本单元。理解生命的奥秘、探索疾病的机制与治疗手段，都要求在单细胞及亚细胞水平上对生命活动与生理现象进行实时、原位与动态的探测[1]。受细胞外微环境与细胞内物质组成差异的共同作用，细胞在新陈代谢、呼吸作用、信息传递、物质运输等生命活动中表现出较大的差异[2]。传统的分子生物学研究手段仅仅能够获得针对群体细胞分析的稳态平均结果，难以在细胞及亚细胞层次上提供生物差异性信息。随着现代生命科学的研究逐渐进入单细胞时代，单细胞分析化学应运而生[3]。单细胞分析是利用高灵敏、高特异性与高通量的分析化学技术，在特定时间空间尺度下对细胞的物质组成及结构的多样性和差异性进行解析，力求精确反映细胞、亚细胞与单分子水平的生理状态与生命过程[4]。随着分析化学、生命科学和生物医学等学科的不断交融与蓬勃发展，单细胞分析化学已经成为一个横跨多学科的前沿科学领域，在现代分析化学研究中占据着举足轻重的地位[5]。

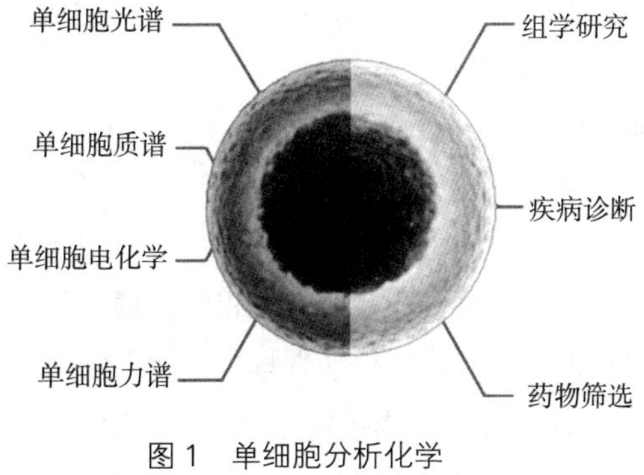

图 1　单细胞分析化学

自 20 世纪 90 年代以来，随着对细胞操控、采样与超高分辨荧光成像等技术的出现，国际上的单细胞分析研究呈现快速发展趋势。其中一个代表为 Betzig、Hell 与 Moerner 因发展 PALM 与 STED 荧光显微镜技术的突出贡献而获得的 2014 年诺贝尔化学奖[6]。这些超分辨技术具有纳米级分辨率，能够在细胞与亚细胞尺度上对荧光标记的生物分子进行时空分布的精确解析，成为单细胞层次上的结构生物学和生物物理学的重要研究手段。同时，多种技术的联用也推动着单细胞分析技术朝着更高分辨率与更高检测通量的方向发展[7]。基于微流控技术与单细胞成像技术的单细胞组学分析与高通量测序，已成为分子生物学研究的强有力工具[8]。2016 年 11 月 *Nature Biotechnology* 杂志以“Focus on Single-cell technology”为题，对单细胞分析技术的最新进展与挑战进行了重点介绍[9]。

近年来，国家自然科学基金委和科技部对单细胞分析方面的研究予以大力支持和积极资助，我国的单细胞分析研究迈入一个高速发展时期。以单细胞为研究对象的诸多先进精密仪器与联用技术的不断涌现，极大地推动了我国单细胞分析研究实力的提升，使得国内外在这一领域研究水平的差距大大缩小。伴随着单细胞分析方法的灵敏度与时空分辨率逐步提高，单细胞分析的应用也在不断拓宽和深入，尤其在细胞生物学异质性、肿瘤医学、免疫治疗、药物筛选及疾病诊断等方面展现了可喜的应用前景。

二、单细胞分析化学研究进展

单细胞分析化学在光谱、质谱、电化学、力谱等分析原理的基础上蓬勃发展，新仪器和新材料的出现使得单细胞分析化学技术在成像时空分辨率、检测灵敏度和选择性方面显著提高[3, 4]。

（一）单细胞光谱成像分析技术

单细胞光谱成像分析技术包括荧光成像、拉曼成像、表面等离激元共振成像、同步辐射 X 射线成像等。单细胞光谱成像分析主要通过不断提高成像技术的时空分辨率和灵敏度，实现单细胞的“可视化”分析[10]，以满足生命科学和医学研究对获取单细胞内物质信息的深度和广度需求的不断增加。

1. 单细胞荧光成像分析

荧光探针技术　利用荧光标记的生化分子对单细胞中靶标物进行特异性识别，并借助荧光信号放大机制，可实现对靶标物质高选择性、高灵敏度的成像分析[11-15]。我国科学家在该领域研究活跃，处于国际领先水平，取得了多项原创性的研究成果。本文篇幅所限，无法全部列出领域内的所有工作，仅能选择一些代表性例子进行简介。

唐本忠等开发了一系列具有聚集诱导发光（AIE）特性的荧光分子，该类分子在稀溶液中不发出荧光，但在聚集态可以发出强荧光。该独特的发光性质，使得 AIE 分子作为荧光探针，被广泛用于细胞成像、小分子及酶活性检测[16, 17]。另外，该团队最近设计合成了具有光激活特性的 AIE 前驱体探针，可特异性针对细胞内脂滴实现高点亮比率成像，具有光激活效率高、特异性好、合成简便等优势[18]。在亚细胞层次成像分析上，彭孝军等设计合成了对不同细胞器（包括线粒体、溶菌体、细胞核、高尔基体、内质网等）具有靶向性的荧光探针。凭借着低细胞毒性、双光子激发和近红外荧光发射等优良性能，这些探针被用于对特定细胞器内的小分子代谢物浓度以及黏度、极性和 pH 等微环境状态进行荧光成像分析[19]。例如，使用比率型荧光探针检测到癌细胞的线粒体极性低于正常细胞[20]，以及使用 IMC 识别基团的荧光探针在肿瘤细胞高尔基体内点亮过表达的 COX-2 酶，实现较深的肿瘤组织层析荧光成像[21]。

谭蔚泓等基于核酸适配体的分子识别，开发了一系列可实现肿瘤细胞特异性成像的高性能荧光探针[15, 22]。他们利用核酸适配体对靶标分子特异性的构型变化和核酸链置换反应，设计了一种多功能的 DNA“纳米爪”，对细胞表面的复杂生物标记物进行逻辑性的分析，诊断结果的信号输出以及自动化的靶向光动力治疗[23]。此外，还在细胞表面建立了基于核酸适配体的逻辑信号回路，实现了对细胞表面多种生物标志物的逻辑运算[24]，为精准分子医学的发展提供了新思路。李景虹等将固定的单细胞中的 microRNA 分子通过核酸聚合酶进行滚环扩增，使每个 microRNA 的扩增产物结合大量荧光分子，产生足够强的荧光信号，从而实现在单细胞中对单个 microRNA 分子进行荧光成像。采用这一技术，选用不同扩增引物，即可对细胞中不同序列的 microRNA 分别进行成像分析[25, 26]。除 microRNA 以外，该方法也同样适用于对单细胞内的 mRNA 分子进行成像分析[27]。李正平等利用核酸茎环引物诱导指数扩增（SPEA）反应，在单细胞中对端粒酶活性实现了超灵敏的荧光分析[28]。

基于纳米颗粒荧光探针的单细胞成像分析，为研究细胞内生物反应过程的化学结构变化提供了崭新的途径。鞠熀先等根据“单激发－双发光共振能量转移（D-LRET）”机制，用近红外光激发的上转换纳米颗粒作为能量供体，不同荧光标记的两种蛋白质糖基作为能量受体，提出了对单细胞表面特定蛋白质上的两种不同聚糖修饰进行同时原位成像的新策略[29]。林君等开发了基于上转换纳米材料的红、绿、蓝等多色单细胞荧光成像方法，具有背景信号极低的特性[30]。在利用该方法研究细胞的内吞行为时发现，纳米材料被细胞所摄取的量随着时间延长而增加，且被共定位于溶酶体中，表明这种纳米材料是以依赖能量的内吞方式被细胞摄取的。李景虹等以氧化石墨烯纳米片作为能量受体和核酸选择性吸附材料，通过荧光标记的核酸适配体对 ATP 和 GTP 的特异性相互作用导致其在石墨烯纳米片上脱附的机制，实现在细胞中对两种代谢物 ATP 和 GTP 的同时原位成像分析[31, 32]。

单分子荧光技术兴起于 1995 年并发展至今[33]，在探针性能、细胞标记、成像技术以及在细胞信号传导、基因表达与调控等方面的应用均取得重要进展[34]。单分子荧光技术凭借其超高灵敏度的优势，将成像技术的观测对象集中在单个或若干个荧光分子的水平，这显著地提高了单细胞荧光成像分析的灵敏度。我国科学家在单分子荧光成像仪器、理论及应用方面均取得了一系列有影响力的成果。

谢晓亮等开发的反射光片层显微镜（RLSM）技术将选择性平板照射与垂直方向的照射物镜和观测物镜结合，组装成为仪器核心，实现了对哺乳动物细胞核中单个荧光蛋白质分子的示踪[35]。利用该项技术，成功观测到了细胞核中转录因子和DNA的动态相互作用，以及RNA聚合酶II在细胞核中的空间分布规律[36]。方晓红等发展了一种基于隐藏马尔科夫模型（HMM）的新算法，通过单分子荧光强度变化来研究信号蛋白亚基组成，并发现单个肾上腺素受体β2AR在4种不同临床药物作用下聚集状态的改变[37]。任吉存等发展了一种暗场发光相关谱（DFSCS）的单粒子光谱技术，将单个纳米粒子的扩散行为与在暗场下光散射的波动关联起来，并建立了相应理论模型描述纳米粒子在均匀发光场内的扩散机理，可用于研究细胞体内作为探针或药物载体的纳米粒子的实时扩散情况[38]。杨朝勇等使用微流控芯片系统产生水凝胶微滴，封装单分子或单个细胞，具有良好的生物相容性，能够对单分子和单细胞进行扩增和培养，且利于在长时间尺度跟踪分子和细胞的状态。通过大量微滴进行快速平行分析，实现在微流控芯片系统中对单分子和单细胞的分析与成像[39, 40]。

超分辨荧光显微技术包括光激活定位显微镜（PALM）、受激发射损耗显微镜（STED）、结构照明显微镜（SIM）、及随机光学重建显微镜（STORM）超分辨显微技术[41]，突破了传统可见光成像的衍射极限，达到显著小于可见光波长的几个到几十个纳米水平的空间分辨率。我国在超分辨荧光显微成像技术领域取得了多方面有特色的研究进展。

在超分辨荧光显微成像仪器构建方面，匡翠方等提出了一种基于虚拟波矢调制的超分辨荧光显微术，使用一个探测器阵列代替传统共聚焦显微镜中的点探测器来获得样品的图像信息，突破了光学显微成像分辨率极限，实现了对活体细胞的非侵入性、实时、三维的立体成像，具有装置简便、信噪比高和相差矫正等优点[42, 43]。陈良怡等发明了超高Z轴分辨率的扫描全内反射角显微镜（va-TIRF）和高分辨率结构光成像显微镜（Super-SIM microscope），用于细胞和分子水平囊泡分泌和胞吞的机制的研究；还构建了双光子三轴扫描光片显微镜（2P3A-DLSM），用于胰岛水平上胰岛素分泌的三维成像和斑马鱼在体胰岛功能检测，其在三维空间Z轴具备高分辨率、超宽视野和微秒级成像速度[44, 45]。

在荧光材料方面，超分辨荧光显微成像技术要求荧光分子或颗粒快速闪烁，在某一个时间段持续扫描并通过计算得到高分辨成像图像，从而定位出荧光分子或颗粒的准确位置。席彭等制备了铥掺杂的上转换荧光纳米材料，可通过808nm与980nm两个波长的近红外激光交替抑制和激活纳米材料的STED高性能荧光，成像分辨率达到28nm，具有近红外激发光生物组织穿透度深、生物相容性好和背景干扰低的优势[46]。另外，该研究组也使用单分散的半导体聚合物荧光纳米粒子作为超分辨荧光显微技术的荧光材料，兼备了光能效率、光稳定性和生物相容性的优势[47]。

在技术联用方面，刘松琴等将超分辨荧光成像技术与质谱技术相结合，首先利用超分辨荧光成像机制精确定位细胞中特定的亚细胞区域，再进行定点离子化的质谱分析，从而成功实现对单细胞中多种小分子目标物空间分布的非标记成像分析[48]。在技术应用方面，张奇伟等利用三维 STORM 超分辨显微技术和荧光分子信标，成功在细胞核中原位观测到了 2.5kb 非重复 DNA 片段，这是目前能观测到的尺寸最小的非重复 DNA 片段[49]。

2. 拉曼光谱成像分析

拉曼光谱能够提供与物质化学结构及组成相关的分子振动指纹信息，具备无损、无标和多组分同时测量的性能[50, 51]。表面增强拉曼光谱（SERS）技术将痕量分析物较弱的拉曼散射信号极大地增强，达到成像仪器可检测的水平，这使得基于拉曼光谱的单细胞成像分析成为可能[52, 53]。

何耀等[54]以银纳米颗粒包覆的硅片为基底，对细胞凋亡过程产生的 SERS 信号进行高灵敏度检测，实现了对单细胞水平的细胞凋亡活动的监控。龙亿涛等开发了反应型 SERS 纳米探针，并用于对单细胞内的内源性气体信号分子 H_2S 和 CO 的高选择性、高灵敏度原位动态监测[55, 56]。黄岩谊等将微流控系统中培养的大量单细胞用受激拉曼散射显微镜（SRSM）进行成像分析，得到了在不同培养条件下多个单细胞中脂滴形貌及含量的统计分布，发现在非脂肪细胞中脂肪富集主要导致脂滴数量的增多[57]。鞠熀先等利用金纳米颗粒的区域可控 SERS 效应，对经药物处理的 MCF-7 癌细胞表面 EpCAM 蛋白上的唾液酸糖基化程度进行了成像分析[58]。徐健等发展了一种稳定同位素标记的拉曼分析法（Raman-SIP），在两种菌落之间研究重水（D_2O）可逆共标记的代谢物反应，对单细胞内含碳代谢物的分析十分有效[59, 60]。王康等在金表面上修饰有机氰化物作为 SERS 探针，通过该有机氰化物与 Fe^{3+} 离子或血红素蛋白相互作用后导致的拉曼光谱变化，实现了单细胞中 Fe^{3+} 离子和血红素蛋白的定量分析[61]。

3. 表面等离激元共振成像分析

表面等离激元共振成像是一种经典的免标记细胞成像技术[62]。基于数十纳米厚度的金属薄膜对具有特定波长和入射角的入射光产生的表面等离子共振吸收效应，利用反射光强度与金属薄膜表面局部折射率之间灵敏、定量的依赖关系，可以对单细胞质量密度分布进行成像[63]。王伟等提出了一种基于表面等离子共振显微镜的膜蛋白 - 配体相互作用的原位分析方法[64]，以免标记方式获取单细胞质量密度分布图像，在单细胞水平上实现了膜蛋白 - 配体原位结合动力学常数及其空间分布，在尽可能真实的环境下研究膜蛋白行为，揭示细胞生物化学环境对其相互作用的影响规律。刘震等提出了表面等离激元免疫夹心法（PISA）[65]，将活体微萃取和表面等离激元光学检测相结合，检测到单个活细胞和活体动物中的低拷贝数蛋白，揭示了癌细胞和正常细胞中碱性磷酸酶和 survivin 等低含量蛋白质的表达差异。何彦等利用等离激元纳米粒子的光稳定性与高亮度的特性，建立了以金纳米颗粒为探针的暗场单颗粒示踪技术，实现了对细胞内吞纳米颗粒的动态过程分析，同时还利用金银核壳结构纳米棒探针发展了对细胞内酶促生成硫化氢的动态监测方法，为细胞内酶活性调控与异质性的研究提供了新的研究手段[66, 67]。龙亿涛等采用表面等离激元共振–荧光联用成像技术，在单细胞层面上实现了野生型和变异型p53蛋白的原位成像，发现变异型p53蛋白在肿瘤细胞内过表达而正常p53蛋白受到抑制[68]。

4. 同步辐射X射线成像分析

同步辐射X射线显微成像技术利用光源本身的超短波长特点，能够对单细胞中的具有X射线响应的金属纳米颗粒进行十纳米级别超高分辨率的三维成像[69]。陈春英等结合了SR–TXM和SR–XANES两种同步辐射X–射线分析技术，对银纳米粒子在单细胞中的三维分布进行高分辨率成像[70]。通过不同价态银元素的信号差异，揭示了银纳米粒子的细胞毒性主要源于产生Ag^+离子及其与细胞内生物分子形成的Ag–O–及Ag–S–组分。樊春海等建立了适用于同步辐射X射线显微技术的细胞结构成像技术，实现了氮化硅薄膜窗口上单细胞的培养，观察到微管蛋白的胞内分布[71]。

（二）单细胞质谱分析

细胞内发生的生物过程常常是多种分子协同作用的结果，发展新的单细胞分析方法，最大限度地满足细胞中多组分同时分析的需求，并能够对细胞事件中的未知成分进行快速定性和精准定量分析，是单细胞生物学研究对分析化学的迫切需求。依据细胞中各种成分的分子量不同，在质谱中能够形成按照质量数排列的谱峰，进一步通过多级质谱分析，即可获得细胞中各种成分的分子信息，实现多组分的同时分析[72, 73]。国外这一领域的研究以Ewing、Spengler和Sweedler等课题组的工作为代表[73]。Dorrestein等对细胞内各种未知成分进行无标记快速鉴定，获得细胞中蛋白质乃至小分子代谢物的“组学”信息[74]。此外，Nolan等建立的单细胞质谱流式技术（Mass Cytometry）[75, 76]和张新荣等开发的稳定金属同位素质谱标记方法[77]，采用同位素内标和稀释技术可实现对细胞内待测分子的精准定量。

张新荣等提出脉冲进样直流电喷雾法[78]，能够在电极溶液非接触的情况下诱发样品电喷雾，提高单细胞质谱分析灵敏度和缩减取样量至pL级别，使得对多种小分子代谢物的检出限达fmol（10^{-15}mol）以下。例如，对肿瘤MCF–7细胞中的重要能量参数ATP/AMP和ADP/AMP浓度比值进行了同时检测，发现细胞增殖时能量物质ATP含量比例明显增多，且ATP通路与葡萄糖的磷酸戊糖途径密切相关[79]。杭纬等发展了高辐射飞秒激光离子化正交时间飞行质谱（fs–LI–O–TOFMS）方法[80]，在单细胞中进行元素分析，能够在细胞内测得ICP–MS较难测出的P、S、Cl等信号偏弱的非金属元素的含量。丰伟悦等展示了一种可以在单细胞中原位定量分析金属纳米粒子含量的激光消融电感应耦合等离子体质谱（LA–ICPMS）方法[81]，以及测量单细胞中Fe、Cu、Zn、Mn、P和S含量及分布的时间分辨电感应耦合等离子体质谱[82]。刘红霞等将微流控芯片系统与基质辅助激光解吸电离质谱（MALDI-MS）相结合[83]，发展了一种高通量快速检测单细胞膜上多种磷脂含量的方法。林金明等开发了探针电喷雾电离质谱（PESI-MS）对细胞样品进行分析[84]。徐静娟等利用同步极化诱导电喷雾电离质谱（SPI-ESI）对单细胞同时进行正负离子两种模式的分析，成功对单细胞中多达 90 余种代谢物实现同时检测。袁必锋等利用液相色谱 - 二级电喷雾电离质谱联用（LC-ESI-MS/MS）方法，对从捕获的循环肿瘤细胞中提取并水解的核酸样品进行分析，可对细胞核酸不同碱基的甲基化进行检测[85]。黄光明等将膜片钳技术和质谱技术相结合，使得神经细胞在进行质谱分析时正常存活，可以在老鼠大脑切片中的单个神经细胞的细胞质中取出纳升水平的样品进行分析，从而实现对神经细胞生理活动中细胞质内物质组成变化的分析[86]。

（三）单细胞电化学分析

在细胞的新陈代谢、呼吸作用、传递信息、物质运输等生命活动中，往往存在着电活性物质的定向有序传递、传导和转移等过程，如葡萄糖酶促代谢过程中的电子转移。电化学技术作为一种简单、快速、灵敏的检测手段，在细胞内的这些生化反应过程的研究中得到了广泛的应用，成为研究细胞生物过程的最有力工具之一。由于细胞的超微体积，电化学分析技术应用于单细胞分析时，使用纳米尺度的电极以保证信号来源于单个细胞[87]。将纳米电极阵列化集成，结合电化学技术的快速、高灵敏特点和无衍射极限限制的优势，还可构建基于电化学的成像分析技术，实现单细胞和亚细胞尺度的实时动态图像分析，以及对细胞表面的物质与信息的传导进行实时原位监测[88]。

黄卫华等利用纳米材料优异的信号放大、光电催化特性以及细胞黏附分子的生物相容性等功能，在单细胞水平上实现了多种信号分子的实时动态监测，并对其释放动力学行为进行了研究[89]。他们研制了纳米铂颗粒修饰的 TiC@C 纳米纤维阵列电极，对生长素（IAA）的检出灵敏度高达 1.0 nmol/L，并实时监测了单细胞内单个原生质体 IAA 的释放，发现其具有量子释放特征，为生长素胞吐释放的运输模式提供了直接证据[90]。他们以 TiO_2 纳米颗粒作为光降解自清洁材料、石墨烯和纳米金颗粒作为高灵敏电化学传感材料，构建了高灵敏、光致自清洁微电极阵列，实现了单细胞 NO 释放的高灵敏检测以及电极的高效循环利用[91]。他们还制作了电化学性能优异和生物相容性良好的高性能柔性可拉伸电化学传感器，并成功用于机械敏感型细胞和组织信号分子的实时监测[92]。江德臣等利用微孔电极阵列中引入纳环结构的方式，使得电极表面可以通过修饰显著提高分析灵敏度，成功检测到单细胞内低浓度过氧化氢的高强度信号[93]。他们还制备了尖端有可活动环电极的纳米毛细管，在电极附近飞升级别体积中高效产生过氧化氢，实现对单细胞内葡萄糖浓度和鞘磷脂酶活性的高灵敏度分析[94]。施国跃等制备了半径 3nm 的金纳米盘电极，实现对大鼠嗜铬细胞受到刺激时释放多巴胺的高空间分辨率电化学监控分析[95]。等离子体的电化学阻抗成像技术（PEIC）利用阻抗和表面电荷密度的定量关系，将交流电势下表面等离子体共振的信息通过计算转换为阻抗信息成像。陶农建和王伟等利用该技术对细胞凋亡、细胞电穿孔过程进行了实时成像研究[96]，显示了其非标记电化学成像技术高灵敏、高时空分辨率的优势。李景虹等实现了无标记的对细胞信号转导过程中的信号分子动态变化进行高时空分辨率的成像分析，获得了对 G 蛋白偶联受体受激初期钙离子信号高度动态及瞬态变化的信息[97]。

（四）单细胞力谱成像分析

生物力学的研究表明机械力对维持生命体内组织与器官的正常形态、结构与功能发挥着举足轻重的作用。在细胞层面上，对环境中机械力信号的感知是包含触觉、听觉、压觉、痛觉等在内的许多重要生理过程的生物学基础，同时机械力信号在细胞调控中也发挥着重要作用。随着生物力学在单细胞分析领域的发展，发展单细胞力谱分析方法的重要性不言而喻[98]。生命活动中的机械力信号具有信号小、变化范围大、传导速度快等特征，使得单细胞力谱分析的发展面临巨大挑战。

随着原子力显微镜（AFM）的出现，通过对 AFM 探针表面进行生物分子的修饰，再将探针在细胞膜表面进行扫描并实施记录相应的力学信号，即可实现细胞膜表面生物分子作用过程中力学信号的检测。杨培慧等用 AFM 实时原位研究了白藜芦醇药物对 MCF-7 癌细胞表皮生长因子受体的抑制作用[99]。白藜芦醇改变了细胞的形状和刚性，发现了表皮生长因子表达抑制作用来源于在细胞表面形成表皮生长因子功能化的尖端突起。刘连庆等也使用 AFM 观察甲胺喋呤药物作用下 4 种不同细胞系的单细胞形貌和黏弹性的变化，发现药物作用下细胞形貌发生变化，且杨氏模量和弛豫时间显著降低，有望作为一种药物评价方法[100]。

近年来发展起来的光镊技术也被应用于单细胞力谱分析[101]。谢晖等使用具有双探针光镊的自制纳米机器人系统，对单细胞进行取和放等操作，通过将两个单细胞老鼠肌原母细胞黏合和拉离，测量了细胞与培养基底以及细胞间的粘合力[102]。由于 AFM 和光镊技术对仪器设备要求较高，且具有操作难度大、检测通量小等不足之处，极大限制了其在细胞力学研究中的应用。为了满足单细胞力谱分析发展的需求，研究人员逐步开发了与光谱成像技术联用的单细胞力谱分析方法。方晓红等研制了超分辨荧光显微技术与 AFM 联用成像系统，实现了 STED 与 AFM 的同步扫描成像，以获得荧光与表面形貌两个维度信息的相关性[103]。

三、单细胞分析化学技术的应用

随着近年来单细胞分析化学技术的蓬勃发展，其时空分辨率、检测灵敏度、检测通量、细胞相容性等方面取得了显著进步。同时，单细胞分析化学在单个细胞的分析深度和群体细胞分析广度两个维度上不断强化，使之成为对于生物和医学研究具有重要推动作用的分析技术，在组学分析、疾病诊断、药物筛选等领域得到广泛应用[104-106]。

（一）基于高灵敏高通量单细胞分析技术的组学研究

在单细胞水平上进行基因组学、转录组学、蛋白质组学分析，可以揭示群体细胞间更深层次的统计信息，而这些重要信息被传统组学分析得到的多个细胞的加权平均值所掩盖[107]。单细胞分析在代谢组学方面近年来的进展较为缓慢，主要由于小分子代谢物种类繁多，且对于较低含量的代谢物普遍适用的分析方法有限[108]。

在基因组学方面，基因测序技术的发展使得测序灵敏度显著增加，测序仪器也越发简单和低成本，促进了单细胞基因组学分析的迅速发展。对单细胞进行全基因组测序时，由于单细胞中通常只有低拷贝 DNA，需要在测序前对全基因组进行扩增以达到测序所需的最低浓度。传统的 PCR 扩增方法是非线性指数扩增，虽然扩增效率很高，但是往往受到不同基因核酸序列差异的影响，从而导致基因扩增程度不一致，产生定量分析严重偏差。谢晓亮等建立了一种称为基于多重退火和循环放大周期（MALBAC）的扩增新方法，并应用于单个人体细胞全基因测序[109, 110]，其优势是扩增产物被环化从而扩增产物不会作为模板导致非线性扩增的发生，显著降低不同序列的扩增偏差，帮助体外受精卵细胞的优选。最近，他们还开发了另一种通过转座子插入的线性扩增（LIANTI）方法，是目前性能最好的全基因组扩增方法[111]。刘笔锋等发展了基于微流控芯片的超高通量测序法，并对单细胞的 DNA 损坏进行分析[112]。该方法根据彗星 DNA 电泳的形状来对 DNA 损坏进行评估，是一种简单、高效且高度可重复的单细胞基因组学测序法。汤富酬等用全基因组重亚硫酸盐测序法对处于不同发育时期的人早期胚胎细胞的基因组进行甲基化特征分析，发现主要的去甲基化发生于发育的 2-cell 期[113, 114]。尽管基因组学分析揭示了细胞基因信息的静态存储状态，但是无法由此得知不同基因被转录为 mRNA 的效率信息。

转录组学分析的对象则是细胞中的 mRNA，反映了细胞在不同状态时基因组表达效率的动态过程。RNA 测序技术诸如 RNA-Seq 等技术的突飞猛进，使单细胞转录组学分析成为现实[115]。例如，单细胞转录组学分析已被 Regev 和 Haber 分别成功应用于免疫细胞[116]和循环肿瘤细胞[117]的研究中，其方法的基础是反转录 PCR 扩增 RNA 成为 cDNA，然后进行类似于基因组学的定量测序分析。由于细胞中 mRNA 种类众多，对全转录组进行测序分析是一大挑战。黄岩谊等将 RNA-Seq 植入微流控芯片，从裂解后的细胞中捕捉带有 PolyA 末端的 mRNA 进行基于 cDNA 的二代测序[118]，可实现单细胞的全转录组学分析，获得每个细胞序列长达 20 万个碱基的读出。

单细胞蛋白质组学分析对灵敏度有较高的要求，目前尚无适宜于组学分析使用的高效可靠的蛋白质扩增方法。尽管可以使用抗体技术对单细胞的蛋白质进行高灵敏度的分析，但是单细胞蛋白质组学分析技术目前通常仅能对一定种类的蛋白质进行同时分析，难以覆盖细胞中的大部分蛋白质种类[119]。单细胞蛋白质组学分析最成熟的方法是荧光激活细胞分选技术（FACS）。通过荧光标记的抗体特异性结合细胞表面的膜蛋白标志物，利用带有分选功能的流式细胞仪实现对含有目标膜蛋白的细胞分析和分选。为提高分析蛋白质的种类，Nolan 等开发了质谱流式细胞技术（CyTOF），该方法利用质谱标签标记的抗体，基于流式细胞仪的高通量进样方式，使用质谱作为检测器，可将分析蛋白质种类从 FACS 的 12 种左右提高到 30 种以上[76]。Heath 发展的单细胞条码芯片法[120]和 Ploegh 发展的微刻法[121]是在显著提高通量的前提下进一步增加了可同时检测的蛋白质种类。刘笔锋等提出了一种对低丰度蛋白质组进行单细胞化学蛋白质组分析的策略，检测到了以往方法难以测得的膜蛋白 GB1 及其剪接变种在单细胞表面的表达情况[122]；同时，他们还利用穿膜活性探针在单细胞的细胞质中对功能性蛋白质进行组学分析[123]。张祥民等将 LC–MS 集成系统用于单细胞蛋白质组学研究，对 100 个细胞实现超高灵敏度检测，发现 800 多种蛋白质的存在[124]。

（二）基于高选择性单细胞分析技术的疾病诊断与药物筛选

单细胞分析技术被广泛应用于疾病诊断，特别是癌症的早期诊断与预警。传统的疾病诊断方法通常难以得到组织样品中大量细胞中极少数病变细胞的信息，单细胞分析技术在疾病的早期就检测出极少数的病变细胞方面具有独特的优势[125]。

循环肿瘤细胞被认为是癌症发生恶性转移的罪魁祸首。循环肿瘤细胞的检测分析，是评估癌症发生转移风险的有效手段，对于预防性用药有重要的指导意义[126]。循环肿瘤细胞在血液中的含量非常小，利用微流控系统在血液中捕获循环肿瘤细胞，高通量的单细胞分析方法对血液中少量的循环肿瘤细胞的膜蛋白、核酸或代谢物进行分析，检测出了循环肿瘤细胞在膜蛋白、转录水平和代谢水平上与同类正常细胞的差异[127]。我国科学家利用单细胞分析化学技术，在该领域做出了一系列具有特色和应用前景的研究工作。

白凡等对从病人血液中捕获的循环肿瘤细胞进行单细胞全基因组测序，发现了肿瘤细胞特征的基因外显子组具有病人自身以及癌症种类的个性化特征[127]，该分析结果对个性化的诊断及用药具有指导意义。方晓红等筛选了多种特异识别非小细胞肺癌 NSCLC 细胞的核酸适体，将其修饰到微流芯片中的纳米线阵列，实现了对血液中稀少肿瘤细胞的有效捕获、释放和基因分析[128]。张祥民等利用以硝酸纤维素膜作为基质的表面增强拉曼光谱成像技术，对从血液中捕获的循环肿瘤细胞进行分析，以非小细胞肺癌 NCI–H1650 细胞作为目标肿瘤细胞，显示出高捕获效率和准确率[129]。施奇惠等组合了荧光标记底物和条码标记抗体微阵列，制作了集成微芯片用以对循环肿瘤细胞的葡萄糖摄入、胞内功能蛋白质以及基因突变进行单细胞分析，研究了不同循环肿瘤细胞的代谢水平、功能蛋白表达和基因突变情况之间的相关性[130, 131]。邵晨等以细胞表面的碳酸酐酶和 CD147 抗原作为组合靶标物，在肾透明细胞癌病人的血液中高效捕获到了循环肿瘤细胞，显著提高了捕获效率和后续单细胞分析的灵敏度[132]。

由于与人体内细胞的环境存在差异，利用培养细胞进行小分子药物筛选时，同类细胞对药物的响应存在显著的差异。杨朝勇等用微孔阵列芯片制作出高通量微胶原蛋白阵列，在三维空间长期培养细胞并可用于高通量单细胞分析[133]，结果发现小分子药物阿霉素对细胞的半抑制增殖浓度远低于半数存活抑制浓度。近年来，癌症免疫疗法受到越来越多的关注与应用。癌症免疫疗法作用于病人后产生大量T细胞，单细胞分析发现通常其中具备癌细胞杀灭能力的T细胞（活性T细胞）仅占极少数。与细胞分选结合的单细胞分析技术，还能够将免疫疗法产生的活性T细胞从样品中提取出来并进行单细胞组学分析，得到足够数量的活性T细胞的多维度生化指标信息。刘澎涛等利用单细胞RNA测序技术对老鼠骨髓前体细胞进行转录组学分析，揭示了先天淋巴样细胞的前体、发育和通路与PD-1表达的关联[134]，为探索基于PD-1和PD-L1的免疫疗法提供了新的视角。

四、总结与展望

作为高灵敏、高特异性与高通量分析技术的代表，单细胞分析化学在未来生命分析化学研究中将长期占据举足轻重的地位。在过去的5年里，我国的科研工作者在单细胞分析化学这一领域已取得长足进步。无论在单细胞操控技术的开发，超分辨成像及联用技术的构建，还是在高灵敏高时空分辨单细胞单分子检测技术的发展与应用等方面，都在逐渐实现对国际研究水平的追赶，甚至在部分领域实现了国际领先。例如，我国学者建立的基于单个卵细胞的高精度全基因组测序技术的基因异常分析，已成功应用于临床医学中检测女方家族遗传疾病，并在世界范围内获得广泛的关注。

尽管中国学者在单细胞分析的各个研究领域都取得了重大进展，在单细胞分析化学日新月异的发展形势下，我国单细胞分析化学仍需在如下几个方面加强努力，以提高中国单细胞分析研究的整体水平。

第一，发展高灵敏、高通用性的单细胞分析技术的新原理与新方法。我国单细胞分析开展较晚和力量分散，目前应用于单细胞分析的仪器与技术，如单细胞操控技术、超分辨成像技术等研究仍与国际先进水平存在较大差异。这要求国内的学者在单细胞分析的新仪器与新技术方面攻克难关，发展属于国内知识原始创新的具有国际影响力的新原理、新技术。

第二，突破现有单细胞组学技术的壁垒，发展能够检测较低丰度且难以信号放大的生物分子的技术。与核酸分子不同，蛋白质、磷脂以及多糖等生物分子往往无法以扩增的方式进行信号放大，这为单细胞蛋白组学与代谢物组学的研究提出了新的挑战与机遇。

第三，发展适用于活细胞内信号通路实时、原位、动态、免干扰的分析手段。细胞的生命活动与生理过程都是在细胞内复杂而有序的信号通路调节下得以实现的。由于细胞内信号分子的含量低、周期短且易受外界干扰，活细胞内信号通路的研究长期以来是化学生物学研究中尚未解决的重大科学问题。现有的单细胞分析技术，如单细胞光谱与质谱分析技术等，仍难以对活细胞中信号分子及其信号转导与调控进行实时动态的研究。因此，发展适用于活细胞信号通路分析研究的新技术具有十分重要的科学意义。

第四，拓展单细胞分析在基础生命科学与重大疾病诊断中的应用，满足国内临床医学研究中的实际需求。在未来的单细胞分析的应用研究中，需要针对具有挑战性的生物医学与生命科学问题形成系统而深入的研究，如核酸测序在人工受精卵技术、神经系统的生物过程抗阻断、免疫应答等，使基础研究的成果更好地服务于人们的生活与解决健康问题。

第五，在新的形势与挑战下，进一步加强跨学科交叉，围绕着重大科学目标开展国内外研究团队的通力协作，在国际单细胞分析研究中的竞争发展中占据一席之地。借鉴国外在单细胞基因测序技术商业化的经验，争取政策支持、发挥市场资源配置力量和积极引导民间资金，有望将单细胞分析的研究推向一个新高度。

参考文献

[1] Spiller D G，Wood C D，Rand D A，et al. Measurement of single-cell dynamics [J]. Nature，2010，465（7299）：736-745.

[2] Snijder B，Pelkmans L. Origins of regulated cell-to-cell variability [J]. Nature Reviews Molecular Cell Biology，2011，12（2）：119-125.

[3] Armbrecht L，Dittrich P S. Recent advances in the analysis of single cells [J]. Analytical Chemistry，2017，89（1）：2-21.

[4] Lin Y，Trouillon R，Safina G，et al. Chemical analysis of single cells [J]. Analytical Chemistry，2011，83（12）：4369-4392.

[5] Bahcall O，Eggleston A K，Le Bot N，et al. Frontiers in biology [J]. Nature，2017，541（7637）：301-301.

[6] von Diezmann A，Shechtman Y，Moerner W E. Three-dimensional localization of single molecules for super-resolution imaging and single-particle tracking [J]. Chemical Reviews，2017，117（11）：7244-7275.

[7] Gooding J，Magnussen O，Fermin D，et al. From single cells to single molecules: General discussion [J]. Faraday Discussions，2016，193: 141-170.

[8] Theberge A B，Courtois F，Schaerli Y，et al. Microdroplets in microfluidics: An evolving platform for discoveries in chemistry and biology [J]. Angewandte Chemie International Edition，2010，49（34）：5846-5868.

[9] Chen X，Love J C，Navin N E，et al. Single-cell analysis at the threshold [J]. Nature Biotechnology，2016，34（11）：1111-1118.

[10] Stender A S，Marchuk K，Liu C，et al. Single cell optical imaging and spectroscopy [J]. Chemical Reviews，2013，113（4）：2469-2527.

[11] Okumoto S. Imaging approach for monitoring cellular metabolites and ions using genetically encoded biosensors [J].

Current Opinion in Biotechnology, 2010, 21 (1): 45–54.

[12] Kikuchi K. Design, synthesis and biological application of chemical probes for bio–imaging [J]. Chemical Society Reviews, 2010, 39 (6): 2048–2053.

[13] Gao X H, Nie S M. Molecular profiling of single cells and tissue specimens with quantum dots [J]. Trends in Biotechnology, 2003, 21 (9): 371–373.

[14] Mei J, Leung N L C, Kwok R T K, et al. Aggregation–induced emission: Together we shine, united we soar! [J]. Chemical Reviews, 2015, 115 (21): 11718–11940.

[15] Tan W, Donovan M J, Jiang J. Aptamers from cell–based selection for bioanalytical applications [J]. Chemical Reviews, 2013, 113 (4): 2842–2862.

[16] Chen M, Li L, Nie H, et al. Tetraphenylpyrazine–based AIEgens: Facile preparation and tunable light emission [J]. Chemical Science, 2015, 6 (3): 1932–1937.

[17] Ding D, Li K, Liu B, et al. Bioprobes based on AIE fluorogens [J]. Accounts of Chemical Research, 2013, 46 (11): 2441–2453.

[18] Gao M, Su H, Lin Y, et al. Photoactivatable aggregation–induced emission probes for lipid droplets–specific live cell imaging [J]. Chemical Science, 2017, 8 (3): 1763–1768.

[19] Zhu H, Fan J, Du J, et al. Fluorescent probes for sensing and imaging within specific cellular organelles [J]. Accounts of Chemical Research, 2016, 49 (10): 2115–2126.

[20] Jiang N, Fan J, Xu F, et al. Ratiometric fluorescence imaging of cellular polarity: Decrease in mitochondrial polarity in cancer cells [J]. Angewandte Chemie International Edition, 2015, 54 (8): 2510–2514.

[21] Zhang H, Fan J, Wang J, et al. An off–on COX–2–specific fluorescent probe: Targeting the Golgi apparatus of cancer cells [J]. Journal of the American Chemical Society, 2013, 135 (31): 11663–11669.

[22] Liang H, Zhang X B, Lv Y, et al. Functional DNA–containing nanomaterials: Cellular applications in biosensing, imaging, and targeted therapy [J]. Accounts of Chemical Research, 2014, 47 (6): 1891–1901.

[23] You M, Peng L, Shao N, et al. DNA "nano–claw": Logic–based autonomous cancer targeting and therapy [J]. Journal of the American Chemical Society, 2014, 136 (4): 1256–1259.

[24] You M, Zhu G, Chen T, et al. Programmable and multiparameter DNA–based logic platform for cancer recognition and targeted therapy [J]. Journal of the American Chemical Society, 2015, 137 (2): 667–674.

[25] Deng R, Tang L, Tian Q, et al. Toehold–initiated rolling circle amplification for visualizing individual microRNAs in situ in single cells [J]. Angewandte Chemie International Edition, 2014, 53 (9): 2389–2393.

[26] Deng R, Zhang K, Li J. Isothermal amplification for microRNA detection: From the test tube to the cell [J]. Accounts of Chemical Research, 2017, 50 (4): 1059–1068.

[27] Deng R, Zhang K, Sun Y, et al. Highly specific imaging of mRNA in single cells by target RNA–initiated rolling circle amplification [J]. Chemical Science, 2017, 8 (5): 3668–3675.

[28] Wang H, Wang H, Liu C, et al. Ultrasensitive detection of telomerase activity in a single cell using stem–loop primer–mediated exponential amplification (SPEA) with near zero nonspecific signal [J]. Chemical Science, 2016, 7 (8): 4945–4950.

[29] Wu N, Bao L, Ding L, et al. A single excitation–duplexed imaging strategy for profiling cell surface protein–specific glycoforms [J]. Angewandte Chemie International Edition, 2016, 55 (17): 5220–5224.

[30] Dai Y, Xiao H, Liu J, et al. In vivo multimodality imaging and cancer therapy by near–infrared light–triggered trans–platinum pro–drug–conjugated upconverison nanoparticles [J]. Journal of the American Chemical Society, 2013, 135 (50): 18920–18929.

[31] Tang L, Wang Y, Li J. The graphene/nucleic acid nanobiointerface [J]. Chemical Society Reviews, 2015, 44 (19): 6954–6980.

[32] Wang Y, Tang L, Li Z, et al. In situ simultaneous monitoring of ATP and GTP using a graphene oxide nanosheet–based sensing platform in living cells [J] . Nature Protocols, 2014, 9 (8) : 1944–1955.

[33] Funatsu T, Harada Y, Tokunaga M, et al. Imaging of single fluorescent molecules and individual ATP turnovers by single myosin molecules in aqueous–solution [J] . Nature, 1995, 374 (6522) : 555–559.

[34] Li G–W, Xie X S. Central dogma at the single–molecule level in living cells [J] . Nature, 2011, 475 (7356) : 308–315.

[35] Gebhardt J C M, Suter D M, Roy R, et al. Single–molecule imaging of transcription factor binding to DNA in live mammalian cells [J] . Nature Methods, 2013, 10 (5) : 421–426.

[36] Zhao Z W, Roy R, Gebhardt J C M, et al. Spatial organization of RNA polymerase II inside a mammalian cell nucleus revealed by reflected light–sheet superresolution microscopy [J] . Proceedings of the National Academy of Sciences of the United States of America, 2014, 111 (2) : 681–686.

[37] Sun Y, Li N, Zhang M, et al. Single–molecule imaging reveals the stoichiometry change of beta (2) –adrenergic receptors by a pharmacological biased ligand [J] . Chemical Communications, 2016, 52 (44) : 7086–7089.

[38] Liu H, Dong C, Ren J. Tempo–spatially resolved scattering correlation spectroscopy under dark–field illumination and its application to investigate dynamic behaviors of gold nanoparticles in live cells [J] . Journal of the American Chemical Society, 2014, 136 (7) : 2775–2785.

[39] Zhu Z, Yang C J. Hydrogel droplet microfluidics for high–throughput single molecule/cell analysis [J] . Accounts of Chemical Research, 2017, 50 (1) : 22–31.

[40] Guan Z, Jia S, Zhu Z, et al. Facile and rapid generation of large–scale microcollagen gel array for long–term single–cell 3D culture and cell proliferation heterogeneity analysis [J] . Analytical Chemistry, 2014, 86 (5) : 2789–2797.

[41] French J B, Jones S A, Deng H, et al. Spatial colocalization and functional link of purinosomes with mitochondria [J] . Science, 2016, 351 (6274) : 733–737.

[42] Li H, Huang Y, Kuang C, et al. Method of super–resolution based on array detection and maximum–likelihood estimation [J] . Applied Optics, 2016, 55 (35) : 9925–9931.

[43] Kuang C, Ma Y, Zhou R, et al. Virtual k–space modulation optical microscopy [J] . Physical Review Letters, 2016, 117 (2) : 028102.

[44] Zong W, Zhao J, Chen X, et al. Large–field high–resolution two–photon digital scanned light–sheet microscopy [J] . Cell Research, 2015, 25 (2) : 254–257.

[45] Zong W, Huang X, Zhang C, et al. Shadowless–illuminated variable–angle TIRF (siva–TIRF) microscopy for the observation of spatial–temporal dynamics in live cells [J] . Biomedical Optics Express, 2014, 5 (5) : 1530–1540.

[46] Liu Y, Lu Y, Yang X, et al. Amplified stimulated emission in upconversion nanoparticles for super–resolution nanoscopy [J] . Nature, 2017, 543 (7644) : 229–233.

[47] Chen X, Li R, Liu Z, et al. Small photoblinking semiconductor polymer dots for fluorescence nanoscopy [J] . Advanced Materials, 2017, 29 (5) : 1604850.

[48] Hua X, Szymanski C, Wang Z, et al. Chemical imaging of molecular changes in a hydrated single cell by dynamic secondary ion mass spectrometry and super–resolution microscopy [J] . Integrative Biology, 2016, 8 (5) : 635–644.

[49] Ni Y, Cao B, Ma T, et al. Super–resolution imaging of a 2.5 kb non–repetitive DNA in situ in the nuclear genome using molecular beacon probes [J] . eLife, 2017, 6e21660.

[50] Krafft C, Schie I W, Meyer T, et al. Developments in spontaneous and coherent Raman scattering microscopic imaging for biomedical applications [J] . Chemical Society Reviews, 2016, 45 (7) : 1819–1849.

[51] Cheng J-X, Xie X S. Vibrational spectroscopic imaging of living systems: An emerging platform for biology and medicine [J]. Science, 2015, 350 (6264): 8870.

[52] Li M, Xu J, Romero-Gonzalez M, et al. Single cell Raman spectroscopy for cell sorting and imaging [J]. Current Opinion in Biotechnology, 2012, 23 (1): 56-63.

[53] Wachsmann-Hogiu S, Weeks T, Huser T. Chemical analysis in vivo and in vitro by Raman spectroscopy-from single cells to humans [J]. Current Opinion in Biotechnology, 2009, 20 (1): 63-73.

[54] Jiang X, Jiang Z, Xu T, et al. Surface-enhanced Raman scattering-based sensing in vitro: Facile and label-free detection of apoptotic cells at the single-cell level [J]. Analytical Chemistry, 2013, 85 (5): 2809-2816.

[55] Li D W, Qu L L, Hu K, et al. Monitoring of endogenous hydrogen sulfide in living cells using surface-enhanced Raman scattering [J]. Angewandte Chemie International Edition, 2015, 54 (43): 12758-12761.

[56] Cao Y, Li D W, Zhao L J, et al. Highly selective detection of carbon monoxide in living cells by palladacycle carbonylation-based surface enhanced Raman spectroscopy nanosensors [J]. Analytical Chemistry, 2015, 87 (19): 9696-9701.

[57] Cao C, Zhou D, Chen T, et al. Label-free digital quantification of lipid droplets in single cells by stimulated Raman microscopy on a microfluidic platform [J]. Analytical Chemistry, 2016, 88 (9): 4931-4939.

[58] Chen Y, Ding L, Song W, et al. Protein-specific Raman imaging of glycosylation on single cells with zone-controllable SERS effect [J]. Chemical Science, 2016, 7 (1): 569-574.

[59] Wang Y, Song Y, Tao Y, et al. Reverse and multiple stable isotope probing to study bacterial metabolism and interactions at the single cell level [J]. Analytical Chemistry, 2016, 88 (19): 9443-9450.

[60] Tao Y, Wang Y, Huang S, et al. Metabolic-activity-based assessment of antimicrobial effects by D_2O-labeled single-cell Raman microspectroscopy [J]. Analytical Chemistry, 2017, 89 (7): 4108-4115.

[61] Hanif S, Liu H, Chen M, et al. Organic cyanide decorated SERS active nanopipettes for quantitative detection of hemeproteins and Fe^{3+} in single cells [J]. Analytical Chemistry, 2017, 89 (4): 2522-2530.

[62] Zeidan E, Kepley C L, Sayes C, et al. Surface plasmon resonance: A label-free tool for cellular analysis [J]. Nanomedicine, 2015, 10 (11): 1833-1846.

[63] Fang Y, Chen S, Wang W, et al. Real-time monitoring of phosphorylation kinetics with self-assembled nano-oscillators [J]. Angewandte Chemie International Edition, 2015, 54 (8): 2538-2542.

[64] Zhang F, Wang S, Yin L, et al. Quantification of epidermal growth factor receptor expression level and binding kinetics on cell surfaces by surface plasmon resonance imaging [J]. Analytical Chemistry, 2015, 87 (19): 9960-9965.

[65] Liu J, Yin D, Wang S, et al. Probing low-copy-number proteins in a single living cell [J]. Angewandte Chemie International Edition, 2016, 55 (42): 13215-13218.

[66] Xu D, He Y, Yeung E S. Direct observation of the orientation dynamics of single protein-coated nanoparticles at liquid/solid interfaces [J]. Angewandte Chemie International Edition, 2014, 53 (27): 6951-6955.

[67] Xiong B, Zhou R, Hao J, et al. Highly sensitive sulphide mapping in live cells by kinetic spectral analysis of single Au-Ag core-shell nanoparticles [J]. Nature Communications, 2013, 4.

[68] Qian R, Cao Y, Long Y T. Dual-targeting nanovesicles for in situ intracellular imaging of and discrimination between wild-type and mutant p53 [J]. Angewandte Chemie International Edition, 2016, 55 (2): 719-723.

[69] Bohic S, Simionovici A, Snigirev A, et al. Synchrotron hard X-ray microprobe: Fluorescence imaging of single cells [J]. Applied Physics Letters, 2001, 78 (22): 3544-3546.

[70] Wang L, Zhang T, Li P, et al. Use of synchrotron radiation-analytical techniques to reveal chemical origin of silver-nanoparticle cytotoxicity [J]. ACS Nano, 2015, 9 (6): 6532-6547.

[71] Zhu Y, Earnest T, Huang Q, et al. Synchrotron-based X-ray-sensitive nanoprobes for cellular imaging [J].

Advanced Materials，2014，26（46）：7889–7895.

［72］ Colliver T L，Brummel C L，Pacholski M L，et al. Atomic and molecular imaging at the single–cell level with TOF–SIMS［J］. Analytical Chemistry，1997，69（13）：2225–2231.

［73］ Passarelli M K，Ewing A G. Single–cell imaging mass spectrometry［J］. Current Opinion in Chemical Biology，2013，17（5）：854–859.

［74］ Petras D，Jarmusch A K，Dorrestein P C. From single cells to our planet–recent advances in using mass spectrometry for spatially resolved metabolomics［J］. Current Opinion in Chemical Biology，2017，36: 24–31.

［75］ Spitzer M H，Nolan G P. Mass cytometry: Single cells，many features［J］. Cell，2016，165（4）：780–791.

［76］ Bendall S C，Simonds E F，Qiu P，et al. Single–cell mass cytometry of differential immune and drug responses across a human hematopoietic continuum［J］. Science，2011，332（6030）：687–696.

［77］ Liu R，Zhang S，Wei C，et al. Metal stable isotope tagging: Renaissance of radioimmunoassay for multiplex and absolute quantification of biomolecules［J］. Accounts of Chemical Research，2016，49（5）：775–783.

［78］ Gong X，Zhao Y，Cai S，et al. Single cell analysis with probe ESI–mass spectrometry: Detection of metabolites at cellular and subcellular levels［J］. Analytical Chemistry，2014，86（8）：3809–3816.

［79］ Gong X，Xiong X，Wang S，et al. Desalting by crystallization: Detection of attomole biomolecules in picoliter buffers by mass spectrometry［J］. Analytical Chemistry，2015，87（19）：9745–9751.

［80］ Gao Y，Lin Y，Zhang B，et al. Single–cell elemental analysis via high irradiance femtosecond laser ionization time–of–flight mass spectrometry［J］. Analytical Chemistry，2013，85（9）：4268–4272.

［81］ Wang M，Zheng L N，Wang B，et al. Quantitative analysis of gold nanoparticles in single cells by laser ablation inductively coupled plasma–mass spectrometry［J］. Analytical Chemistry，2014，86（20）：10252–10256.

［82］ Wang H，Wang B，Wang M，et al. Time–resolved ICP–MS analysis of mineral element contents and distribution patterns in single cells［J］. Analyst，2015，140（2）：523–531.

［83］ Xie W，Gao D，Jin F，et al. Study of phospholipids in single cells using an integrated microfluidic device combined with matrix–assisted laser desorption/ionization mass spectrometry［J］. Analytical Chemistry，2015，87（14）：7052–7059.

［84］ Chen F，Lin L，Zhang J，et al. Single–cell analysis using drop–on–demand inkjet printing and probe electrospray ionization mass spectrometry［J］. Analytical Chemistry，2016，88（8）：4354–4360.

［85］ Huang W，Qi C B，Lv S W，et al. Determination of DNA and RNA methylation in circulating tumor cells by mass spectrometry［J］. Analytical Chemistry，2016，88（2）：1378–1384.

［86］ Zhu H，Zou G，Wang N，et al. Single–neuron identification of chemical constituents，physiological changes，and metabolism using mass spectrometry［J］. Proceedings of the National Academy of Sciences of the United States of America，2017，114（10）：2586–2591.

［87］ Zhang A，Lieber C M. Nano–bioelectronics［J］. Chemical Reviews，2016，116（1）：215–257.

［88］ Spira M E，Hai A. Multi–electrode array technologies for neuroscience and cardiology［J］. Nature Nanotechnology，2013，8（2）：83–94.

［89］ Li Y T，Zhang S H，Wang L，et al. Nanoelectrode for amperometric monitoring of individual vesicular exocytosis inside single synapses［J］. Angewandte Chemie International Edition，2014，53（46）：12456–12460.

［90］ Liu J T，Hu L S，Liu Y L，et al. Real–time monitoring of auxin vesicular exocytotic efflux from single plant protoplasts by amperometry at microelectrodes decorated with nanowires［J］. Angewandte Chemie International Edition，2014，53（10）：2643–2647.

［91］ Xu J Q，Liu Y–L，Wang Q，et al. Photocatalytically renewable micro–electrochemical sensor for real–time monitoring of cells［J］. Angewandte Chemie International Edition，2015，54（48）：14402–14406.

［92］ Liu Y L，Jin Z H，Liu Y H，et al. Stretchable electrochemical sensor for real–time monitoring of cells and tissues［J］.

Angewandte Chemie International Edition, 2016, 55 (14): 4537–4541.
[93] Zhuang L, Zuo H, Wu Z, et al. Enhanced electrochemical nanoring electrode for analysis of cytosol in single cells [J]. Analytical Chemistry, 2014, 86 (23): 11517–11522.
[94] Pan R, Xu M, Jiang D, et al. Nanokit for single-cell electrochemical analyses [J]. Proceedings of the National Academy of Sciences of the United States of America, 2016, 113 (41): 11436–11440.
[95] Liu Y, Li M, Zhang F, et al. Development of Au disk nanoelectrode down to 3 nm in radius for detection of dopamine release from a single cell [J]. Analytical Chemistry, 2015, 87 (11): 5531–5538.
[96] Fang Y, Wang H, Yu H, et al. Plasmonic imaging of electrochemical reactions of single nanoparticles [J]. Accounts of Chemical Research, 2016, 49 (11): 2614–2624.
[97] Lu J, Li J. Label-free imaging of dynamic and transient calcium signaling in single cells [J]. Angewandte Chemie International Edition, 2015, 54 (46): 13576–13580.
[98] Kashef J, Franz C M. Quantitative methods for analyzing cell-cell adhesion in development [J]. Developmental Biology, 2015, 401 (1): 165–174.
[99] Zhang L, Yang F, Cai J Y, et al. In-situ detection of resveratrol inhibition effect on epidermal growth factor receptor of living MCF-7 cells by atomic force microscopy [J]. Biosensors & Bioelectronics, 2014, 56: 271–277.
[100] Li M, Liu L, Xiao X, et al. Effects of methotrexate on the viscoelastic properties of single cells probed by atomic force microscopy [J]. Journal of Biological Physics, 2016, 42 (4): 551–569.
[101] Heller I, Hoekstra T P, King G A, et al. Optical tweezers analysis of DNA-protein complexes [J]. Chemical Reviews, 2014, 114 (6): 3087–3119.
[102] Xie H, Yin M, Rong W, et al. In situ quantification of living cell adhesion forces: Single cell force spectroscopy with a nanotweezer [J]. Langmuir, 2014, 30 (10): 2952–2959.
[103] Ruan H, Yu J, Yuan J, et al. Nanoscale distribution of transforming growth factor receptor on post-Golgi vesicle revealed by super-resolution microscopy [J]. Chemistry-an Asian Journal, 2016, 11 (23): 3359–3364.
[104] Stegle O, Teichmann S A, Marioni J C. Computational and analytical challenges in single-cell transcriptomics [J]. Nature Reviews Genetics, 2015, 16 (3): 133–145.
[105] Chattopadhyay P K, Gierahn T M, Roederer M, et al. Single-cell technologies for monitoring immune systems [J]. Nature Immunology, 2014, 15 (2): 128–135.
[106] Heath J R, Ribas A, Mischel P S. Single-cell analysis tools for drug discovery and development [J]. Nature Reviews Drug Discovery, 2016, 15 (3): 204–216.
[107] Junker J P, van Oudenaarden A. Every cell is special: Genome-wide studies add a new dimension to single-cell biology [J]. Cell, 2014, 157 (1): 8–11.
[108] Zenobi R. Single-cell metabolomics: Analytical and biological perspectives [J]. Science, 2013, 342 (6163): 1243259.
[109] Hou Y, Fan W, Yan L, et al. Genome analyses of single human oocytes [J]. Cell, 2013, 155 (7): 1492–1506.
[110] Zong C, Lu S, Chapman A R, et al. Genome-wide detection of single-nucleotide and copy-number variations of a single human cell [J]. Science, 2012, 338 (6114): 1622–1626.
[111] Chen C, Xing D, Tan L, et al. Single-cell whole-genome analyses by linear amplification via transposon insertion (LIANTI) [J]. Science, 2017, 356 (6334): 189–194.
[112] Li Y, Feng X, Du W, et al. Ultrahigh-throughput approach for analyzing single-cell genomic damage with an agarose-based microfluidic comet array [J]. Analytical Chemistry, 2013, 85 (8): 4066–4073.
[113] Guo H, Zhu P, Yan L, et al. The DNA methylation landscape of human early embryos [J]. Nature, 2014, 511 (7511): 606–610.

[114] Guo H, Zhu P, Guo F, et al. Profiling DNA methylome landscapes of mammalian cells with single-cell reduced-representation bisulfite sequencing [J] . Nature Protocols, 2015, 10 (5) : 645-659.

[115] Saliba A E, Westermann A J, Gorski S A, et al. Single-cell RNA-seq: Advances and future challenges [J] . Nucleic Acids Research, 2014, 42 (14) : 8845-8860.

[116] Shalek A K, Satija R, Adiconis X, et al. Single-cell transcriptomics reveals bimodality in expression and splicing in immune cells [J] . Nature, 2013, 498 (7453) : 236-240.

[117] Ting D T, Wittner B S, Ligorio M, et al. Single-cell RNA sequencing identifies extracellular matrix gene expression by pancreatic circulating tumor cells [J] . Cell Reports, 2014, 8 (6) : 1905-1918.

[118] Streets A M, Zhang X, Cao C, et al. Microfluidic single-cell whole-transcriptome sequencing [J] . Proceedings of the National Academy of Sciences of the United States of America, 2014, 111 (19) : 7048-7053.

[119] Lu Y, Yang L, Wei W, et al. Microchip-based single-cell functional proteomics for biomedical applications [J] . Lab on a Chip, 2017, 17 (7) : 1250-1263.

[120] Shi Q H, Qin L D, Wei W, et al. Single-cell proteomic chip for profiling intracellular signaling pathways in single tumor cells [J] . Proceedings of the National Academy of Sciences of the United States of America, 2012, 109 (2) : 419-424.

[121] Love J C, Ronan J L, Grotenbreg G M, et al. A microengraving method for rapid selection of single cells producing antigen-specific antibodies [J] . Nature Biotechnology, 2006, 24 (6) : 703-707.

[122] Xu F, Zhao H, Feng X, et al. Single-cell chemical proteomics with an activity-based probe: Identification of low-copy membrane proteins on primary neurons [J] . Angewandte Chemie International Edition, 2014, 53 (26) : 6730-6733.

[123] Chen D, Fan F, Zhao X, et al. Single cell chemical proteomics with membrane-permeable activity based probe for identification of functional proteins in lysosome of tumors [J] . Analytical Chemistry, 2016, 88 (4) : 2466-2471.

[124] Chen Q, Yan G, Gao M, et al. Ultrasensitive proteome profiling for 100 living cells by direct cell injection, online digestion and nano-LC-MS/MS analysis [J] . Analytical Chemistry, 2015, 87 (13) : 6674-6680.

[125] Bendall S C, Nolan G P. From single cells to deep phenotypes in cancer [J] . Nature Biotechnology, 2012, 30 (7) : 639-647.

[126] Song Y, Tian T, Shi Y, et al. Enrichment and single-cell analysis of circulating tumor cells [J] . Chemical Science, 2017, 8 (3) : 1736-1751.

[127] Ni X H, Zhuo M L, Su Z, et al. Reproducible copy number variation patterns among single circulating tumor cells of lung cancer patients [J] . Proceedings of the National Academy of Sciences of the United States of America, 2013, 110 (52) : 21083-21088.

[128] Shen Q L, Xu L, Zhao L B, et al. Specific capture and release of circulating tumor cells using aptamer-modified nanosubstrates [J] . Advanced Materials, 2013, 25 (16) : 2368-2373.

[129] Zhang P, Zhang R, Gao M, et al. Novel nitrocellulose membrane substrate for efficient analysis of circulating tumor cells coupled with surface-enhanced Raman scattering imaging [J] . ACS Applied Materials & Interfaces, 2014, 6 (1) : 370-376.

[130] Zhang Y, Tang Y, Sun S, et al. Single-cell codetection of metabolic activity, intracellular functional proteins, and genetic mutations from rare circulating tumor cells [J] . Analytical Chemistry, 2015, 87 (19) : 9761-9768.

[131] Yang L, Wang Z, Deng Y, et al. Single-cell, multiplexed protein detection of rare tumor cells based on a beads-on-barcode antibody microarray [J] . Analytical Chemistry, 2016, 88 (22) : 11077-11083.

[132] Liu S, Tian Z, Zhang L, et al. Combined cell surface carbonic anhydrase 9 and cd147 antigens enable high-

efficiency capture of circulating tumor cells in clear cell renal cell carcinoma patients[J]. Oncotarget, 2016, 7(37): 59877–59891.

[133] Guan Z C，Jia S S，Zhu Z，et al. Facile and rapid generation of large–scale microcollagen gel array for long–term single–cell 3D culture and cell proliferation heterogeneity analysis [J] . Analytical Chemistry，2014，86 (5) : 2789–2797.

[134] Yu Y，Tsang J C H，Wang C，et al. Single–cell RNA–seq identifies a PD–1 (hi) ILC progenitor and defines its development pathway [J] . Nature，2016，539 (7627) : 102–106.

撰稿人：向 宇 熊 斌 何 彦 李景虹

化学反应动力学研究进展

一、引言

在过去的几十年中，化学反应动力学研究取得一系列重大成就：不断揭示各类化学反应的本质，弄清了大气臭氧层破坏的机理，揭示了重要表面催化反应的机制，推进了化学激光的发展，发现了 C60 分子等。1990 年以来，有 7 项诺贝尔化学奖的获奖内容与化学反应动力学密切相关。这些化学反应动力学研究的里程碑充分表明了化学反应动力学是化学学科中异常重要和活跃的前沿研究领域，并指明了化学反应动力学研究的发展趋势：动力学研究逐渐深入到了原子分子以及量子态的层次；分子动态过程的研究由基态的动力学扩展到了激发态的动力学研究；化学反应动力学的研究也由单势能面的绝热动力学深入到了多势能面的非绝热动力学研究；反应过程的时间尺度的研究也从微秒、纳秒走向飞秒、阿秒。另一方面，动力学研究的发展也从简单的气相体系扩展到更复杂的体系，如与大气、燃烧、表面、生物等学科紧密相关的重要体系，为与能源和环境等重要科学技术相关的化学反应过程的研究提供了重要的实验和理论基础。

二、国内外化学反应动力学发展

化学反应动力学以量子力学为理论基础，以分子束、激光等技术方法为主要实验手段，从微观角度研究基元反应，以其对化学反应的深刻认识推动着化学学科不断向深度和广度发展，物理学与化学相互交叉是推动化学反应动力学发展的动力。20 世纪 80 年代中期，李远哲等发展了交叉分子束技术，实现了在振动态分辨水平上对化学反应动力学的研究，并因此获得诺贝尔化学奖。自那时起，量子动力学理论以及新实验方法的广泛运用极大地推动了化学反应动力学研究领域的发展，30 年来，有多项与化学反应动力学直接相关的研究成果获得诺贝尔化学奖：1986 年，Herschbach 等以分子反应动力学的奠基性研究获得诺贝尔化学奖。他们将交叉分子束技术应用于化学反应动力学研究，实现了在单次碰撞条件下的化学反应即真正的基元反应动力学，使得反应动力学相关研究领域得到了蓬勃发展，极大地推动了化学学科的发展。1995 年，Crutzen 等因对臭氧分子损耗反应动力学的研究澄清了大气臭氧层破坏的机理，使得世界各国采取统一行动，阻止了臭氧层损耗的持续发展，因而获得了诺贝尔化学奖。1996 年，Smalley 等在以超声分子束激光光谱技术研究半导体等材料的团簇结构和形成动力学机制的机器上，意外地发现了 C60 分子，开创了纳米科学发展的新纪元，从而获得诺贝尔奖。1999 年，Zewail 将飞秒激光光谱技术应用于化学反应动力学研究，为化学家提供了直接观测化学反应动力学过程和过渡态的技术手段，开创了飞秒化学，从而获得了诺贝尔化学奖。2002 年，Fen、Tanaka 开创性地发展了气相生物大分子的实验研究方法，为生物大分子结构和反应机理的精确实验研究提供了新的途径，因此获得诺贝尔化学奖。2007 年，德国化学家 Ertl 以表面化学反应机理的研究成果获得诺贝尔化学奖，他通过描述氢在金属表面的吸附作用、氨合成的分子机理和固体表面的催化过程，奠定了现代表面化学的基础理论和方法学。

此外，1992 年 Marcus 由于化学反应中的电子转移理论而获诺贝尔化学奖；1998 年 Kohn，Pople 因建立了量化计算方法使得化学科学更加精确化而获奖；2013 年 Karplus、Levitt 及 Warshel 因为发展了大分子动力学理论而获得诺贝尔化学奖。这一系列化学发展中的里程碑式的成就充分说明化学反应动力学是化学学科中异常重要和活跃的前沿领域。

利用现代技术，化学反应动力学研究在许多方向上获得了重要发展和突破。例如，Neumark 等发展和利用负离子光电离能谱实验方法对反应过渡态开展的一系列研究，Zare 等利用 PHOTOLOC 技术在 $H+D_2$ 反应中发现了一些新的现象，刘国平等利用交叉分子束离子成像技术对不少多原子反应进行了研究，取得了许多有意义的结果。

我国的化学反应动力学研究于 20 世纪 70 年代后期开始起步，中国科学院大连化学物理研究所、中国科学院化学研究所、中国科学技术大学等机构相继开展了化学反应动力学方面的实验和理论研究，1987 年国家计委批准建设的分子反应动力学国家重点实验室于 1992 年开始正式开放运行。进入 21 世纪以来，以分子反应动力学国家重点实验室为主要研究基地的我国化学反应动力学研究取得了一系列具有国际水平的研究成果，他们利用自行研制的、高分辨率的交叉分子束仪器，结合反应力学理论方法的发展，在量子过渡态结构以及相关反应动力学等方面的研究取得了突破性的进展。在化学反应动力研究领域所取得的研究成果多次发表在国际顶级学术期刊上，并两次获得国家自然科学奖二等奖（2008 年和 2014 年），两次入选中国十大科技进展新闻（2006 年和 2007 年）。 张存浩院士因其在化学反应动力学等领域开创性的工作及所取得的成就获得了 2013 年度国家最高科学技术奖。这一切都说明我国反应动力学领域的研究已经达到了国际领先的地位，特别是理论与实验的紧密结合已成为反应动力学领域里的一个标杆和典范。

三、我国化学反应动力学近期研究进展

在原子和分子的层次上研究和理解化学反应机理的课题都可以归为化学反应动力学领域。自量子力学诞生以来，尤其是近十几年来，随着实验技术和计算方法以及计算机能力的提高，对所有包括化学反应在内的一切过程的研究都逐步深入到原子分子的层面，所以一方面各研究领域的发展都开始涉及化学反应动力学，另一方面化学反应动力学领域自身的发展也已经影响到其他研究领域。本节将就化学反应动力学研究领域近年来的发展做一个简单梳理，因反应动力学研究已经渗透到化学学科的各个方向，本文的总结并没有囊括近年来化学反应动力学研究的全部成果。

（一）基元化学反应动力学

微观层次的原子和分子的运动遵循着量子力学的规律，因此只有在量子态分辨的水平上研究化学反应动力学才能提供化学过程的完整描述。实际发生的各种化学反应过程非常复杂，为了寻求化学反应的基本规律和机理，用基本的物理概念清晰明确地描述化学反应，就必须深入研究基元化学反应的动力学。测量单次碰撞条件下分子间量子态分辨的基元反应散射微分截面，并依此给出反应过程中的微观过程和机制，是化学反应动力学的最重要也是最基本的研究内容。

三原子基元化学反应是最简单的化学反应，特别是含氢（氘）的三原子体系，如 H/F/Cl+H_2/D_2/HD 等，由于比较容易实现全量子的势能面和动力学计算，一直都是反应动力学领域的基本研究案例，对这些反应的认识深度代表着反应动力学领域的研究水平。F+H_2/D_2/HD 是最重要的化学激光体系之一，而 Cl+H_2 的反应是氯与碳氢化合物化学作用的原型反应，也是大气化学、环境化学和光化学中的一个基本过程。2001 年以来，杨学明等对上述含氢三原子反应体系进行了一系列深入的系统研究[1-3]在 2013 年之前，他们主要关注的是处于振转基态的氢分子的反应，然而分子振动对化学反应有着极其重要的影响，长期以来一直是化学反应动力学研究领域的一个重要课题。但 H_2 是没有极性的分子，只能通过 Raman 激发制备其振动激发态，而 Raman 激发的效率一般很低，激发态 H_2 的反应动力学研究从未在交叉分子束实验中得以实现。

2013 年杨学明、肖春雷等通过自主研发窄线宽的 OPO 激光，在利用 Stark-induced adiabatic Raman Passage（SARP）技术高效制备振动态激发分子方面取得了重大进展，对 HD 分子从（v=0，j=0）到（v=1，j=0）的激发取得了高于 90% 的效率[4]。但是由于该技术需要高度聚焦的激光束，限制了总激发效率，致使该技术无法在交叉分子束实验中得到应用。通过不断尝试，他们掌握了利用受激 Raman 激发在分子束中高效制备振动激发态氢分子的技术，使得在交叉分子束中研究 HD（v=1）散射动力学成为可能。利用该实验技

术，他们对 F+HD（v=1）反应进行研究，发现在后向散射信号随碰撞能的变化曲线上存在振动现象。为解释实验发现，张东辉等进一步提高了 F+H_2 体系势能面的精度。在新的势能面上，理论与实验取得了高度吻合。理论研究证实实验所观察到的振动现象是由束缚在产物 HF（v'=4）绝热振动曲线上的两个共振态所引起的。研究发现，HF（v'=4）绝热振动曲线在反应物端与 HD（v=1）态相关联，因而它们只能通过 HD 的振动激发来探测，而不能通过平动能的增加而进入[5]。这项研究表明，对于化学反应，分子振动激发不仅提供能量，也能开启新的反应通道，使我们能观察到在基态反应中所无法观察到的共振现象，这对提高我们对化学反应本质机理的认识有着非常重要的意义。

2015 年，肖春雷、杨学明等利用该技术研究了 Cl+HD（v=1）→ DCl+H 反应。前人的研究工作表明这一反应与 F+H_2 反应体系有很大的差异，在 HD 处于基态时没有任何共振的迹象。他们利用后向散射能谱（BSS）技术发现，在后向散射的 DCl 产物信号随碰撞能的变化曲线上存在着明显的振荡现象，但前向散射信号非常小。为解释这些实验现象，孙志刚、张东辉等重新构建了 Cl+HD（v=1）→ DCl+H 反应高精度的势能面，在此基础上开展了精确的量子动力学计算。理论研究确认了该反应中共振态的存在，从而首次在 F+H_2 体系以外的三原子反应中发现了反应共振态。与以前在 F+H_2 体系中发现的 Feshbach 共振态不同，新发现的共振态兼有 Feshbach 共振态和 Shape 共振态的性质，因而寿命只有 20 fs 左右，大大短于 F+HD 反应共振态的寿命（100fs）。理论分析表明，由于 H 与 DCl 的相互作用，过渡态区域 D–Cl 化学键在第二振动激发态（v_{DCl}=2）的绝热势能曲线上明显被“软化”，使得该绝热势能曲线在反应过渡态区域形成一个明显势阱，这与 HD 基态反应中过渡态区域明显存在的势垒有很大的差别。由于 Cl+HD（v=1）→ DCl+H 反应主要是沿着该绝热势能曲线进行，共振态对其有重要影响，从而使该化学反应显现出明显的化学反应共振特征。研究还发现共振显著提升该化学反应的反应速率常数并且极大地影响了产物的振转态分布，因此对于认识该化学反应有着重要的意义。进一步的理论分析表明，此类化学键“软化”现象是由于反应过渡态附近的非谐性所导致的，而几乎所有的化学反应的过渡态附近都存在非常大的非谐性，因而往往能在振动激发态绝热势能面上能造成一定的势阱，并有可能支持共振态，所以化学反应共振态在反应物振动激发态反应中很可能是一个普遍现象[6]。该研究还能帮助我们认识燃烧化学等过程中普遍存在的分子激发振动态反应的动力学真面目。

即便是简单的三原子分子的化学反应动力学研究也会有非常重大的意义。地球原始大气中氧气的起源机制一直是人们关注的课题。原始大气中含有大量的 CO_2，CO_2 如何生成氧分子就成了该问题的一个重要焦点。田善喜研究组利用自主研制的负离子速度成像装置研究了 CO_2 分子电子贴附解离（Dissociative Electron Attachment，DEA）动力学过程[7]，发现其中有稳定 O_2 的产生，即 e^-（15.9~19.0eV）+CO_2 → C^-+O_2。其反应截面较“三体复合反应”和“直接光解”的截面大。这是一个以前未被发现的“氧气起源”新途径。

SO_2 光化学在大气化学和地球化学中具有关键作用。南京大学谢代前等基于高精度的量子化学从头算方法，构建了 SO_2 单重态 $^1B_1$ 和 $^1A_2$ 激发态的非绝热势能面，以及绝热的基态 $^1A_1$ 与三重态激发态 $^3B_1$ 和 $^3A_2$ 的势能面[8]。对于 $^1B_1$ 和 $^1A_2$ 的非绝热势能面，他们使用基于电子波函数系数的准非绝热化的方法来处理两个态之间的锥形交叉耦合，并采用 Chebyshev 传播方法计算了 SO_2 的 B 带电子吸收谱，计算结果与实验吻合。理论结果表明，$^1B_1$ 和 $^1A_2$ 态的锥形交叉耦合是造成 B 带电子吸收谱表现出丰富的振荡特征的主要原因。为了研究 SO_2 光激发过程中振电耦合的非绝热动力学性质，他们采用了 3- 态（$^1B_1$，$^1A_2$，$^3B_1$）模型，假定 SO_2 分子在光激发后快速弛豫到 $^1A_2$ 上的低的振动激发态，然后用含时波包方法模拟了不同同位素（^{32}S，^{33}S，^{34}S，^{36}S）取代的 SO_2 分子从 4 个低振动态（000，100,010,001）非绝热转换到三重态的几率。研究表明，对同位素 $^{36}SO_2$，从（000）、（100）和（001）振动态出发都具有最大的衰减速率，而 $^{33}SO_2$ 的（010）态也具有大的衰减速率，从而合理解释了最新实验观测到的 ^{36}S 和 ^{33}S 的反常大的同位素效应。理论和实验共同揭示了 S 同位素在 SO_2 激发态动力学中的非质量同位素效应，可作为地球及其他行星科学中重要的示踪器[9]。

范德华相互作用对于许多化学和生物过程如生物大分子三级结构的形成、蛋白质晶体的电子隧穿等非常重要，人们普遍以为范德华阱在化学反应中是不可避免的。2017 年，边文生课题组在对 C（^1D）+D_2 反应的研究工作中却发现弱的范德华力在这个势阱控速反应的入口谷处形成了范德华鞍，这种结构在低碰撞能下表现出与范德华阱完全不同的动力学作用。他们提出范德华鞍的概念并给出了一般定义，揭示了范德华鞍引起的新反应机理和范德华鞍在势阱控速反应中的一般意义[10]。

多原子分子化学体系比三原子更为普遍，而且还有许多三原子反应所没有的动力学现象。然而多原子体系反应动力学和激发态分子反应动力学的研究无论在理论还是实验方面都远比简单三原子反应体系困难。2013 年，张东辉等充分利用对称性对 $H+CH_4 \rightarrow H_2+CH_3$ 反应体系进行减维处理，由此对 $H+CD_4 \rightarrow HD+CD_3$ 反应进行的态 – 态微分截面的计算结果与实验结果高度吻合，从而首次在理论上获得了一个多原子体系精确的态 – 态微分截面[11]。

精确的势能面是量子动力学计算的基础，张东辉等发展和利用了一系列有效的方法，极大地提高了利用神经网络（NN）方法构造势能面的拟合精度，成功构造了 OH_3 和 HO+CO 这两个四原子体系最为精确的势能面[12, 13]，也构造了 H/F/Cl+CH_4 体系的势能面。在此基础上，刘舒和张东辉等首次在理论上对 $H+H_2O$ 初始基态、第一对称和反对称伸缩振动激发态的反应进行了全维态 – 态量子动力学研究。计算发现对称和反对称激发态反应表现的非常相似，更重要的是产生的 OH 都只有很小一部分在 v=1 态上，分布比例和基态与激发态的相对反应性非常接近，从而证实了 local mode 图像，以及不反应的 OH 键在反应中作为旁观者[14]。该研究团队对 $H'+CH_4 \rightarrow CH_3H'+H$ 取代反应及其同位素类似物进行了精确的量子动力学研究。该反应是最简单的经由背面攻击瓦尔登翻转机理实现的反应。研究发现，反应的阈值能量远大于势垒高度，并且反应显示出不同的同位素效应：阈值能量以上的反应截面具有很强的正向二级同位素效应，即 $H'+CH_4 \rightarrow CH_3H'+H$ 反应的反应性显著大于 $H'+CHD_3 \rightarrow CD_3H'+H$ 反应；而反应的热速率常数表现出反向二级动力学同位素效应，即 $H+CH_4$ 反应的速率略小于 $H+CHD_3$ 反应[15]。

（二）表面化学反应动力学

在许多过程中，化学反应是在具有催化功能的界面或者表面上发生的。在原子分子层次上研究表面相关的化学反应过程对于弄清表面反应的机理非常重要。相对于气相系统，即使是单个吸附在表面上的分子的化学体系都极其复杂，涉及表面和体相的大数量的原子，包含了表面电子激发、分子吸附和脱附、电荷输运、能量的传递和耗散等复杂过程。

自 1972 年日本科学家 Fujishima 和 Honda 发现 TiO_2 的光催化解离水的作用以来，TiO_2 在光催化方面的应用和研究得到了日益广泛的关注。前人发现，在 TiO_2 光催化分解水过程中，甲醇的加入能够提高产氢效率，同时甲醇本身也能光催化产氢。但是甲醇光催化产氢的化学反应机理并不十分清晰。杨学明团队利用自行研制的基于高灵敏度质谱的表面光化学装置，系统地研究了单分子层（ML）甲醇覆盖的金红石 TiO_2（110）表面在紫外光照射后的反应动力学过程。早期结果表明甲醇在光照过程中通过 O–H 键及 C–H 键的断裂形成甲醛，解离出来的大量氢原子转移到旁边的桥氧原子（BBO）上形成 BBO–H，没有形成氢气[16]。他们进一步研究发现在 TiO_2 表面升温过程中，一部分氢原子会夺取表面的桥氧原子先以水（H_2O）的形式从表面脱附出来，产生表面氧空位，只有少量的氢原子会复合成氢气脱附出来。随着表面氧空位浓度的增加，桥氧上剩余的氢原子则更容易结合成氢气分子脱离[17]。他们又比较了波长 266nm 和波长 355nm 光照条件下甲醇分子光致解离的量子产率。发现 266nm 比波长 355nm 光照产率高两个数量级左右，而 TiO_2 对 266nm 的吸收效率仅为 355nm 的两倍[18]。这表明光子能量（及其所激发的电子 – 空穴对的能量）对光催化效率的有着重要的影响。这一结果对广为接受的认为光催化反应速率主要取决于光照产生的有效电子 – 空穴对的数目的光催化模型提出了挑战。

激发态电子结构在 TiO_2 光催化过程的电荷转移中起着非常重要的作用，因此研究激发态电子结构将有助于我们理解光催化过程的微观机理。之前人们认为费米能级以上 2.4 eV 处的电子激发态是吸附质诱导的。杨学明团队通过扫描激发光波长，用双光子光电子能谱（2PPE）的方法重新研究了 TiO_2 的激发态电子结构，实验结果表明这个电子激发态并不是吸附质导致的，而是 TiO_2 本身所固有的，并且与 Ti 离子的还原密切相关[19]。周传耀等通过变波长 2PPE 进一步证实 TiO_2（110）费米能级以上 2.5 ± 0.2eV 处的电子激发态是一个与 Ti^{3+} 相关的固有电子态。刘利民和 Selloni 等应用基于杂化泛函（HSE06）的密度泛函理论计算证实了实验结果，并且明确了带隙态的 dxy 属性和激发态的 dxz/dyz/dz^2 属性[20]。

TiO_2 有金红石（rutile）、锐钛矿（anatase）和板钛矿三种晶体结构，rutile 型 TiO_2 得到的研究最多，但普遍认为 anatase 表面的光催化产氢效率要高于 rutile 表面。杨学明团队细致地研究了甲醇分子在 anatase-TiO_2（101）表面的光化学反应。结果表明，甲醇分子在 anatase-TiO_2（101）表面的光化学行为和在 rutile-TiO_2（110）表面很相似。但是，由于 anatase-TiO_2（101）表面结构的特异性，桥氧原子比 rutile-TiO_2（110）表面更稳定，在升温过程中，甲醇解离产生的桥氧氢原子不会夺取桥氧原子生成水（H_2O）从表面脱附出来，进一步分析表明大约有 40% 左右的桥氧氢原子以氢气形式脱附出来，这就解释了 anatase-TiO_2（101）表面产氢效率较高的原因[21]。

在表面化学动力学理论方面，研究者们力争从量子力学第一原理出发计算和阐述极其复杂的表面化学反应机制。水及其同位素分子在金属表面的解离吸附是一个典型和具有实际意义的量子动力学研究范例。谢代前等研究了 HOD 在 Cu（111）表面的选键解离动力学。结果表明，HOD 局域伸缩振动模式的激发可以有效地增强激发键的断裂，这是由于反应途径上的“后”过渡态以及水分子的较慢的分子内振动能量重整所致[22]。在六维近似下，H_2O/HOD/D_2O 在反应中质心位置和方位角固定在过渡态处，忽略了分子在表面的质心平移运动和方位角影响。刘天辉、傅碧娜、张东辉等发展了七维的量子动力学模型，把方位角这一自由度包括在动力学计算中。研究发现，把方位角固定在过渡态所处的角度（也就是六维模型）显著高估 H_2O 的解离吸附概率，因而六维模型是很不精确的。同时，他们还发现对六维不同方位角概率的平均可以较好地给出七维吸附概率，称该方法为方位角平均法[23]。这项研究不仅提高了人们对 H_2O 解离吸附动力学的认识，并提出了一个能比较精确处理多原子分子在固体表面散射的方位角问题的方法，对气相 - 表面动力学理论研究的发展有着重要的意义。他们紧接着又完成了 H_2O 在金属表面解离全维（九维）量子动力学方法的发展，并进行了 H_2O 在 Cu（111）表面解离动力学的计算，从而首次实现了对一个三原子分子在固体表面量子散射的全维研究。研究发现，H_2O 的振动激发，无论是伸缩振动还是弯曲振动，都能显著增加解离概率。伸缩振动能最为有效，其次是弯曲振动能，它们都比平动能更有效促进反应的进行，而反对称伸缩振动只比对称伸缩振动的效益高一点。他们的研究还发现其提出的质心位置平均方法能很好的给出全维的结果，进一步验证了质心位置平均方法的可靠性[24]。

李微雪等开展了表面结构敏感性对合成气转化反应活性的影响的理论研究。煤、生物质、页岩气来源的合成气，经费托合成制液体燃料是满足能源需求的重要途径。钴基催化剂因其较高的长链烷烃选择性和活性、稳定性，受到长期关注。费托合成反应是一个低温有利的强放热反应，寻找低温、高活性的钴基催化剂是目前研究的关键，需要深入研究相应催化反应的结构敏感性。长期以来，人们发现合成气转化活性显著依赖于钴催化剂的晶相结构：当六角密堆晶相 HCP-Co 的含量较高时催化活性较高，而当面心立方密堆

FCC-Co 含量较高时催化活性则相对较低。通过第一性原理计算的方法，他们发现 HCP 晶相比 FCC 晶相金属钴对 CO 活化具有更高的本征催化活性。同时，在 HCP-Co 上，CO 倾向于直接解离过程；在 FCC-Co 上，CO 需要通过氢助的路径进行解离（CO+H → HCO，HCO → CH+O）。HCP-Co 和 FCC-Co 的显著不同主要是由于 HCP-Co 和 FCC-Co 具有不同的晶体结构对称性和形貌，导致在 HCP-Co 存在不同于 FCC-Co 的大量特异性的活性位。通过晶相和形貌来控制催化反应的活性为设计稳定、高效的高活性催化剂提供了一种新的方向[25]。

（三）团簇与催化反应机制

随着世界范围内富含甲烷的页岩气、天然气水合物、生物沼气等的大规模发现与开采，以储量相对丰富和价格低廉的天然气替代石油生产液体燃料和基础化学品成为了学术界和产业界研究和发展的重点。具有四面体对称性的甲烷分子是自然界中最最稳定的有机小分子，它的选择活化和定向转化是一个世界性难题，被誉为是催化，乃至化学领域的“圣杯”，长期以来一直是国内外科学家研究的主题。包信和等基于“纳米限域催化”的新概念，创造性地构建了硅化物晶格限域的单铁中心催化剂，成功实现了甲烷在无氧条件下选择活化，一步高效生产乙烯、芳烃和氢气等高值化学品的研究成果。根据该催化过程的动力学行为推测该转化过程为甲烷在催化剂表面产生甲基，继而甲基在气相中偶联生成乙烯、苯和萘等产物。樊红军等从理论上说明该过程是否为自由基机理主要是探索是否存在合理的从甲基到乙烯、苯和萘的转化机理，并且该机理所推测的反应性质与实验观察是否吻合。实验上唐紫超等利用新型的常温催化反应器与真空紫外单光子电离分子束质谱对催化反应中的甲基自由基和中间体进行探测。该反应的实验条件、产品组成及分布与催化剂最佳优化条件相一致。他们准确地探测到甲基自由基和中间体，所得到的产物分布、温度特性、转化率等与理论预测非常吻合，为验证甲烷无氧制乙烯催化机理提供了重要的实验证据[26]。

2013 年，何圣贵等建立了用于团簇光反应研究的两级反射式飞行时间质谱方法，成功应用于甲醇光反应机理研究[27]。他们还建立了激光溅射制备 - 四极杆选质 - 离子阱约束并反应 - 高分辨率反射式飞行时间质谱检测的团簇反应装置[28]，该装置成功应用于一系列团簇反应，在单原子化学反应性研究方面取得了进展。他们开展了大尺寸镧氧团簇正离子（La_2O_3）$_N^+$（N=1~8）和钪氧团簇负离子（Sc_2O_3）$_N$$O^-$（$N$=1~18）与甲烷、丁烷等小分子的反应研究，发现其碳氢活化的反应性具有显著的尺寸依赖性，其中（La_2O_3）$_N^+$（N=2 和 5）以及（Sc_2O_3）NO^-（N=4 和 12）的活性高于相邻尺寸的团簇。密度泛函理论研究很好地解释了实验中发现的尺寸效应，表明团簇的反应中心为氧自由基（O^-），其局部的电荷和自旋分布控制了团簇反应性的尺寸效应。他们继而开展了 CO 在纳米尺寸 TiO_2 和 ZrO_2 团簇上的氧化研究。TiO_2 与 ZrO_2 纳米粒子是常见的催化材料（催化剂或催化剂的载体），催化材料表面的活性氧是导致 CO 等分子低温氧化的关键物种，活性氧中的原子氧自由基

（O^-）非常活泼，难以被捕获表征，他们将氧自由基制备到 TiO_2 和 ZrO_2 纳米尺寸团簇上。质谱实验发现，在不同载体上氧自由基的反应性显著不同：在（TiO_2）$_N$（N=3~25）上生成气态的 CO_2，在（ZrO_2）$_N$（N=3~25）上产生吸附态的 CO_2，这一发现在分子水平上解释了如下重要事实：在CO低温催化氧化中，TiO_2 材料表现出高稳定性，而 ZrO_2 却快速失活[29]，研究工作被选为 JACS 封面并被作为研究亮点进行了报导[30]。何圣贵等还对钛氧团簇担载的 Au 原子在 CO 氧化过程中所起的作用进行了研究[31]，他们在研究中还发现在钒氧团簇阴离子担载的金二聚体 Au_2 的辅助下，O_2^{2-} 物种可以直接或间接氧化 CO[32]。

就像 C60 的发现过程一样，对团簇动力学的研究还会发现新的物种分子。李隽、王来生等通过结合光电子能谱特征实验、全局结构搜索和高精度量子化学理论计算，首次在气相中观察到由双链交织而成、具有完美 D_{2d} 对称性的空心笼状 $B_{40}^{-/0}$ 全硼富勒烯团簇（All-Boron Fullerenes），并将其命名为硼球烯（Borospherenes）。硼球烯 B40 是继 C60 之后第二个从实验和理论上完全确认的无机非金属笼状团簇，标志着硼球烯实验和理论研究的开端[33]。

氧化态是化学中最重要的基本概念之一，是元素的固有性质，它反映了元素在化合物中及其在反应过程中电子的得失能力。2014 年，周鸣飞等采用脉冲激光溅射 – 超声分子束载带技术在气相条件下制备四氧化铱正离子，并利用串级飞行时间质谱 – 红外光解离光谱装置成功测得气相四氧化依离子［IrO_4］$^+$ 的红外振动光谱，结合量化计算证实四氧化铱离子中的铱处于 +9 价态，从而首次在实验上确定了元素 +9 价态的存在[34]。在此基础上，他们进一步对过渡金属和镧系金属的高氧化态开展了研究，通过低温基质隔离红外光谱和串级飞行时间质谱 – 红外光解离光谱实验获得了 PrO_2^+ 络合物以及 PrO_4 中性分子的红外振动光谱，结合理论计算证实了其中的 Pr 处于 +5 价态，从而确认了镧系金属元素最高氧化态可以达到 +5 价[35]。这一成果对镧系金属氧化价态的拓展及成键特性的理解具有重要科学意义，也为进一步宏观合成该类高氧化价态化合物及其在化学反应体系中的应用提供了基础。

（四）复杂及生物大分子体系动力学

在现实世界中，化学过程大多不是孤立或者原子数目较少的简单的体系，多数是处于和周围环境不可分的状态（如液相或固相中的化学过程），或者所涉及的原子数目非常巨大（如生物大分子）。对这样复杂化学体系的机理性研究是化学反应动力学必须面对的挑战。在过去的几年里，随着实验技术的发展和计算机能力的提高，我国复杂体系化学动力学研究也取得了一系列重要进展。

配位键是金属有机材料中最基本的结构，对分子激发态的结构和性质影响巨大。金属有机超分子配合物可以形成一系列特殊的框架结构，空腔大小达到纳米量级，表现出很好的主客体性能，可以与其他物质发生多种非键相互作用，金属有机配体作用使得这一类物质具有特殊的电子结构。由于这一系列特殊结构与性质，使得金属有机超分子配合物常常表现出独特的光化学和光物理特性。研究这些超分子配合物的激发态动力学性质对于人们理解和应用其光化学和光物理特性有非常重要的理论和现实意义。韩克利等采用时间分辨光谱实验与量子化学理论计算相结合的方法研究了一系列不同结构的超分子体系的激发态动力学性质，深入分析了重金属原子和配位结构在分子激发态动力学中的作用[36]。

内源性过氧化亚硝酰根（$ONOO^-$）已被确认是一种生物体内的强氧化剂。过氧化亚硝酰根的这种化学性质使得它成为生物所患各种疾病的主要致病因子，过氧化亚硝酰根甚至可以作为一种潜在的候选抗癌药物。韩克利等首次实现了对水溶液和细胞内的氧化还原循环对过氧化亚硝酰 / 谷胱甘肽的近红外荧光检测。该工作实现了在分子、细胞和活体三个层次上对具有生理氧化还原活性的物种的原位、实时、动态荧光成像分析[37]。该研究成果在一定程度上揭示了生物体内氧化还原过程对生物生理和病理的影响，为进一步的生物医学研究提供了有力的实用工具。

线型三奎烷是一种重要的自然产物，其合成过程一直是学术界研究的热点。韩克利等采用密度泛函理论对用 *N*– 氮丙啶亚胺化合物作为原料生成线型三奎烷的反应进行了理论研究，结果表明该反应要经过四个步骤：氮丙啶三元环裂解、1，3– 偶极环加成、脱氮和分子内［3+2］环加成。脱氮会生成单重态和三重态的三亚甲基甲烷双自由基中间体，单重态三亚甲基甲烷双自由基的分子内［3+2］环加成遵循协同的反应机理且具有较低的能量，是占优势的反应路径。此外，研究还发现单重态三亚甲基甲烷双自由基有两种环加成模式，分别经过两个过渡态生成两种立体选择性不同的产物，两个过渡态的能量差与产物的立体选择性结果一致[38]。该项研究揭示了该类反应的详细反应机理，确认了最低能量反应路径和反应的决速步，阐明了立体选择性产生的原因。

π 堆积体系在自然界中广泛存在，比如有机半导体、液晶和 DNA 双螺旋结构等。由于体系中存在较强的 π–π 耦合，因此体系中电荷迁移较为容易。利用这一特性，π 堆积体系常应用于电子器件中，如有机半导体晶体管、DNA 分子导线等。不同于传统的半导体体系，有机 π 堆积体系的电荷迁移服从跃迁机制，因而常规的半导体理论方法无法用于描述其电荷迁移行为。而且由于有机分子的多样性，有机 π 堆积体系也种类繁多。如何从上百万种可能中找到所需性能的有机 π 堆积体系是这个领域的关键问题。一个能够准确定量化预测有机 π 堆积体系电荷迁移的理论方法是解决这一关键问题的核心。邓伟侨、韩克利团队基于前期工作，提出进一步完善的理论方法，做到仅根据 π 堆积体系的晶体结构就能预测该体系的载流子迁移率，得到的预测结果与实验结果吻合得很好[39]。这个成果为使用计算机大规模筛选所需性能的 π 堆积体系打下坚实基础。

在生物大分子领域，李国辉等与实验合作者密切合作，针对组蛋白 H3 第 4 位赖氨酸的甲基转移酶 MLL 修饰底物过程的动态学分子机制，提出和设计了结合增强型采样技术的分子动力学模拟方案，以及利用 QM/MM/MD 方法计算底物结合和催化过程自由能的研究思路。动态学模拟发现，在有辅助蛋白存在情况下，MLL 酶活性口袋区域关键残基的动态学特性有很大变化：辅助蛋白的存在导致了活性口袋区域的关键残基运动更加倾向于一致性，适合于稳定底物；没有辅助蛋白存在时关键残基运动趋于无规则，不利于底物结合。对 MLL 甲基化酶不同家族成员的整个甲基化修饰过程而言，底物结合要比底物催化更具有选择性，整个修饰过程的限速步在于底物结合[40]。

即使采用化学反应动力学传统的实验技术也可以对复杂生物现象做出机理性探索。郭源等用高分辨宽带和频振动光谱研究了气 / 液界面 Ca^{2+} 对鞘磷脂（ESM）单分子膜的结构和取向的影响，澄清了 Ca^{2+} 与 ESM 相互作用的机理。Ca^{2+} 首先与 PO_2^- 基团结合，导致 PO_2^- 部分脱水，极性头基部分发生再取向。随着 Ca^{2+} 不断加入，PO_2^- 基团与鞘氨醇骨架上的 3–OH 之间的分子内氢键也遭到破坏，使得鞘氨醇骨架变得更加有序，同时骨架末端甲基基团的取向由指向液面变为指向空气相[41]，从而在分子水平上给出了 Ca^{2+} 与 ESM 膜相互作用的基本物理图像，有助于深入理解神经细胞信号传导的分子机理，了解生物体内电解质对神经传导影响的机制。夏安东、王江云等通过基因编码的方法将一系列电子受体苯丙氨酸类似物引入绿色荧光蛋白，利用飞秒瞬态吸收光谱研究了绿色荧光蛋白中的光致电子转移过程。他们测量到电子转移发生在皮秒范围（接近光系统 I 中最快的电子转移步骤）。利用晶体结构研究测量了发色团到电子受体之间的距离，揭示了该电子转移过程是距离依赖的过程（与光系统 I 一致）[42]。

三、未来发展趋势和对策

在过去的 5 年内，我国化学动力学领域的研究者们在化学反应动力学传统领域基元化学反应动力学取得长足进展的同时，从理论和实验两方面努力将研究方向扩展到了激发态反应、多势能面的非绝热过程、表面化学反应、复杂体系（液相、固相等）、大分子（包括团簇和生物分子）体系动态学等方面，并取得了令人瞩目的研究成果。今后几年，随着化学动力学的实验技术和理论方法以及计算机能力的进一步提高和扩展应用，特别是国家重大科研仪器设备专项“基于可调极紫外相干光源的综合实验研究装置”项目目前已经顺利启动，将于 2018 年投入科研工作，我国的化学动力学的研究水平将得到极大的提高，使我们有能力面对化学动力学领域来自学科自身发展和国家能源、环境以及国防需求的挑战。

基元化学反应动力学是化学动力学理论的试金石和一切复杂化学反应动力学研究的出发点和基础。人们将把在三原子反应上取得成功的理论与实验紧密结合的研究拓展到多原子反应，在前所未有的精度和广度上研究多原子体系的反应动力学问题，特别是在燃烧过程和化学激光中扮演着重要角色的基元化学反应。实验上利用和发展各种分子束技术，尤其是发展基于极紫外相干光源的新一代通用型交叉分子束装置、激光受激拉曼泵浦、受激辐射泵浦以及红外激发泵浦技术、高能原子分子束技术。理论上发展多原子反应体系高精度势能面构造方法和精确的量子动力学理论方法。通过理论与实验的紧密结合，研究振动激发对反应动力学的影响、高反应能条件下的反应动力学机理，发现多原子反应的新机理与规律，把气相反应动力学研究推向一个全新的高度，实现在定量水平对复杂分子反应速率的计算，为燃烧反应、激光化学、大气化学、星际化学的研究提供理论保障。

在许多化学过程中，化学反应是在具有催化功能的团簇分子表面上发生的，近年来备受关注的雾霾环境污染问题也和团簇的形成机制密切相关。我们将发展和利用基于飞行时间质谱、四极质谱和离子阱等质谱与光谱联用技术，尤其是基于红外自由电子激光和极紫外相干光源的红外多光子解离光谱、紫外－红外双共振电离/解离光谱、基于极紫外相干光源的红外光解－真空紫外电离凹陷谱等技术，开展选质量催化相关团簇的结构和反应动力学研究。通过对由气态硫酸、水分子等大气分子经由氢键或化学键作用形成的不同大小的分子簇的组成及光谱结构测定，了解气体分子凝聚产生气溶胶及雾霾的成核机制。

表界面催化研究的关键是表界面催化反应动态学，利用催化剂的表界面结构和电荷分布特性来调控基元化学反应，实现降低特定基元反应的活化能，达到高活性和高选择性的目标。化学反应动力学前沿研究将以提高传统非石油碳资源（包括煤炭、天然气、页岩气等）转化利用效率和太阳能的有效利用为导向，针对在实际反应条件下的表界面催化反应及光催化反应的特点，发展基于高亮度极紫外相干光源（大连极紫外相干光源）的新一代催化产物原位动态高灵敏探测的实验方法，以及先进的具有高空间和时间分辨性能的原位动态结构和组分表征技术，发展基于第一性原理的可靠的表界面基元反应动力学理论与计算方法，以及包含时间和压力等条件的连接微观和宏观的催化材料、催化反应模拟和预测技术，在原子分子层次上确定催化反应的活性中间体，揭示其构效关系和动态变化规律，为新一代催化剂设计提供理论基础。针对天然气、合成气的高效催化转化，开展结合气相和表界面的基元反应动力学研究，为这些重要催化工程的工业化提供理论基础和科学支撑。

参考文献

[1] 戴东旭，杨学明．基元化学反应态态动力学研究[J]．化学进展，2007，19（11）：1633–1645.

[2] 戴东旭，杨学明．化学反应过渡态的结构和动力学[J]．中国科学 B 辑：化学，2009，39（10）：1089–1101.

[3] 戴东旭，张东辉，杨学明．化学反应动力学[J]．中国科学院院刊，2011，（Z1）:80–92.

[4] Wang T，Yang T，Xiao C，et al. Highly efficient pumping of vibrationally excited HD molecules via stark–induced adiabatic raman passage[J]. The Journal of Physical Chemistry Letters，2013，4（3）：368–371.

[5] Wang T，Chen J，Yang T，et al. Dynamical resonances accessible only by reagent vibrational excitation in the F+HD → HF+D reaction[J]. Science，2013，342（6165）：1499–1502.

[6] Yang T，Chen J，Huang L，et al. Extremely short–lived reaction resonances in Cl+HD（v=1）→ DCl+H due to chemical bond softening[J]. Science，2015，347（6127）：60–63.

[7] Wang X D，Gao X F，Xuan C J，et al.Dissociative electron attachment to CO_2 produces molecular oxygen[J].

Nature Chemistry，2016，8（3）：258–263.

[8] Xie C，Hu X，Zhou L，et al. Ab initio determination of potential energy surfaces for the first two UV absorption bands of SO_2 [J]. The Journal of Chemical Physics，2013，139（1）：014305.

[9] Whitehill A R，Xie C，Hu X，et al. Vibrionic origin of sulfur mass–independent isotope effect in photoexcitation of SO_2 and the implications to the early earth' s atmosphere [J]. Proceedings of the National Academy of Sciences of the United States of America，2013，11044: 17697–17702.

[10] Shen Z, Ma H, Zhang C, et al. Dynamical Importance of van der Waals saddle and excited potential surface in C（1D）+D_2 complex–forming reaction [J]. Nature Communications，2017，8: 14094.

[11] Liu S，Chen J，Zhang Z，et al. Communication: A six–dimensional state–to–state quantum dynamics study of the H+CH_4 → H_2+CH_3 reaction（J =0）[J]. The Journal of Chemical Physics，2013，138: 011101.

[12] Chen J，Xu X，Xu X，et al. A global potential energy surface for the H_2+OH ↔ H_2O+H reaction using neural networks [J]. The Journal of Chemical Physics，2013，138: 154301.

[13] Chen J，Xu X，Xu X，et al. Communication: An accurate global potential energy surface for the OH+CO → H+CO_2 reaction using neural networks [J]. The Journal of Chemical Physics，2013，138: 221104.

[14] Liu S and Zhang D H. A local mode picture for H atom reaction with vibrationally excited H_2O: a full dimensional state–to–state quantum dynamics investigation [J]. Chemical Science，2016，7（1）：261–265.

[15] Zhao Z，Zhang Z，Liu S，et al. Dynamical barrier and isotope effects in the simplest substitution reaction via Walden inversion mechanism [J]. Nature Communications，2017，8: 14506.

[16] Guo Q，Xu C，RengZ，et al. Stepwise Photocatalytic Dissociation of Methanol and Water on TiO2（110）[J]. Journal of the American Chemical Society，2012，134（32）：13366–13373.

[17] Xu C，Yang W，Guo Q，et al. Molecular Hydrogen Formation from Photocatalysis of Methanol on TiO_2（110）[J]. Journal of the American Chemical Society，2013，135（28）：10206–10209.

[18] Xu C，Yang W，Ren Z，et al. Strong photon energy dependence of the photocatalytic dissociation rate of methanol on TiO_2（110）[J]. Journal of the American Chemical Society，2013，135（50）：19039–19045.

[19] Mao X，Lang X，Wang Z，et al. Band–gap states of TiO_2（110）：major contribution from surface defects [J]. The Journal of PhysicalChemistry，2013，4（22）：3839–3844.

[20] Wang Z，Wen B，Hao Q，et al. Localized excitation of Ti^{3+}ions in the photoabsorption and photocatalytic activity of reduced rutile TiO_2 [J]. Journal of American Chemical Society，2015，137（28）：9146–9152.

[21] Xu C，Yang W，Guo Q，et al. Molecular hydrogen formation from photocatalysis of methanol on anatase–TiO_2（101）[J]. Journal of the American Chemical Society，2014，136（2）：602–605.

[22] Jiang B，Xie D，Guo H，et al. Vibrationally mediated bond selective dissociative chemisorption of HOD on Cu（111）[J]. Chemical Science，2013，4（1）：503–508.

[23] Liu T，Zhang Z，Fu B，et al. A seven–dimensional quantum dynamics study of the dissociative chemisorption of H_2O on Cu（111）：effects of azimuthal angles and azimuthal angle–averaging [J]. Chemical Science，2016，7（3）：1840–1845.

[24] Zhang Z，Liu T，Fu B，et al. First–principles quantum dynamical theory for the dissociative chemisorption of H_2O on rigid Cu（111）[J]. Nature Communication，2016，7: 11953.

[25] Liu J X，Su H Y，Sun D P，et al. Crystallographic dependence of CO activation on cobalt catalysts: HCP versus FCC[J]. Journal of the American Chemical Society，2013，135（44）：16284–16287.

[26] Guo X，Fang G，Li G，et al. Direct，Nonoxidative conversion of methane to ethylene，aromatics，and hydrogen [J]. Science，2014，344: 616–619.

[27] Yuan Z，Zhao Y X，Li X N，et al. Reactions of $V_4O_{10}^+$cluster ions with simple inorganic and organic molecules [J]. International Journal of Mass Spectrometry，2013，354–355: 105–112.

[28] Yuan Z, Li Z Y, Zhou Z X, et al. Thermal reactions of (V_2O_5) nO^- (n=1–3) cluster anions with ethylene and propylene: oxygen atom transfer versus molecular association [J]. The Journal of Physical Chemistry, 2014, 118(27): 14967–14976.

[29] Ma J B, Xu B, Meng J H, et al. Reactivity of atomic oxygen radical anions bound to titania and zirconia nanoparticles in the gas phase: low–temperature oxidation of carbon monoxide [J]. Journal of the American Chemical Society, 2013, 135 (8): 2991–2998.

[30] Hellemans A. Finally caught in the act: atomic oxygen radical anions essential in the oxidation of carbon monoxide [J]. Journal of the American Chemical Society, 2013, 135 (14): 5229–5229.

[31] Li X N, Yuan Z, He S G. CO oxidation promoted by gold atoms supported on titanium oxide cluster anions [J]. Journal of the American Society, 2014, 136 (9): 3617–3623.

[32] Yuan Z, Li X N, He S G. CO oxidation promoted by gold atoms loosely attached in $AuFeO_3^-$ cluster anions [J]. The Journal of Physical Chemistry Letters, 2014, 5 (9): 1585–1590.

[33] Zhai H J, Zhao Y F, Li W L, et al. Observation of an all–boron fullerene [J]. Nature Chemistry, 2014, 6: 616–619.

[34] Wang G, Zhou M, Goettel J T, et al. Identification of an iridium–containing compound with a formal oxidation state of IX [J]. Nature, 2104, 514:475–477.

[35] Zhang Q, Hu S X, Qu H, et al. Pentavalent lanthanide compounds: formation and characterization of praseodymium(V) oxides [J]. Angewandte Chemie International Edition, 2016, 55 (24): 6896–6900.

[36] Chen J S, Zhao G J, Cook T R, et al. Photophysical properties of self–assembled multinuclear platinum metallacycles with different conformational geometries [J], Journal of the American Chemical Society, 2013, 135 (17): 6694–6702.

[37] Yu F, Li P, Wang B, et al. Reversible near–infrared fluorescent probe introducing tellurium to mimetic glutathione peroxidase for monitoring the redox cycles between peroxynitrite and glutathione in vivo [J], Journal of the American Chemical Society, 2013, 135 (20): 7674–7680.

[38] Qiao Y, Han K L. Theoretical investigations toward the tandem reactions of N–aziridinyl imine compounds forming triquinanes via trimethylenemethanediyls: mechanisms and stereoselectivity [J], Organic & Biomolecular Chemistry, 2014, 12: 1220–1231.

[39] Deng W Q, Sun L, Huang J D, et al. Quantitative prediction of charge mobilities of π–stacked systems by first–principles simulation [J]. Nature Protocols, 2015, 10: 632–642.

[40] Li Y, Han J, Zhang Y, et al. Structural basis for activity regulation of MLL family methyltransferases [J]. Nature, 2016, 530 (7591): 447–452.

[41] Feng R J, Lin L, Li Y Y, et al. Effect of Ca^{2+} to sphingomyelin investigated by sum frequency generation vibrational spectroscopy [J]. Biophysical Journal, 2017, 112 (10): 2173–2183.

[42] Lv X, Yu Y, Zhou M, et al. Ultrafast photoinduced electron transfer in green fluorescent protein bearing a genetically encoded electron acceptor [J]. Journal of the American Chemical Society, 2015, 137 (23): 7270–7273.

撰稿人：戴东旭　杨学明

资源化学研究进展

一、引言

资源化学是化学学科的一个新研究领域，一个年轻的分支学科。近数十年来，随着化学学科分支学科不断推陈出新。如人们熟悉的生物化学被化学生物学取代，经典的环境化学被绿色化学超越等，资源化学是这些化学学科分支学科不断推陈出新过程中的其中一个分支学科。化学生物学、绿色化学等新化学分支学科由于首先由国外学者提出，因此毫无疑问的、很快为国内化学界同行所接受。不同于化学生物学、绿色化学等新化学分支学科，资源化学由田伟生在二十世纪九十年代提出后在国内主流化学界备受争议。“资源化学”的学术术语就曾被质疑，质疑“资源化学”来源什么英文词语。为此作者不得不专门给出“Resource Chemistry”英文术语以回复争议。值得庆幸的是国家自然科学基金委员会等基金组织、相关化学化工企业和有关化学学术期刊在资源化学在争议声中给予了大力支持和鼓励。1990 年，国家自然科学基金委批准了“甾体皂甙元与三萜皂甙元资源合理利用研究”项目，此后又继续批准了“甾体皂甙元资源合理利用研究”项目。2000 年，浙江医药集团公司仙居制药厂资助成立了“资源化学与甾体化学联合实验室”。1999 年《科学新闻周刊》刊登了田伟生撰写的“资源化学研究大有可为”文章，2010 年，化学进展特邀刊登“资源化学研究进展”综述论文。中国化学会有机化学学科委员会、化学教育学科委员会，兰州大学、北京大学等大学，中国科学院昆明植物研究所、长春应用化学研究所等科研院所先后邀请田伟生进行了四十余次学术报告，专门介绍资源化学概念与研究进展。最近在中国化学会支持下由西北大学成功地承办了首届资源化学学术研讨会，并且确定了第二届会议承办单位。有了这些支持才使资源化学领域的研究工作得以坚持，才使资源化学这一年轻的化学分支学科得以生存、成长壮大。

资源化学是一个被唤醒的古老的学科。二十世纪八十年代末，田伟生刚获得独立开展有机化学学科研究工作的机会，作为一个准备投身化学科学研究工作的人必须首先认真确定自己领导团队的研究目标、方向与领域。“我为什么要做研究”“我能研究什么问题”，带着这些基本问题，面对社会可持续发展中“资源日益匮乏和环境污染加剧”严重问题和社会公众对化学及其化学工业的指责，基于化学，特别是有机合成化学的发展历史和现状，作者认识到：社会公众对化学及其化学工业的指责是要求化学家担当起更多地社会职责，社会可持续发展中的“资源日益匮乏和环境污染加剧”问题应该是化学家面临的挑战性研究任务。故田伟生选择了旨在解决社会可持续发展中的“资源日益匮乏和环境污染加剧”问题作为自己研究团队的研究目标。为了便于与学术同行及社会公众交流，作者把自己选择的研究领域定义为“资源化学”，即专门从事研究资源利用过程中的化学问题。在资源化学研究过程中作者进一步认识到：尽管资源化学的学术术语被我们正式提出至今不到三十年，但是资源化学是一个事实上已经存在的古老自然科学学科。讲其古老是因为它的实践活动一直伴随着人类利用物质资源的全过程。是资源利用导致了化学及其相关分支学科诞生，同样是资源利用过程出现的问题唤醒了资源化学。人类利用物质资源的过程中逐步了解了物质的组成、性质和转化并产生化学学科，化学学科的诞生促进了人类利用资源的效率和速度。但是由于人类利用资源理念的历史局限，过度粗放型利用资源导致了资源浪费和环境污染的问题，从而影响到人类社会可持续性发展。转化资源利用模式，用精准利用资源模式替代粗放型利用资源模式是解决资源浪费和环境污染的问题的必然之路。如何实现精准利用资源物质？在分子水平上利用资源是精准利用资源的最佳方式和途径，但是这需要化学家研究给出相应资源型化合物特定的化学性质及其可实际应用的技术。当前许多物质资源的组成分子已经被化学家研究清楚，越来越多的物质分子已经被化学家发现和发明，这些为在分子水平上利用资源提供了基础。资源化学的学科研究任务就是进一步深入了解清楚资源型化合物的各种化学和生物学性质，为合理、精准利用资源提出科学依据和相关技术。

资源化学是解决资源浪费和环境污染的利器，同时也为化学学科发展带来新机遇的分支学科。化学学科及其分支学科产生于人类利用资源的实践活动，但是随着学科深入发展，化学学科研究越来越纯学术化，越来越脱离人类社会实践活动。社会公众对化学及其化学工业的问责，社会可持续发展中“资源日益匮乏和环境污染加剧”问题，化学学科发展中找不到研究目标的困惑促使化学家从纯学术研究的理想天堂回归到人类社会现实。化学家在深入研究资源型化合物的性质并形成相应技术就能精准利用资源，实现精准利用资源就能减少资源浪费，没有资源浪费也就从源头消除了环境污染。毫无疑问，资源化学是解决资源浪费和环境污染的利器。近数十年来，人类在化学物质的发现和发明方面能力得到了充分的提高。以处在化学中心的合成化学而言，早在上世纪六十年代，诺贝尔化学奖评审委员会在表彰著名化学家伍德沃德在有机合成领域为人类做出的贡献时就宣称：合成化学的主人是自然界，若有第二当属伍德沃德。九十年代，著名化学家科瑞总结的有机合成中“逆合成分析”使有机合成不再神秘。只要有足够的研究经费和时间，人类可以合成任何想合成的分子不再是梦想！在化学反应的研究中，从经典的官能团转化研究已经发展到实现不活泼 C–H，C–C 的立体选择性转化。化学家下一步研究什么？化学学科研究在许多国家已经不被看好！但是，在资源型化合物利用中还存在许多挑战性的任务。资源型化合物结构复杂且存在多种官能团，化学家在实验室理想状态下进行的许许多多理想反应不能很好地应用于资源型化合物。如何使化学在人类利用资源中，在社会可持续发展中发挥更多作用这是资源化学的任务，而这也为化学学科研究工作带来了新机遇。

在此专题报告中作者将着重介绍资源化学概念和作者自己在资源化学研究中取得的成果。报告内容包括三个部分，即资源化学概念，资源化学实践与成果和资源化学学科发展展望。

二、资源化学概念[1]

广义上讲，资源化学就是研究物质资源的化学组成及其反应和应用的专门科学学科。尽管资源化学的学术名称被我们正式提出时间不久，但是资源化学的实践活动却一直伴随着人类利用资源活动的全过程，资源化学应该是一门最古老的自然科学学科。人类在利用资源的过程中逐步了解了物质资源的组成、组成成分的物理化学和生物学性质导致化学及其相关分支学科的相继诞生。例如，为了专门研究来源于有机体中的有机化合物时产生了无机化学与有机化学；为了专门分析物质组成和分析方法建立了分析化学；为了展开对高分子量物质的专门研究而设立的高分子化学；专门研究天然物质的化学被定义为天然产物化学，天然产物化学又被进一步细分为植物化学、动物化学、生物化学、天然海洋化学等；有机合成化学专门研究天然的和非天然的分子的合成及合成方法学的科学；专门针对用途的化学研究有医药化学、制药化学、农药化学、香料化学、染料化学以及食品化学等。鉴于大宗数量资源所涉及的矿产资源、石油资源、煤炭资源以及林产资源的化学研究已经建立了相应的化学分支学科，如矿产化学、冶金化学、石油化学、煤炭化学、林产化学等。与此同时，由化学化工工业及其相关工业生产的化学结构、化学物理性质和生物学性质多样化的资源型化合物，特别是资源型有机化合物的深入研究工作并未真正得到化学家普遍重视。因此，我们提出并希望给予重视的资源化学是指资源型化合物的化学，特别是资源型有机化合物的化学，而非广义的资源化学。

大宗型资源如煤炭、石油、矿产，它们的化学组成与结构、性能与使用已经得到了充分的研究并且得到实际应用。由于规模效应缘故，它们的利用已经趋于合理、精准，大化工企业即便产生了废弃物也便于进行后治理。与大宗型资源利用情况不同，化合物资源的特点规模小、种类多、结构复杂、化学、物理和生物学性质千变万化。由于化合物资源研究的难度高、加上它们作为资源型化合物属性的可变性大，故长期以来很难得到人们重视。正因为此化合物资源宝贵财富在利用过程中缺乏科学依据，许多化合物资源得不到合理利用。粗放型利用化合物资源过程中不仅造成资源浪费，而且导致严重环境污染。特别值得注意的是：由于进行与化合物资源利用有关的企业多数情况下其企业的综合能力有限（资金、技术和环保意识等）、生产规模小、生产地点分散等客观原因使化合物资源利用过程中产生的环境污染根本无法得到后治理。人们很习惯把环境污染的恶名戴在化学化工企业的头顶上，而产生环境污染的企业是属于那些利用化合物资源的企业。但这些企业是与药品、农药、精细化工产品生产密切相关的企业。由于我们不可能通过简单地让企业关门停产就能解决问题，解决此问题的根本出路是政府管理部门大力支持科研院所和大学加强资源化学学科领域研究为相关企业提供合理、精准利用化合物资源的科学依据和技术，从源头解决问题。

化合物是保持物质基本属性的分子，资源型化合物仅仅是指那些数量可规模化、价格低廉、方便易得，甚至被当作废弃物抛弃的化学分子。化合物由天然来源和人工合成获得，以此资源性化合物同样被分为天然资源性化合物和人工合成的资源型化合物。前者主要来源为动植物、微生物等生物体组成成分和体内代谢产物，如纤维素、甲壳素、木质素、蛋白质、各种天然糖、羟基酸、氨基糖、氨基酸、萜类和甾体、生物碱等；后者基本为化学、制药和精细化工工业产品生产过程中的副产物、中间体和产品。资源型化合物和非资源型化合物处在动态变化之中，随其规模、价格和需求的变化而变化。

资源化学研究内容包括两部分：①资源型化合物的发现与开发；②资源型化合物的反应及其在具有重要应用背景的目标化合物合成中的应用。

资源型化合物的发现与开发是一项挑战性任务。发现和开发适合于具有重要应用背景的目标化合物分子合成的资源型化合物对于在读化学的研究生、青年学者、甚至长期工作在实验室仅从事某一研究领域的专家来说确实是一件挑战性任务。因为在实验室从事化学研究的学者一般很少考虑研究成本问题，而化学书籍和科学期刊中给出的化合物信息中同样很少介绍它们的规模、价格。即使化学试剂手册中给出了试剂的价格，但也很难代表它们真实的价值。尽管如此，从众多被发现和发明的化合物中确定资源型化合物的设想通过化学家的努力还是可以实现的。生物多样性是自然界的基本规律，生物多样性导致了生物相互之间的互补性，从而实现了生物界的和谐共存。在结构和生物活性多样性的天然产物中同样存在着十分有趣的互补性，在自然界当发现一些生物活性高、数量微小的分子时候，总会发现还存在一些与前者结构类同的生物活性低但数量上成规模的化合物。如此在从自然界获得某目标分子的结构和生物活性信息后，人们总是可以在自然界寻找出高效合成所需分子及其家族化合物的合成前体，即资源型化合物。依据此，化学家通过综合来自天然产物化学家发表研究结果以及通过与相应作者深入交流，结合天然产品加工专家提供的信息而方便地确定那些天然化合物可以成为天然资源型化合物。成功的实例如甾体皂甙元、巴卡亭Ⅲ（baccatin Ⅲ）、青蒿酸等资源型化合物的发现与应用。十九世纪，化学家相继从人体和动物体发现了性激素、孕激素和皮质激素，并了解了它们在人体和动物体内的重要生理作用。在人们希望大量获得这些甾体化合物并使它们成为药物的动力推动下，经过探索研究化学家发现可以从天然界大规模开发获得一些甾体分子、如甾体皂甙元等资源型化合物作为合成药物的原料。上世纪七十年代，美国化学家发现存在于紫衫树皮中的紫衫醇能有效地治疗乳腺肿瘤。但是，紫衫树皮中紫衫醇的含量仅为万分之几，根本无法满足临床治疗需求。有趣的是化学家进一步研究在紫衫树枝叶中发现了适合于合成紫衫醇的前体巴卡亭Ⅲ，它的天然含量要比紫衫醇高出十倍。资源型化合物巴卡亭Ⅲ的发现与开发不仅保障了紫衫醇类药物的高效生产，并且有效地保护紫衫树种。青蒿素是中国化学家发现的当今最有效的抗疟疾药物原料，于其共存于青蒿植物中的青蒿酸无疑是青蒿素的最佳合成原料，青蒿酸生物合成技术的工业化应用更是为高效合成青蒿素合成提供了保障。此外，有机合成化学已经有近二百年的发展史，合成化学的许多成果已经被成功地应用于生产并形成了化学、制药和精细化工工业产品行业。根据文献报道研究成果结合化学、制药和精细化工工业产品信息完全可以确定许多人工合成的资源型化合物。

资源型化合物的反应及其在具有重要应用背景的目标化合物合成中的应用是资源化学研究工作的中心内容。一些资源性化合物如氨基酸、羟基酸、单糖和单萜的反应及其在药物和生物活性分子合成中应用已有文献综述，作者课题组也对天然资源型化合物甾体皂甙元和人工合成资源型化合物氟烷基磺酰氟的反应和合成进行了系统研究，但是还有许多大家熟知的资源性化合物的化学有待研究，如糠醇、氨基葡萄糖等。由于化石燃料和电能的广泛应用，在农村农作物杆秸不必继续作为燃料使用。目前一些地方焚烧处理农作物杆秸，不仅浪费资源而且污染环境。利用现有技术从作物杆秸获取糠醇和糠醛，它们作为资源性化合物完全可以被用作为药物、农药和精细化工产品原料。甲壳素在规模上是仅次于纤维素的生物物质，其单体氨基葡萄糖是合成许多天然产物以及抗生素的资源型化合物。柠檬苦素类化合物存在于柑橘类水果的皮与籽中，它们结构奇特，数量可观属于萜类生物活性分子合成的资源性化合物。在我国利用甾体皂甙元合成甾体药物过程中，每年废弃手性资源型化合物达上千吨，如何利用它们合成医药、农药和生物活性目标化合物是十分值得研究的课题。三萜类天然化合物广泛存在于许多中药中，如甘草、人参、党参、柴胡、远志等。甘草成分甘草酸被用于食品、烟草和药品加工生产。为了获取甘草酸，我国西北和华北地区的植被一直受到严重破坏。已知来源于植物楤木等植物中的资源型化合物齐墩果酸与甘草酸的化学结构十分近似，若能转化齐墩果酸成为甘草酸，不仅可以解决甘草酸来源问题，还可从根本上解决因乱挖甘草造成的植被破坏问题。如何将齐墩果酸高效转化成为甘草酸对化学家来说是一个挑战性的课题，也属于当今有机化学热点研究领域 C–H 官能团中的一个实实在在的问题。被我国剑麻加工工业废弃的资源型化合物剑麻皂甙元可以作为制药工业的宝贵原料，但是一直成为污染环境的污染物。用其代替薯蓣皂甙元合成甾体药物不仅可以减少环境污染，还可以降低甾体药物生产成本。薯蓣皂甙元是资源型化合物，也是我国生产甾体药物的主要原料，它在薯蓣属植物根茎中含量约3%左右。提取薯蓣皂甙元的生产工艺在用酸水解薯蓣皂甙元皂甙时一直把同时存在的占总量90%以上的淀粉、多糖和纤维素当作废弃物投放到环境，不仅浪费了资源，还严重污染了环境。由于以化合物资源为导向的资源化学研究领域还未真正被重视，产业界普遍缺乏合理利用化合物资源的科学依据和技术，因此造成我国的化学化工及其相关行业比大化工行业的环境污染问题更为严重。精准利用资源需要化学家花大力气对资源型化合物的反应和合成应用展开研究，作者相信资源化学在提高资源利用精准度和社会可持续发展中一定大有可为。

资源化学研究目的旨在通过资源型化合物反应和应用的研究，为提高资源精准利用，减少资源利用过程中的资源浪费和环境污染提供科学依据；为解决资源浪费和环境污染问题提供策略和实用技术。

三、资源化学实践与成果

作者在提出资源化学概念后，带领自己的研究团队围绕着天然资源型化合物甾体皂甙元和非天然资源型化合物氟烷基磺酰氟的反应及其在有机合成中应用方面进行了系统研究。本章节将通过我们在资源化学实践中取得的部分研究成果进一步说明资源化学研究工作的必要性和重要性。

（一）天然资源型化合物甾体皂甙元的反应及其在有机合成中应用研究进展

资源型化合物包括天然资源型化合物和非天然型化合物，甾体皂甙元被我们选择作为天然资源型化合物的研究对象。甾体皂甙元是一类含螺甾基本结构的化合物，在天然植物中则以与糖链接的皂甙形式存在。甾体皂甙主要存在于薯蓣科、百合科、玄参科、菝葜科、蒺藜科和龙舌兰科等植物中，目前人工种植的黄姜是提取薯蓣皂甙元的主要植物原料。选择天然资源型化合物甾体皂甙元作为研究对象的主要原因为：它们是我国甾体药物生产的基本原料。我国薯蓣皂甙元年生产量最高达四千吨以上，剑麻皂甙元年产量约在五百吨以上且大部分作为环境污染物被废弃。其次是甾体皂甙元的生产历史超过半个世纪，其生产工艺十分成熟，因此甾体皂甙元的价格极为便宜，最低价格每公斤仅一百五十多元人民币。并且由于从薯蓣植物根茎中提取薯蓣皂甙元时仅利用了其资源的3%左右，其余的淀粉、纤维素和鼠李糖等尚未被利用，故其生产成本仍然存在降低空间。最重要的原因是由于甾体皂甙元的生产技术过于陈旧，不仅存在严重环境污染问题，而且存在资源浪费问题。甾体皂甙元是中国的特色化工产品，其利用过程中的问题不受国际上专家关注，故解决甾体皂甙元资源型化合物的研究工作只能由中国化学家自己承担。

甾体皂甙元裂解物的洁净氧化降解反应。我国甾体药物生产基本上以薯蓣皂甙元作为起始原料，利用甾体皂甙元合成甾体药物首先需要氧化降解其为孕甾烯酮醇。如图1

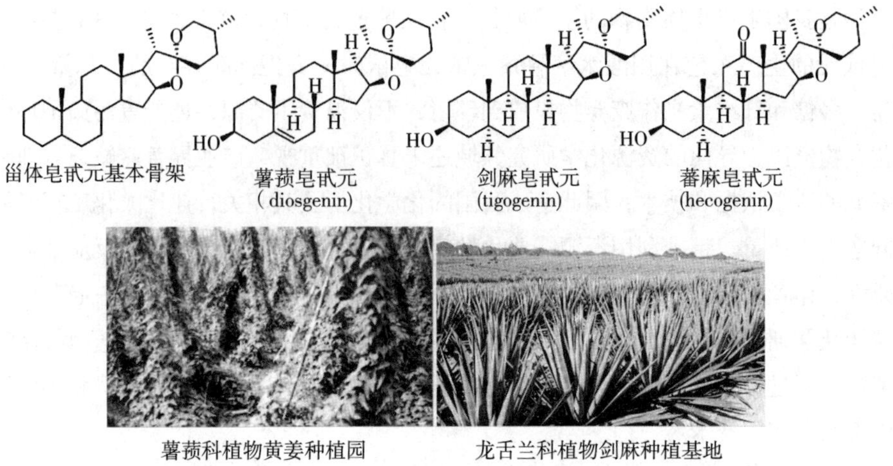

图 1　甾体皂甙元的化学结构以及含甾体皂甙元植物的种植基地

所示，实现这一转化的基本方法是依据美国化学家 Marker 研究给出的氧化降解反应[2]。Marker 反应在 20 世纪 40 年代被应用于工业生产，我国从五十年代起一直沿用至今。转化甾体皂甙元成为孕甾烯酮醇的关键反应是其裂解产物的氧化反应。Marker 反应采用三氧化铬作为氧化试剂氧化降解甾体皂甙元裂解物，每氧化降解一吨甾体皂甙元将会产生约四吨含金属铬盐的废弃物。按照我国以往每年消耗四千吨以上甾体皂甙元计算，将有上万吨以上的含金属铬盐的废弃物产生。由于这些环境污染物分散在边缘山区的小化工厂中，它们的后续处理实际上无法进行。此外，在含金属铬废弃物中还含占降解产品近三分之一（大约一千吨）的手性分子被作为废弃物处理，造成甾体皂甙元资源利用过程中的严重浪费。

针对甾体皂甙元资源利用过程中的这一重大环境环境污染问题，我们系统地研究了用双氧水和氧气代替铬酐氧化降解甾体皂甙元的反应，研究发现用 30% 双氧水商品可以代替铬酐氧化降解甾体皂甙元裂解物成为相应的孕甾烯酮醇[3]。特别值得注意的是我们发展的反应不仅解决了甾体皂甙元铬酐氧化降解的金属铬环境污染问题，还可以方便地回收氧化降解产生的手性分子进一步转化成为宝贵的手性试剂，变废为宝，提高了甾体皂甙元的利用率，实现了原子经济性利用地甾体皂甙元资源的目标。基于这一反应的甾体皂甙元洁净氧化降解技术在完成一百公斤规模实验后于最近两年已经被转让给两家生产企业。毫无疑问，随着此技术的产业化推广应用，不仅将从源头解决我国甾体皂甙元资源利用过程中的资源浪费和环境污染的问题，也将促进我国甾体药物工业和手性药物的发展。

双氧水作为氧化剂直接氧化降解甾体皂甙元的反应。在图 2 反应中从甾体皂甙元制备孕甾烯酮醇的反应时，首先需要通过高温、高压裂解反应转化甾体皂甙元成为相应的甾体皂甙元裂解产物。直接氧化降解甾体皂甙元的反应将有可能为工业部门提供更安全和便捷的甾体皂甙元资源利用技术，为此我们进一步研究了双氧水在各种酸性条件下氧化降解

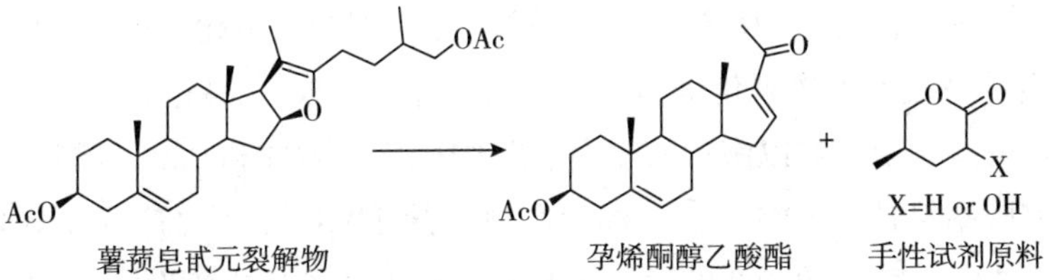

依据反应	氧化剂	产品	资源利用度	环境污染状况
Marker 反应	CrO_3	孕烯酮醇乙酸酯	75%（理论上）	严重环境污染
我们的反应	30%H_2O_2	孕烯酮醇乙酸酯和手性试剂原料	100%（理论上）	无环境污染

图 2　薯蓣皂甙元氧化降解中 Marker 反应和我们发现的反应比较

甾体皂甙元的反应。研究发现：30% 双氧水在甲酸等介质中能以优异收率转化甾体皂甙元成为相应的 Baeyer–Villiger 氧化产物（见图 3），即孕甾 –16，20– 醇类化合物和 4– 甲基戊内酯[4]。还发现卤素（如溴和碘）能够改变反应的区域选择性给出相应的非正常的 Baeyer–Villiger 氧化产物，即孕甾 –16，22– 酸内酯类化合物和 3– 甲基丁内酯[5]。前一个反应已完成数十公斤规模试验，后一反应在试验室已反复进行了上百克规模的试验，它们均适合工业化生产应用。过氧三氟乙酸可直接氧化假甾体皂甙元，给出雄甾 –16，17– 醇类化合物。它们为高效利用甾体皂甙元资源提供了新的科学依据，为高效设计和合成甾体药物以及潜在应用价值的生物活性甾体分子提供了新机遇。

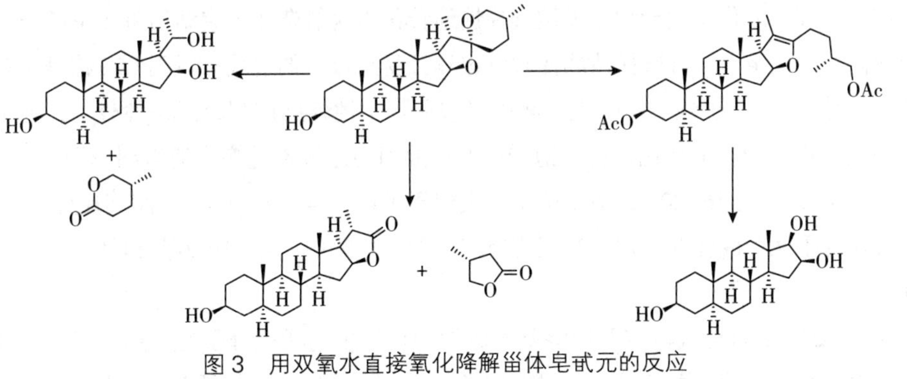

图 3　用双氧水直接氧化降解甾体皂甙元的反应

甾体皂甙元双氧水直接氧化降解产物的反应。用双氧水代替铬酐氧化降解假甾体皂甙元得到的孕甾酮醇类化合物可以直接用于甾体药物合成，而甾体皂甙元双氧水直接氧化产物孕甾 –16，20– 醇和孕甾 –16，22– 酸内酯等在有机合成鲜有文献报道。为此，我们进一步研究了它们的反应（图 4）。孕甾 –3，16，20– 三醇分子中三个仲羟基由于所处空间位置不同，它们对不同氧化剂表现出不同反应性能。与高价碘试剂，如 Dess – Martin 氧化试剂反应，可以选择性氧化其 C16 羟基[6]。孕甾三醇的微生物氧化与化学氧化具有互补性，其 C20 羟基更容易被微生物氧化得到 16- 羟基孕甾 -3，20- 二酮。孕甾三醇在溴化氢乙酸溶液（或乙酰溴）中反应以满意的收率给出 C16 溴代，C3 和 C20 乙酰化产物，乙酰化溴代产物最好收率可达到 89%[6, 7]。溴代产物碎裂可得到 C16（17）- 甾烯化合物，16- 溴代孕甾 -3，20- 二醇乙酸酯在碱性条件下可经 Grob 碎裂反应给出雄甾 -16- 烯化合物，它为甾体肌肉松弛剂和哺乳动物信息素合成提供方便[8, 9]。

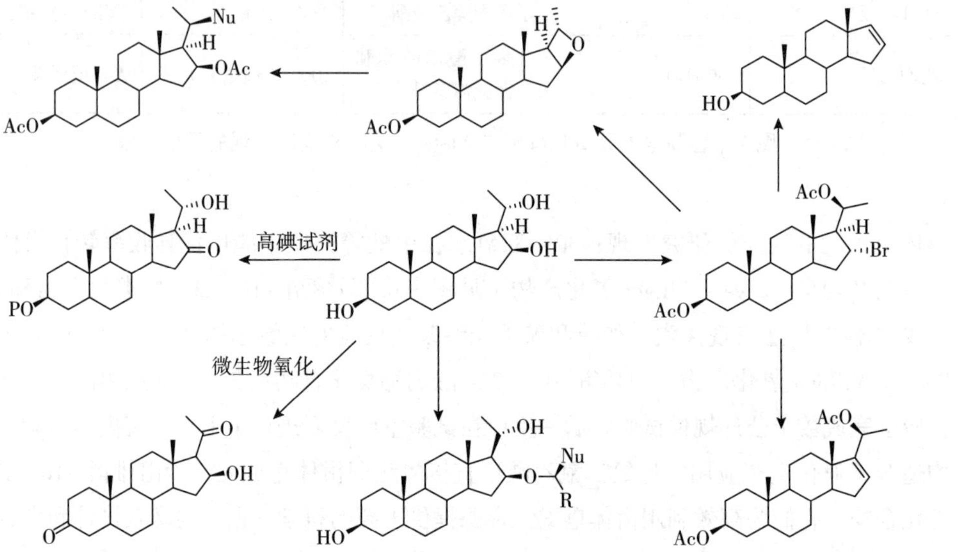

图 4　孕甾 -3，16，20- 三醇的反应

孕甾三醇乙酰化溴代产物还可被选择性地转化为含氧杂环丁烷结构单元孕甾分子[10]。该中间体在酸性条件下可选择性地在 C20 位置接受亲核试剂进攻，得到 C20 构型翻转的类天然甾体分子[11]。手性 2，4- 戊二醇曾经被用作手性辅助试剂合成手性分子，但是它在反应后被转化成为非手性化合物不饱和戊酮，故因使用成本过高未能得到应用。孕甾 -3，16，20- 三醇分子的 16，20- 二醇结构单元类似于手性 2，4- 戊二醇，可用其代替手性 2，4- 戊二醇作为手性合成辅助试剂，其反应所产生的孕烯酮醇化合物正好是甾体药物合成关键中间体。

孕甾 -22- 醛是合成具有胆固醇基本骨架的甾体生物活性分子的常用合成中间体，但其存在合成原料昂贵（如大豆甾醇或麦角甾醇）、自身化学稳定性差（容易异构化）、反应立体选择性低等问题。与此相反，孕甾 -16，22- 酸内酯合成原料便宜且制备方便，化学性质稳定且有优异的反应选择性。如图 5 所示，孕甾 -22- 醛与异戊炔基锂反应其立体选择仅 1∶1.5，而孕甾内酯经加成 / 还原反应的立体选择性高达 1∶12。这些恰好克服了利用孕甾 -22- 醛作为关键合成中间体合成天然胆甾类生物活性分子时存在的问题[12]。此外，由于孕甾内酯 C16 位存在羟基基团也为目标分子 D 环修饰带来方便。

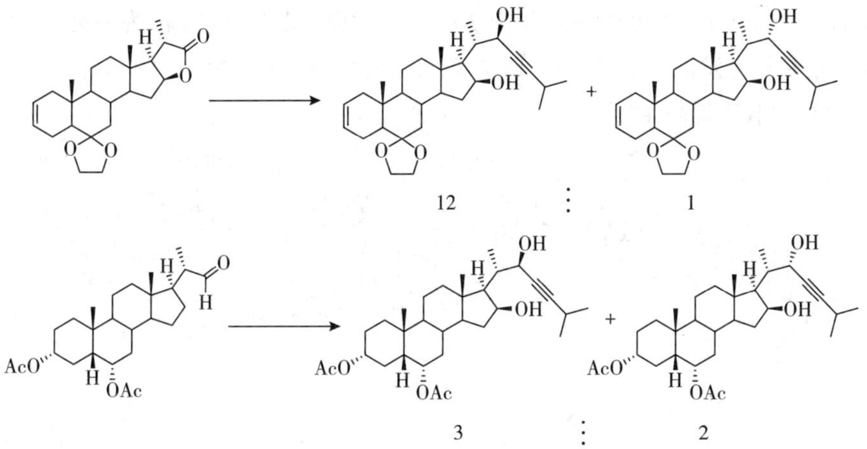

图 5　甾体内酯加成 / 还原反应立体选择性与甾体 –22– 醛的比较

甾体皂甙元双氧水直接氧化降解产物在甾体药物和天然产物合成中应用。在我们专门研究资源型化合物甾体皂甙元的反应之前，合成甾体药物和天然甾体分子一般都采用甾体皂甙元的降解产物孕甾烯酮醇或去氢表雄酮作为关键合成中间体。双氧水直接氧化降解甾体皂甙元得到的孕甾 –16，20– 醇和孕甾 –16，22– 酸内酯为甾体药物和生物活性天然甾体分子新合成策略和路线设计带来了机遇。研究表明用孕甾 –16，20– 醇和孕甾 –16，22– 酸内酯代替孕甾烯酮醇或去氢表雄酮作为关键合成中间体后确实显著提高了目标化合物的合成效率。

以猪外信息素合成为例，用孕甾三醇作为关键合成中间体，从薯蓣皂甙元出发经孕甾三醇的乙酰化溴代和碎裂等四步反应即可合成猪外信息素雄甾 –16– 烯 –3– 酮（见图 6）[8]。而文献以去氢表雄酮为关键合成中间体的方法，从薯蓣皂甙元出发，经 Marker 氧化降解反应和 Beckemann 降解反应等十步反应方可合成得到目标分子[13]。

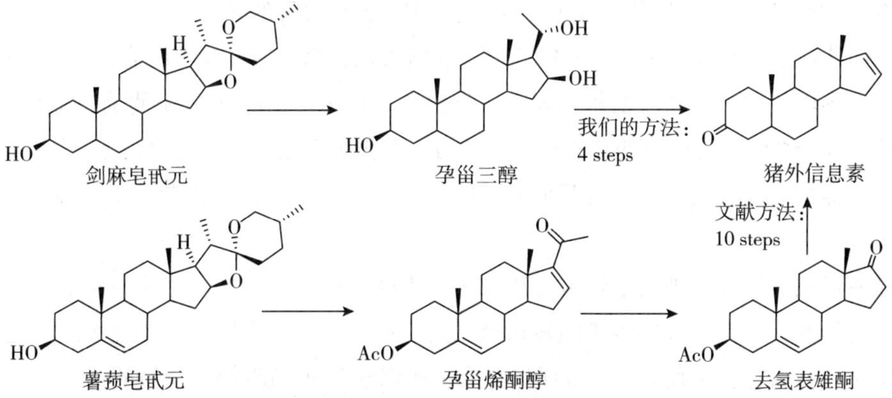

图 6　利用孕甾三醇合成猪外信息素的合成效率比较

吡嗪双甾体是一类化学结构新颖、抗肿瘤活性显著的海洋天然产物。我们以孕甾内酯作为关键中间体，从蕃麻皂甙元出发经 19 步反应，以 7.2% 总收率的合成完成了其代表化合物 cephalostatin 1 北片断的合成（见图 7）。比文献报道以孕甾烯酮醇为关键合成中间体，从相同起始原料出发的 35 步反应，1.6% 总收率的传统合成策略减少了 16 步反应（接近 50%），并使总收率提高 4.5 倍[14]。

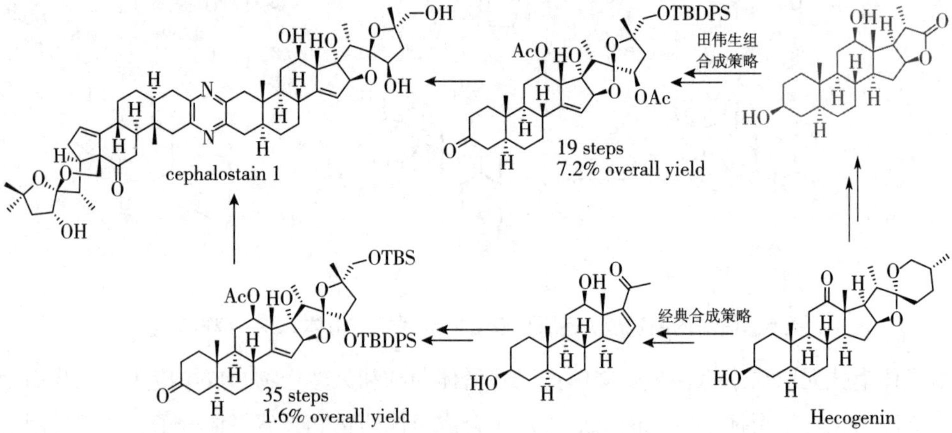

图 7　利用孕甾内酯合成 cephalostatin 1 北片段的合成效率比较

以孕甾三醇和孕甾内酯等作为关键合成中间体，我们完成了苦楝甾醇（azedarachol）[15]、白前甙元 C 和 D（glaucogenins C and D）[16,17]、17*S*–潘库溴铵（17*S*–pancuronium bromide）[18]、倍他米松（betamethasone）[19]、吡嗪双甾体 cephalostatin 1[14]、油菜甾醇内酯（brassinolide）[20–22]、海洋甾醇 clathsterol[23–25]、certonarsterol D_2、D_3 和 N_1[20] 和甾体生物碱 clionamine D[26] 等具有显著生物活性的天然甾体分子和甾体药物的高效合成。分别从孕甾三醇和孕烯酮醇出发我们还完成了 saundersiosides 的形式合成[27]。

苦楝甾醇是从日本产苦楝根皮分离得到昆虫拒食剂，活性测试显示害昆虫 *Ajrotis sejetum* Denis. 的幼虫有拒食作用。我们以孕甾三醇为原料，经其 C3– 羟基消除、双羟化和 C20– 羟基反转并选择性酯化等 12 步反应完成了苦楝甾醇的合成[15]。

白前甙元是一类双裂孕甾烷天然产物，其糖甙化合物普遍存在于白前、白薇、徐长卿以及马兰等中药中，生物活性研究表明该类化合物对 α – 正链 RNA 病毒有抑制活性，有可能开发成为抑制植物病毒的新农药。以孕甾三醇作为合成原料，经单线态氧烯反应和二价铁参与的烷氧基过氧化物碎裂两个关键反应我们完成了双氢白前甙元 C[16] 和白前甙元 D[17] 的合成。

潘库溴铵类肌肉松弛剂已经被应用作为手术辅助药物，早期研究认为其 17*S*– 异构体有更好的活性，但是合成困难未能被选择开发。以孕甾三醇为原料，经溴代乙酰化反应、碎裂反应和消除等反应我们可以方便地完成了肌肉松弛剂 17*S*- 潘库溴铵的合成[18]，为开发新型潘库溴铵类肌肉松弛剂药物提供了机会。

倍他米松是一种常用的糖皮质激素药物，从孕甾三醇出发，经溴代乙酰化、甲基铜锂试剂的亲核取代、烯醇硅醚环氧化等反应完成了倍他米松的关键中间体的高效合成[19]，

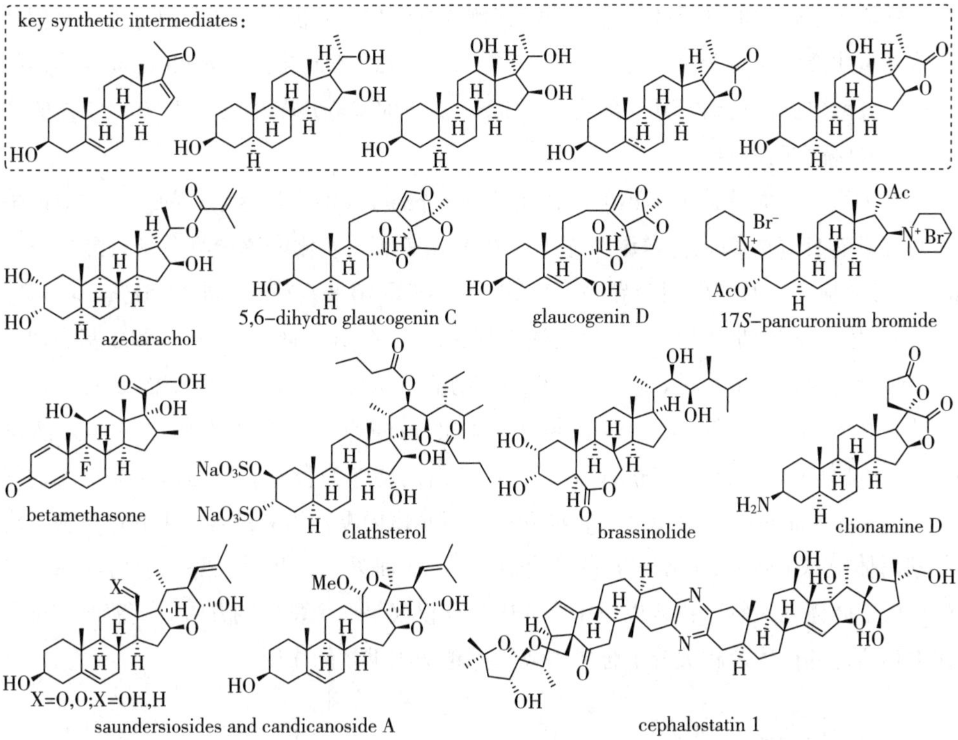

图 8　利用孕甾三醇和孕甾内酯合成的部分甾体药物和天然产物分子

为糖皮质激素药物新合成路线设计与新技术进一步开发提供了科学依据。在倍他米松合成过程中我们还发现了甾体环氧烯丙醇的串联重排加成反应。

吡嗪双甾体 cephalostatin 1 是一类化学结构新颖的、具有显著抗肿瘤活性的海洋天然产物，也是至今为止化学结构最为复杂，在合成上最具有挑战性的天然产物之一，它的合成吸引了许多合成化学家的关注[28]。从蕃麻皂甙元降解得到的孕甾四醇和孕甾内酯出发，我们分别以 27 步反应，6.2% 收率合成了其南片段，以 19 步反应，7.2% 收率合成了其北片段，然后经四步反应进行南 / 北片段连接完成了 cephalostatin 1 的合成[14]，提供一个高效实用的合成此类天然产物的方法。我们还从蕃麻皂甙元降解内酯出发合成了另一类结构独特吡嗪双甾体 ritterazine N 的类似物[29]。

油菜甾醇内酯是首先被从油菜花粉中分离得到的甾体植物生长激素，后发现它及其类似物广泛存在与植物体内。由于其对农作物的增产作用和其从自然界直接获取的困难性（在干燥油菜花粉中含量约百万分之一），油菜甾醇内酯类化合物合成一直受到合成化学家的关注。我们从薯蓣皂甙元降解得到的甾体内酯出发，经二十步反应，以 7% 总收率完成油菜甾醇内酯的克级规模合成，为油菜甾醇内酯类植物生长激素的基础研究提供了必须的样品[21, 22]。

海洋甾醇 clathsterol 是从红海海绵 *Clathria* 中分离得到的一种甾醇硫酸盐，对 HIV-1

逆转录酶具有很好的抑制活性。为了确定其侧的立体化学并发展高效合成方法，我们从剑麻皂甙元氧化降解所得孕甾内酯出发完成了 C22、C23- 不同构型的 clathsterol 异构体的合成，通过波谱数据比较确定其立体化学为 22S、23S 构型。并以 21 步反应，11% 总收率完成 clathsterol 硫酸盐的合成[25]。

自非洲海域的海绵体分离出的甾体生物碱 clionamine D，具有罕见的 γ - 螺双内酯结构，并且具有良好的细胞自噬调节作用。从剑麻皂甙元氧化降解产物孕甾内酯出发、经 Schenck 烯反应、以及三价锰介导的［3+2］环加成等 10 步反应，我们以 51% 总收率完成它的合成[26]。通过初步生物活性研究还发现它的 3- 位异构体和 3- 羟基和羰基类似物具有相似生物活性，初步判断 clionamine D 的生物活性与其 γ - 螺双内酯结构有关。

孕甾内酯作为天然甾醇合成关键中间体的策略也被姜标等人用于海洋天然甾醇 certonarsterol D_2 的合成[30]；Morzycki 等人用孕甾内酯作为合成 F 环为糖结构单元的螺甾分子“glycospirostanes”的原料[31]；俞飚等人以孕甾内酯作为关键合成中间体完成了海星多羟基甾体皂甙 linckoside A 和 B 的全合成[32]。显而易见，孕甾三醇和孕甾内酯等作为关键合成中间体在天然产物合成应用研究中目前仅仅是一个良好的开端，随着时间推移它们在生物活性的甾体及相关分子合成中或发挥越来越重要的作用。

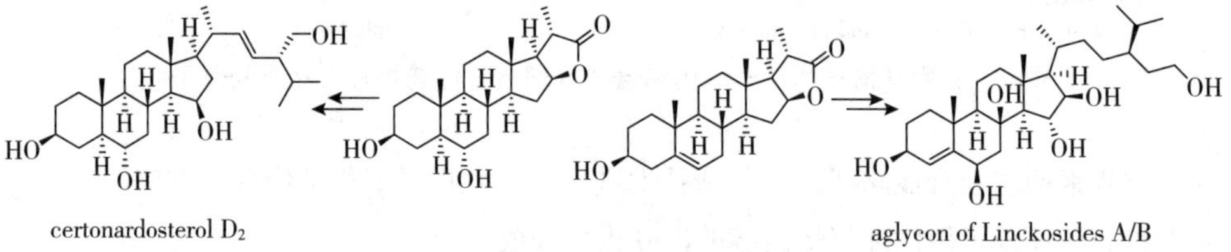

图 9　海洋甾醇 Certonarsterol D_2 和 Linckoside A 和 B 甙元的合成

甾体皂甙元氧化降解所获手性试剂及其在有机合成中的应用。如上所述，基于双氧水氧化甾体皂甙元反应而发展的甾体皂甙元洁净氧化降解技术已经开始向生产企业转让。这些技术在生产企业全部推广实施后，每年将有可能生产出近千吨带手性甲基侧链的手性试剂和手性合成原料，如何使这些从废弃物中获得的宝贵手性试剂和原料应用于工业生产、服务于社会是我们研究工作的追求目标之一。

尽管文献中已经报道了许多合成带甲基侧链的手性分子的方法，然而已知的手性试剂中，带甲基侧链的手性试剂品种少，价格昂贵。如 Aldrich 试剂公司销售的最常见的 2- 甲基 -3- 羟基丙酸甲酯，每克售价 450 元，作为工业原料每公斤也需要 6 万元（甾体皂甙元氧化降解主产品孕烯酮醇每公斤也仅仅 800 元左右）。由此可见我国在甾体皂甙元资源利用过程中不仅存在严重环境污染，同时存在巨大资源浪费。若按照 2- 甲基 -3- 羟基丙酸甲酯工业原料价格计算，我国每年在甾体皂甙元资源利用过程中被废弃的手性试剂价值约在上百亿人民币！在从甾体皂甙元氧化降解废弃物中回收到手性化合物 4*R*- 甲基 -5- 羟基戊酸内酯和 4*R*- 甲基 -2，5- 二羟基戊酸 - δ - 内酯后，我们系统地研究了它们的反应，将它们转化成为一些便于保存和方便使用的手性试剂，图 8 列举部分手性试剂的化学结构[33, 34]。这些手性试剂和原料为手性药物、农药和昂贵的香精香料等的高效合成提供了物质基础。

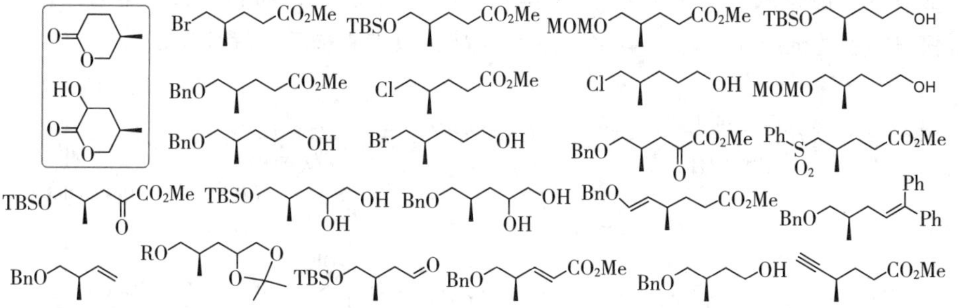

图 10　由戊酸－δ－内酯衍生的部分手性试剂

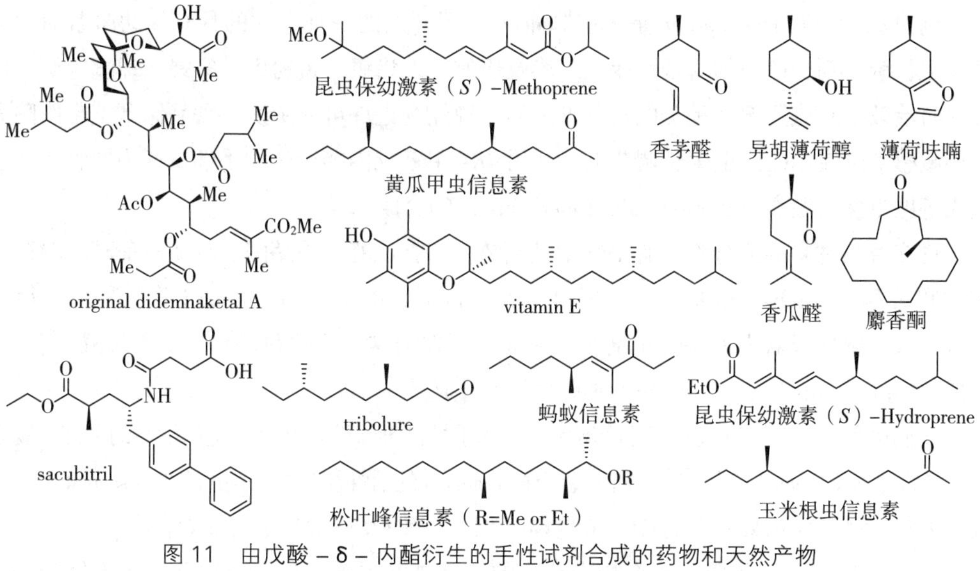

图 11　由戊酸－δ－内酯衍生的手性试剂合成的药物和天然产物

传统农药已经在防治病虫害促进农作物丰产方面发挥了巨大的作用，但是它们对土壤和水资源造成的严重污染也越来越受到人们关注。发展高效的、无环境污染的新农药是社会对化学家提出的挑战性任务。昆虫信息素是存在于昆虫体内的内源性激素物质，能够调控昆虫群体的各种行为。昆虫信息素应该是十分有效地防治各种虫害的药剂，但由于其在昆虫体内含量微少导致从天然来源获取它们十分困难。与此同时，由于合成所需的手性原料和试剂品种少、价格昂贵也限制这些昆虫信息素的高效合成。为了解决传统农药带来的环境污染问题，发展高效的合成各种不同昆虫的信息素的方法，我们利用上述已经发展手性试剂完成了一些光学活性的昆虫信息素，如松叶蜂信息素[35, 36]、玉米根虫信息素[37]、昆虫保幼激素[38, 39]、黄瓜甲虫信息素[40]、蚂蚁信息素[41]、谷盗聚集信息素 tribolure[42]

等的化学合成。

松叶蜂广泛分布于欧洲、亚洲以及北美洲，是危害针叶林的主要害虫。它的合成受到许多合成化学家的关注[37]。其中，Mori 等人从光学活性香茅醇合成松叶蜂时需经过十二步反应合成才能得到关键中间体 4– 甲基 – 溴代十二烷[43]，而利用从甾体皂甙元氧化降解所获得的手性试剂 5– 溴代 –4– 甲基戊酸甲酯，只需三步反应就可合成。

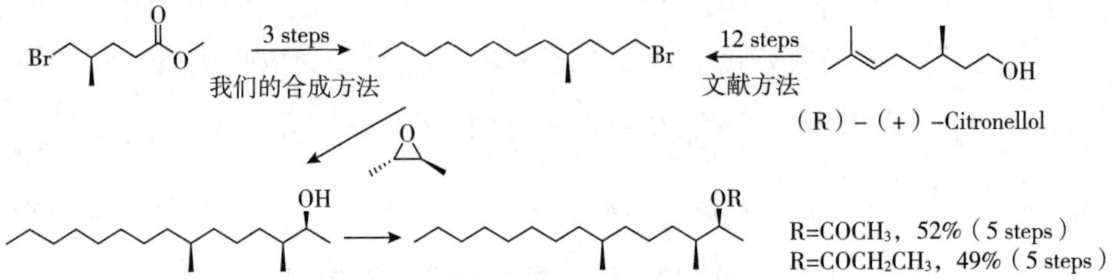

图 12　松叶蜂性信息素合成方法效率比较

昆虫保幼激素 Hydroprene 是美国 Zoecon 公司开发的一种有效的具有保幼激素活性的昆虫生长调节剂，该药剂对烟芽夜蛾、埃及伊蚊、大蜡螟、黄粉甲、家蝇、蟑螂等均有良好的防治效果。该药剂还是美国目前唯一的一种经登记许可在厨房、食品储藏室、医院和一些敏感部位使用的昆虫生长调节剂。利用 4*R*– 甲基戊内酯衍生的手性试剂为原料，我们完成了昆虫保幼激素 hydroprene 以及 methoprene 的合成[37–39]。

许多香精香料的成分都含有手性甲基侧链，如香瓜醛、香茅醛、异胡薄荷醇、薄荷呋喃、麝香酮、Citralis Nitrile®、Tropional®、Firsantol®、Rose oxide、Cedrol 等。利用从甾体皂甙元氧化降解所获得的手性试剂我们完成了麝香酮[44]、薄荷呋喃[45]、香瓜醛[45]、香茅醛[45]和 Citralis Nitrile®[46]等香料手性分子合成。

我们也利用从甾体皂甙元氧化降解所获得的手性试剂于药物和具有生物活性的天然产物合成。如维生素 E（vitamin E）[47–49]、抗心衰药物 Entrestro 的活性组分之一沙库必曲（sacubitril）[50]、海洋天然产物 didemnaketal A 的 C1—C8 核心片段[51]和 C9—C23[52]等高效合成。

从甾体皂甙元氧化降解废弃物中获取的手性试剂也被其他研究团队用于天然产物合成。马大为等人合成在 halipeptins A—D 海洋环肽合成中利用了 4R– 甲基戊内酯[53]。涂永强等人在合成海洋天然产物 didemnakatol A[54]中向我们实验室购买了相应的手性试剂。

手性甲基侧链的结构单元广泛存在于生物活性天然产物、医药、农药和香料分子结构中，以往由于制备困难限制这些分子的合成与应用。随着我们甾体皂甙元洁净氧化反应在工业生产的推广应用，手性甲基侧链的结构单元的试剂和原料将能够实现大规模生产，届时价廉的且多样化手性试剂和原料的提供必然促进手性药物、农药和精细化工产品的发展。

甾体皂甙元 E/F 的开环反应及其在甾体天然产物分子合成中应用。胆固醇是存在于人类、哺乳动物、昆虫等动物体内最基本甾体化合物，它们是动物体内细胞膜的组织部分，也是各种甾体激素物质的合成前体。植物体内的植物甾醇化学结构类似于胆固醇，真菌体内的麦角甾醇化学结构也胆固醇十分相近。已知天然界存在上千种具有胆固醇基本骨架的生物活性分子，它们是生物体内的重要活性物质，也是发展新医药、新农药的先导分子。甾体皂甙元的E/F螺环开环产物不仅具有胆固醇相同的基本骨架，而且其已经存在的官能团为合成具有胆固醇基本骨架的各类天然产物分子提供了方便。利用甾体皂甙元完整骨架合成具有胆固醇基本骨架的各类天然产物分子的关键是需要了解甾体皂甙元的E/F螺环的反应性能。

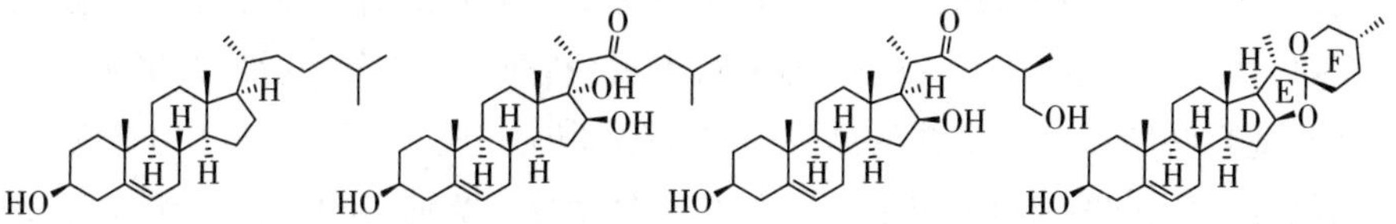

胆固醇　　OSW-1甙元　　皂甙元EF环开环产物　　薯蓣皂甙元

图13　胆固醇、天然产物OSW-1甙元与甾体皂甙元结构比较

有机反应的普适性研究近年来在我国有机化学界受到越来越多的关注和宠爱，这是因为所有官能团或结构单元的转化反应都受到反应底物的严格限制。反应应用范围严格受制于反应底物，许多常见反应不适合于资源化合物。为了精准利用资源型化合物，故它们的反应及其应用研究必须给予重视。

甾体皂甙元E/F螺环为缩酮官能团，已知缩酮很容易酸水解成为相应链状酮醇化合物，然而在我们研究中发现甾体皂甙元E/F螺环缩酮在酸水解条件下根本无法得到相应的链状酮醇化合物。设想通过保护缩酮开环后的羰基或羟基促进其开环反应，为此我们研究了甾体皂甙元的开环缩硫酮化和开环乙酰化反应。在研究甾体皂甙元的开环硫缩酮化反应中发现在路易斯酸存在下单硫醇与甾体皂甙元反应给出F环开环的26-硫代产物；当与双硫醇反应时则给出F环开环、分子内氧化还原以及硫缩醛化的26-硫缩醛产物；当与硫化氢反应时直接给出了硫代甾体皂甙元[55, 56]。这一反应为合成硫代甾体皂甙提供了有效的方法，它也被应用于利用薯蓣皂甙元完整骨架合成具有显著抗肿瘤生物活性的天然产物OSW-1的甙元[57, 58]。16-羟基甾体皂甙元在相同反应条件下与硫醇反应给出相应的16-单或双硫缩酮产物[59, 60]，它们经乙酰化反应得到相应的16-烯基硫醚产物。这些反应已被应用于名贵中药重楼活性成分甙元pennogenin的合成[61]。此外，16-硫缩酮基团存在提高了胆甾-22-酮还原的立体选择性，通常给出C22S-构型还原产物。

路易斯酸催化的甾体皂甙元溴代（及卤代）反应随甾体皂甙元结构不同而给出不同结果。16-羟基甾体皂甙元给出的是26-溴代E/F螺环开环产物，5，6-二溴代-16-羟基薯蓣皂甙元溴代给出相应的26-溴代E/F螺环开环产物用锌粉脱溴时可直接给出16，22-二氧代胆固醇，也可选择性脱溴给出16，22-二氧代-26-溴代胆固醇[60]。前者应用于合成OSW-1甙元[62]，后者应用于甾体生物碱的合成[63]。甾体皂甙元直接溴代给出的主要产物是26-溴代E/F螺环开环产物[64, 65]，它为C26为氮取代的甾体生物碱甙元合成提供了方便[66]。12- 氧代甾体皂甙元（如蕃麻皂甙元）反应则可以得到 16,26- 二溴代产物[64]，这为从甾体皂甙元完整骨架合成天然产物 jaborosalactones 类化合物提供了方法[67]。在甾体皂甙元 26- 硫代、卤代开环反应的基础上，我们考察了其 26- 胺化开环反应。研究发现在路易斯酸存在下，磺酰胺与甾体皂甙元反应可以获得胺化开环反应结果，此反应也被应用于甾体生物碱甙元，如中药龙葵生物碱合成[68]。

为了实现甾体皂甙元 E/F 螺环开环我们早在 20 世纪 90 年代就研究了其乙酰化开环反应[69]。由于反应时间未能很好控制我们得到了 E 环开环后进一步乙酰化反应产物[70]，此反应在随后被 Sandoval-Ramirez 等人通过控制反应时间在 15 分钟后得以实现[71, 72]。这一反应特别适合呋咱型甾体皂甙的高效合成，但是仍然不适合其 D 环结构修饰，原因是 C16- 乙酰基脱去后即刻环化回到甾体皂甙元 E/F 螺环。为此，我们进一步研究了甾体皂甙元的乙酰化开环 / 还原或腈基化反应[64]，反应产物适合于进一步进行其 D 环结构修饰。

甾体皂甙元 F 环可以方便通过还原打开，利用 26- 酰基碳正离子的分子内活化作用可以实现甾体皂甙元 E 环的开环[73]，但是反应伴随的重排反应的发生而导致收率显著下降。对此反应我们通过添加卤代试剂抑制了重排反应给出一个新的溴代分子内酰化开环反应[74]。此方法操作方便、反应收率高，适合于 C25- 位为 R 或 S 两种构型立体化学的天然产物合成。这一反应被成功地应用于天然甜味剂 osladin 甙元[75]、中药重楼活性成分皂甙元 pennogenin[64]、中药小百部活性成分 aspafiliosides E 和 F 的甙元[76, 77]、天然抗菌活性甾体 boophiline 以及甾体生物碱 solasodine、solanidine、demissidine[75, 78]等的高效合成以及线虫生长调节剂 dafachronic Acid A 的形式合成。

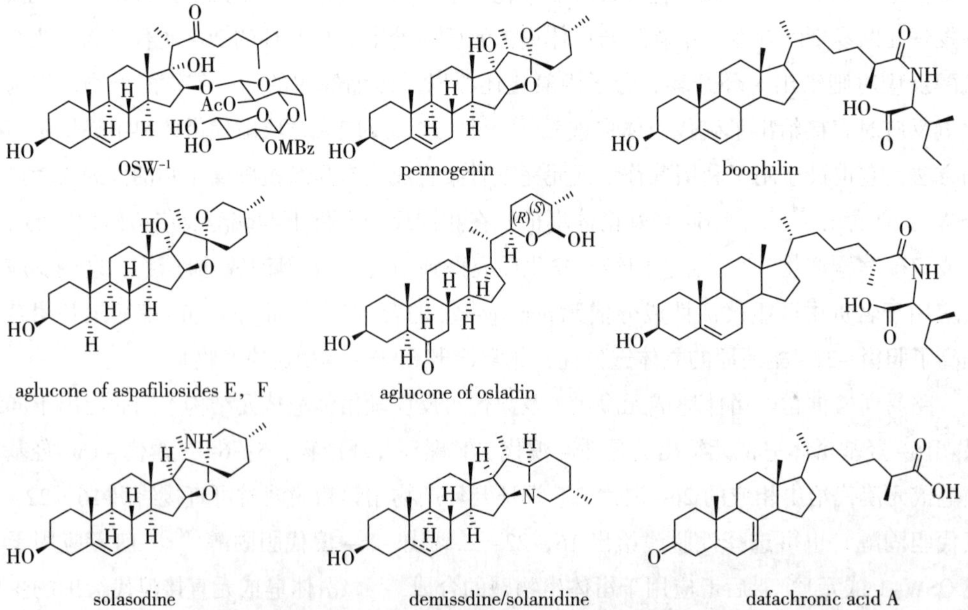

图 14　利用甾体皂甙元完整骨架合成的部分天然甾体分子

海洋天然产物 saundersiosides 是一类具有 24（23→22）–移–胆甾烷结构和显著的抗肿瘤活性的胆甾烷分子。最近我们研究发现甾体皂甙元乙酰化开环产物经二醋酸碘苯引发的 Favorskii 重排可以高效转化成为天然产物 saundersiosides 的基本骨架[79]。这一反应为甾体皂甙元资源利用和 saundersiosides 等海洋天然产物高效合成进一步提供了基础。

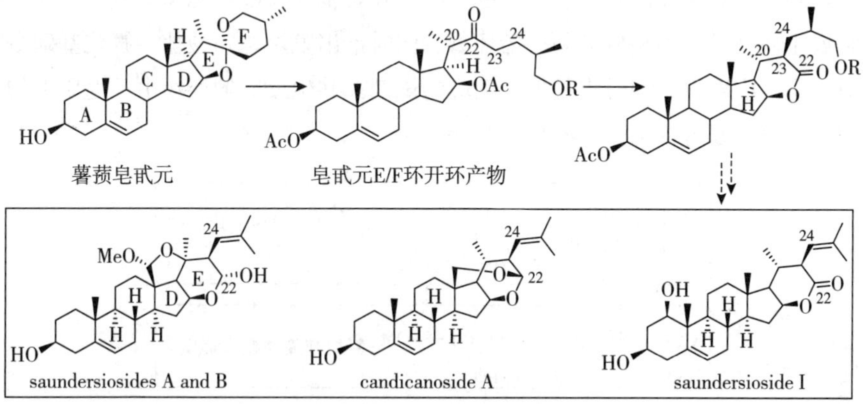

图 15　Saundersioside 基本骨架合成

我国天然资源型化合物薯蓣皂甙元的生产和利用长达半个多世纪，由于粗放性利用造成了严重资源浪费和环境污染问题。针对上述问题我们经过近三十年系统研究，比较准确了解了甾体皂甙元的主要反应性能，为"原子经济性"和精准型地利用甾体皂甙元资源提供了科学依据和实用技术。

（二）非天然资源型化合物氟烷基磺酰氟的反应及其在有机合成中应用研究进展

含氟有机化合物在天然界极为罕见。氟烷基磺酰氟是合成含氟表面活性剂的原料，它们通常通过烷基磺酰氯电解氟化或由四氟乙烯经过调聚反应而得。前者主要有全氟辛烷磺酰氟和全氟丁烷磺酰氟，后者由末端分别由氯代和氢原子取代的八氟乙氧基乙烷磺酰氟。它们经水解反应可以获得系列相应的磺酸盐，作为高效表面活性剂广泛应用于电镀、纺织等工业领域。烷基磺酰氟水解反应是一个十分优秀的"原子经济性"反应，"原子经济性"反应是我们在许多反应研究（如前述甾体皂甙元的反应）中追求的最高目标。但是对于一个已经满足了"原子经济性"原则的反应，如烷基磺酰氟水解反应，是否还有潜力可发掘？这是提高我们精准利用资源型化合物的新目标！借鉴于自然界宏观的生物共生的现象，联想到微观化学反应也属于相互共生的真实状况，我们认为通过"共生反应"（对应英文为 symbiotic reaction），即让两个以上的反应在一个反应体系进行，可以从反应器具、溶剂、试剂等多个方面提高反应效率，减少反应副产物产生。与通常只期望单一反应产物的反应不同，共生反应的最大挑战是产物的分离，故在设计共生反应时应提前考虑产物的分离问题。

氟烷基磺酰氟水解与烯烃环氧化的共生反应。烯烃环氧化反应无论在实验室还是在工业生产上都是一个十分常见的反应。环己烯环氧在工业上有许多用途，它的生产或经过金属催化的环氧化、或卤醇化、或催化氧化。一般情况下，环己烯不能直接与双氧水发生反应。当使环已烯、氟烷基磺酰氟、双氧水和无机碱在一个反应体系时，氟烷基磺酰氟水解与烯烃环氧化的共生反应顺利发生。在反应中氟烷基磺酰氟首先与双氧水反应生成过氧磺酸，它氧化环己烯成为相应的环己烯环氧，同时给出氟烷基磺酸盐。氟烷基磺酰氟在此共生反应即是反应原料，同时也是另一反应的试剂。共生反应的产物由于性质上的显著差别，它们在分离上没有带来额外困难[80, 81]。

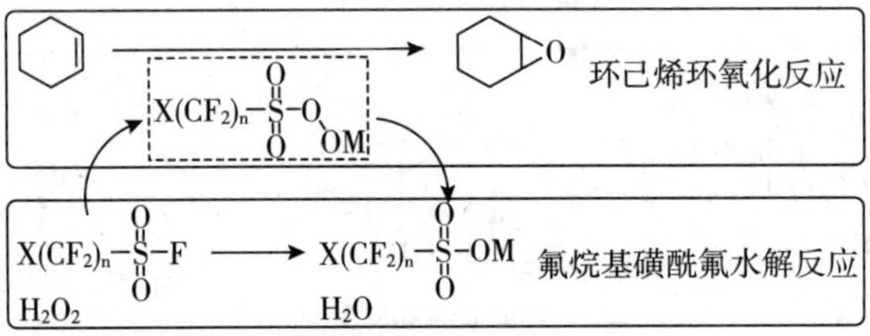

图 16　氟烷基磺酰氟水解与烯烃环氧化共生反应

氟烷基磺酰氟水解与羰基化合物转化的共生反应。二十世纪九十年代初，美国默克公司在甾体 5α-还原酶及其抑制剂研究成果基础上成功开发了首个新型的治疗老年前列腺肥大药物菲那甾胺并且上市。当时由于知识产权问题促使我国考虑发展我国自己的同类新药，中国药科大学等单位承担了国家防治老年疾病新药爱普列特研制。

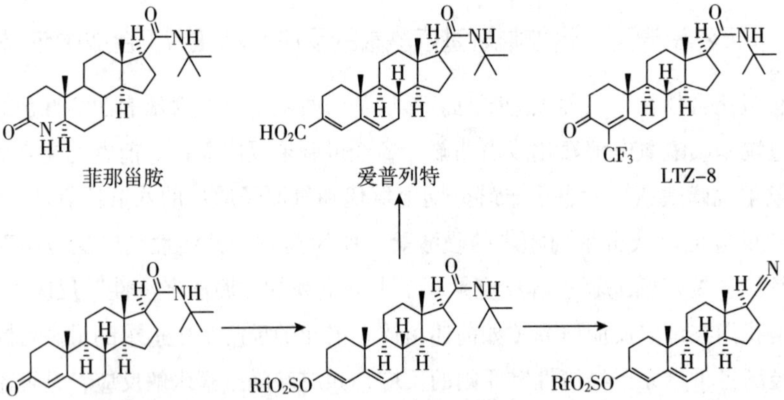

图 17　氟烷基磺酰氟与羰基化合物反应及其应用

无论如何，项目一开始就在爱普列特原料药合成中遇到了难题。按照文献方法，爱普列特结构中的不饱和酸可以经过孕甾酮烯醇三氟甲基磺酸酯的钯催化偶联反应合成，此反应在实验室属于一个常见反应。但是，由于试剂价格昂贵（当时三氟甲基磺酸酐尚未国产化，价格每公斤上千美金以上）、反应操作困难（极其容易潮解），而且存在反应副产物多等问题使其不能适合于工业生产[6]。在此项目面临下马的情况下，我们接受了项目组的请求和挑战。按照项目组提出的所需试剂价格便宜、易得（国产）和反应适合工业生产的要求，在调研后选择用国产的八氟乙氧基乙基磺酰氟（当时每公斤价格 200–300 人民币）作为三氟甲基磺酸酐代替试剂进行研究。在系统研究了八氟乙氧基乙基磺酰氟等氟烷基磺酰氟与羰基化合物反应后我们获得了具有知识产权保护的爱普列特新合成技术[82–84]。此反应保障了国家新药爱普列特研制项目顺利进行，此后项目承担单位主动邀请中科院上海有机化学研究所成为项目参加单位。2002 年，爱普列特作为同类药物的第二个新产品首次由我国研发成功上市。目前产品年销售额已突破一亿元人民币，总销售已超过十亿。在此基础上，我们进一步设计、合成并通过活性测试发现了新型甾体 5α–还原酶抑制剂 LTZ–8[85–88]。

氟烷基磺酰氟水解与高烯丙醇重排的共生反应。氟原子的特殊性能使其取代分子中其他原子或基团后显著改变分子原有性质。在研究氟烷基磺酰氟与甾醇的反应中我们发现了氟烷基磺酰氟引发的 19–羟基甾醇的高烯丙醇重排反应，反应为设计合成新型雌激素药物拮抗剂带来了机遇，也被应用于海洋天然产物 spiniferin 1 合成及其立体结构确定[89–95]。它也可以用于 cyclocitrinol 天然产物合成[96, 97]。

氟烷基磺酰氟水解与邻二醇环氧化的共生反应。氟烷基磺酰基是一个比普通磺酰基，如甲磺酰基或对甲苯磺酰基更好的离去基团。它与羟基反应时可原地形成碳正离子，当其近邻位置上同时存在有另外的亲核基团时，可即刻引发分子内反应。当氟烷基磺酰氟与邻二醇反应时可高效地给出相应的环氧化合物，而不容易形成双磺酸酯副产物。此反应被成功地应用于中药黄皮果活性成分黄皮酰胺[98, 99]和抗心衰药物 Entrestro 的活性组分之一沙库必曲的合成[50]。

氟烷基磺酰氟水解与其他反应的共生反应有：与羟基基团氟代反应的共生反应[100]；与邻位胺基醇和 *N*–硫代酰基保护的邻位胺基醇环化给出相应的氮杂环丙烷和噻唑啉反应的共生反应[101]；与羧酸酯化、酰胺化和酸酐化反应的共生反应[102]；与肟重排的共生反应不仅适合于酸敏感反应底物，其产物为腈基而非酰胺[103]；与 α，β–不饱和酮肟的 Beckmann 重排反应的共生反应给出在合成上具有重要价值的 *N*–烯基酰胺[104]，他们是黄皮酰胺类天然产物和甾体肌肉松弛剂的关键合成中间体；与醛肟脱水的共生反应[105]；氟烷基磺酰氟醇解生成的磺酸酯作为烷基化试剂可以原地进行亲电烷基化反应[106]。

虽然我们在研究氟烷基磺酰氟的反应中提出了“共生反应”概念，并以此进一步提高了氟烷基磺酰氟资源的利用度，但是共生反应绝不局限于与氟烷基磺酰氟水解的共生反应。甾体皂甙元氧化降解同时得到孕甾烯酮醇和 5–甲基己内酯的反应是共生反应，孕甾 –16，20–二醇缩酮的亲核加成反应可以得到孕甾烯酮醇和醛的手性加成产物也是共生反应。在昆虫拒食剂苦楝甾醇的合成中，合成中间体孕甾 –2–烯 –16，20–二醇的合成我们采用的是磺酸酯消除和脱缩酮保护基的共生反应，即利用了消除反应产生的磺酸作为脱脱缩酮保护基的试剂。

综上所述，针对天然资源型化合物甾体皂甙元利用过程中存在的资源浪费和环境污染问题，按照“原子经济性”利用资源型化合物的原则，依据甾体甾甙元反应的新成果，我们发展了用双氧水代替铬酐氧化降解甾体皂甙元及其裂解物的新方法和新技术，实现了理论上“百分之百”利用资源型化合物的设想，关键技术已转让给相关企业。针对非天然资源型化合物氟烷基磺酰氟水解反应已经属于“原子经济性”反应的特点，提出了“共生反应”概念，为进一步提高源型化合物氟烷基磺酰氟的利用价值开辟了新思路。我们研究的共生反应及其在药物和天然产物合成中应用的成果，如新药爱普列特已经显示出应有经济效益和社会效益。

四、资源化学学科发展展望

为了更好开展资源化学研究工作，2000 年我们研究团队与企业共同成立了首个“资源化学与甾体化学联合实验室”。田伟生 2003 年应邀在上海师范大学兼职期间倡导并开展资源化学研究，2004 年该校接受资源化学理念并成功申请到上海教委与教育部共建的“资源化学重点实验室”。通过了解已知国内目前已建立与资源化学有关的实验室分别有：北京化工大学的“化工资源有效利用国家重点实验室”；西南大学的“药用资源化学研究所”；西安近代化学研究所的“氟氮化工资源高效开发与利用国家重点实验室”；中国科学院兰州化学物理研究所的“中国科学院西北特色植物资源化学重点实验室”；中国科学院新疆理化技术研究所的“中国科学院干旱区植物资源化学重点实验室”；青海民族大学的“青藏高原植物资源化学研究省级重点实验室”；成都理工大学的“矿产资源化学四川省高校重点实验室”；中科院青海盐湖研究所的“青海省盐湖资源化学重点实验室”和天津科技大学的“天津市海洋资源与化学重点实验室”。我国已经建立的资源化学实验室涵盖了生物资源化学、矿石资源化学和化石资源化学，相信上述资源化学实验室的建立和建设能够为开展资源化学研究提供了良好保障和支持。与此同时，我们研究团队不仅得到了企业的支持，也得到了国家自然科学基金委、中科院和上海有机化学研究所以及上海科委的支持。

二十世纪九十年代初，在我们研究团队开始系统地、专门开展资源型化合物甾体皂甙元和氟烷基磺酰氟的反应及其应用研究。与此同时，国内外许多研究团队也以各种不同形式开展与资源化学相关的研究。由于篇幅限制，本文在此只能举例简单进行介绍。

早在二十世纪八十年代，为了提高天然产物合成效率，Hanessian 就提出了“手性元法”合成复杂天然产物的概念[107, 108]。手性羟基酸，如苹果酸、乳酸、酒石酸等来源方便，量大且价格低廉，是一类重要的资源型化合物。它们的反应及其在有机合成中应用研究文献数在 2000 年前就达到了 2500 多篇，Gawronski 等人撰写的《在有机合成中的酒石酸和苹果酸》一书详细地介绍了酒石酸和苹果酸作为手性合成元、手性配体的反应及其应用[109]。厦门大学黄培强等从九十年代起一直坚持羟基酸的反应及其合成应用研究取得了系列成果，详细情况可见厦门大学黄培强研究团队的网页介绍。

手性氨基酸多数已有工业化产品，直接或经过转化利用氨基酸合成药物和天然产物的研究一直未间断。它们不仅作为天然产物合成原料，还被衍生成各种手性配体用于不对称合成，或直接衍生成为手性催化剂，Coppola 等人早在八十年代就已经总结过氨基酸作为手性元用于天然复杂分子合成的状况[110, 111]。最近氨基酸的 C-H 官能团化研究成果为氨基酸及其肽类物质利用提供了更多机会。

葡萄糖等各种六碳糖和五碳糖价廉易得，它们分子中的手性羟基可以直接或通过高选择的转化反应用于许多手性药物和天然产物合成[112]。吴毓林等也曾经以糖作为手性源完成了一些重要的天然产物合成[113]。与糖转化相关化合物如糠醛、Levoglucosenone 等大批量生产和应用也受到关注[114]。

天然萜类化合物，单萜、倍半萜、二萜、三萜等资源十分丰富，利用萜类化合物作为手性元合成复杂天然产物同样受到合成化学家关注[115]。

木质素是一类含苯丙醇基本结构单元的天然高分子聚合物，是植物纤维细胞壁的基本组成成分之一，能起到增强植物机械强度和抵抗微生物侵蚀的作用。它们也是造纸工业环境污染物的主要成分。如何降解木质素成为苯丙醇类基本化工原料一直是化学家关注的研究题目，但是问题还未能真正得以解决[116-118]。无论如何，芳香族化合物的去芳构化反应已经被许多合成化学家用于天然产物合成[119]。

以氨基葡萄糖为基本结构单元的甲壳素在自然界的存量仅次于纤维素，从甲壳素水解制备氨基葡萄糖已经工业化，但是要真正解决其工艺中存在的环境污染问题还需要深入研究。另外，如何利用氨基葡萄糖作为手性合成元的研究还不多见[120, 121]。

制药工业和化学化工工业能够生产数目众多的、结构多样性有机分子，它们或是目标分子、或是合成中间体、或是副产物，由于规模化使它们的价格低廉且易于获得。例如氯霉素生产已有六十多年历史，它及其副产物右旋体价格极其便宜，利用氯霉素作为资源型化合物的研究同样受到国内外化学家关注。陈芬儿等人利用氯霉素衍生的手性配体完成了生物素等手性药物合成[122]。更多的研究工作可见王海峰博士学位中引用的文献[123]。氯霉素也可以作为手性原料用于药物、农药和天然产物合成。

通过对已知分子基本骨架的官能团转化、结构重排和引入新官能团是发展新药物、新农药和新材料最高效的途径。对于天然产物而言，同类天然产物的特点是具有相同或相关基本结构，差别仅仅表现在特定位置上氧化态以及立体化学不同。我们也注意到在自然界同类化合物中，其含量似乎与生物活性成反比。紫衫醇及其类似物的顺利生产是资源化学研究的一个成功例子。存在于紫衫树叶中的巴卡亭 III 含量高于紫衫树皮中紫衫醇十倍之多，利用巴卡亭 III 合成紫衫醇解决了紫衫醇药用原料缺乏问题，避免了因扒紫衫树皮提取紫衫醇导致紫衫树种灭绝的危险[124].

海洋天然产物考替甾亭 A 是一个有可能发展成为抗肿瘤药物的复杂甾体分子，为了发展考替甾亭 A 的高效合成方法，资源型化合物雌酚酮、考的松等甾体药物被作为合成原料，两者均为有数十年生产历史的药品，其方便易得、价格低廉。甾体 C/D 环结构单元，合成方法成熟，且有商品供给[125]，也被用于考替甾亭 A 等天然产物合成。最近，Baran 等在前人工作基础上重新研究了通过光化 / 碎裂反应进行甾体 -19- 甲基的羟基化反应，并应用此反应从考的松合成了多羟基甾体分子 ouabagenin[126]。Baran 等还研究了甾体 -C12- 羟基化反应及其在多羟基甾体分子合成中应用[127]。翟宏斌等用来源方便的山道年作为合成原料给出了合成天然产物苦艾素最简洁高效的方法[128]。最近杨震等也以易得的香紫苏醇为原料合成了天然产物 hispiodanin A[129]。从已有的资源型化合物合成所需目标分子毫无疑问是一种发展高效、实用的途径。但是鉴于资源型化合物多数结构复杂，以及对它们的反应性能知之甚少，或虽经前人研究但未能真正清楚了解，故对大多数有机化学家来说仍然是一个挑战性问题。

我国有比较丰富的化合物资源，资源化学研究在我国会大有所为。针对我国氟矿和氟化合物资源结合工业需求，中科院上海有机化学研究所的几代研究人员一直坚持氟有机化学研究，陈庆云等从氟化学工业基本原料四氟磺内酯发展的三氟甲基化试剂为含三氟甲基的药物和农药提供了高效方法[130]。氟利昂（C_lCF_2H）是应用量最大、应用范围最广的一类传统制冷剂，近年来，由于国际环保政策调整，它作为制冷剂的用量逐年减少，为了开发其的新用途，最近张新刚把氟利昂发展成为一种新型的二氟甲基化试剂[131]。

近年来，国际上有机化学界重新启动化学惰性的 C–H 键官能化反应，如 C–H 氧化[132]、C–H 胺化[133]、C–H 磺酰胺化[134]、C–H 氟化[135, 136]、C–H 甲基化[137]、C–H 键转化成为其他 C–C 键[138]等化学转化的研究成果使利用资源型化合物的策略和方式比经典的仅仅通过官能团转化的方式有了更多的选择。

通过化学惰性的 C–H 官能化反应对药物、类药物分子、或天然产物的后修饰（在合成最后阶段进行官能团化或多样化）的合成策略近来在国际有机合成界受到重视[134, 135, 139, 140]，这些研究工作理念与我们提出的资源化学可为同工异曲。不同之处是我们是专门针对某特定资源型化合物系统研究其反应及其应用，旨在使资源型化合物资源得到充分、合理利用，对用什么反应不是特别在意；而其它研究者则以药物、类药物分子、或天然产物的后修饰为目标而进行特定反应的研究，不在乎所用反应底物。共同之处都是为了充分利用已有化合物资源（药物、类药物分子、或天然产物）高效合成目标分子，但是毫无疑问资源化学研究对于资源合理利用的针对性更强，其实际应用价值更高。资源利用是社会可持续性发展中的永久性问题。在当前“资源匮乏、环境污染”困扰社会可持续性发展的情况下，在化学家发现、发明的化合物数目剧增而化学家自己被禁锢于文献上的一些基本化学反应和转化方法学而找不到重大化学研究课题的情况下，资源化学毫无疑问给化学家提出一个挑战性任务！最近在中国化学会支持下，西北大学成功承办了我国首届资源化学学术研讨会，会议还确定了第二届会议承办单位。资源化学作为化学学科的一个分支学科在我国经历了三十年的历练，已经得到社会、企业和学术界的普遍认可。我们确信：资源化学学科如同其他新兴学科一样，一定会在社会可持续发展的需求中迅速发展壮大！

参考文献

[1] 田伟生，科学新闻周刊，1999，5，13.

[2] Marker R E，Rohrmann E STEROLS. LXXXI. Conversion of sarsasa-prgenin to Pregenanediol [J]. Journal of the American Chemical Society 1939，61（12）: 3592-3593.

[3] 田伟生，刘闪闪，沈军伟．一种新的孕烯酮醇化合物合成方法．中国专利，CN1146574C，2004.

[4] 田伟生．一种降解甾体皂甙元成为孕甾醇的方法及其用途．中国专利，CN1061985C，2001.

[5] 田伟生，李民，殷海峰，等．内酯化合物、合成方法及其用途．中国专利，CN1120844C，2003.

[6] 朱臻．5α-还原酶抑制剂的合成及工艺研究与甾体药物合成原料探索．中国药科大学博士学位论文，南京，1996.

[7] Lin J R，Wang J，Chen L J，et al. Studies on the regioselectivity of acetylation-bromination in pregnanetriol [J]. Acta Chimica Sinica 2006，1265-1268.

[8] 田伟生，许启海，李伯玉．一种合成猪外激素的方法．中国专利，专利号：ZL200710037402.9，2009.

[9] Lin J R，Zhou N Y，Xu Q H，et al. The fragmentation reaction of 16R-bromopregnane- 3S，20S-diol [J]. Tetrahedron Letters 2007，48（29）: 4987-4989.

[10] Lin J R，Jiang S F，Zhou N Y，et al. Controllable Reaction of 16R-Bromopregnane- 3S，20S-diol Diacetate with Bases [J]. Acta Chimica Sinica 2008，2637-2645.

[11] Han J, Lin J R, Jin R H, et al. 16S,20S-环氧孕甾-3S-醇乙酸酯的合成及其区域选择性环氧开环取代反应[J]. Acta Chimica Sinica 2011，69（19）: 2272-2280.

[12] 王涛．甾体内酯的加成反应及其在天然产物合成中的应用。中国科学院上海有机化学研究所博士学位论文，上海，2004.

[13] Cox P J，Turner A B. Synthesis，x-ray structure and molecular mechanics studies of the boar taint steroid（5α-androst-16-en-3-one）[J]. Tetrahedron 1984，40（16）: 3153-3158.

[14] Shi Y，Jia L Q，Xiao Q，et al. A Practical Synthesis of Cephalostatin 1 [J]. Chemistry - An Asian Journal 2011，6（3）: 786-790.

[15] Zhang D S，Shi Y，Tian W-S. Semisynthesis of Azedarachol from Pregnanetriol，a Degradative Product of Tigogenin [J]. Chinese Journal of Chemistry 2015，33（6）: 669-673.

[16] Gui J H，Wang D H，Tian W S. Biomimetic synthesis of 5，6-dihydro-glaucogenin c: Construction of the disecopregnane skeleton by iron（ii）-promoted fragmentation of an α-alkoxy hydroperoxide [J]. Angewandte Chemie International Edition 2011，50（31）: 7093-7096.

[17] Gui J H，Tian H L，Tian W S. Synthesis of glaucogenin D，a structurally unique disecopregnane steroid with potential antiviral activity [J]. Organic Letters 2013，15（18）: 4802-4805.

[18] 周楠琰．16R-溴代孕甾-3S，20S-二醇的碎裂反应及其在甾体肌肉松弛剂合成中应用，上海师范大学硕士学位论文，2008.

[19] Wang S S, Shi Y, Tian W S A Formal Synthesis of Betamethasone [J]. Chinese Journal of Chemistry 2015, 33（6）: 637-642.

[20] 沈凯圣. 油菜甾醇内酯和 Certonardsterol D2、D3 和 N1 的合成研究. 中国科学院上海有机化学研究所博士学位论文，上海，2005.

[21] Zhu W，Wang H，Fujioka S，et al. Homeostasis of Brassinosteroids Regulated by DRL1，a Putative Acyltransferase in Arabidopsis [J]. Molecular Plant 2013，6 (2): 546–558.

[22] Xin P，Yan J，Li B，et al. A Comprehensive and Effective Mass Spectrometry–Based Screening Strategy for Discovery and Identification of New Brassinosteroids from Rice Tissues [J]. Frontiers in Plant Science 2016，7: 1786.

[23] 冯锋. Clathsterol 的合成研究. 中国科学院上海有机化学研究所博士学位论文，上海，2005.

[24] Cong R G，Zhang Y H，Tian W S. A concise synthesis of the steroidal core of clathsterol [J]. Tetrahedron Letters 2010，51 (30): 3890–3892.

[25] Zhou T，Feng F，Shi Y，et al. Synthesis Toward and Stereochemical Assignment of Clathsterol: Exploring Diverse Strategies to Polyoxygenated Sterols [J]. Organic Letters 2016，18 (9): 2308–2311.

[26] Wang S S, Shi Y, Tian W S. Highly Efficient and Scalable Synthesis of Clionamine D [J]. Organic Letters 2014, 16(8): 2177–2179.

[27] Cheng S L，Jiang X L，Shi Y，et al. Concise Synthesis of the Core Structures of Saundersiosides [J]. Organic Letters 2015，17 (10): 2346–2349.

[28] Lee S，LaCour T G，Fuchs P L. Chemistry of Trisdecacyclic Pyrazine Antineoplastics: The Cephalostatins and Ritterazines [J]. Chemical Reviews 2009，109 (6): 2275–2314.

[29] Shi Y，Jiang X L，Tian W S. Synthesis of 12，12′–azo–13，13′–diepi–Ritterazine N [J]. The Journal of Organic Chemistry 2017，82 (1): 269–275.

[30] Jiang B, Shi H P, Tian W S, et al. The convergent synthesis of novel cytotoxic certonardosterol D2 from diosgenin [J]. Tetrahedron 2008，64 (3): 469–476.

[31] Czajkowska D，Morzycki J W，Santillan R，et al. Synthesis of "glycospirostanes" via ring–closing metathesis [J]. Steroids 2009，74 (13–14): 1073–1079.

[32] Zhu D P，Yu B Total Synthesis of Linckosides A and B，the Representative Starfish Polyhydroxysteroid Glycosides with Neuritogenic Activities [J]. Journal of the American Chemical Society 2015，137 (48): 15098–15101.

[33] 田伟生，丁凯，黄悦. 带手性侧链的 – 羟基酸衍生物，中国专利 专利号 CN100548965C，2009.

[34] 田伟生，沈军伟，沈伟，等. 一类手性甲基 1，3– 官能团合成子的合成方法 CN100378042C，2008.

[35] Wang Z K，Tian W S，Pan X F. A concise synthesis of the sex pheromones of Pine sawflies [J]. Chinese Journal of Organic Chemistry 2007, 866–869.

[36] Wang Z K，Xu Q H，Tian W S，et al. Stereoselective synthesis of (2S，3S，7S) –3，7– dimethylpentadec–2–yl acetate and propionate，the sex pheromones of pine sawflies [J]. Tetrahedron Letters 2007，48 (42): 7549–7551.

[37] 王子坤. 基于甾体降解废弃物（R）–4– 甲基 –6– 戊内酯合理利用的昆虫信息素等天然产物合成. 兰州大学博士学位论文，兰州，2007.

[38] Zhang S J, Dong H D, Gui J H, et al. Stereoselective synthesis of the insect growth regulator (S) – (+) –hydroprene through Suzuki – Miyaura cross–coupling [J]. Tetrahedron Letters 2012，53 (15): 1882–1884.

[39] 何承宇. 基于甲基侧链手性源合成昆虫保幼激素（S）–methoprene 等活性分子的研究，中国科学院上海有机化学研究所硕士学位论文，上海，2016.

[40] Shen W，Hao X，Shi Y，et al. Synthesis of (6R，12R) –6，12–Dimethylpentadecan–2–one，the Female–Produced Sex Pheromone from Banded Cucumber Beetle Diabrotica balteata，Based on a Chiron Approach. [J]. Nat Prod Commun 2015，2155–2160.

[41] 沈伟.（R）–4– 甲基戊内酯在若干含手性甲基天然产物分子合成中应用研究，中国科学院上海有机化学研

究所博士学位论文，上海，2009.
[42] Wang C，He C Y，Shi Y，et al. Synthesis of Tribolure，the Common Aggregation Pheromone of Four Tribolium Flour Beetles [J] . Chinese Journal of Chemistry 2015，33（6）: 627–631.
[43] Mori K，Tamada S，Matsui M. Stereocontrolled synthesis of all of the four possible stereoisomers of erythro–3，7–dimethyl– pentadec–2–yl acetate and propionate，the sex pheromone of the pine sawflies [J] . Tetrahedron Letters 1978，19（10）: 901–904.
[44] Shen J W，Shi Y，Tian W S. Synthesis of（R）–（–）–Muscone from（R）–5–Bromo–4–methylpentanoate: A Chiron Approach [J] . Chinese Journal of Chemistry 2015，33（6）: 683–687.
[45] 张永霞，利用甾体甾甙元氧化降解废弃物合成薄荷呋喃及其相关香料的研究，中国科学院上海有机化学研究所硕士学位论文，上海，2008.
[46] 程利利，手性试剂（R）–4– 甲基 – γ – 丁酸内酯的制备及其在 Citralis Nitrile 香料合成中的应用，上海师范大学硕士学位论文，2008.
[47] Huang X G，Li T，Lin J R，et al. Asymmetric synthesis of vitamin E [J] . Chinese Journal of Organic Chemistry 2006，1353–1361.
[48] 李涛，利用甾体甾甙元降解废弃物合成维生素 E 等天然产物，中国科学院上海有机化学研究所博士学位论文，上海，2007.
[49] Li T，Huang X G，Lin J R，et al. Total synthesis of（2R，6R）–2，6，10– trimethylundecan–1–ol with（R）–5–methyl–delta–valerolactone from the industrial waste [J] . Acta Chimica Sinica 2007，65（12）: 1165–1171.
[50] Wang Y，Chen F E，Shi Y，et al. Multigram scale，chiron–based synthesis of sacubitril [J] . Tetrahedron Letters 2016，57（52）: 5928–5930.
[51] Zhang S J，Shi Y，Tian W S. Synthesis of C1 – C9 Domain of the Nominal Didemnaketal A [J] . Chinese Journal of Chemistry 2015，33（6）: 663–668.
[52] 张顺吉，甾体甾甙元降解废弃物在 didemnakatol A 等天然产物合成的应用，中国科学院上海有机化学研究所博士学位论文，上海，2011.
[53] Yu S Y，Pan X，Lin X，et al. Total synthesis of halipeptin A: a potent antiinflammatory cyclic depsipeptide [J] . Angewandte Chemie International Edition 2004，44（1）: 135–138.
[54] Zhang F M，Peng L，Li H，et al. Total Synthesis of the Nominal Didemnaketal A [J] . Angewandte Chemie International Edition 2012，51（43）: 10846–10850.
[55] Tian W S, Guan H P, Pan X F. Mercaptolysis of the E/F rings of steroidal sapogenins: A concise synthesis of Delta(20（22）)–furostene–26–thioethers [J] . Chinese Journal of Chemistry 2003，21（7）: 784–788.
[56] Wang J，Wu J J，Tian W S. BF3 · Et2O Promoted Sulfuration of Steroidal Sapogenins [J] . Chinese Journal of Chemistry 2015，33（6）: 632–636.
[57] 彭小文，利用薯蓣皂甙元完整骨架合成 OSW–1 甙元的研究，中国科学院上海有机化学研究所博士学位论文，上海，1999.
[58] Xu Q H，Peng X W，Tian W S. A new strategy for synthesizing the steroids with side chains from steroidal sapogenins: synthesis of the aglycone of OSW–1 by using the intact skeleton of diosgenin [J] . Tetrahedron Letters 2003，44（52）: 9375–9377.
[59] 兰泉，Cephalostatin 1 “北片段” 的合成研究，中国科学院上海有机化学研究所博士学位论文，上海，2005.
[60] 王静，利用薯蓣皂甙元完整骨架合成 OSW–1 甙元，上海师范大学硕士学位论文，2007.
[61] Tian W S，Xu Q H，Chen L，et al. Synthesis of pennogenin utilizing the intact skeleton of diosgenin [J] . Science in China Series B: Chemistry 2004，47（2）: 142–144.
[62] Chen L J，Xu Q H，Huang H，et al. Synthesis of 5（6）–dihydro–OSW–1 by using the intact skeleton of tigogenin

[J]. Tetrahedron Letters 2007, 48 (19): 3475–3477.

[63] Tang X M, Xu Q H, Wang J, et al. Synthesis of Solasodine by Using the Intact Skeleton of Diosgenin [J]. Acta Chimica Sinica 2007, 65 (20): 2315–2319.

[64] 吴晶晶，甾体皂甙元的开环反应及其在 Pennogenin 等天然产物合成中应用，中国科学院上海有机化学研究所博士学位论文，上海，2014.

[65] 田伟生，吴晶晶，高冉，等. 一种 26- 卤代呋咱化合物，合成方法及用途，中国专利 专利号：ZL 201410770644.9，2016.

[66] Wu J J, Shi Y, Tian W S. Facile synthesis of solasodine based on a mild halogenation–ring opening reaction of spiroketals in steroidal sapogenins [J]. Tetrahedron Letters 2015, 56 (10): 1215–1217.

[67] Chen L X, He H, Qiu F. Natural withanolides: an overview [J]. Natural Product Reports 2011, 28 (4): 705–740.

[68] Wu J J, Gao R, Shi Y, et al. Direct amination of EF spiroketal in steroidal sapogenins: an efficient synthetic strategy and method for related alkaloids [J]. Tetrahedron Letters 2015, 56 (47): 6639–6642.

[69] Tian W S, Guan H P, Pan X F. Lewis acid mediated ring F opening–acetylation of steroid sapogenin [J] Chinese Chem. Lett., 1994, 5, 1013

[70] Yang Q X, Tian W S, Pan S. Restudies on the Lewis Acid Catalytic Acetylation opening Reaction of Steroidal Sapogenin [J]. Acta Chimica Sinica 2004, 62 (21): 2171–2176.

[71] Fern á ndez–Herrera M A, S á nchez– S á nchez L, Pinto B M, et al. Synthesis of the steroidal glycoside (25R)–3β, 16β–diacetoxyl–12, 22–dioxo–5α–cholestan–26–yl β–D glucopyranoside and its anti–cancerproperties on Hela, CaSki and ViBo cells [J]. Eur.J. Med. Chem. 2010, 45, 4827–4837.

[72] Fern á ndez–Herrera M A, S á nchez– S á nchez L, Pinto B M, et al. Synthesis of 26–hydroxy– 22–oxocholestanic frameworks from diosgenin and hecogenin and their in vitro antiproliferative and apoptotic activity on human cervical cancer CaSki cells [J]. Bioorganic & Medicinal Chemistry 2010, 18 (7): 2474–2484.

[73] Gonzalez A G, Francisco C G, Freire R, et al. New sources of steroid sapogenins .26. Stereoselective opening of ring e in furostan sapogenins – efficient route to 16, 22R, 26–hydroxy steroids [J]. Tetrahedron Letters 1974, (49–5): 4289–4292.

[74] 张晓菲，利用薯蓣皂甙元完整骨架合成 Osladin 甙元，上海师范大学硕士学位论文，2014.

[75] Zhang X F, Wu J J, Shi Y, et al. Formal synthesis of osladin based on an activation relay process [J]. Tetrahedron Letters 2014, 55 (33): 4639–4642.

[76] Wu J J, Shi Y, Tian W S. Synthesis of the aglycon of aspafiliosides E and F based on cascade reactions [J]. Chemical Communications 2016, 52 (9): 1942–1944.

[77] Wu J J, Shi Y, Tian W S. Synthesis of the aglycon of aspafiliosides E and F via a spiroketal–forming cascade [J]. Tetrahedron Letters 2017, 58 (10): 923–925.

[78] Zhang Z D, Shi Y, Wu J J, et al. Synthesis of Demissidine and Solanidine [J]. Organic Letters 2016, 18 (12): 3038–3040.

[79] Jiang X L, Shi Y, Tian W S. Constructing 24 (23 → 22) –abeo–Cholestane from Tigogenin in a 20 (22 → 23) –abeo–Way via a PhI (OAc) 2–mediated Favorskii Rearrangement [J]. The Journal of Organic Chemistry 2017, 82 (8): 4402–4406.

[80] 严兆华，氟代烷基磺酰氟 RfSO2F 在有机合成化学中的应用，中国科学院上海有机化学研究所博士学位论文，上海，1999.

[81] Yan Z H, Tian W S. Poly (per) fluoroalkanesulfonyl fluoride promoted olefin epoxidation with 30% aqueous hydrogen peroxide [J]. Tetrahedron Letters 2004, 45 (10): 2211–2213.

[82] 田伟生，一种甾体烯醇多氟烃基磺酸酯化合物及其衍生物、用途和制备方法，中国专利，专利号：

CN1055930C, 2000.

[83] Zhu Z, Tian W S, Liao Q J. A practical procedure for chemo- and regioselective conversion of steroid 3-ketones into the corresponding enol sulfonates using 3-oxa-octafluoro pentanosulfonyl fluoride [J]. Tetrahedron Letters 1996, 37 (47): 8553-8556.

[84] Tian W S, Zhu Z, Liao Q J, et al. A practical synthesis of 3-substituted Δ3, 5 (6) -Steroids as new potential 5α-reductase inhibitor [J]. Bioorganic & Medicinal Chemistry Letters 1998, 8 (15): 1949-1952.

[85] Fei X S, Tian W S, Chen Q Y. Synthesis of 4-trifluoromethylsteroids: A novel class of steroid 5α-reductase inhibitors [J]. Bioorganic & Medicinal Chemistry Letters 1997, 7 (24): 3113-3118.

[86] Fei X S, Tian W S, Chen Q Y. New, convenient route for trifluoromethylation of steroidal molecules [J]. Journal of the Chemical Society, Perkin Transactions 1 1998, (6): 1139-1142.

[87] Xin J, Tian W S, Sun Y, et al. The screening and anti-prostatic hyperplasia activity assessment of novel steroid 5alpha-reductase inhibitors LTZ-8 and others [J]. Chinese Journal of Pharmacology and Toxicology 2003, 17 (5): 395-400.

[88] Fei X S, Tian W S, Ding K, et al. New, Convenient Route For Trifluoromethylation Of Steroidal Molecules [J]. Organic Syntheses 2010, 87: 126-136.

[89] 雷震，氟烷基磺酰氟与 19- 羟基睾酮反应的研究，中国科学院上海有机化学研究所硕士学位论文，1997.

[90] Tian W S, Lei Z, Chen L, et al. Some new reactions of poly (per) fluoroalkanesulfonyl fluorides with steroidal molecules [J]. Journal of Fluorine Chemistry 2000, 101 (2): 305-308.

[91] Chen L, Ding K, Tian W S. Total synthesis of (+/-) -dihydrospiniferin-1 via a polyfluoro alkanosulfonyl fluoride induced tandem carbonium ion rearrangement reaction [J]. Chemical Communications 2003, (7): 838-839.

[92] Ding K, Sun Y S, Tian W S. Total Synthesis of (+/-) -Spiniferin-1 via a Polyfluoro- alkanosulfonyl Fluoride Induced Homoallylic Carbocation Rearrangement Reaction [J]. Journal of Organic Chemistry 2011, 76 (5): 1495-1498.

[93] Sun Y S, Ding K, Tian W S. Total syntheses of (Sp) - (+) - and (Rp) - (-) -spiniferin-1, a pair of unusual natural products with planar chirality [J]. Chemical Communications 2011, 47 (37): 10437-10439.

[94] Tian W S, Ding K, Sun Y S, et al. Total Synthesis and Establishment of the Stereochemistry of Spiniferin-1, a Rare Planar Chiral Marine Natural Product with a 1,6-Methano[10]annulene Skeleton [J]. Synthesis 2013, 45(04): 438-447.

[95] He L, Xiang H, Zhang L Y, et al. Novel estrogen receptor ligands and their structure-activity relationship evaluated by scintillation proximity assay for high-throughput screening [J]. Drug Development Research 2005, 64 (4): 203-212.

[96] Du L, Zhu T, Fang Y, et al. Unusual C25 Steroid Isomers with Bicyclo [4.4.1] A/B Rings from a Volcano Ash-Derived Fungus Penicillium citrinum [J]. Journal of Natural Products 2008, 71 (8): 1343-1351.

[97] Liu T, Yan Y, Ma H, et al. Synthesis of Cyclocitrinol Skeleton via a Carbocation Rearrangement [J]. Chinese Journal of Organic Chemistry 2014, 34 (9): 1793-1799.

[98] Yan Z H, Wang J Q, Tian W S. A concise total synthesis of (-) -dehydroclausenamide utilizing the novel formation of cis-epoxide as the key step [J]. Tetrahedron Letters 2003, 44 (52): 9383-9384.

[99] Yan Z H, Wei M, Tian W S, et al. The dehydro-epoxidation of methyl (2S, 3R) -2, 3-dihydroxyl-3-benzypropronate with fluoroalkanosulfonyl fluorides [J]. Acta Chimica Sinica 2011, 69 (10): 1269-1272.

[100] Yan Z H, Lai K, Tian W S, et al. Deoxyfluorination of peracetylated pyranose hemiacetal with fluoroalkanesulfonyl fluoride [J]. Acta Chimica Sinica 2012, 70 (11): 1322-1326.

[101] Yan Z H, Guan C B, Yu Z X, et al. Fluoroalkanosulfonyl fluorides-mediated cyclodehydration of β-hydroxy sulfonamides and β-hydroxy thioamides to the corresponding aziridines and thiazolines [J]. Tetrahedron Letters

2013，54（43）: 5788–5790.
[102] Yan Z H，Tian W S，Zeng F，et al. 5H–3–oxa–Octafluoropentanesulfonyl fluoride: a novel and efficient condensing agent for esterification，amidation and anhydridization [J]. Tetrahedron Letters 2009，50（23）: 2727–2729.
[103] Gui J，Wang Y，Tian H，et al. Perfluoroalkylsulfonyl fluoride–mediated abnormal Beckmann rearrangement of steroid 17–oximes with acid–labile groups [J]. Tetrahedron Letters 2014，55（30）: 4233–4235.
[104] Yan Z H，Xu Y，Tian W S. A new and concise way to enamides by fluoroalkanosulfonyl fluoride mediated Beckmann rearrangement of α，β–unsaturated ketoximes [J]. Tetrahedron Letters 2014，55（52）: 7186–7189.
[105] Yan Z H，Tian H，Zhao D D，et al. Perfluoroalkanosulfonyl fluoride: A useful reagent for dehydration of aldoximes to nitriles [J]. Chinese Chemical Letters 2016，27（1）: 96–98.
[106] Yan Z，Jin H，Yu X，et al. One–Step C– or O–Benzylation of 1，3–Dicarbonyls with Benzyl Alcohols Promoted by Perfluoroalkanosulfonyl Fluoride [J]. Chinese Journal of Organic Chemistry 2017，37（1）: 196–202.
[107] Hanessian S. Total synthesis of natural products: The chiron' approch，Pergamon，Oxford，1983.
[108] Hanessian S. The enterprise of synthesis: from concept to practice [J]，J. Org. Chem. 2012，77，6657–6688.
[109] Gawroski J，Gawronska K. Tartaric and malic acids in synthesis: a source book of building blocks，ligands，auxilianries and resolving agents，John Wiley & Sons Inc，New York，1999.
[110] Coppola G M，Schuster H F. Asymmetric synthesis – construction of chiral molecules using amino acids，John Wiley & Sons Inc New York，1987.
[111] Reetz M T. Synthesis and diastereoselective reactions of N，N–dibenzylamino aldehydes and related compounds [J]. Chemical Reviews 1999，99（5）: 1121–1162.
[112] 孔繁祚. 糖化学. 科学出版社，北京，2005.
[113] 吴毓林，姚祝军. 天然产物合成化学 – 科学与艺术的探索，科学出版社，北京，2006.
[114] Bozell J J edited Chemicals and materials from renewable resources，ACS symposium series 784，America chemical society，Washington，DC 2001.
[115] Ho T L，Enantioselective synthesis: Naturalproducts from chiral terpenes，New York，Wiley，1992.
[116] 蒋挺大，木质素（第二版），化学工业出版社，北京，2008.
[117] 刘嘉川，刘温霞，杨桂花，等. 造纸植物资源化学，科学出版社，北京，2012.
[118] Rahimi A，Ulbrich A，Coon J J，et al. Formic–acid–induced depolymerization of oxidized lignin to aromatics [J]. Nature 2014，515（7526）: 249–252.
[119] Roche S P，Porco J A，Jr. Dearomatization Strategies in the Synthesis of Complex Natural Products [J]. Angewandte Chemie–International Edition 2011，50（18）: 4068–4093.
[120] 蒋挺大，甲壳素，化学工业出版社，北京，2003.
[121] 施晓文，邓红兵，杜予民. 甲壳素和壳聚糖材料及应用，化学化工出版社，2015.
[122] Wang H，Yan L，Wu Y，et al. Chloramphenicol base chemistry. Part 10（1）: Asymmetric synthesis of alpha–hydroxy chiral alcohols via intramolecular Michael additions of gamma–hydroxy–alpha，beta–unsaturated enones with chloramphenicol base derived bifunctional urea organocatalysts [J]. Tetrahedron 2017，73（19）: 2793–2800.
[123] 王海峰. 基于氯霉素骨架的催化剂和配体的不对称反应研究及其在药物合成中应用，复旦大学博士学位论文，上海，2017.
[124] Denis J N，Greene A E，Guenard D，et al. A highly efficient，practical approach to natural taxol [J]. Journal of the American Chemical Society 1988，110（17）: 5917–5919.
[125] 吴毓林，等. 天然产物全合成荟萃 – 抗生素及其他，669–692，科学出版社，北京，2012.

[126] Renata H, Zhou Q, D ü nstl G, et al. Development of a Concise Synthesis of Ouabagenin and Hydroxylated Corticosteroid Analogues[J]. Journal of the American Chemical Society 2015, 137(3): 1330–1340.

[127] See Y Y, Herrmann A T, Aihara Y, et al. Scalable C–H Oxidation with Copper: Synthesis of Polyoxypregnanes[J]. Journal of the American Chemical Society 2015, 137(43): 13776–13779.

[128] Zhang W, Luo S, Fang, et al. Total Synthesis of Absinthin[J]. Journal of the American Chemical Society 2005, 127(1): 18–19.

[129] Li F, Tu Q, Chen S, et al. Bioinspired Asymmetric Synthesis of Hispidanin A[J]. Angewandte Chemie International Edition 2017, 56(21): 5844–5848.

[130] Chen Q Y, Wu S W. Methyl fluorosulphonyldifluoroacetate; a new trifluoromethylating agent[J]. Journal of the Chemical Society, Chemical Communications 1989, (11): 705–706.

[131] Feng Z, Min Q Q, Fu X P, et al. Chlorodifluoromethane–triggered formation of difluoromethylated arenes catalysed by palladium[J]. Nat Chem 2017, advance online publication.

[132] Osberger T J, Rogness D C, Kohrt J T, et al. Oxidative diversification of amino acids and peptides by small–molecule iron catalysis[J]. Nature 2016, 537(7619): 214–219.

[133] Park Y, Kim Y, Chang S. Transition Metal–Catalyzed C–H Amination: Scope, Mechanism, and Applications[J]. Chemical Reviews 2017, advance online publication.

[134] Dai H X, Stepan A F, Plummer M S, et al. Divergent C–H Functionalizations Directed by Sulfonamide Pharmacophores: Late–Stage Diversification as a Tool for Drug Discovery[J]. Journal of the American Chemical Society 2011, 133(18): 7222–7228.

[135] Fier P S, Hartwig J F. Synthesis and Late–Stage Functionalization of Complex Molecules through C–H Fluorination and Nucleophilic Aromatic Substitution[J]. Journal of the American Chemical Society 2014, 136(28): 10139–10147.

[136] Wang J, S á nchez–Roselló M, Aceña J L, et al. Fluorine in Pharmaceutical Industry: Fluorine–Containing Drugs Introduced to the Market in the Last Decade (2001–2011)[J]. Chemical Reviews 2014, 114(4): 2432–2506.

[137] Barreiro E J, K ü mmerle A E, Fraga C A M. The Methylation Effect in Medicinal Chemistry[J]. Chemical Reviews 2011, 111(9): 5215–5246.

[138] Colby D A, Bergman R G, Ellman J A. Rhodium–Catalyzed C–C Bond Formation via Heteroatom–Directed C–H Bond Activation[J]. Chemical Reviews 2010, 110(2): 624–655.

[139] Cernak T, Dykstra K D, Tyagarajan S, et al. The medicinal chemist's toolbox for late stage functionalization of drug–like molecules[J]. Chemical Society Reviews 2016, 45(3): 546–576.

[140] Wencel Delord J, Glorius F. C–H bond activation enables the rapid construction and late–stage diversification of functional molecules[J]. Nat Chem 2013, 5(5): 369–375.

撰稿人：田伟生　史　勇　严兆华

ABSTRACTS

Comprehensive Report

Report on Advances in Chemistry

In the past two years, chemistry in China has made remarkable progress both in scientific research and education. The quantity and quality of the academic papers published continued at the international leading position. This report summarizes the progress made by our country's chemical workers during the first half of 2015 to 2017, including five major branches of inorganic chemistry, organic chemistry, analytical chemistry, macromolecular chemistry, nuclear chemistry and radiochemistry as well as seven fields of expertise and eight interdisciplinary and other specialized fields. The report quoted more than 850 references, which can more accurately reflect the progress of our chemistry.

The comprehensive report has four sections: introduction, development and research of higher chemistry education, important progresses of Chinese chemistry, and development trends and perspectives.

Chinese organic chemists continued their excellent performance in the past two years. In terms of quantity, China has been among the top two contributors of research publications for most of the high-ranking academic journals, indicating the chemical research in China has reached a high level. New progresses have been made in the fields of synthetic methodologies and natural product synthesis; organofluorine chemistry and natural product chemistry have been already in the world's leading positions.

Organic reactions and synthetic methodologies have been, and continues to be, the main arena for organic chemists. A number of researchers have already created sub-fields and kept excelling therein. Liu Guosheng and co-workers reported a copper-catalyzed radical relay pathway for enantioselective conversion of benzylic C–H bonds into benzylic nitriles, affording with high enantioselectivity products that are key precursors to important bioactive molecules. This is the first research paper in *Science* magazine in organic chemistry field from China. Organic synthesis has experienced a dramatic increase, the field has reached an awe-inspiring level, aiming to provide large quantities of complex natural products with a minimum amount of labor and material expenses. Dozens of complex natural products were synthesized by Chinese researchers. Li Ang's group accomplished the total syntheses of more than a dozen of challenging natural products including aflavazole, aspidodasycarpine, et al. Direct introduction of fluorine atom or fluorine-containing group into organic molecules has developed as a popular research field in organofluorine chemistry, and Chinese researchers have created a competitive strength in this field and been recognized by international community. Research mainly focused on developing fluorine-related reagents and methodologies. Reagents and reactions for deoxyfluorination, monofluoromethylation, difluoromethylation, trifluoromethylation, difluoromethylthiolation, and trifluoromethylthiolation were developed. In the field of natural product chemistry, more than 6000 new natural products were isolated and reported. Among the compounds isolated, 349 compounds have new skeletons and more than 1800 compounds exhibit bioactivities. Notable natural products include phainanoids A-F, a new class of potent immunosuppressive triterpenoids, and perforalactone A, a new 20S quassinoid with insecticidal activity against *Aphis medicaginis Koch* and antagonist activity at the nicotinic acetylcholine receptor of Drosophila melanogaster.

In the field of inorganic chemistry, fruitful results in the following fields were achieved: (i) basic scientific field; (ii) new energy utilization and storage, new materials (such as flexible artificial machine, super capacitor, self-repairing materials, ferroelectric materials, high efficient adsorption material) ; (iii) catalysis; (iv) cancer treatment. A number of projects were awarded the 2nd class Natural Science prizes, including "Macro scale fabrication of nanostructure units and functionalization of macro scale assembly" (Yu Shuhong et al, University of Science and Technology) , "Structure design and functional construction of metal organic framework based on fluorescence sensing" (Qian Guodong et al, Zhejiang University) , "Basic physical and chemical of organic research on effect transistors" (Hu Wenping et al, Institute of Chemistry, Chinese Academy of Sciences) , "Basic studies on controlled assembly and composite of chemically modified graphene" (Shi Gaoquan et al, Tsinghua University) , "Construction and synergistic mechanism of high performance composite electrode materials for energy storage" (Huang Yunhui et al, Huazhong University of Science and Technology) .

In the field of physical chemistry, chemical dynamics is a frontier sub-discipline that studies the dynamics and laws of chemical reactions at microscopic levels. Exploring and understanding the dynamical nature of chemical processes plays an extremely important role in promoting the development of chemistry. In the past two years, a series of advances have been made in the study of chemical dynamics in China. Tian Shanxi found an oxygen generation pathway in the dissociative electron attachment process of CO_2, thus revealing the new mechanism of oxygen origin in the earth primordial atmosphere. Xie Daiqian et al. found that for the photolysis of phenol molecules in the vibrational ground state of the first electronic excited state, the adiabatic model is not correct for describing the dynamics near the conical intersection. In the next few years, with the further development and improvement of the experimental and theoretical methods of chemical dynamics and the ability of computer, the research level of chemical dynamics in China will be greatly improved, so that we can face the challenges in the field of chemical dynamics from the discipline development itself and the national demands in the areas of energy, environment and defense.

The summary of recent research progresses of electrochemistry includes published articles and awards, recent advances in the last three years, and its development trend and prospect. With the developing of the economy and the increasing of the environmental pollution, it is necessary to develop renewable sources to replace the limited resources of petroleum and other resources. Electrochemical energy conversion and storage are still the focus of international research on Electrochemistry. In the past three years, Chinese scientists have achieved fruitful results in the energy electrochemistry. And in the fuel cells and water electrolysis, scientists have made a series of remarkable progress in studying noble metal catalyst with low content, noble metal catalysts replacement, the establishment of active sites model, catalytic mechanism and stability enhancement. The general trend of the electrochemical power source is developing toward high specific energy and safe, it develops the high voltage, large capacity cathode materials, and high capacity anode materials, and on the contrary, it produces strong safety solid-based and water-based Li ion batteries. Chinese scholars have their advantages in the material design and preparation in electrochemical energy conversion and storage fields. Fundamental research in electrochemistry mainly focuses on applying the electrochemical in situ technique, consisting of in situ spectroscopy, optical techniques, and in situ transmission electron microscopy (TEM) and so on, to study the electrochemical processes and the interfacial changes. They rationally design high-performance electrode materials based on the atomic structure, molecular level and nanometer scale, the reaction mechanism and the structure-activity relationship of the electrode materials.

Both ionic liquids (ILs) and supercritical fluids are considered as green solvents with tunable properties. ILs have unusual properties including wide liquid range, negligible volatility, weakly coordinating properties, high thermal and chemical stability, high solubility for both organic or inorganic substance, and high designability. Supercritical CO_2 ($scCO_2$) is readily available, inexpensive, nontoxic, nonflammable, and can be easily recaptured and recycled after use. Especially, the physical and chemical properties of $scCO_2$ can be easily adjusted by varying operating pressure and temperature. The thermodynamic studies on ILs and $scCO_2$ systems are of great importance from the viewpoints of fundamental research and their practical applications in chemical reaction, material synthesis, gas adsorption, separation, etc. This report reviews the main progress in the chemical thermodynamics of green solvent systems in recent three years in China, including the physicochemical properties of IL systems, thermodynamic properties of CO_2/water involved systems and the non-equilibrium thermodynamics of the transfer-reaction process in IL aqueous solutions.

Biophysical chemistry is an emerging interdiscipline. It advances very fast worldwide and has made significant contribution to our understanding of life phenomenon. In China, we have also enjoyed the research progress and expansion of the scientific scope in biophysical chemistry. However, in comparison with the cutting edge of the world, biophysical chemistry in China still has a long way to go.

In the field of analytical chemistry, the number researchers grew rapidly, and the research level elevated constantly in 2016-2017. Great progresses have been made in the fields of bioanalysis and biosensing, together with big developments in the fields of single molecular and single cellular analysis, *in vivo* bioanalysis, bioanalysis based on functional nucleic acids, biomolecular recognition, *in vivo* imaging, nanoanalysis and interdiscipline combination analysis. In 2016, professor Fan Chunhai et al won the National Natural Science Award of China (the Second Class) for their work on mechanism, regulation and bioanalytical application of biomolecular interface action process. In 2017, professor Tang Benzhong in Hong Kong University of Science and Technology and his cooperators' project on aggregation induced emission passed the first and final round evaluation of National Natural Science Award of China (the Second Class) . Professor

Yang Xiurong from Changchun institute of applied chemistry, Chinese academy of sciences won the "scientific innovation and advancement certificate" on the first occasion and only 254 people in China were awarded this honor.

Chromatography, as a branch of separation and analysis science, has played an increasingly important role both in the basic research of natural science and in the development of national economy. In the past three years, chromatographic research in our country has made significant progress in many aspects including sample pretreatment, chromatographic stationary phase and column technology, multidimensional and integrated devices, innovative chromatography instruments and devices, as well as the separation and analysis of complex samples (proteomics, metabolomics and Chinese medicine etc.) . According to the document index, since 2010, the number of SCI papers published by Chinese chromatographic scientists surpassed that of United States, and has always maintained a leading position. From 2015 to March of 2017, the number of SCI papers published in this field accounted for nearly 26.43% of the total, which is improved in comparison to that of the previous three years (2013 to May 2015, 25% for our country) . In addition, the number of articles published in first class journals of this field such as *Journal of Chromatography A*, *Analytical Chemistry* and *Nature*-branded primary research journals are increasing year by year. It shows that the development of chromatography in China has been developing well in recent years. The overall level has reached the international advanced level, and some research directions have reached the international leading level. However, with the development of science and technology, chromatography technology is facing many challenges, especially the development of life science, materials science and environmental science requires further chromatographic techniques for more complete functions, analysis speed, higher separation ability and higher degree of automation.

The progress in polymer science in China from 2015-2017 was summarized. An overview on the achievements of polymer chemistry, functional polymer including solar cell based on polymers, biomedical polymers, and polymer physics was provided. Chinese polymer Scientists have made great contributions in following areas. New methods for polymer synthesis such as amine-yne click polymerization, slow chain-walking copolymerization of ethylene with polar monomers, multicomponent polymerization, and COFs, polymer solar cell based non-fullerene electron accepter instead of C_{60}. In the field of shape memory, discovered the selective actuation method for controlling multi-shape recovery using radiofrequency waves, proposed the concept of ultra-soft shape memory polymer and synthesized such a material. Based on the photoresponsive linear liquid crystal polymer (LCP) , fabricated light-controllable self-driven tubular microactuators, in which complex liquids were precisely propelled under light irradiation without any aid of lines and pumps. This kind of tubular micro actuators presents a conceptually novel way to propel liquids by capillary force arising from photo-induced asymmetric deformation. It is notable that Chinese scientists have carried out the systematical research on crystalline CO_2-based polycarbonates prepared enantioselective copolymerization from CO_2 and epoxides and had made a great breakthrough.

Nuclear chemistry and radiochemistry are important parts of nuclear science and engineering, covering several research fields such as nuclear fuel cycle, nuclear chemistry, radioanalytical chemistry, nuclear pharmaceutical chemistry and labeled compounds, environmental radiochemistry, management of radioactive waste, application of radiation chemistry, and so on. At present, solvent extraction is the primary technology in reprocessing and high-level liquid waste separation. Separation of lanthanides and actinides has been the focus and difficulty of reprocessing research till today. Developing novel extractants is among the most important research aspect, the selection of an extractant of good characters determines the success of extraction and separation. In addition, many countries in the world are working on the research of the materials for uranium extraction from seawater, and excellent materials of the adsorbent is the chief target of separation uranium from seawater and wastewater with low concentration uranium. At present, scientists of all over the world such as China, the United States and Japan have chosen the adsorbent that has oxime group in molecular structure. In the field of radiopharmaceuticals, the frontiers research in fundamental area are new PET drugs labeled with nuclides Ga-68, Cu-64, Zr-89 and new SPECT drugs labeled with nuclides Tc-99m and I-123 .The multimodal imaging medicines is also an important development direction, and accurate treatment of tumor by drugs labeled such as Lu-177, Y-90 always is hot area.

Environmental chemistry, as a discipline of studying the existence, behavior, effect, and control principle of toxic substances in the environment, is an important branch of chemistry and also the core component of environmental science. China has experienced rapid economic development, and at the same time there are many emerging environmental problems. Recently, the atmospheric fine particles and ozone, persistent organic pollutants, micro/nano plastics and resistance gene have attracted more and more attention, gradually becoming hot research areas of environmental chemistry. In recent years, the study on analytical method, environmental fate, and risk assessment of organic pollutants added to the "Stockholm Convention" provides important support for ensuring the environmental safety and implementation of Stockholm Conventions in China. The establishment of China's atmospheric Hg emission inventory and Hg flows in China is also helpful for fulfilling the new signed "Minamata Convention on mercury". The recent advances in key mechanism of combined pollution formation in regional atmosphere, application and effect of environmental nanomaterials, environmental computational chemistry and predictive toxicology, environmental toxicology and health, effectively promote our understanding of environmental pollution and its remediation.

Green Chemistry, also known as Environmentally Benign chemistry or Environmentally Friendly Chemical, is a multidisciplinary and interdisciplinary research. Its kernel concept is to use the principles of chemical technology to reduce or eliminate the use and production of chemical products which are harmful to human health or environment. Green Chemistry is an update and development of traditional chemical research methods, focuses on the prevention of pollution from the molecular lever that can use resources reasonably and effectively. To meet the challenge of sustainable development, it aims at solving the fundamental contradiction between environmental pollution and economic developing. Based on the main feature of atom economy, the reaction and reaction process of Green Chemistry highlights its complete understanding of environmental protection: using non-toxic raw materials, under the condition of non-toxic reaction, few by-products or even "zero emission", and finally producing environment-friendly chemical products. In the past three years, researchers in China have made a series of breakthroughs in the field of Green Chemistry, and published many papers in the international top journals. Ma Ding *et al.* from Peking University have reported the syntheses of a new platinum-molybdenum carbide nanostructure as the low-temperature and high-efficiency catalysts for aqueous phase methanol reforming, which was published in *Nature*. The synergy between the highly dispersed platinum center and the molybdenum carbide substrate accelerated the methanol-reforming reaction at the interface between platinum and molybdenum carbide substrate, giving rise to high hydrogen production activity in aqueous phase methanol reforming reaction. Not long afterward, they constructed layered gold clusters on the molybdenum carbide substrate to create an interfacial catalyst system for the ultra-low-temperature water-gas shift reaction, which was reported by Science. These series of work have attracted wide attention from both domestic and foreign academia and industry. The direct production of lower olefins from syngas by Sun Yuhang et al. from Shanghai Advanced Research Institute of Chinese Academy of Sciences was reported by *Nature* and selected as the main research progress of National Natural Science Foundation of China in 2016. In 2017, the same research group reported the direct conversion of carbon dioxide into liquid fuels with high selectivity over a bifunctional catalyst in *Nature Chemistry*, which was considered a breakthrough in the field of carbon dioxide conversion. These achievements show that the development of Green Chemistry in China is gaining momentum and some of research results have achieved an international leading level.

Theoretical chemistry constitutes a significant part of modern chemistry, and it plays an indispensable role in molecular property prediction and reaction mechanism investigations. Over the past few years, theoretical chemistry groups have been expanding rapidly in China, and high-level researches have been conducted. Prominent progresses are made in the areas of electronic structure theory, chemical reaction dynamics, and statistical mechanics, as well as their applications in life and material sciences. These lead to influential publications in Science, PNAS, JACS, and invited reviews in Acc Chem Res, Chem Soc Rev, and so on. Prof. Liu Wenjian won the 2017 Fukui Medal from the Asia-Pacific Association of Theoretical and Computational Chemists (APATCC) Prof. Gao Yiqin won the 2016 Pople Medal from APATCC. In 2015, the 15th International Congress of Quantum Chemistry was organized successfully in China for the first time, which also shows the rising impact of Chinese theoretical chemistry in the world.

Modern crystal chemistry is interdisciplinary, which focus on the crystal structure and performance principle. The development of crystal chemistry is of great importance in the disciplines such as materials science, energy science, environmental science, synthetic chemistry, biochemistry, geochemistry and mineralogy. According to the incomplete statistics, there were more than 3, 000 academic papers in this field published by Chinese chemists in 2015-2017. In recent years, Chinese chemists have achieved a leading position in the study of Metal-Organic Frameworks (MOFs) which is one of the frontiers of crystal chemistry. Their work has broadened the functions and application areas of MOFs, deepened the understanding of the relationship between macro-material performances and micro-structure, and developed some new systems and synthesis methods. At the same time, the design and structure of organometallic supramolecular materials, chiral metal organic macrocycles, cluster complexes, covalent organic framework (COFs) , hydrogen-bonded organic frameworks (HOF) materials research and the preparation of novel non-linear optical (NLO) crystal materials have also made new progress. Research progress and highlights achieved in zeolites and related porous materials in 2014-2017 are summarized. The trends in the field of zeolites and porous materials are predicted.

Colloid and Interface Chemistry deals with colloidal dispersion system and interface phenomena, and has wide application in various fields, such as energy, chemical manufacture, biology, and environmental science. Recently, due to the fast development in the fields of advanced functional materials, bionics, and biomedicine, more and more attention has been paid to the molecular assembly and the material arrangement. In the past two years, many original research achievements had been made by the colloid and interface chemists of China, and have aroused more and more interests from the international scholars, which significantly increase the international impact of Chinese colloid and interface chemistry. These achievements can be generalized as follows: (1) fabrication of novel amphiphilic ordered self-assemblies and their applications in biomedicine, especially with respect to the supramolecular assembly, one-dimensional ordered assembly, organic-inorganic composite assembly, ordered thin films with patterns on surfaces; (2) application of colloid and interface chemistry in the preparation of micro- and nano- functional materials, including the morphological controlled inorganic materials, organic-inorganic composite materials, noble metal nanomaterials, and functional gels; (3) novel application of colloid and interface chemistry in biosensors; (4) novel methods of colloid and interface chemistry. As a theoretical and practical subject, the development of modern economics and society will provide wide space for colloid and interface chemistry. It can be expected that in future Colloid and Interface Chemistry will been paid more attention to the basic physical chemistry problems, such as the fabrication and theory of novel amphiphilic ordered self-assemblies, and the theoretical guide for the interface structure and function control of micro- and nano-materials. Alternatively, the continuous penetration of new methods into colloid and interface system will resulted in new crossing points of the subject, which will powerfully promote the development of colloid and interface chemistry.

In recent years, the revision of national high school chemistry curriculum standard, which aims to develop students' key competence, has been completed. And in new chemistry curriculum, chemical core literacy is constructed, which includes five elements: *Macroscopic and Microscopic Analysis*, *Ideas of Change and Balance*, *Evidence Reasoning and Model Understanding*, *Experimental Inquiry and Innovation Consciousness*, *and Scientific Spirit and Social Responsibility*. The practice of chemical teaching reform and its research achievements have been steadily promoted both at home and abroad. The results from China have been published in some international journals, such as, *Journal of Chemical Education*, *Chemistry Education Research & Practice*. There are some regular competitions and academic exchanges activities, for instance, *Conference on the Implementation and Achievement of the New Chemistry Curriculum*. Chinese chemistry researchers take an active part in the annual conference of IUPAC, EASE, ESERA and other international chemistry/science education research institute. More importantly, the first international chemistry/science education conference, *2015 International Conference of East-Asian Association for Science Education*, was held in Beijing Normal University on

October 2015. In short, Chinese researchers have emendated the national high school chemistry curriculum standards, and make the classroom teaching reform continue to establish the chemical education system that accords with Chinese culture. Following the national development strategy, those researchers exert themselves to build a platform for academic exchange activities at home and abroad, in order to promote the export and sharing of the chemical education research results with Chinese characteristics to represent China story and take the responsibility of a big nation in chemistry education.

Written by Tian Weisheng, Hao Linxiao, Zhu Yvjun

Reports on Special Topics

Report on Advances in Nonlinear Optical Crystalline Materials

Second-order nonlinear optical (NLO) materials can halve the wavelength of light (or double the frequency) by second harmonic generation (SHG) process. Technologically, these materials are used in semiconductor manufacturing, photolithography, laser systems, atto-second pulse generation, and advanced instruments, etc. Thus, they are of current interests and great importance in modern laser science and technology. In the past decades, Chinese scientists made significant contributions to the development of NLO materials such as the establishment of anionic group theory, the discovery of commercial NLO crystals β-BaB_2O_4 (BBO) and LiB_3O_5 (LBO), as well as the solely available deep-UV NLO crystal $KBe_2BO_3F_2$ (KBBF). In particular, BBO and LBO were hailed as "China band" crystals and promoted the rapid development of China's related industrial chain on NLO materials and laser techniques. Moreover, based on deep-UV solid-state lasers produced by KBBF crystal, Chinese scientists have successfully developed a series of unique instruments such as ultrahigh resolution photoemission spectrometry and photoelectron emission microscopy. These instruments have played a significant role in many advanced scientific fields including grapheme, high temperature superconductivity, topological insulator, and wide bandgap semiconductor. These outstanding works make China in a leading role in the field of NLO materials. Since 2013, Chinese scientists have published a number of high-impact papers on NLO materials in top journals such as J. Am. Chem. Soc., Angew. Chem.

Int. Ed., Adv. Mater., and Nat. Commun.. Besides, some of them are invited to publish related reviews on famous reviews journals, including Chem. Soc. Rev., Coord. Chem. Rev., and Acc. Chem. Res., and to make conference reports and keynote reports, as well as to serve as editors or editorial board members. In addition, member of Chinese Academy of Sciences Chuang-Tian Chen won one of the top prizes of International Organization for Crystal Growth—Laudise prize in 2013. It can be concluded that the leading role of China in the field of NLO materials has been further consolidated in recent years. In this report, we summarize typical achievements from 2013 to 2016 in the development of NLO materials in the deep-UV (wavelengths below 200 nm), UV-vis-near-IR, as well as middle and far IR spectral regions. Accordingly, we propose several suggestions as following: 1. Innovate new preparation methods of NLO materials beyond the traditional methods to enrich the material sources and structures, which will help break the performance bottleneck of known NLO materials. 2. Develop rapid and accurate characterization techniques, since the current techniques are still too rough to give very accurate measurement results on polycrystalline samples. On the other hand, although measurements using bulk single crystals can provide accurate results, they always need years of efforts to growth. 3. Establish new structure-property relationship theories on NLO materials, especially on IR NLO materials. The anionic group theory has proved to be very effective in deep-UV and UV NLO materials, but it is difficult to predict the performances of IR NLO materials. 4. Support researches on bulk crystal growth and device preparation. To date, great efforts have been paid on the discovery of new NLO materials, but these studies are usually limited to polycrystalline samples and there are relatively much less concerns on the bulk crystal growth and subsequent device preparation.

Written by Junhua Luo, Zheshuai Lin, Maochun Hong

Report on Advances in Biomineralization and Bioinspired Synthesis of Inorganic Materials

It has been hundreds of millions of years since minerals were used by living creatures for various purposes. Biominerals, as well as their formation process, are quite different from non-biological minerals. While the inorganic components of these biominerals are common and abundant in nature, they exhibit extraordinary performance due to their unique hierarchical structures. The properties of biominerals are also highly desired for artificial materials. Therefore, the study of bioinspired synthesis of inorganic materials is progressing fast in recent years. Besides, there is an urgent demand for new materials with surpassing performance because of the rapid development of our country in many fields. The study of biominerals is inspiring for the research of new materials. For example, the precise control of shape and morphology of crystals and the hierarchically ordered structures at multiscales in biominerals are instructive for fabricating macroscopic materials.

In this Report, we will summarize and analyze the development and newest achievements in this emerging field. The first chapter is about biomineralization and biominerals such as mollusk shell, bone and tooth. The fine structures, the toughening mechanisms and the phase transformation are the focuses in the chapter. The second chapter is about the biomimetic mineralization of calcium carbonate, calcium phosphate, silicon, iron oxides and etc. We discuss the effect of additives, the role of metastable phases, mesocrystals and the controlled synthesis of these minerals. The applications of these artificial minerals are also introduced in the chapter. The third chapter is about the bioinspired synthesis of inorganic materials. Bioinspired synthesis is divided into three parts, i.e., the structural bioinspiration, functional bioinspiration and componential bioinspiration.

In summary, the research of biomineralization and bioinspired synthesis of inorganic materials are valuable both theoretically and practically. In 2005, *Science* journal released a special issue entitled "*Design for Living*" that discussed the strategies for fabricating advanced bioinspired materials. In 2015, *Nature* journal released a supplement issue entitled "biomaterials", where the importance of biomimetic strategies for designing new materials is emphasized. We thus anticipate unexpected and counter-intuitive materials will be discovered based on bioinspired synthesis. Moreover, we conclude with a brief survey and perspectives of the challenges and opportunities in this active field, and propose potential research topics in the coming years.

Written by Mao Libo, Meng Yufeng, Gao Huailing, Yu Shuhong

Report on Advances in Organic Photovoltaic Technology

As an important clean-energy technology, organic solar cells (OSCs) have attracted wide attention from all over the world, which have many advantages in making large-area solar panels by low-cost roll-to-roll solution processing methods. After many years of development, OSCs have become the frontier science field involving organic chemistry, materials science, semiconductor physics and other interdisciplines. Power the power conversion efficiencies (PCEs) is one of the most important evaluation indexes of OSCs, which is proportional to the three parameters, the open circuit voltage (*VOC*) , short circuit current (*JSC*) and fill factor (FF) . The past several years have witnessed many significant progresses of OSCs, and the PCEs have been boosted to over 13%, demonstrating their great potential in practical applications.

Advances in the design of novel photovoltaic materials and application of highly efficient device structures have enabled the rapid development of OSCs. Donor and acceptor are two main key materials in OSCs, and both of them experienced rapid progresses recently. For polymer donors, from early homopolymer like polythiophene to typical donor-acceptor (D-A) structure polymers, many excellent donors have been developed and applied in fabricating OSCs. In addition, organic small molecular donors, such as thiophene- or benzodithiophene-based molecules, also showed very good photovoltaic performance. For acceptors, fullerene derivatives like C60 or C70-phenylbutanoic acid methyl ester ($PC_{61}BM$, $PC_{71}BM$) , C60 or C70-indene adduct ($IC_{61}BA$ or $IC_{71}BA$) were the most widely used electron acceptor material and achieved many important results in the field of OSCs. Recently, in order to overcome the drawbacks of fullerene derivatives like high cost, difficult modification of absorption and molecular energy levels, many highly efficient non-fullerene acceptors have been exploited for OSCs. Nowadays, the PCEs of non-fullerene-based OSCs have surpassed their fullerene counterparts, achieving the highest PCE of over 13%. Bulk heterojunction comprising donor and acceptor mixture as active layer is the most popular device structure of OSCs. The combination of donor and acceptor materials, modification of interlayer, and application of tandem structures are all of great importance to achieve excellent photovoltaic performance for OSCs. For example, in comparison with binary OSCs, ternary

OSCs comprised of two donors or two acceptors can utilize more solar photons, benefiting to obtain a higher *JSC*. Tandem solar cell can achieve an additive *VOC* of the two sub cells. Along with the emerging of highly efficient non-fullerene acceptors, device optimization meet more opportunity, and which is expected to get higher performance OSCs.

Researchers in China have made many outstanding contributions to the development of the OSCs field and achieved many important results. In this research progress, we discussed the latest advances of OSCs in the view of molecular design, device fabrication and working mechanism, and we highlighted the results achieved by China research groups. According the to the main contents, this research progress can be divided to five parts, electron-donating materials, electron-accepting materials, interlayer materials, high-efficiency OSCs fabrication and working mechanism. After a brief summary in the end, we present some inevitable questions about the development of commercial OSCs, such as stability and manufacturability. From the aspect of material design and device optimization, we think these problems can be solved in the near future.

Written by Hou Jianhui, Yao Huifeng

Report on Advances in C-H activation

Traditional organic synthesis relies on the transformation of functional groups, which are not common in organic molecules. Besides, most of them require multi-steps tosynthesize. In contrast, relatively unreactive chemical bonds including C-H bonds are widely existed in organic compounds. Thus, the cleavages and transformations of various C-H bonds were very important in organic synthesis. Due to high dissociative energy, lack of strong coordination site and difficulty to distinguish various C-H bonds, reactivity and selectivity are oft-discussed great issues in C-H activation. As a result, C-H activation was considered as the grail of chemistry.

Since the concept of C-H activation proposed by Bergman and Graham in the 1960s, this area has achieved great progress in the last decades. In the just past three years, C-H activation also has developed very well. Based on the functionalized pattern of metal and substrate, C-H activation could be classified as four patterns: oxidative addition of C-H bond to low-valent metal, electronic substitution of electron-rich substrate, σ-bond metathesis, metal mediated homolysis of C-H bond. In recent years, another kind of pattern called CMD (concerted metalation-deprotonation) has also attracted more and more attention.

Up to now, the most practical strategy for C-H activation was still using directing group. Notably, as the development of this area, mild condition for C-H activation was well developed, for example, using oxygen or air as oxidant was widely existed in many reports. Compared to sp^2 C-H functionalization, transition-metal catalyzed sp^3C-H activation was far behind. The two most important strategies for such transformations were nitrene insertion and radical pathway. Activation of methane also achieves great progress, such as transformation to benzene and ethene.

Organoboron compounds play an important role in fields ranging from materials science to biochemistry to organic synthesis. They could be served as very useful synthons. Recently boronation of methane was realized, which stands out as one of the notable landmarks in the progress of C-H activation.

Oxidative coupling of two C-H bonds, which also means cross-dehydrogenative coupling (CDC) was ideal synthetic route to construct new carbon-carbon bond. In recent years, CDC process could be seen in all kind of aromatic compounds. Compared to sp^2 C-H/sp^2 C-H coupling, sp^2 C-H/sp^3 C-H coupling was very limited. Most of them were concentrated on active C-H bond. Very recently, palladium catalyzed cross-coupling of inert sp^2 C-H bond and sp^3 C-H bond was realized by using acridine as ligand.

The utilization of electric current or light were consider as a potentially ideal energy for chemical transformation. In recent years, C-H activation by electrochemical or photo-redox process was also well developed, providing novel synthetic planning for the synthesis of natural products, drug molecules, and functional materials. Frustrated Lewis pair (FLP) , which means a compound or mixture containing a Lewis acid and a Lewis base that, because of steric hindrance, cannot combine to form a classical adducts, also show powerful applications in selective C-H functionalization.

Due to the limit page, this issue mainly illustrated the development of C-H activation involving metal-intermediates in last three years. To further introduce the perspective of C-H functionalization, we will also purpose the new strategies and viewpoint about this area in the part of outlook.

Written by Fang Huayi, Shi Zhangjie

Report on Advances in Aggregation-Induced Emission

Conventional fluorophores emit strongly in solution but their emission is greatly weakened or even totally quenched in the aggregate state. This thorny obstacle of aggregation-caused quenching (ACQ) severely undermines the performance of many leading fluorophores in the aggregate state. Aggregation-induced emission (AIE) , conceptually coined by Tang *et al.* in 2001, stands for a unique phenomenon that a series of non-planar luminogenic molecules that are non-luminescent or weakly emissive in solution state are induced to emit strongly in the aggregate or solid states. This important concept of AIE challenges and overturns the general belief of ACQ, and has exerted a great impact on the area of photo-physics and luminescent materials, etc. AIE has paved new avenues towards highly emissive solid-state functional materials and provided an array of high-tech applications.

Nowadays, AIE has become a hot research topic as evidenced by the exponential growth of AIE-related publications and citations. AIE-related researches have been developed into an emerging field that is spearheaded by Chinese scientists and followed by researchers from about 1100 universities and institutes in more than 60 countries and regions. AIE was ranked No. 3 by Thomson Reuters in the top 10 research fronts in Chemistry and Materials sciences in 2013 and listed as No. 2 as a "key hot research front" in 2015 as co-reported by National Science Library of Chinese Academy of Science and Thomson Reuters. The AIE dots (the particles generated from AIE materials) were also reported by a *Nature* News Feature article as one of the four key materials for "nanolight revolution".

The researches on AIE theories, materials and techniques are currently enjoying a very bright prospect. Based on numerous experimental measurements and theoretical calculations, a comprehensive AIE working mechanism of restriction of intramolecular motion (RIM) , inducing restriction of intramolecular rotation (RIR) and restriction of intramolecular vibration (RIV) , has been figured out, which has been widely accepted and used in various AIE systems by researchers all over the world.

At the same time, a huge wave of robust AIE luminogens (AIEgens) with emission colors covering from deep blue to near infrared and excellent fluorescence quantum yields up to 100% are prepared. These AIEgens have been widely applied in opto-electronic devices, chemical sensors, fluorescence bioprobes, biomedicine, and so forth. In addition, guided by AIE mechanism, diverse solid-state luminescent materials with unique properties, such as room temperature phosphorescence of organic AIEgens, clusteroluminescence of non-aromatic systems, have been explored. These interesting and "uncommon" emission phenomena may trigger new revolutions in photo-physics, chemical and materials sciences, and related technologies. In view of the rapid development of this promising research area of AIE, this report introduces the recent advances on AIE researches, including molecular design, AIE working mechanism, applications of AIEgens as smart materials and in opto-electronic and biologic fields. It is anticipated that the report will help researchers to have a clear picture of the present situation and future direction of AIE researches.

Written by Tang Benzhong, Dong Yongqiang, Peng Qian, Chi Zhenguo, Shi Heping, Qian Jun, Zhao Zujin, Qin Anjun

Report on Advances in Single Cell Analytical Chemistry

Cells are the basic building blocks of biology, which play critical roles in understanding the mystery of life and investigating the mechanism of diseases and their cures. Cells in the same tissue or organ may even be quite different from each other in metabolism, respiration, signal transduction or mass transportation, because of the environment variation from each to cell. To study this type of difference, single cell analytical chemistry has been established in recent years. The successful techniques for the analytical chemistry of single cells are extremely sensitive, selective, high-throughput, as well as high in spatial and temporal resolutions. These techniques, based on optical spectroscope, mass spectrum, electrochemical analysis and force imaging, are able to precisely illustrate the single-cellular, sub-cellular and single-molecular information of biological systems. The development of chemistry, biology and medical science is more and more interactive, making single cell analytical chemistry become an area across multiple disciplines and a frontier topic of modern analytical chemistry. The November 2016 issue of the highly impacted journal *Nature Biotechnology* was entitled "Focus on single-cell technology", suggesting the broad interest and urgent need of this research field. The most well-known state-of-art technique for single cell analysis is super-resolution microscope, whose three inventors were indeed awarded the 2014 Nobel Prize in Chemistry.

In the recent five years, Chinese researchers have made outstanding achievement in the analytical chemistry of single cells, with the funding support from both Natural Science Foundation of China and Ministry of Science and Technology of China. Examples include methods and techniques for single cell manipulation, super-resolution imaging, ultra-fast tracking, and single molecule detection. This paper summarizes the above achievement and also discussed the perspective of this field in the near future. The first chapter is the introduction of the field, followed by the second chapter focusing on the progress of the field in four directions including optical, mass spectral, electrochemical and mechanical techniques. The third chapter describes the application of single cell analysis in "omics" (such as genomics, transcriptomics and proteomics), disease diagnosis, and drug screening. In the final chapter, we provide an outlook of the field and the challenges remaining for our Chinese research community in the field.

Written by Xiang Yu, Xiong Bin, He Yan, Li Jinghong

Report on Advances in Chemical Reaction Dynamics

Chemistry is one of the main methods and tools for human beings to understand and change the material world. Understanding and studying the nature and regularity of chemical reaction is an important content of chemistry. Chemical Reaction Dynamics utilizes the new experimental and theoretical methods developed by modern physics to study the dynamical behavior of the reaction system at the molecular level, in order to explain and predict the macroscopic kinetics law of the reaction system. In the past few decades, a series of significant achievements have been made in the study of chemical reaction dynamics. Since 1990, seven Nobel Prize-winning awards were closely related to chemical reaction dynamics. These milestones in the study of chemical reaction dynamics show that chemical reaction dynamics is an extremely important and active frontier area in chemistry. This report highlights the progress made by chemical reaction dynamics researchers in China since 2013 on the elementary chemical reaction dynamics, surface chemical reaction dynamics, cluster and catalytic reaction mechanism and complex and biomolecular dynamics.

Xueming Yang's team observed the dynamical resonances accessible only by reagent vibrational excitation in the F+HD (v=1) reaction for the first time. They also discovered the reaction resonance state resulted by the chemical bond "softening" in the transition state region of Cl+HD (v=1) →DCl+H reaction. Shanxi Tian's group found a passway to O_2 production in the dissociative electron attachment (DEA) process of CO_2, thus revealing the new mechanism of oxygen origin in the Earth's primordial atmosphere. Daiqian Xie et al. found that for the photolysis of phenol molecules in the vibrational ground state of the first electronic excited state, the adiabatic model is not correct for describing the dynamics near the conical intersection. Wensheng Bian's group found that the weak long-range forces cause van der Waals saddles in thel C (^{1}D) +D_2 complex-forming reaction that have very different dynamical effects from van der Waals wells at low collision energies. Shu Liu and Donghui Zhang reported the first full-dimensional state-to-state study for the H+H_2O reaction, and confirmed the local mode picture of the OH bond in this reaction. They also performed accurate quantum dynamics studies on the simplest Walden inversion reaction of H+CH_4→CH_3H+H and its isotopic analogues.

The experimental investigation in Xueming Yang's group provided strong evidence that molecular hydrogen from photocatalysis of methanol on TiO_2 (110) is produced via a thermal recombination reaction of hydrogen atoms. They also observed the strong photon energy dependence of photocatalytic dissociation rate of methanol on TiO_2 (110) . This result raises doubt about the widely accepted photocatalysis model on TiO_2, which assumes that the reaction of the adsorbate is only dependent on the number of electron-hole pairs created by photoexcitation. Xueming Yang and Chuanyao Zhou et al. revealed that the electronic excited state at 2.5 eV above the TiO_2 (110) Fermi level is an intrinsic electronic state associated with Ti^{3+}. Using a six-dimensional quantum model, Daiqian Xie et al. showed that excitations in local stretching modes of the HOD molecules on Cu (111) selectively enhance cleavage of the excited bond. Bina Fu and Donghui Zhang reported the first full-dimensional quantum dynamics study for the dissociative chemisorption of H_2O on rigid Cu (111) surface with all the nine molecular degrees of freedom fully coupled, based on an accurate full-dimensional potential energy surface.

Mingfei Zhou et al determined the existence of oxidation state of IX experimentally for the first time by measuring the Infrared photodissociation spectra of the [IrO_4] $^+$cations. By combining infrared spectroscopic and advanced quantum chemistry studies of PrO_4 and PrO_2^+complexes, they also showed that these species have the unprecedented Pr^V oxidation state. In cooperation with the experts of catalysis, Zichao Tang and Hongjun Fan theoretically and experimentally validated the radical mechanism of the dehydrogenation coupling reaction of methane on lattice-confined single iron sites embedded within a silicide matrix. Shenggui He's group found that promoted by the gold dimer in $Au_2VO_3^-$ and $Au_2VO_4^-$ clusters, the CO can be directly or indirectly oxidized by O_2^{2-} species.

Weiqiao Deng and Keli Han's team have perfected the theory and method to predict the carrier mobility of the system based on the crystal structure of π-stacked systems which widely exist in nature. Li Guohui et al collaborating with the experimental researchers, proposed and implemented new enhanced sampling technique and QM/MM computational scheme for enzyme catalysis study , with these development the difference and mechanism of enzymatic catalysis of MLL family were studied and uncovered quantitatively. Through femtosecond transient absorption measurement, Andong Xia et al. showed that photoinduced electron transfer (PET) from the GFP chromophore to FNO2Phe occurs within 11 ps. Yuan Guo et al. investigated the interactions between Ca^{2+}ions and egg sphingomyelin (ESM) Langmuir monolayers at the air/water interface by sub-wavenumber high-resolution broadband sum frequency generation vibrational spectroscopy, clarified the mechanism of interaction between $Ca^{2+}$and sphingomyelin monolayers.

In summary, over the past 5 years, besides the great progress in the elementary chemical reaction dynamics, researchers of chemical reaction dynamics in China extended the research content from both theory and experiment to the excited state reaction, the nonadiabatic process of multi-potential energy surfaces, reaction dynamics on surface, complex system dynamics (liquid, solid et al) and macromolecular (including cluster and biomolecular) system dynamics, and have made remarkable achievements. In the next few years, with the further development and improvement of the experimental and theoretical methods of chemical dynamics and the ability of computer, the research level of chemical dynamics in China will be greatly improved, so that we can face the challenges in the field of chemical dynamics from the discipline itself development and national energy, environment and defense demand.

Written by Dai Dongxu, Yang Xueming

Report on Advances in Resource Chemistry

Resource chemistry is an emerging subdiscipline of chemical science, its concept and term was firstly introduced by author W.-S. Tian in the early 1990s. This special report consists of three parts with 140 references, encompassing the concept, practice and future outlook of resource chemistry.

In the first part, this report firstly comes back a background in resource chemistry and then introduces academic definitions, research contents, and ultimate goals of resource chemistry. Resource waste and environmental pollution have been the severe problems which have to be faced by human society development. What can chemists do to alleviate these concerns? First and foremost, the society requires the chemists to provide the relevant scientific basis and technology for the accurate and efficient utilization of resources. Although the generalized resource chemistry involves the chemical problems of all resource utilization processes, which include mineral, fossil and biological resources, the resource chemistry in this special report is only definited as chemistry of the resource compounds generated from mineral, fossil and biological resources. The research contents of resource chemistry include the discover and exploitation of the resource compounds as well as their reaction and synthetic application. The ultimate goal of resource chemistry is to provide the scientific basis and technology for the accurate and efficient utilization of resources and the elimination of environmental pollution.

The second part of the report summarizes our own efforts in resource chemistry, which have aculminated in about 100 references. A substantial portion of our program is dedicated to the study of steroidal sapogenins, a type of abundant natural resource compounds and basic industrial starting materials for the production of steroidal drugs. In order to utilize the steroidal sapogenins in an atomic economical way, we have developed a new method to oxidatively degrade pseudosapogenins (e.g., pseudo- tigogenin and pseudodiosgenin) into the corresponding pregnenolone acetates, using 30% H_2O_2 instead of the conventionally employed CrO_3, thereby avoid environmental pollution generated from CrO_3, as oxidizing agent. Moreover, this new method also improves the utilization efficiency of steroidal sapogenins as it allows the recycling of chiral (R) -4-methyl-δ-valero- lactone from the reaction waste stream. Direct oxidization of steroidal sapogenins with peracids (generated in situ from 30% H_2O_2 and formic acid) produces pregnane-16, 20-diols and (R) -4-methyl-δ-valerolactone. Interestingly, we found that iodine can change the regioselectivity of steroidal sapogenins oxidization to afford abnormal Baeyer-Villiger oxidation product: steroidal 16-hydroxyl-22- lactones and (R) -3-methyl-γ- butyrolactone. Furthermore, we explored the reactions and synthetic applications of these resource compounds generated from steroidal sapogenins such as pregnane-16, 20-diols, 16-hydroxyl-22-lactones, (R) -4-methyl-δ-valerolactone, and (R) -3-methyl -γ-butyrolactone. Pregnane-16, 20-diols and 16-hydroxyl-22-lactones were used in the efficient synthesis of steroidal drugs and natural products with potent bioactivities such as cephalostatine 1, azedarachol, glaucogenins C and D, 17S-pancuronium bromide, betamethasone, brassinolide, clathsterol, certonarsterol (D2, D3, and N1) , clionamine D, and saundersiosides. (R) -4-Methyl-δ-valerolactone and (R) -3-methyl-γ- butyrolactone were used as the starting materials for the synthesis of chiral-methyl containing insect hormone, flavor and fragrance molecules, chiral drugs as well as natural products (e.g., hydroprene, methoprene, sex pheromones of pine sawflies and southern corn rootworm, tribolure, muscone, menthofuran, citronellal, Citralis Nitrile, vitamin E, sacubitril, pennogenin, and didemnaketal A) . On the other hand, the ring-opening reactions of E/F-ring of steroidal sapogenins were also investigated, allowing the utilization of its intact skeleton in the preparations of the natural products with the basic structure of cholesterol. These selective ring-opening reactions, including bromination, amino, thio- and bromo-lactonization ring-openings, are successfully applied in the highly efficient synthesis of the aglycone of natural product OSW-1, pennogenin, osladin, aspafiliosides E and F as well as boophiline, solasodine, solanidine and demissidine.

Fluoroalkanosulfonyl fluorides are commercially available in bulk and serve as representative examples of non-natural resource compounds. They are employed as starting materials for production of fluoroalkanesulfonic acid salts, valuable surfactants. To further bolster the practical value of fluoroalkanesulfonyl fluorides as well as other resource compounds, Tian to put forth the concept of symbiotic reactions. to maximize the values of these. we designed the symbiotic reaction processes wherein the hydrolysis of fluoro- alkanosulfonyl fluoride are merged with other reactions in tandem. In the designed symbiotic reaction, fluoroalkanosulfonyl fluoride not only serves as the starting material of one reaction, the hydrolysis of fluoroalkanosulfonyl fluoride, but also become an essential reagent for another reaction. Fluoroalkanesulfonyl fluoride has been widely used as hydroxyl- activating reagent in the preparation of fluorinated compounds from alcohols. They have been used to prepare cis-epoxides from chiral vicinal diols, and to access aziridines/thiazolines from β-hydroxy sulfonamides/ thioamides. In addition, they could also be enlisted to trigger the homoallylic carbocation rearrangement of 19-hydroxymethyl steroid, acid-free Backmann rearrangement, esterification and amidation reactions. These reactions have been applied in the efficient syntheses of some natural products, such as (-) -dehydroclausenamide and (±) -spiniferin. Symbiotic reaction is defined as a process wherein more than one chemical transformations are choreographed in the same pot and to produce more than one valuable products concurrently. We believe that, aside from our examples presented herein, more types of symbiotic reactions can be realized through rational design.

The final part of the report provides an abbreviated picture of other chemists' achievements in areas related to resource chemistry, covering about 40 references. This will be followed by our outlook for resource chemistry. In recent years, a number of laboratories have been established in China to investigate problems pertaining to resource chemistry. Recently, the Chinese Chemical Society also held its inaugural symposium on resource chemistry. The C-H functionalization reactions and late-stage functionalization of the resource compounds have become "hot" area in chemistry. Resource chemistry is an emerging scientific field with potentials to solve the longstanding problems of environmental pollution and resource wastage. We envision that as the chemical community faces ever-increasing social demand on efficiency and sustainability, the concept of resource chemistry will be understood and embraced by many. Thus, we are optimistic that the goals we set forth will eventually be realized—chemistry will no longer be associated with fear-inducing pollution or environmental hazards. Instead, the society will cherish her numerous achievements in the betterment of mankind.

Written by Tian Weisheng, Shi Yong, Yan Zhaohua

索引